D0292054

CONTROL MECHANISMS AND PROTEIN SYNTHESIS

S. D. WAINWRIGHT

Control mechanisms & protein synthesis

COLUMBIA UNIVERSITY PRESS
New York and London 1972

QH604 .W34

S. D. Wainwright is Research Professor in the Biochemistry Department of the Faculty of Medicine at Dalhousie University in Halifax, Nova Scotia, Canada. He is Medical Research Associate of the Medical Research Council of Canada.

Copyright © 1972 Columbia University Press
Library of Congress Catalog Card Number: 72-2348
ISBN: 0-231-03432-6
Printed in the United States of America

To Martin, Jacques, Francis, Howard, and Dave,

without whose dedicated example and patient guidance this book could not have been written.

INDIANA UNIVERSITY LIBRARY NORTHWEST

For it is my express purpose in writing this work to enable diligent readers to teach themselves.—Claudius Galenus, Second Century B.C.

Read not to contradict and confute, nor to believe and take for granted, . . . but to weigh and consider.—Francis Bacon, 1597.

PREFACE

This book is based upon a course of lectures offered in the graduate program of the Biochemistry Department of the Faculty of Medicine at Dalhousie University, and it is written with the same philosophy and objectives. It is my hope that this volume will serve as a sufficient summary of developments in the field to this date so that future students, and others, may devote their studies primarily to advances yet to be made. I have therefore willingly responded to Dr. Borek's instruction to "give lots of references" to the original literature. Nevertheless, it is neither practicable nor desirable to attempt to cite all relevant published articles. Therefore, by way of apology to workers whose valuable contributions have not been cited, it seems appropriate to state explicitly the principal factors which have guided me in selecting references.

My overriding concern has been the potential usefulness of the reference to my own students. Therefore, with a few important exceptions, I have sought to include only articles written in English or French, in publications readily available in the library of this Medical School and my own collection. In my experience, collections of articles in publications of proceedings of symposia are, in practice, not readily available to groups of students, and wherever possible I have cited alternate references. Finally, I have tried to limit documentation of individual points to three or four references, except for areas of very rapid development or controversy.

I am indebted to many friends and colleagues for their

generous and valuable assistance in the preparation of this volume. In particular, I am most grateful to my wife, Lillian, as a collaborator in my own studies, for advice and criticism on everything contained herein, and for moral support and understanding throughout the endeavor. Very special thanks are also due my sons David and Peter, who have shown tolerance, understanding, and encouragement beyond their years in all my moods; past and present students, who have served as guinea pigs in the evolution of the course on which this book is based; and to my colleague Dr. S. J. Patrick for encouragement and criticism of the entire volume. I am grateful for valuable criticism of the drafts of individual chapters to the following friends: Drs. S. T. Bayley, E. Borek, M. Brossard, S. Dubiski, M. J. Fraser, R. S. Gilmour, R. H. Hall, C. W. Helleiner, J. G. Kaplan, A. T. Matheson, J. M. Neelin, M. R. Pollock, V. L. Seligy, M. Sung, J. R. Trevithick, R. W. Turkington, H. J. Vogel, and E. W. Yamada.

I wish to thank Dr. R. L. Perlman and Dr. M. Herzberg for generously supplying me with Figures 4.5 and 5.3, respectively. All other figures were drawn for me by Mr. H. MacLennan of the Audio Visual Service of the Faculty of Medicine, Dalhousie University, to whom I am indebted for his skill and patience. My thanks are also extended to the authors and publishers acknowledged, for their kind permission to reproduce the figures indicated. I thank Mr. F. F. Foy of the Editorial Service of our Faculty of Medicine for checking cited references which are no longer readily available to me; Mrs. L. Goldstein, Mrs. M. McGaffney, Mrs. J. Pringle, and Mrs. M. Wade for typing various sections of the manuscript; Mr. R. J. Tilley and the editorial staff of Columbia University Press for their assistance; and Dr. E. Borek, series editor, for persuading me to set pen to paper.

I wish also to thank Mrs. M. Cannon, Miss S. A. Hamlin, Mrs. O. P. Jenkins, Miss N. N. McGrath, and Mr. G. Sheppard for their valuable contributions to our studies, which have also made it possible for me to have the necessary time available to write this volume. These studies have been supported throughout by the Medical Research Council. Various phases have also been supported in part by the Banting Foundation, Canada Department of Health and Welfare, Dalhousie Medical Research Fund, and National Cancer Institute of Canada.

S. D. Wainwright

Halifax, Nova Scotia, Canada
June 1971

CONTENTS

An explanation of this is that, in all the productions of Nature or of art, what already exists potentially is brought into being only by what exists actually.—Aristotle, Fourth Century B.C.

Lastly, the innermost causes and reasons for the creations of these various elemental beings is hidden from our understanding and knowledge. But at the end, all things shall be revealed to us, the greatest and the least, and the reason for all things.—Theophrastus von Hohenheim (Paracelsus), Sixteenth Century.

CHAPTER 1: **INTRODUCTION: THE SYSTEM REGULATED**

One of the most important biological events is the fertilization of a normal human ovum by a normal sperm. The ensuing fusion of the haploid nuclei of the two gametes to constitute a diploid zygote initiates the lengthy and complex developmental processes which normally lead to the birth of a healthy human infant, destined to yet further growth and maturation. One single cell gives rise to millions of cells of various types, arranged in the complex structures of the organs, and capable of performing a diversity of metabolic functions. Some are destined never to divide again, while others not only will divide but also are not yet committed to become a specific type of cell.

The importance of the event does not lie specifically in the conception of another human being. Similar and equally spectacular consequences follow the formation of a normal zygote in all species of metazoa. Rather, the significance of the event resides in the fact that it is not a general property of all cells to originate such an exquisitely organized assembly of highly diverse progeny. A single cell of the bacterium *Escherichia coli* introduced into a nutrient medium containing all the necessary ingredients will rapidly multiply and give rise to a culture containing several million cells. However, this population of cells has no organized structure and (excepting the consequences of spontaneous mutations) the progeny cells are identical in form and properties with the original cell. Nor can this highly dramatic difference between the progeny of these two types of cell be summarily dismissed as reflecting an invariant common difference in property between all metazoan cells and all microbial cells, or even all bacterial cells. Established lines of metazoan cells growing in tissue culture usually show considerably more similarity to bacterial cultures than to cultures of explanted organs of the type from which they were originally derived. In the case of tumor cells of the ascites type the similarity to bacterial cultures extends even to the absence of any degree of organized structure in the cell population. Conversely, many types of bacterial cell (e.g., *Bacillus subtilis*) will, under appropriate conditions, undergo a form of differentiation and give rise to spores.

There is indeed a profound difference between cells of higher forms (eucaryotes) and of bacteria (procaryotes) in the degree of complexity of the assembly of their nuclear genes in ordered structures, or organelles. The double set of genes of the diploid metazoan cell is organized into two homologous sets of highly complex chromosomes and confined by a membrane within a discrete nucleus. On the other hand, the single set of nuclear genes of a haploid bacterial cell comprises a single, relatively simple chromosome lying in intimate contact with the cell cytoplasm.[1] As we shall repeatedly note, the more complex organization of the metazoan genome is associated with some differences in the de-

[1] Under favorable conditions bacterial cells may contain more than one nucleus per cell. Further, many strains of bacterial cell contain genes which may either be integrated into the chromosome or may exist in the cytoplasm independently of the chromosome as entities known as episomes.

Episomes will be discussed briefly in Chapter 3. For a comprehensive treatment the reader is referred to Campbell (1969).

tailed processes involved in the synthesis of protein and in the control thereof. Nevertheless, there is no comparable difference in organization of the genome of a fertilized mammalian egg, cells of the intact embryo which develop from it, or of a culture of cells taken from that embryo. Divergencies in the properties observed in the fertilized egg, the embryonic liver, and cultures of embryonic liver cells are due neither to differences in structure of the cell nuclei nor to differences in the types of genes present in the nuclei. Rather they reflect the many and various subtle mechanisms which determine what portion of the genome of an individual cell is expressed and thereby influences the properties of the cell; to what extent it is expressed, and at what stage in the life cycle of the cell and of the organism of which it is part. That is, mechanisms involved in the interaction of the microenvironment and heredity at the gene or molecular level.

Until relatively recently the nature of these mechanisms, and to some extent even their existence, was largely a matter of conjecture. Extensive morphological and biochemical studies had yielded detailed descriptions of the course of development of a number of animal species. Careful and detailed experiments had led to development of the concepts of inducers, organizers, and cortical layers. Yet, despite intensive searches spanning decades, the nature of these entities eluded (and still largely eludes) identification. Thus, for example, it was possible to formulate schemes whereby the course of embryonic development could be interpreted in terms of complex patterns of induced enzyme synthesis (see Chapter 3). Yet at the time there was no compelling reason to believe that this particular phenomenon occurred in cells of higher forms. Nor was the nature of the underlying mechanism controlling the phenomenon of induced enzyme synthesis as observed with microbial cells itself understood. Indeed, we had little understanding even of the processes involved in the synthesis of protein beyond the existence of a massively documented relationship between synthesis of protein on one hand and the synthesis or "metabolic turnover" of RNA on the other.

The nature of that relationship escaped us until Jacob and Monod (1961[a]) formulated their concepts of messenger ribonucleic acid (mRNA) as the intermediary through which nuclear genes exert their control of the types and quantities of proteins formed by a cell. Their elegant hypothesis suggested mechanisms by which both: (a) "*genetic information*" encoded within a nuclear gene to specify the structure of a unique protein could be "*translated*" in the synthesis of that protein in the cell cytoplasm; and (b) regulation of the rate of formation of that protein could be exerted by the influence of regulatory genes upon the production of the corresponding messenger RNA—"*transcription.*"[2] At the same time, Nirenberg and Matthaei (1961) demonstrated that cell-free preparations from the bacterium *Escherichia coli* would synthesize polyphenylalanine when supplemented with polyuridylic acid as an artificial "template RNA." This finding led directly to the "cracking of the genetic code" and, in turn, more profound understanding of the nature of gene mutations and of the course of evolution (see Jukes 1966, Yčas 1969). In addition, it permitted a more rapid direct analysis of the processes involved in the biosynthesis of protein than had been available in the studies pioneered by Siekevitz (1952) and by Zamecnik and Keller (1954).

[2] The term "translation" was introduced into the jargon of the field in the course of early approaches to "cracking the genetic code." Thus, for example, Crick (1959) observed that the problem was one of "translating from one language to another; that is from the 4-letter language of the nucleic acids to the 20-letter language of the proteins." Recognition of its role as the intermediary in gene expression led to evolution of the related concept of messenger RNA as a gene "transcript" in which a "genetic message encoded in the deoxyribonucleotide script of the nucleic acid is reiterated, or transcribed, in the ribonucleotide script of the same language." Thus *the process of gene transcription represents a "transfer of genetic information"* from DNA into RNA by transcription of the DNA message.

Subsequent general description of the synthesis of RNA upon a DNA "template" as transcription has created a potential ambiguity. Therefore, in the following pages I shall use the term transcription *from DNA* to refer to the transfer of genetic information in the synthesis of mRNA, and the term transcription *of RNA* to designate the synthesis of RNA on a DNA "template," as appropriate.

In the period since the publication of those two reports we have witnessed an enormous increase in our knowledge of the components and reactions involved in the biosynthesis of proteins. We have seen the hypotheses of Jacob and Monod repeatedly put to critical test, to be validated in some cases, and refined or modified as necessary in others. In the process, we have gained considerable insight into some of the mechanisms serving to regulate some aspects of the synthesis of some proteins. Nevertheless, many regulatory processes are but little understood, and other as yet unsuspected regulatory mechanisms may still remain to be disclosed. In the following chapters I shall attempt to assess the current status of our understanding of the types and nature of the known regulatory processes. However, before proceeding to this analysis I must first describe the components of the system subject to regulation, "the protein-synthesizing machinery" of the cells. There are some differences in detail between procaryotes and eucaryotes, and between the intracellular organelles of eucaryotes and the surrounding cytoplasm. Therefore, where there are such differences, it is most convenient to consider the system found in bacteria as representative and then note the major differences. The bacterial system has recently been extensively reviewed (Lengyel and Söll 1969, Lipmann 1969, Lipmann et al. 1967). Components of the system identified to date in *E. coli* are listed in Table 1.1. Lists of components identified in other species of bacteria are less complete but similar.[3]

THE POLYRIBOSOME AS A DYNAMIC ASSEMBLY LINE FOR PROTEIN SYNTHESIS

A polypeptide chain is synthesized by the stepwise addition of single amino acid residues to a progressively lengthening "nascent" peptide chain. Progressive elongation of this nascent chain involves a series of reactions which result in formation of a new peptide bond. The latter reaction occurs

[3] See footnote 5.

TABLE 1.1. Identified components of the "protein-synthesizing system" of E. coli

1. Ribosomes (70S)
   (a) Small ribosome (30S) subunits consisting of one species of RNA (16S) and some 20 proteins, including initiation and dissociation factors.
   (b) Large ribosome (50S) subunits consisting of one large species (23S) and one small species (5S) of RNA and some 30 proteins, including peptidyl transferase.
2. Messenger RNA.
3. Initiation factor $\mathbf{F}_1$, $\mathbf{F}_2$ and a family of factors $\mathbf{F}_3$.
4. Elongation factors **Tu**, **Ts**, and **G**.
5. Release factors $\mathbf{R}_1$ and $\mathbf{R}_2$ plus factor **S**.
6. Dissociation factor.
7. Transfer RNAs; approximately 40 species, including "major" and "minor" species for 20 acids.
8. Aminoacyl-tRNA synthetases; approximately 20.
9. Methionyl-$\text{tRNA}_{\text{F}}^{\text{Met}}$ transformylase.
10. Peptide deformylase.
11. Aminopeptidase.
12. Amino acids, ATP, GTP, glutathione, and inorganic ions.
13. Peptidyl hydrolase, an accessory enzyme.

within a ribonucleoprotein particle known as a *ribosome* (Blobel and Sabatini 1970, Hawtrey 1969, Rich, Eikenberry, and Malkin 1966).

The order of amino acid residues in the polypeptide chain (*primary structure*) is specified by the sequence of nucleotide bases in a *messenger RNA* (mRNA), which serves as a "template" against which the "incoming" amino acid residues are aligned in turn. The sequence of bases in the messenger RNA is determined by the sequence of nucleotide base pairs in the deoxyribonucleic acid (DNA) of the *corresponding structural gene* from which it is transcribed. Each amino acid residue of the polypeptide is represented by a *codon* consisting of a triplet of nucleotide bases in the messenger RNA and a corresponding triplet of base pairs in the DNA of the gene (see Figure 1.1). The individual amino acids are aligned correctly by means of "*adaptor*" ribonucleic acids or *transfer ribonucleic acids* (tRNAs). Each amino acid can be esterified to the 3′-terminal adenosine residue of one, or more, amino acid-specific transfer RNA(s) by the action of specific *amino-acyl-tRNA synthetase,* or "charging," enzymes. Individual aminoacyl-tRNAs then associate with the corresponding codons of the mRNA through interaction between the codon and a complementary *anticodon* triplet of nucleotides in the tRNA. The correlations between individual amino acids and

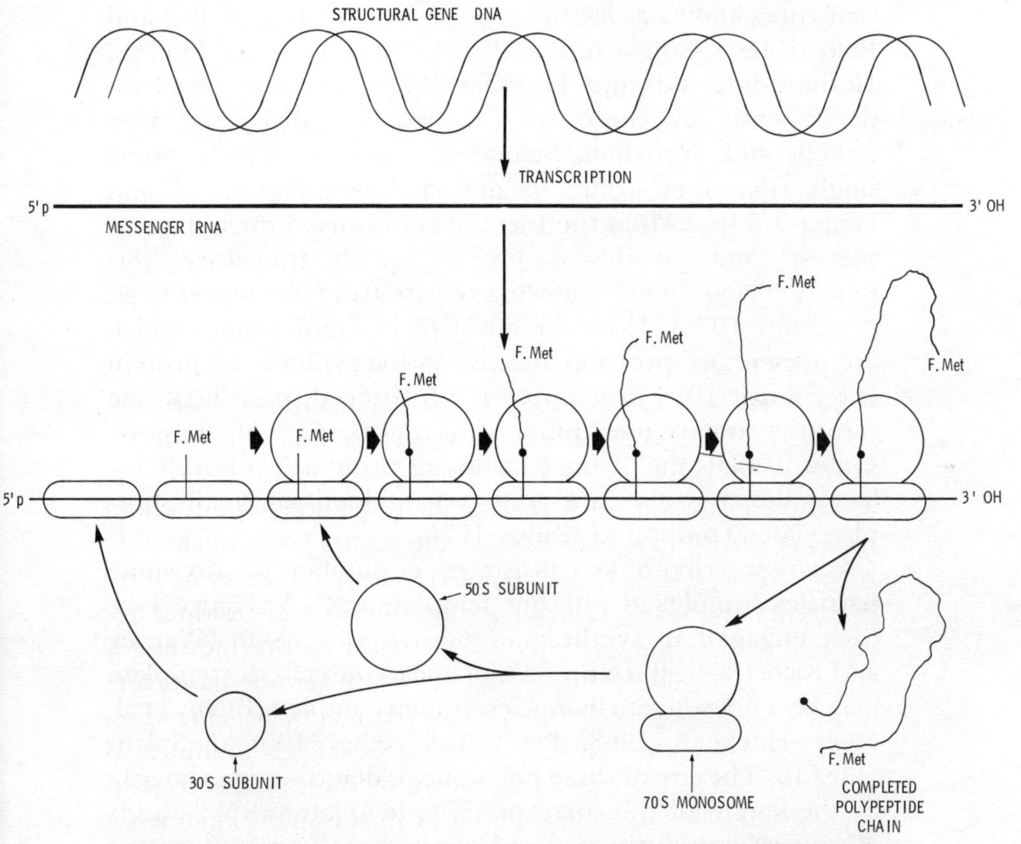

FIGURE 1.1. Schematic representation of a polysome and the ribosome subunit cycle

In this diagram emphasis is placed upon the flow of ribosome subunits through the subunit cycle. Detailed reactions are illustrated in Figures 1.3 and 1.4. In procaryotes subunits attach to "nascent" mRNA chains during the process of "transcription" (see Chapter 5). It is not yet known whether 70S monosomes dissociate before or after release (see page 35).

corresponding codons—the "*genetic code*"—and complementary anticodons have been treated in detail in another volume of this series (Jukes 1966) and will not be considered further here.

Few of the ribosomal particles appear to be present as single ribosomes (or *monosomes*) in the cytoplasm of cells actively synthesizing protein. The majority appear to be in structures known as *polyribosomes* or *polysomes* (e.g., Kabat and Rich 1969, Mangiarotti and Schlessinger 1966, Phillips, Hotham-Iglewski, and Franklin 1969), first discovered independently by Gierer (1963), Warner, Rich, and Hall (1962), and Wettstein, Staehelin, and Noll (1963). Some single ribosomes would be expected (see Figure 1.1 and Figure 1.3 [p. 21]) in the form of transitory "initiation complexes," and possibly others as equally transitory "termination ribosomes" following release from polysomes (e.g., Algranati 1970). However, most of the monosomes which are present are probably inactive in the synthesis of protein (e.g., Kabat 1970, Kaempfer 1970). Indeed, in at least one case, the monosomes differ from the particles of the polysomes in that one of the constituent proteins is phosphorylated, and they also lack an as yet unidentified small phosphorylated compound (Kabat 1970).

Each polyribosome consists of a number of ribosome particles complexed with one template mRNA (Figure 1.1), each engaged in synthesis of one peptide chain (Warner and Rich 1964[a]). In the case of eucaryote cells the template may be a messenger ribonucleoprotein complex (Burny et al. 1969, Henshaw 1968, Perry and Kelley 1968[a], Spirin 1969[a]). The size of these polysomes is determined primarily by the length of the corresponding template mRNA and is in turn related to the size and number of species of protein specified by the mRNA (Kiho and Rich 1965, Rich, Warner, and Goodman 1963).

Synthesis of a polypeptide chain commences at the N-terminal end, and the nascent protein attached to the polysomes consists largely of N-terminal sequences of incomplete peptide chains of various lengths (Canfield and Anfinsen 1963, Dintzis 1961, Luck and Barry 1964, Naughton and Dintzis 1962). The N-terminal amino acid of the peptide chain corresponds to the 5′-terminal portion of the corresponding messenger RNA and the 3′-terminal section of the coding strand of the DNA in the genome (Guest and Yanofsky 1966, Robertson and Zinder 1968). Thus "translation of the genetic message" commences at the 5′-end of the mRNA. Similarly, in the case of an mRNA which specifies more than one polypeptide chain (a *polycistronic mRNA,* see Chapter 5) translation of the "message" normally commences from "initiation sites" which represent the 5′-termini of segments corresponding to individual peptide chains. Further discussion of the translation of such polycistronic mRNAs must, however, be deferred to later chapters.

EFFICIENCY OF THE PROTEIN-SYNTHESIZING MACHINERY OF THE CELL

The synthesis of an entire polypeptide chain is a very rapid process and remarkably free of errors. Synthesis of the polypeptide chain subunit of the β-galactosidase enzyme of the bacterium *Escherichia coli* has been reported to be complete in 80 seconds at 37°C, which represents a rate of 15 peptide bonds formed per second (Lacroute and Stent 1968). Minimum rates determined at 37°C for other systems include 15 bonds per second for the enzymes of the tryptophan biosynthetic pathway of *E. coli* (Rose, Mosteller, and Yanofsky 1970) and approximately 2 to 3 bonds per second for hemoglobin synthesized by suspensions of reticulocytes (Kabat 1968, Lingrel and Borsook 1963). Discrepancies between these various values probably stem from the assumptions necessary to derive them from the experimental data on which they are based, although the possibility of real differences arising from the effect of specific mechanisms regulating the rates of synthesis of particular proteins (see Chapter 9) cannot be dismissed. The assembly of polypeptide chains in most cell-free extracts used for *in vitro* studies to date is considerably less efficient than in the corresponding intact cells, but is nevertheless relatively rapid. Adamson, Herbert, and Godchaux (1968) have reported a cell-free system from reticulocytes to be almost as efficient as the intact cells. However, more representative of the *in vitro* systems now in use are the reports of rates ranging from 25 to 30 peptide bonds formed per *minute* to 3 bonds per second at 31 to 32°C (Webster and Zinder 1969, Wilhelm and Haselkorn 1970) for systems derived from *E. coli.*

The frequency of insertion of an incorrect amino acid into the growing peptide chain has been estimated to be no greater than 1 in 3000 within intact cells (Loftfield 1963). A number of highly purified specific proteins have been reported to exhibit a *microheterogeneity* in which two different amino acid residues occupy the same position in the primary sequence of the protein (e.g., von Ehrenstein 1966). However, in the cases extensively studied to date this appears to reflect the presence of two different alleles of a structural gene, or even duplication of a gene (see Chapter 2), within the genome of the cell. This "*fidelity of translation*" is determined primarily by the specificity of the various aminoacyl-tRNA synthetase enzymes which "recognize" and select the amino to be attached to any one given adaptor tRNA. In turn, the specificity of the synthetase enzyme may be partly restricted as a consequence of formation of a complex between the enzyme protein and the corresponding substrate tRNA (Baldwin and Berg 1966, Loftfield and Eigner 1965, Loftfield, Hecht, and Eigner 1963, Printz and Gross 1967). The extreme fidelity of translation normally obtained with intact cells can be altered in the presence of certain antibiotics (Davies 1966, Gorini, Jacoby, and Breckenridge 1966).

THE RIBOSOMES

Our knowledge of the processes involved in the synthesis of polypeptide chains comes largely from studies with cell-free bacterial extracts supplemented with either a synthetic polyribonucleotide, or the RNA of certain small bacterial viruses (*bacteriophages* or *phages*),[4] as a messenger RNA. Studies with such "exogenously programmed" systems have shown that the synthesis of a polypeptide chain occurs in three distinct stages: namely, (a) formation of a series of *initiation complexes,* (b) synthesis of peptide bonds and *chain elongation,* and (c) *chain termination* and release from the polysome complex. These studies have also shown that the first of the stages involves the formation of ribosome particles from smaller ribonucleoprotein particles. On the other hand, studies made primarily with intact cells have shown that in leaving the polysome complex individual ribosomes dissociate again into the constituent subunits. Thus the synthesis of the polypeptide chains specified by a given mRNA is paralleled by a "*ribosome cycle*" in which ribosome subunits associate and dissociate (Figure 1.1). In the process of this ribosome cycle the subunits of the ribosome monomers freely exchange (Guthrie and Nomura 1968, Kaempfer, Meselson, and Raskas 1968). It is, therefore, necessary to commence with a brief consideration of the structural features of ribosomes which are prominent in discussion of the mechanisms of protein synthesis.

The ribosome particles which have been most intensively studied are those of the bacterium *E. coli* (Kurland 1970, Nomura 1970, Spirin and Gavrilova 1969). They are representative of the ribosomes of most bacteria which have been studied. Some differences in particular physicochemical properties have been observed for ribosomes derived from other species, and notably thermophilic and halophilic species (e.g., Bayley and Kushner 1964). These differences

[4] The bacteriophages **R17**, **f2**, **MS2**, and **M12** comprise a group of closely related viruses, which are considered to be formally equivalent to a series of mutants derived from a single "prototype" virus. Bacteriophage **Q*β*** is a similar, but unrelated, virus. Each contains only three cistrons specifying, respectively, viral coat protein, viral RNA-"replicase," and "maturation" protein (e.g., Gussin 1966, Horiuchi, Lodish, and Zinder 1966, Horiuchi and Matsuhashi 1970). The sequence of these cistrons has been formally demonstrated to be identical in two strains (Jeppesen et al. 1970, Konings et al. 1970).

will not be considered further beyond noting that they may be partially responsible for differences in specificity of responses between extracts of *E. coli* and of some other species of bacteria to a single template RNA (Lodish 1969, Lodish and Robertson 1969) discussed below.

The ribosome monomer of *E. coli* is an ellipsoidal particle of molecular weight about 2.65 x 10^6 and sedimentation constant (S_{20},w) approximately 70S, consisting of about 64 percent ribonucleic acid and 36 percent protein, together with small quantities of polyamines and metal ions. This 70S monomer consists of two smaller subunits of similar overall chemical composition (Hill, Rossetti, and Van Holde 1969, Spahr 1962, Tissières et al. 1959). The larger of the subunits has a molecular weight of 1.55 x 10^6 and sedimentation constant (S_{20},w) of about 50S. The smaller subunit has a molecular weight of 9 x 10^5 and sedimentation constant (S_{20},w) of approximately 30S. Electron micrographs of the 70S ribosome monomer show a prominent groove in the region of contact between the 30S and 50S subunits (Huxley and Zubay 1960).

The 30S subunits contain one molecule of RNA of molecular weight and sedimentation constant of about 5.5 x 10^5 and 16S, respectively, together with some 20 or 21 proteins (Craven et al. 1969, Nomura 1970, Tissières et al. 1959). The combined molecular weights of one molecule of each of the distinct species of protein exceeds the quantity of protein present in any one 30S ribosome subunit. Further, the concentrations of some of the proteins are somewhat variable and are invariably less than one molecule per 30S particle. It therefore seems probable that each subunit contains a core of structural proteins and that enzymic components required for polypeptide formation transiently associate with the subunit only as required (Craven et al. 1969, Kaltschmidt and Whittmann 1970, Kurland 1970, Nomura 1970). Nevertheless, the possibility of a family of different 30S subunits cannot be entirely dismissed, for there is some evidence of possible heterogeneity of structure in the RNA component (Lane and Tamaoki 1969, Nichols and Lane 1966, 1967, Young 1968). This subunit contains sites of attachment of messenger RNA and, through the latter, of attachment of aminoacyl-tRNAs and peptidyl-tRNAs (Kaji, Suzuka, and Kaji 1965, Takanami and Okamoto 1963). In addition, it contains the sites of association of a number of the enzymic components of the system involved particularly in the initiation and chain termination stages of polypeptide synthesis (Miller, Zasloff, and Ochoa 1969, Parenti-Rosina, Eisenstadt, and Eisenstadt 1969).

The 50S subunit contains one molecule of a large RNA of molecular weight and sedimentation constant of, respectively, about 1.1×10^6 and 23S, together with one molecule of smaller RNA of molecular weight about 40,000 and S value approximately 5S. As in the case of the 30S subunit, there is some evidence of possible heterogeneity in the larger RNA molecule (Nichols and Lane 1967, Schaup, Best, and Goodman 1969, Young 1968). The primary structure of the 5S RNA has been completely determined, and two major species differing in only one nucleotide base have been found within cells of a single species of *E. coli* (Brownlee, Sanger, and Barrell 1967, 1968). About 30 proteins, including the enzyme peptidyl transferase, are found in the 50S subunit. There is more uncertainty regarding the precise number than in the case of the 30S subunit, but all (or almost all) of the proteins appear to be structural components of the particle (Kaltschmidt and Whittmann 1970, Nomura 1970, Traut et al. 1969).

This 50S subunit contains two major sites of attachment of aminoacyl-tRNAs (Gilbert 1963, Warner and Rich 1964[b]). One is known as the *donor* or *peptidyl* site, and is also the site of attachment of the growing "nascent" peptide chain. This site will bind aminoacyl-tRNA only at unphysiologically high concentrations of magnesium ions (Cannon 1967). The smaller 5S RNA appears to be involved in the structure of this site, as well as in stabilizing the entire ribosome against inactivation by ribonuclease enzyme (Siddiqui and Hosokawa 1968, 1969). A second site, known as the *acceptor* or *aminoacyl* site, is a region at which the ribosome subunit complexes with the "acceptor" terminal -pCpCpA_{OH} grouping of the aminoacyl-tRNA aligned against the codon specifying the next amino acid to be inserted into the growing peptide chain (Cannon, Klug, and Gilbert 1963, Hishizawa, Lessard, and Pestka 1970, Pestka 1969[a]). Evidence of possible similar additional *decoding* sites corresponding to the second amino acid to be inserted into the growing chain has been reported (Culp, McKeehan, and Hardesty 1969[b], Leder et al. 1969, Schreier and Noll 1970, Swan et al. 1969). However, it now seems unlikely that the 50S subunit contains more than two sites of attachment of aminoacyl-tRNAs under physiological conditions (Roufa, Doctor, and Leder 1970, Roufa, Skogerson, and Leder 1970).

INITIATOR TRANSFER RNA AND INITIATION SEQUENCES IN MESSENGER RNAs

Study of the reactions involved in formation of the first peptide bond stems from the discovery by Marcker and Sanger (1964) of formation of N-formyl-methionyl-$tRNA_F^{Met}$ in whole cells and in extracts of *E. coli.* It is now believed that all, or virtually all, proteins of *E. coli,* and other bacteria, are synthesized *in vivo* from an N-terminal formyl-methionine residue (e.g., Lengyel and Söll 1969).[5] Cells of *E. coli* treated

[5] As indicated, there is very strong circumstantial evidence to support this belief in the case of *E. coli.* However, recent studies indicate that an alternative mechanism for initiation of protein synthesis may be available to streptococcal cells in which dihydrofolate reductase activity is blocked by trimethoprim (Pine, Gordon, and Sarimo 1969, Samuel, D'Ari, and Rabinowitz 1970), and in the halophile *Halobacterium cutirubrum* (White and Bayley 1971). Indeed, studies with trimethoprim-treated *E. coli* do not formally exclude the hypothesis that initiation of protein synthesis by an alternative mechanism requires the participation of a labile component which itself can be synthesized only from an N-terminal formyl-methionine residue. Therefore, the possible existence of alternative "substitute," or minor, mechanisms of protein initiation in bacteria cannot be entirely dismissed.

with trimethoprim to eliminate the ability to formylate methionyl-tRNA are almost totally unable to synthesize any protein (Eisenstadt and Lengyel 1966, Shih, Eisenstadt, and Lengyel 1966). Further, synthesis of any polypeptide *in vitro* in the presence of low concentrations of magnesium ions is dependent upon the presence of N-formyl-methionyl-tRNA (or a system capable of formylating methionyl-tRNA) and the resulting product peptide has an N-terminal formyl-methionine residue (see Jukes 1966, Lengyel and Söll 1969). In addition, the presence of an -ApUpG methionine codon in a synthetic polyribonucleotide serves to "phase" the *in vitro* synthesis of polypeptide to yield one unique sequence of amino acid residues (Ghosh, Söll, and Khorana 1967, Sundararajan and Thach 1966).

Other *in vitro* studies demonstrated that N-formyl-methionyl-tRNA was bound to the *donor peptidyl* site of the 50S ribosome subunit. Further, by addition of puromycin to the system it could be displaced from that site in formation of N-formyl-methionyl-puromycin, as though formyl-methionine were a peptide (Bretscher 1966, Leder and Bursztyn 1966). This permitted recognition that one of the essential features of the initiation process was that the donor site of the 50S subunit was occupied by an *N-acylated aminoacyl-tRNA,* in which the primary α-amine group of the aminoacyl-tRNA was "blocked" and unionized. In turn, this led to the use of "artificial" *N-acylaminoacyl-tRNAs* in the *in vitro* systems for more convenient study of the early initiation reactions (see Lipmann 1969).

Studies with synthetic polyribonucleotide polymers as artificial mRNAs have shown that the codon -ApUpG may specify either an initiator formyl-methionine residue or a methionine residue at an internal position of a polypeptide chain (e.g., Ghosh, Söll, and Khorana 1967). Discrimination between initiator and internal methionine codons of natural mRNAs must therefore be based upon features of the mRNAs additional to the presence of the initiator codon. A

C-TERMINUS OF COAT PROTEIN

—Ala—Asn—Ser—Gly —Ileu-Tyr·COOH

····GCA·AAC·UUC·GGU·AUC·UAC·UAA·UAG

FIGURE 1.2. Nucleotide sequence of a termination sequence, initiation sequence, and intervening segment of "untranslated" codons in the RNA of bacteriophage R17

From the data of Nichols (1970) and Steitz (1969) UAA, UAG Terminator codons, AUG Initiator codons.

number of recent studies indicate that the -ApUpG codons which serve as the sites of initiation of polypeptide are defined by the secondary structure of "*initiation sequences*" of nucleotides, portions of which are not themselves translated as corresponding peptides.

Steitz (1969) isolated polynucleotide segments of the (messenger) RNA of the bacterial virus **R17** from each of the sites of initiation of synthesis of RNA-dependent RNA-polymerase (or "replicase") (see Figure 1.2), viral coat protein, and A-protein (or "maturation protein"), respectively. Each segment contained an -ApUpG initiator codon followed by appropriate codons to specify the N-terminal amino acid sequence of the corresponding protein. In each sequence the -ApUpG codon was preceded by a "nonsense" -UpGpA terminator codon and one or three "untranslatable" codons. It seemed unlikely that these particular -UpGpA codons did serve as "signals" for termination of peptide chain synthesis from the viral genome, for the base sequences preceding them did not correspond to the C-terminal amino acid sequences of proteins coded by the viral genome. However, this common feature of the initiation sequences of the RNA served to emphasize that one feature which serves to distinguish an initiator codon from that of an internal methionine residue is the "context" of preceding codons.

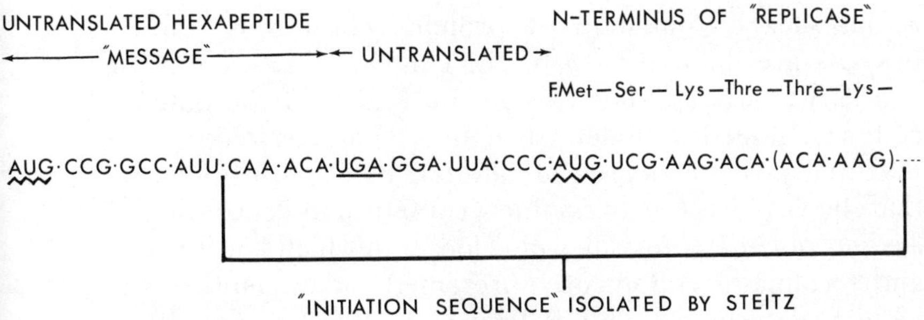

UNTRANSLATED HEXAPEPTIDE "MESSAGE" ← UNTRANSLATED → N-TERMINUS OF "REPLICASE"

F.Met—Ser — Lys—Thre—Thre—Lys—

AUG·CCG·GCC·AUU·CAA·ACA·UGA·GGA·UUA·CCC·AUG·UCG·AAG·ACA·(ACA·AAG)·...

"INITIATION SEQUENCE" ISOLATED BY STEITZ

Subsequently, Nichols (1970) isolated a fragment from RNase T_1 digests of virus **R17** RNA corresponding to the C-terminal sequence of the coat protein and overlapping the N-terminal sequence of the replicase (Figure 1.2). In this fragment, the base sequence corresponding to the end of the coat protein is followed by two terminator codons in tandem. These are followed by an -ApUpG codon and a base sequence which *could* code for the synthesis of a hexapeptide. However, this segment of the genome is not normally translated in formation of peptide (Capecchi and Klein 1970, Steitz 1969).

Similar initiation sequences have also been isolated for the coat protein cistrons of the bacterial viruses **Qβ** (Hindley and Staples 1969) and **f2** (Gupta et al. 1970). The untranslated portions of the various isolated initiation sequences are different in each case and contain no common sequence of nucleotides. Further, unequal numbers of polypeptide chains are synthesized on the three segments of the viral RNA, both *in vivo* and *in vitro* (e.g., Lodish 1969, Nathans et al. 1966, Ohtaka and Spiegelman 1963). Indeed, the secondary structure of the RNAs precludes efficient initiation of peptide chain synthesis from some potential sites of initiation (Bassel 1968, Gesteland and Spahr 1969, Lodish 1970[a], Nichols 1970). Thus, it seems probable that the actual sites of initiation of protein synthesis in natural mRNAs are defined both by the context of bases and the *secondary structure of the RNA* in the region of the initiator codons. Indeed, "model" structures characterized by the maximal possible degree of internal base pairing indicate that the viral RNA may assume a configuration consisting of a series of consecutive "hairpin" loops, in which key initiator and terminator codons are "presented" at the ends of the loops (Adams et al. 1969, Billeter et al. 1969, Nichols 1970, Steitz 1969).

However, such a configuration may reflect constraints necessary for "packing" the viral RNA within the capsid of a mature virus particle and may not be a characteristic feature of the structure of all natural mRNAs.[6] There is now some direct experimental evidence that the individual cistronic portions of polycistronic mRNAs transcribed from bacterial operons are separated by *"intercistronic divides"* consisting of nucleotide bases which are not normally translated in polypeptide synthesis (Rechler and Martin 1970). Further, nascent chains of phage T4-specific mRNA containing up to 100 base residues do not appear to contain any sites for the initiation of protein synthesis (Revel, Herzberg, and Greenshpan 1969). However, initiation sequences have not yet been isolated from bacterial or eucaryote mRNAs.

FORMATION OF SPECIFIC INITIATION COMPLEXES AND ROLE OF INITIATION FACTORS F_3

The first stage in the initiation process is the formation of a series of *initiation complexes,* through which an N-formylmethionyl-tRNA is introduced into the donor site of a ribosomal-mRNA complex. Formation of these complexes requires a source of guanosine triphosphate (GTP) and involves at least three distinct *initiation factors* designated

[6] Preliminary data indicate that functional mRNA transcribed from the genome of bacteriophage T4 in virus-infected cells of *E. coli* has extensive nonrandom secondary structure (Salser 1970).

$\mathbf{F}_1$, $\mathbf{F}_2$, and $\mathbf{F}_3$.[7] These are loosely associated with the 30S ribosome subunit (Miller, Zasloff, and Ochoa 1969, Parenti-Rosina, Eisenstadt, and Eisenstadt 1969) of ribosomes associated with nascent mRNA chains (Revel et al. 1968). Recent studies indicate that factor $\mathbf{F}_3$ is actually a family of related proteins which differ in specificity for different mRNA initiation sequences (Miller and Wahba 1970, Revel et al. 1970).

The overall sequence of reactions currently appears to be essentially as depicted in Figure 1.3. However, details of the reactions involved are far from clear. Indeed, there is some controversy and even apparent conflict in the experimental data regarding the precise function of the initiation factors. This apparent discrepancy reflects differences in the experimental systems studied, and particularly in the structures of the mRNA components. As will be considered below, it also seems probable that a further factor contributing to the observed differences has been the use of ribosome subunits derived from *uninfected* cells.

The first step in the sequence is association of a 30S ribosome subunit with mRNA in a manner which results in an initiator ApUpG codon being presented for complex formation with N-formyl-methionyl-$tRNA_F^{Met}$, to yield the first distinct initiation complex. This is designated initiation complex I in Figure 1.3. This step involves a highly specific mechanism for recognition of the initiator codon region(s) of the mRNA. Details of the mechanism are currently the subject of intense study and speculation. However, it appears that the specificity of the recognition process is derived in part from the structure of the 30S ribosome subunit, and in part from the actions of the initiation factors.

[7] Unfortunately, there is no unanimity in the nomenclature used to describe these initiation factors. The majority of workers appear to accept the designations $\mathbf{F}_1$, $\mathbf{F}_2$, and $\mathbf{F}_3$, and this practice will be followed here. In a second nomenclature, derived from studies of Ravel and Gros (1967), the corresponding fractions are designated A, C, and B, respectively. In yet another recent report they are designated $\mathbf{F}_{I}$, $\mathbf{F}_{II}$, and $\mathbf{F}_{III}$ (Dubnoff and Maitra 1969).

RNA of the virus **f2** will serve as the template for *in vitro* synthesis of all three proteins coded by the viral genome in a cell-free system derived from the natural host, *E. coli.* On the other hand, only the maturation protein is initiated and synthesized in a system derived from *Bacillus stearothermophilus* (Lodish 1969, Steitz 1969), and the **f2** RNA is inactive as an mRNA in systems derived from *Bacillus subtilis* (Lodish and Robertson 1969). This variation in response to a single mRNA between systems derived from different species of bacteria appears to be due entirely to differences in structure of their 30S ribosome subunits (Lodish 1970[b]).[8]

Complex formation between the RNA of bacterial RNA-viruses and 30S ribosome subunits from uninfected *E. coli* cells appears to require specifically the initiation factor $\mathbf{F}_3$ (Dubnoff and Maitra 1971, Iwasaki et al. 1968, Sabol et al. 1970). In contrast, complex formation between the "lysozyme-mRNA" specified by DNA of bacteriophage T4 and 30S ribosome subunits from uninfected cells appears to require specifically factor $\mathbf{F}_2$ (Greenshpan and Revel 1969, Herzberg, Lelong, and Revel 1969). However, when factor $\mathbf{F}_1$ and N-formyl-methionyl-$tRNA_F^{Met}$ are added to permit formation of an initiation complex active in protein synthesis, it is observed that the extent of discrimination between "correct" and "improper" initiation codons is less than when

[8] Lodish has argued from this observation that the initiation factors play no role in recognition of mRNA initiation sequences. However, this conclusion is premature and apparently at variance with the experimental data.

FIGURE 1.3. Schematic representation of the initiation of protein synthesis in E. coli

(Facing page) Regions designated D and A represent the donor and acceptor sites of the 50S ribosome subunits.

Sites of attachment of the initiation factors $\mathbf{F}_1$, $\mathbf{F}_2$, and $\mathbf{F}_3$ are not known. The stage of release of factor $\mathbf{F}_3$ is not known. It is assumed to be released after translocation of the N-formylmethionyl-tRNA into the donor site of the ribosome.

The symbol used to designate N-formyl-methionyl-tRNA does not reflect the actual structure but is intended to indicate that it has a highly ordered structure which includes base-paired regions.

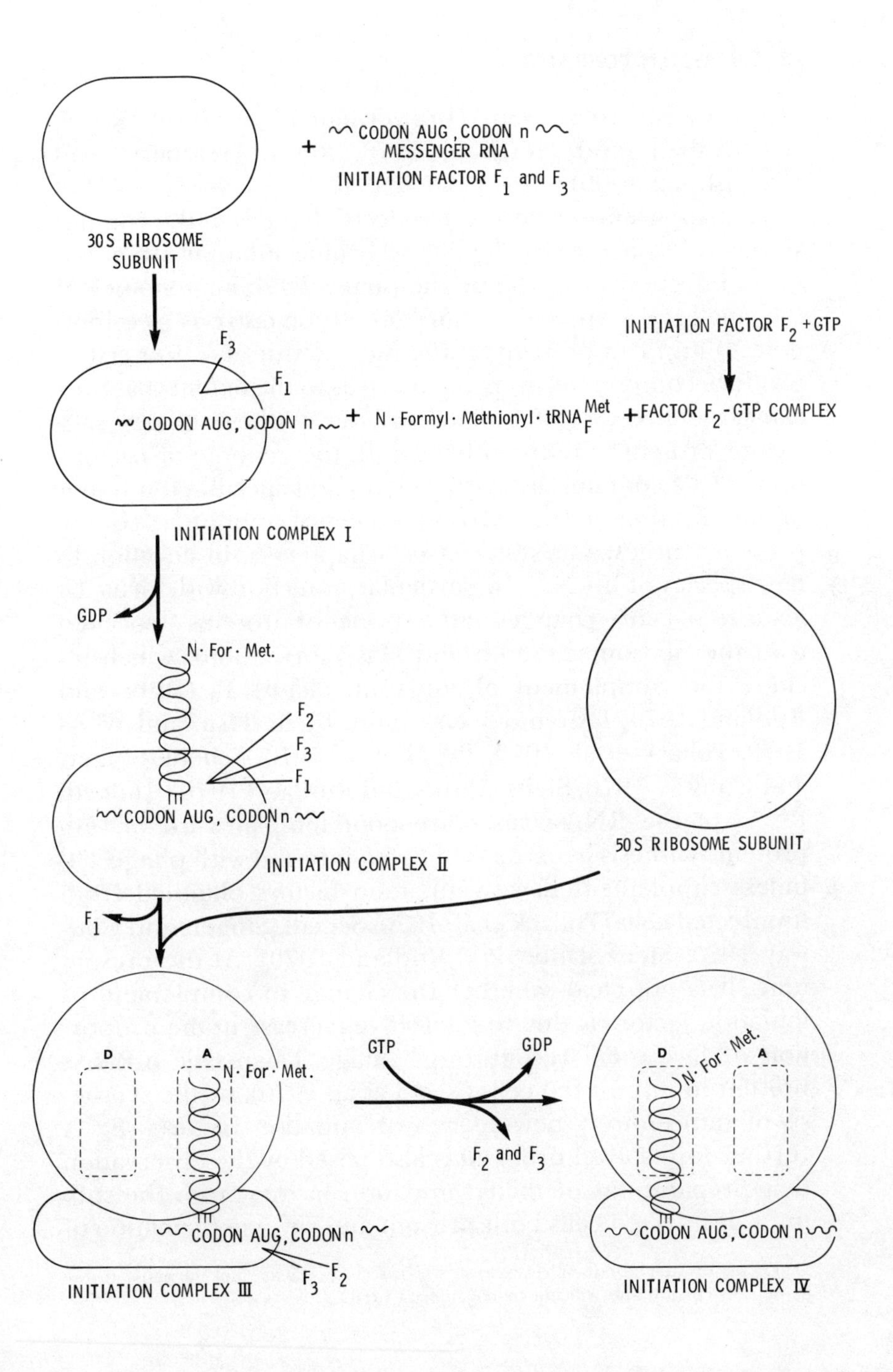

30S RIBOSOME SUBUNIT
+ CODON AUG, CODON n
MESSENGER RNA
INITIATION FACTOR F_1 and F_3
INITIATION FACTOR F_2 + GTP
F_3
F_1
CODON AUG, CODON n
+ N · Formyl · Methionyl · tRNA$_F^{Met}$
+ FACTOR F_2 - GTP COMPLEX
INITIATION COMPLEX I
GDP
N · For · Met.
F_2
F_3
F_1
CODON AUG, CODON n
INITIATION COMPLEX II
50S RIBOSOME SUBUNIT
F_1
D
A
N · For · Met.
GTP
GDP
F_2 and F_3
CODON AUG, CODON n
F_3
F_2
INITIATION COMPLEX III
INITIATION COMPLEX IV

factor F_3 is also present (Brawerman et al. 1969, Revel, Greenshpan, and Herzberg 1970, Revel, Herzberg, and Greenshpan 1969).

This apparent discrepancy reflects changes in the composition and structure of the 30S ribosome subunits attendant upon infection of *E. coli* by the phage T4. The genomes of the small RNA-viruses contain only three cistrons (see footnote 4) and can only direct the metabolism of a host cell to produce components of progeny virus to the extent that they can serve directly as templates for the synthesis of new species of protein.[9] On the other hand, the genome of bacteriophage T4 contains at least 70 genes and specifies a number of modifications in the cell complement of components of the protein-synthesizing system (see Chapter 9), in addition to new species of mRNA. In particular, infection with virus T4 leads to selective changes in the species of proteins associated with the ribosomes (Smith and Haselkorn 1969), which include the complement of initiation factors F_3 (Dube and Rudland 1970, Hsu and Weiss 1969, Klem, Hsu, and Weiss 1970, Pollack et al. 1970, Revel et al. 1970, Schedl, Singer, and Conway 1970, Steitz, Dube, and Rudland 1970). Indeed, RNAs of the RNA-viruses are poor templates for *in vitro* protein synthesis in extracts of *E. coli* infected with phage T4, unless supplemented with initiation factors obtained from uninfected cells (Pollack et al. 1970, Schedl, Singer, and Conway 1970, Steitz, Dube, and Rudland 1970). At the present time, it is not clear whether the change in complement of initiation factors is due to a selective increase in the proportion of factor F_3 "recognizing" phage T4-specific mRNAs present in uninfected cells (Revel et al. 1970), or the synthesis of (an) entirely new species of initiation factor(s) F_3. A further unresolved problem is also posed by the observation that preparations of mixed initiation factors from the subunits of infected cells both prevent and reverse formation of

[9] Effects of attachment of viral capsids to infected cells are not germane to the present discussion and will not be considered further.

initiation complexes containing host cell mRNAs (Dube and Rudland 1970, Klem, Hsu, and Weiss 1970). It remains to be determined whether these effects can be attributed solely to competition between factor $\mathbf{F_3}$-mRNA complexes for "binding site" of the ribosome subunit, or reflect the presence of a more specific inhibitor.

Nevertheless, despite the observed differences between them, all of the experimental systems are dependent upon initiator factors $\mathbf{F_3}$ for complete discrimination between the -ApUpG codons of initiation sequences and other -ApUpG codons present in the mRNA (e.g., Brawerman et al. 1969, Wahba et al. 1969). It therefore seems probable that these factors "recognize" some distinctive feature of the secondary structure of "initiation sequences," in addition to the exposed initiator codon. Factors $\mathbf{F_3}$ can complex directly with both the 30S ribosome subunit and the small viral RNAs (Wahba et al. 1969), and may serve to align the initiation sequence in register with the initiator tRNA-binding site of the ribosome subunit. Present evidence indicates that factors $\mathbf{F_3}$ are released from the ribosome 30S subunits either during association with the 50S subunits or shortly thereafter (Revel et al. 1968, Revel, Herzberg, and Greenshpan 1969).

EVOLUTION OF INITIATION COMPLEXES AND ROLES OF INITIATION FACTORS F_1 AND F_2

The primary role of initiation factor $\mathbf{F_2}$ appears to be in the binding of N-formyl-methionyl-$tRNA_F^{Met}$ by the 30S ribosome subunit. This reaction requires GTP for formation of a normal complex, although GTP can be partly replaced by the analogue guanylyl-methylene disphosphonate (GMP-PCP) in the partial reaction of binding formyl-methionyl-tRNA. In contrast to factor $\mathbf{F_3}$, initiation factor $\mathbf{F_2}$ alone does not directly complex with mRNAs (Revel, Herzberg, and Greenshpan 1969, Wahba et al. 1969). Rather, factor $\mathbf{F_2}$ can complex with both the 30S ribosome subunit and GTP or GDP. The factor is stabilized by complex formation with

GTP and essential sulfhydryl groups are protected (Chae, Mazumder, and Ochoa 1969, Herzberg, Lelong, and Ravel 1969, Lelong et al. 1970, Mazumder, Chae, and Ochoa 1969), presumably through an induced change of conformation. It seems probable that stimulation of the binding of N-formyl-methionyl-tRNA to the 30S ribosome subunit by factor $\mathbf{F}_2$ is due to a similar induced change of conformation. Factor $\mathbf{F}_2$ is probably released from the ribosome 30S subunit during, or shortly after, association with the 50S subunit (Revel et al. 1968, Revel, Herzberg, and Greenshpan 1969). Further analysis of the roles of this factor is complicated by the recent resolution of two distinct components (Herzberg, Lelong, and Revel 1969, Kolakofsky, Dewey, and Thach 1969).

Initiation complex I is converted to complex II (Figure 1.3) by combination with initiation factor $\mathbf{F}_1$ (Hershey, Dewey, and Thach 1969). The mechanism of action of this factor is unclear and appears to be primarily involved in association of complex II with the 50S ribosome unit to yield complex III. However, it appears to stimulate the activities of both factors $\mathbf{F}_2$ and $\mathbf{F}_3$, and is required for maximal efficiency in recognition of initiation sequences (e.g., Brawerman 1969, Revel, Herzberg, and Greenshpan 1969, Sabol et al. 1970).

Unformylated methionyl-tRNA$_F^{Met}$ will not participate in this sequence of reactions at low magnesium concentrations. Some complex can be formed at high magnesium concentrations. However, the process is inhibited somewhat by initiation factors and the complex formed is highly labile. Similarly, N-acetyl-phenylalanyl-tRNA can form a complex at low magnesium concentration with 30S ribosome subunits, but the process is markedly inhibited by initiation factors, and the complex is highly labile. The combined effect of the selective action of the initiation factors and instability of complexes other than those formed with N-formyl-methionyl-tRNA$_F^{Met}$ serves to ensure accuracy of translation of the genetic message in peptide synthesis (Grunberg-Manago et al. 1969, Ohta and Thach 1968, Rudland et al. 1969).

Complex II then associates with a 50S ribosome subunit to yield complex III, in which N-formyl-methionyl-$tRNA_F^{Met}$ is aligned with the initiator codon -ApUpG of the mRNA and attached to the *acceptor* site of the 50S ribosome subunit of an entire 70S ribosome (Hershey and Thach 1967, Sarkar and Thach 1968). Initiation factor $\mathbf{F}_1$ is released in the formation of complex III (Hershey, Dewey, and Thach 1969). The N-formyl-methionyl-$tRNA_F^{Met}$ is then transferred, or *translocated,* to the peptidyl donor site of the ribosome, in a reaction which probably also involves hydrolysis of GTP (Anderson et al. 1967, Hershey, Dewey, and Thach 1969, Mukundan et al. 1968, Thach and Thach 1971). This translocation process must necessarily involve some transposition of the messenger RNA with respect to the ribosome such that the codon specifying the second amino acid of the sequence is aligned with the acceptor site of the ribosome. Details of this process are completely unknown, but interesting model sequences have been proposed by Bretscher (1968[a]), Culp, McKeehan, and Hardesty (1969[b]), Hill (1969), Schreier and Noll (1971), Spirin (1969[b]), and Woese (1970).[10]

The final initiation complex IV (Figures 1.3, 1.4) is now able to accept a molecule of another type of aminoacyl-tRNA and participate in the reactions of chain elongation.

CHAIN ELONGATION AND ROLES OF TRANSFER FACTORS

There is no further requirement for initiation factors in the association of aminoacyl-tRNA with initiation complex IV or for formation of the first peptide bond. Rather the process requires *transfer factors* essential to the cycle of reactions which comprise the process of chain elongation (Erbe, Nau, and Leder 1969, Guthrie and Nomura 1968, Salas et al. 1967). Indeed, at high magnesium concentration the first peptide bond may be formed in the absence of the transfer factors too (Pestka 1969[b]).

[10] In the strictest sense Woese's model is not one of translocation but of "allosteric transition," or change in conformation.

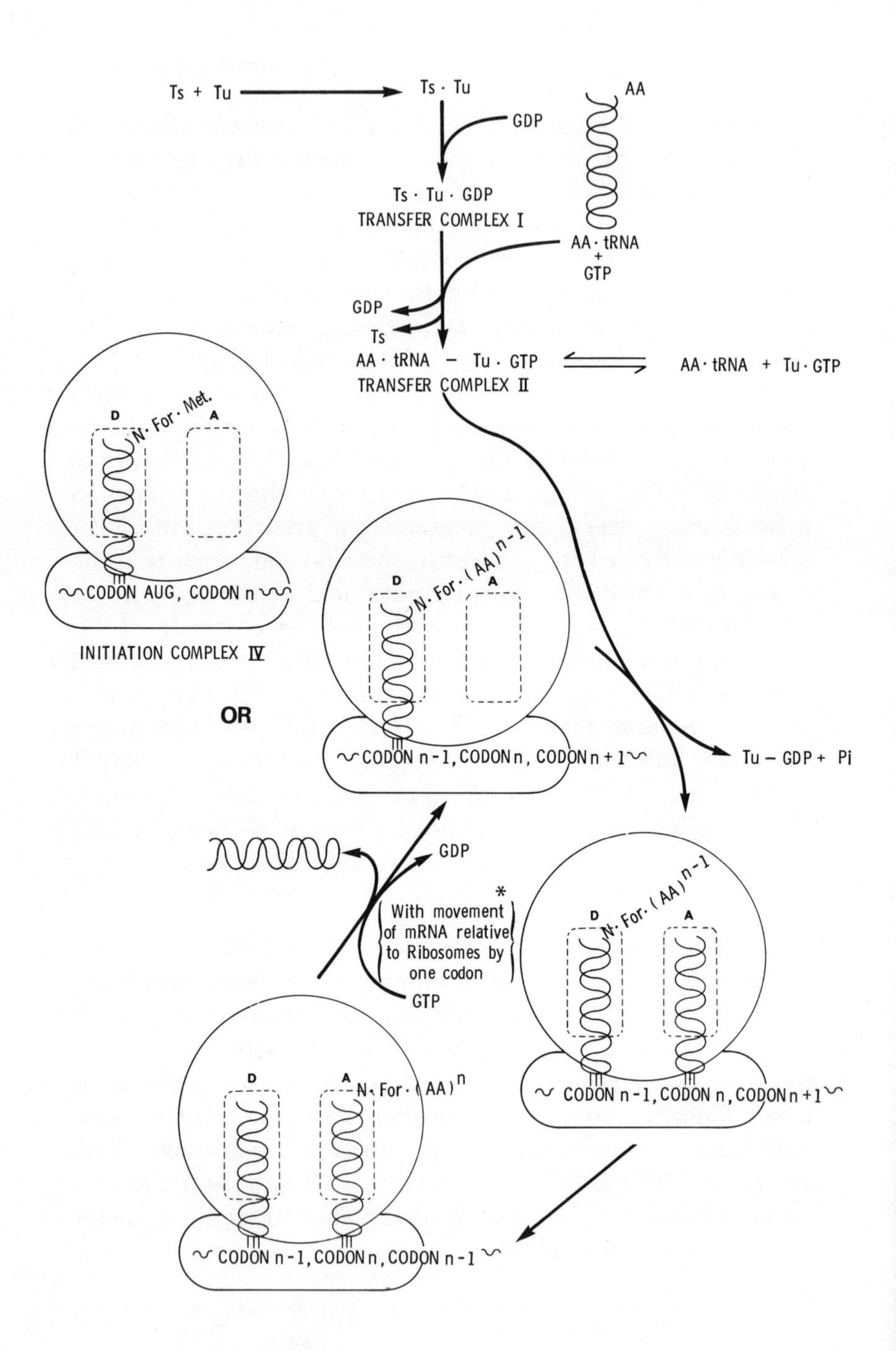

Ts + Tu
Ts · Tu
GDP
AA
Ts · Tu · GDP
TRANSFER COMPLEX I
AA · tRNA
+
GTP
GDP
Ts
AA · tRNA − Tu · GTP
TRANSFER COMPLEX II
AA · tRNA + Tu · GTP
D
A
N · For · Met.
CODON AUG, CODON n
INITIATION COMPLEX IV
OR
N · For · (AA) n-1
CODON n-1, CODON n, CODON n+1
Tu − GDP + Pi
GDP
*
With movement of mRNA relative to Ribosomes by one codon
GTP
N · For · (AA) n-1
CODON n-1, CODON n, CODON n+1
N · For · (AA) n
CODON n-1, CODON n, CODON n-1

Elongation of the peptide chain commences from initiation complex IV; or artificial homologues prepared with an N-acyl-aminoacyl-tRNA or a peptidyl-tRNA in lieu of N-formyl-methionyl-tRNA (Nakamoto and Kolakofsky 1966, Nishizuka and Lipmann 1966). Three distinct transfer factors are required for elongation of the chain, designated as **Ts**, **Tu**, and **G**,[11] respectively (Kaziro and Inoue 1968, Lucas-Lenard and Lipmann 1966, Parmeggiani 1968). A number of details of the reactions of the elongation process remain to be clarified, but the overall sequence appears to be essentially as depicted in Figure 1.4, based on Erbe, Nau, and Leder (1969). Shorey and co-workers (1969) have recently formulated a more extensive schema to depict additional side reactions which can occur *in vitro*.

The early reactions of the sequence leading to elongation of the peptide chain still require substantial clarification. As isolated, transfer factors **Tu** and **Ts** form a complex with each other (Lucas-Lenard and Lipmann 1966, Miller and Weissbach 1969). Interaction of this complex with GDP yields the first recognized "*transfer complex I*" (Figure 1.4) of the reaction sequence (Allende et al. 1967, Ravel, Shorey,

[11] Unfortunately, as with the initiation factors, there is no agreed scheme of nomenclature for the transfer factors of bacteria. The convention followed here is that arising out of the discovery of these factors by Nathans and Lipmann (1961). A second system of nomenclature for the transfer factors of *E. coli* (Ravel et al. 1968) designated these components, respectively, **FI-B**, **FI-A**, and **FII**, but they have recently (Shorey et al. 1969) been changed to **FIs**, **FIu**, and **FII**. The corresponding factors from *B. stearothermophilus* have been designated $\mathbf{S_1}$, $\mathbf{S_3}$, and $\mathbf{S_2}$, respectively (Skoultchi et al. 1968).

FIGURE 1.4. Schematic representation of the elongation of peptide chains in E. coli

(Facing page) Regions designated D and A represent the donor and acceptor sites of the 50S ribosome subunits.

Sites of attachment of factors **Tu** and **Ts** and complexing with GTP are not known. As the stage of release of factor **Ts** is not known, it is shown retained to account for requirement for the factor (see text).

* Completion of each cycle of reactions of chain elongation is represented by an increment in the value of n to $n + 1$.

and Shive 1967).[12] Despite intensive study, both the nature of the reactions involved and the identity of transfer complex I are still uncertain. Addition of GDP to the complex **Tu-Ts** appears to disrupt the association of the two factors **Tu** and **Ts** with formation of a **Tu**-GDP complex. A similar **Tu**-GDP complex can be formed in the absence of factor **Ts**, which appears to catalyze formation of the **Tu**-GDP complex (Ertel et al. 1968[a], 1968[b], Gordon 1968). Miller and Weissbach (1969) have therefore suggested that the initial sequence of reactions may be:

$$\mathbf{Ts} + \mathbf{Tu} \rightarrow \mathbf{Ts} - \mathbf{Tu}$$

$$\mathbf{Ts} - \mathbf{Tu} + \text{GDP} \rightarrow \mathbf{Tu} - \text{GDP} + \mathbf{Ts}$$

However, factor **Ts** also appears to act catalytically in the further conversion of transfer complex I into transfer complex II (Shorey et al. 1969, Skoultchi et al. 1968, Waterson, Beaud, and Lengyel 1970, Weissbach, Miller, and Hachmann 1970).[13] In addition, transfer complex I is retained by a Millipore filter whereas the complex **Tu**-GDP is retained only in the presence of transfer factor **Ts**. It therefore seems probable that factor **Ts** is a component of transfer complex I, but only loosely complexed and readily dissociated from it under the conditions of chromatographic analysis used by Weissbach's group (Miller and Weissbach 1969). The initial reaction sequence could then be depicted as:

$$\mathbf{Ts} + \mathbf{Tu} \rightarrow \mathbf{Ts} - \mathbf{Tu} \xrightarrow{\text{GTP} \rightarrow \text{GDP} + \text{Pi}} \mathbf{Ts} - \mathbf{Tu} - \text{GDP} \rightarrow \text{Complex II}$$

$$\downarrow$$

$$\mathbf{Ts} + \mathbf{Tu} - \text{GDP}$$

Interaction of transfer complex I with GTP and an aminoacyl-tRNA yields transfer complex II (Cooper and

[12] The experiments summarized here were performed with preparations of GTP and the various intermediates were originally believed to contain GTP. However, the actual substrate appears to be GDP (Cooper and Gordon 1969, Lockwood, Hattman, and Maitra 1969, Weissbach, Miller, and Hachmann 1970). I have therefore depicted the reactions as indicated, while retaining the original references to investigations of the roles of the elongation factors.

[13] It has recently been reported that factor **Ts** can be replaced by phospho·enol·pyruvate kinase plus its substrate (Weissbach, Redfield, and Hachmann 1970).

Gordon 1969, Ertel et al. 1968[a], Gordon 1967, Ravel, Shorey, and Shive 1967). The precise nature of complex II remains to be determined. It does not contain transfer factor **Ts** (Ravel, Shorey, and Shive 1968, Shorey et al. 1969, Skoultchi et al. 1970). Furthermore, complex II appears to show ready exchange of the aminoacyl-tRNA component, which indicates that the **Tu**-GTP portion can readily associate with aminoacyl-tRNA in the absence of factor **Ts** (Shorey et al. 1969, Weissbach, Miller, and Hachmann 1970). This property enhances our difficulties in understanding the role of factor **Ts**. It seems possible that combination with both factor **Ts** and aminoacyl-tRNA leads to a conformational change in the structure of unstable **Tu**-GTP complex; that the reaction of complex I with aminoacyl-tRNA serves to displace factor **Ts** from the complex with change of conformation in the complex **Tu**-GTP; or that the apparent roles of factor **Ts** in the structure and activity of complex I are artifacts of the experimental systems studied.

The transfer factors **Tu** and **Ts** exhibit a specificity complementary to that of the initiation factors. They will form complexes of Type II with most, if not all, aminoacyl-tRNAs *except methionyl-tRNA*$_{\mathrm{F}}^{\mathrm{Met}}$. However, they will not form complexes with N-formyl-methionyl-tRNA$_{\mathrm{F}}^{\mathrm{Met}}$, N-acetylphenylalanyl-tRNA, or deacylated tRNAs (Marcker, Clark, and Anderson 1966, Ono et al. 1968, Ravel, Shorey, and Shive 1967). In addition, the analogous factor(s) of wheat embryo have been shown not to complex with denatured aminoacyl-tRNA (Jerez et al. 1969). Thus, the factors have a specificity for both the charged α-amine group of the aminoacyl moiety and the structure of the tRNA. This dual specificity serves to complement that of the initiation factors in discriminating between the N-terminal initiating and internal positions of the peptide chain; particularly in the case of codons specifying methionine and valine.[14]

[14] The valine codons -GpUpG and, to a lesser extent -GpUpA, have activity as initiator codons in model *in vitro* systems (Ghosh, Söll, and Khorana 1967). This property has not been satisfactorily accounted for. There is no evidence that these codons serve as initiator codons *in vivo*.

Transfer complexes of type II readily transfer the aminoacyl-tRNA component to the acceptor site of an initiation complex of type IV (Haenni and Lucas-Lenard 1968, Lucas-Lenard and Haenni 1968, Ravel 1967), with release of complex **Tu**-GDP (Gordon 1969, Ono et al. 1969[a], 1969[b], Shorey et al. 1969). This reaction may be the site of action of recently discovered stimulatory factor(s) which enhance the binding of aminoacyl-tRNA or complex II to the ribosomes (Brot et al. 1970, Kan, Golini, and Thach 1970). The ribosome complex is now fully "loaded," and formation of the first peptide bond is catalyzed by the peptidyl-transferase enzyme component of the 50S ribosome subunit (Erbe, Nau, and Leder 1969, Haenni and Lucas-Lenard 1968, Monro, Cerná, and Marcker 1968). In this process, the N-formylmethionine residue (or other "peptide-equivalent") is transferred to the α-amine grouping of the aminoacyl-tRNA (Takanami 1963) occupying the acceptor site of the ribosome to form a dipeptidyl-tRNA, and leaving deacylated tRNA on the donor site.

Further chain elongation requires the ribosome complex to be "readied" to receive a new molecule of aminoacyl-tRNA. This requires: (a) the deacylated tRNA to be released from the donor site of the ribosome; (b) the peptidyl-tRNA to be transferred, or *translocated,* from the acceptor site of the ribosome to the donor site of the ribosome; and (c) the messenger RNA to be transposed relative to the ribosome to be in correct alignment with the two binding sites of the ribosome.

The deacylated tRNA appears to be actively displaced in the process of translocation (Kuriki and Kaji 1968, Lucas-Lenard and Haenni 1969), and is not released after formation of dipeptidyl-tRNA in the absence of factor **G** (Roufa, Skogerson, and Leder 1970). Ishitsuka and Kaji (1970) have recently reported the requirement for an additional "tRNA-releasing factor" of unknown action, which cannot profitably be considered further at this time. Translocation requires the presence of GTP and transfer factor **G**, or *translocase* (Erbe, Nau, and Leder 1969, Felicetti, Tocchini-Valentini, and DiMatteo 1969, Haenni and Lucas-Lenard 1968, Pestka 1968). The complex factor **G** with ribosomes has GTP-ase activity, and in the translocation reaction the GTP is hydrolyzed and GDP released (Erbe, Nau, and Leder 1969, Haenni and Lucas-Lenard 1968, Nishizuka and Lipmann 1966, Ono et al. 1969[b]). It has been suggested that the energy derived from this hydrolysis serves in the realignment of messenger RNA and ribosome (Nishizuka and Lipmann 1966, Lipmann 1969).

The movement of messenger RNA with respect to the ribosome is formally required in our current concepts both of the nature of the genetic code and of the process of protein biosynthesis, and it is the subject of a number of ingenious hypothetical models (Bretscher 1968[a], Culp, McKeehan, and Hardesty 1969[b], Hill 1969, Spirin 1969[b], Schreier and Noll 1971, Woese 1970). It is therefore unfortunate that there is very little direct experimental evidence bearing directly on this point. In one study, Takanami, Yan, and Jukes (1965) examined the composition of the fragments of virus **f2** RNA firmly bound to the ribosomes (and hence resistant to RNase action) in an *in vitro* system. The composition found after incorporation of amino acids had been allowed to take place was different from that immediately after formation of ribosome-**f2** RNA complex. Other types of experiment consistent with the hypothesis include those in which there may be sequential translation of mRNA *in vitro;* or in which ribospmes "cleared" of nascent peptide by puromycin are allowed to resume protein synthesis after removal of the puromycin. However, a definitive demonstration of the correctness of our beliefs would appear to require analysis of the sequence of protected nucleotides (Steitz 1969) under conditions where only one amino acid residue can be introduced into the growing nascent peptide chain, because of manipulations of the supply of initiation and transfer factors (Erbe, Nau, and Leder 1969, Haenni and Lucas-Lenard 1968).

Upon completion of the translocation reaction the ribosome complex again has the structure of an initiation complex of type IV (Figure 1.4) with an N-formyl-dipeptidyl-tRNA in the donor site of the ribosome. Thus the sequence of reactions outlined serves to elongate the nascent chain at the donor site by one amino acid residue and to restore the structure of the complex to its original state in readiness for a further cycle of reactions. Each successive cycle of reactions elongates the nascent peptide chain by one residue until the polypeptide chain is completed. (This is represented in Figure 1.4 by an increment in the value of n to $n + 1$.)

CHAIN TERMINATION AND RELEASE FACTORS

Completion of the chain is due to alignment of a *terminator codon,* usually -UpApA (see Jukes 1966, Yčas 1969), with the acceptor site of the ribosome. Until recently, this conclusion rested entirely on data obtained with model systems, or with "*amber*" and "*ochre*" mutant strains of bacteria and bacteriophages. These studies demonstrated that the presence of such a terminator codon in a template RNA leads to termination of peptide chain elongation [15] and its release from the ribosome complex (e.g., Bretscher 1968[b], Webster and Zinder 1969). More recent studies, to be considered below, have established that the "nonsense" codons are the normal "signals" for chain termination and reinforced the belief that the ochre codon -UpApA is the one most frequently serving this role.

Analyses of the efficiencies with which nonsense-suppressor genes will support development of selected amber

[15] This statement is not strictly correct in the case of bacterial strains containing *amber-suppressor* or *ochre-suppressor* genes (see Chapter 9). However, even in such strains the frequency with which the "no amino-acid position" is filled by a specific amino acid (the *efficiency of suppression*) is usually low. In the case of the most frequent terminator codon (UpApA) this efficiency of suppression is invariably very low (see Garen 1968).

and ochre mutant strains of bacteriophages have indicated that chain termination *may* also be determined by the "context" of bases surrounding the terminator codon in a "*termination sequence*" (Kuwano, Ishizawa, and Endo 1968, Salser 1969).[16]

The base sequence of one natural termination sequence in the RNA of bacteriophage **R17** has recently been determined (Nichols 1970). This sequence (Figure 1.2) can assume a "hairpin" configuration, with the key terminator codon -UpApA (Capecchi and Klein 1970) exposed, analogous to that proposed for the initiation sequence (Steitz 1969) present in the same RNA fragment. Perhaps of greater interest is the fact that the -UpApA terminator codon is immediately followed by a second, different terminator codon (-UpApG), which could function as a "safety valve" to ensure correct chain termination in strains of bacteria containing ochre-suppressor genes. Direct evidence for other termination sequences is currently limited to demonstrations of untranslated sequences of bases beyond the site of chain termination (e.g., Goodman et al. 1970, Gupta et al. 1970). However, indirect evidence indicates that the presence of two consecutive terminator codons is not an invariable feature of termination base sequences (Rechler and Martin 1970).

Studies with model systems have shown that release of terminated peptide chains is an active process, rather than the passive consequence of an abortive transfer of the peptide to an unoccupied "acceptor" site of the ribosome. Analysis with systems containing highly purified components has shown that peptide chain elongation ceases at the terminator codon, but the peptide chain remains attached to the ribosome complex. Release of the chain requires the presence of protein "*release factors*" and transfer factor **T**, but not that of transfer factor **G** or any species of tRNA (Bretscher 1968[b],

[16] These analyses require the assumption that minor differences in amino acid sequences between completed polypeptide chains coded by the mutant phage genomes and that of the wild-type phage, respectively, do not affect susceptibility to proteolysis, or the efficiency of assembly into viral particles.

Capecchi 1967, Caskey et al. 1968). Two different factors have been resolved: $\mathbf{R}_1$ specific for the terminator codons -UpApA and -UpApG, and $\mathbf{R}_2$ specific for the codons -UpApA and -UpGpA (Scolnick and Caskey 1969, Scolnick et al. 1968).

The mechanism of the termination reaction has not yet been determined. It is catalyzed by peptidyl transferase enzyme and appears to involve formation of a release factor-terminator codon-ribosome complex (Capecchi and Klein 1969, Scolnick et al. 1970, Vogel, Zamir, and Elson 1969). Thus the release factors may function in part as structural analogues of aminoacyl-tRNAs. Activity of the release factors is stimulated by a further protein factor designated **S** (Milman et al. 1969). This factor is distinct from the transfer and initiation factors. However, it has similar properties to transfer factor **Tu**, and may function in a manner similar to that of the transfer and initiation factors in formation of a factor-terminator codon-ribosome complex (Capecchi and Klein 1969, Goldstein and Caskey 1970, Goldstein et al. 1970). Further details of the process have yet to be elucidated.

Microbial cells do contain hydrolase enzyme(s) which cleave the bond between N-acyl-aminoacyl or peptidyl groups and tRNAs, with the exception of N-acylmethionyl-$tRNA_F^{Met}$ (Cuzin et al. 1967, Kössel and Rajbhandary 1968, Vogel, Zamir, and Elson 1968). However, it seems improbable that this enzyme participates in normal chain release. Enzyme substrates are markedly less susceptible to the action of the enzyme when bound to ribosomes (Menninger, Mulholland, and Stirewalt 1970). Further, the release process occurs in simulated chain termination reactions in the absence of added peptidyl-tRNA hydrolase (Goldstein et al. 1970, Menninger, Mulholland, and Stirewalt 1970, Scolnick et al. 1970). In addition, the ribosome-bound fraction of the enzyme appears to be associated predominantly with the 30S subunit (Kössel 1970). It therefore seems probable that this enzyme may serve to prevent fortuitous "improper" peptide chain initiation from aminoacyl-tRNAs other than N-formyl-methionyl-tRNA. The growing nascent peptide chain is presumably protected against any action of this enzyme sterically by virtue of its location within a "pocket" of the 50S ribosome subunit (Blobel and Sabatini 1970).

Deformylation of the polypeptide chain is catalyzed by a specific aminohydrolase enzyme. This enzyme is highly active on N-formyl-methionyl-peptides and the formylated proteins made by *in vitro* systems, but is inactive on free N-formyl-methionine or N-formyl-methionyl-tRNA. It also has lesser activity with artificial N-acyl-methionyl-peptides and N-formyl-aminoacyl-peptides (Adams 1968, Fry and Lamborg 1967, Livingston and Leder 1969, Weissbach and Redfield 1967). The resulting deformylated polypeptide chain may then have the N-terminal methionyl group, and possibly other, residues removed by the action of specific amino-peptidase enzymes to yield the polypeptide chain of final *primary structure* (Natori and Garen 1970). (Discussion of the acquisition of higher orders of structure is deferred to Chapter 9.) Both the deformylation and the removal of N-terminal methionine residues appear to be relatively slow processes in comparison with that of chain elongation (Pine 1969) and probably occur primarily after release of the polypeptide chain from the polysome complex.

THE RIBOSOME CYCLE AND DISSOCIATION FACTOR(S)

The ribosome bearing the completed polypeptide chain appears to be released from the polysome virtually simultaneously with the release of free peptide and deacylated tRNA (Webster and Zinder 1969). It dissociates into the two subunits, which can then participate in a further cycle of reactions of protein synthesis (Guthrie and Nomura 1968, Kaempfer, Meselson, and Raskas 1968). There is, however, some uncertainty at the present time whether the two ribosome subunits are released directly from the polysomes or are dissociated from intermediary free 70S monosomes.

There appear to be few free monosomes in cells of *E. coli* which are actively synthesizing protein (e.g., Kaempfer 1970, Mangiarotti and Schlessinger 1966, Phillips, Hotham-Inglewski, and Franklin 1969), and it is believed that these are probably inactive in protein synthesis.[17] Free monosomes accumulate under some unfavorable experimental conditions, and subsequent dissociation then requires the activity of a *dissociation factor* (Algranati, Gonzalez, and Bade 1969, Kohler, Ron, and Davis 1968, Subramanian, Davis, and Beller 1969). However, it is not known whether these monosomes are formed by an improper reassociation of dissociated subunits (Kaempfer 1970), and possibly subsequent to phosphorylation of one of the component proteins (see Kabat 1970).

There was an initial period of considerable confusion regarding the properties and identity of the dissociation factor (e.g., Albrecht et al. 1970, Gonzalez, Bade, and Algranati 1969, Miall, Kato, and Tamaoki 1970, Subramanian, Ron, and Davis 1968). However, it now seems to be established that the dissociation factor is initiation factor $\mathbf{F_3}$ and that GTP is not required in the dissociation reaction (Dubnoff and Maitra 1971, Sabol et al. 1970, Subramanian and Davis 1970).

PROTEIN SYNTHESIS IN THE CYTOPLASM OF EUCARYOTE CELLS

In any comparison of the protein-synthesizing systems of procaryotes and eucaryotes it is necessary to distinguish between the cytoplasmic system of the latter and the intracellular organelles. The organelles account for only a quantitatively small fraction of the total protein synthesized by most, but not all, eucaryote cells. Nevertheless, these organelles may play a major role in the processes of differentiation (e.g., see S. Nass 1969) and will be considered separately later.

[17] In contrast, cells of *B. subtilis* and *B. licheniformis* appear to contain fairly large pools of 70S monosomes which accumulate and dissociate relatively slowly (Kelley and Schaechter 1970, Van Dijk-Salkinoja and Planta 1970).

Our knowledge of the detailed reactions of protein synthesis in eucaryote cells is considerably less than for bacteria. Properties of some of the identified components of the system (Table 1.2) differ in detail, including the size and properties of the ribosomes and component RNAs. Recent studies indicate that the mRNA component of eucaryote polysomes may be associated with a protein component of unknown function (Burny et al. 1969, Henshaw 1968, Perry and Kelley 1968[a], Spirin 1969[a]). The possibility that this is an artifact of the isolation procedures cannot be dismissed (e.g., Blobel 1971). However, such isolated ribonucleoproteins preferentially reassociate with the smaller ribosome subunits (Pragnell and Arnstein 1970). The major difference between the system of the cytoplasm of eucaryote cells and that of bacteria is that the former contains no N-formyl-methionyl-tRNA; nor does this aminoacyl-tRNA derivative participate in the *in vitro* synthesis of polypeptide by cell-free cytoplasmic systems derived from eucaryotes.

However, despite differences in detail, the reactions appear to be essentially the same as in bacteria, except that

TABLE 1.2. Identified components of the "protein-synthesizing system" of reticulocytes

1. Ribosomes (80S) (see Gould 1970, Holder and Lingrel 1970, Kabat 1970).
   (a) Small (40S) subunits consisting of one species of RNA (18S) and some 20 proteins, including initiation factors.
   (b) Large (60S) subunits containing a large species (28S) of RNA and approximately 30 proteins.
2. Messenger RNA or ribonucleoprotein (e.g., Burny et al. 1969, Pragnell and Arnstein 1970).
3. Initiation factors $\mathbf{M_1}$, $\mathbf{M_2}$ and $\mathbf{M_3}$ (Prichard et al. 1970, Shafritz and Anderson 1970).
4. Elongation factor **TF-I** and **TF-II** (Arlinghaus, Shaeffer, and Schweet 1964, Moon, Collins, and Maxwell 1970).
5. Release factor (Goldstein, Beaudet, and Caskey 1970).
6. Transfer RNAs; not analyzed in detail.
7. Aminoacyl-tRNA synthetases; not analyzed in detail.
8. Aminopeptidase (Yoshida, Watanabe, and Morris 1970).
9. Amino acids, ATP, etc. (Adamson, Herbert, and Godchaux 1968).

polypeptide chain synthesis is not initiated from an N-terminal formyl-methionine residue.[18] Indeed, the basic architectural component, the polysome, was first demonstrated in eucaryote cells (Geirer 1963, Warner, Rich, and Hall 1962, Wettstein, Staehelin, and Noll 1963). Mammalian systems also furnished the first clear demonstration of elongation of nascent peptide chains from the N-terminal end (Dintzis 1961, Naughton and Dintzis 1962). Similarly, transfer factors were also first clearly identified in a reticulocyte system and shown to act in a cyclical sequence in the stepwise addition of single amino acid residues (Arlinghaus, Shaeffer, and Schweet 1964). On the other hand, requirements for initiation factors (Heywood 1970[a], Prichard et al. 1970, Shafritz and Anderson 1970) and release factor(s) (Goldstein, Beaudet, and Caskey 1970) analogous to those of bacteria have only recently been established. As yet, no requirement for a dissociation factor has been shown.

The sequence of reactions involved in the elongation of the peptide chain in eucaryote cytoplasmic systems appears to be entirely homologous with that in bacteria. Aminoacyl-tRNA is inserted into the acceptor site in a reaction requiring GTP and catalyzed by a transfer factor **TF-I** or "binding enzyme," consisting of two components (Arlinghaus, Shaeffer, and Schweet 1964, Culp, McKeehan, and Hardesty 1969[b], Kloppstetch, Schumann, and Klink 1969, Siler and Moldave 1969). Formation of the peptide bond is catalyzed by a peptidyl transferase enzyme component of the large ribosome subunit, and peptidyl-tRNA is translocated into the donor site of the ribosome by a second transfer factor **(TF-II)** analogous to factor **G** of bacteria (Culp, McKeehan, and Hardesty 1969[b], McKeehan and Hardesty 1969, Siler and Moldave 1969, Skogerson and Moldave 1968). Release of the completed peptide chain and dissociation of ribosomes into subunits appear to be energy-dependent processes re-

[18] Formation of the initiation complex also appears to be stimulated by the elongation factor **TF-1** (Shafritz, Laycock, and Anderson 1971).

quiring the action of release factor(s), and possibly other soluble factors (Goldstein, Beaudet, and Caskey 1970, Lawford 1969, Morris 1966).

PEPTIDE CHAIN INITIATION IN EUCARYOTES

Ribosome subunits undergo a cycle of association and dissociation during the course of protein synthesis, as in the case of bacteria (Hoerz and McCarty 1971, Hogan and Korner 1968, Kaempfer 1969). The first reaction in the sequence is association of the smaller ribosome subunit with mRNA or mRNA-protein (Heywood 1970[b], Leader, Wool, and Castles 1970, Pragnell and Arnstein 1970).[19] It appears to require an initiation factor analogous to the bacterial initiation factors $\mathbf{F}_3$ (Heywood 1970[a], Leader, Wool, and Castles 1970, Prichard et al. 1970). Further, present evidence indicates at least some measure of species and tissue specificity in the cell complement of initiation factors which permit discrimination between different species of mRNAs (Heywood 1969, 1970[a], 1970[b], Mathews 1970, Naora and Kodaira 1969).

Evidence available until very recently tended to indicate that eucaryote cells do not initiate synthesis of all types of polypeptide chain from a unique species of initiator tRNA. As noted above, the cytoplasm contains no N-formylmethionyl-tRNA, and this species of acylated-tRNA does not play any role in the *in vitro* synthesis of polypeptide by cytoplasmic systems from eucaryotes. Further, inspection of the N-terminal sequences of proteins isolated from eucaryotes reveals no clear preferential location of one particular amino acid in the N-terminal position.

[19] "Depleted" subunits released by *in vitro* systems "exhausted" by "pre-incubation" (and unable to continue polypeptide synthesis) will spontaneously reform 80S particles. These particles will complex directly with polyuridylic acid and support synthesis of polyphenylalanine, but differ in properties from normal monosomes (Falvey and Staehelin 1970[a], 1970[b]). In these respects they resemble the subunits of salt-extracted ribosomes, which lack factors essential for chain initiation with natural mRNAs (Hamada et al. 1968, Yang, Hamada, and Schweet 1968).

It now seems probable that initiation of most, although possibly not all, proteins of eucaryotes is entirely analogous to that in bacteria, and requires the obligate association of a species of (nonformylated) methionyl-tRNA with an -ApUpG initiator codon. The cytoplasm of all eucaryotes examined contains at least two major species of methionine-specific tRNAs. Only one of the corresponding species of methionyl-tRNA (Met-$tRNA_{F*}$) can be converted to N-formyl-methionyl-tRNA by the transformylase enzyme of *E. coli*. In model systems for *in vitro* polypeptide synthesis, containing synthetic polyribonucleotides as artificial mRNAs, this species of methionyl-tRNA only transfers methionine to the N-terminal position of the polypeptide formed. On the other hand, the other species of methionyl-tRNA (Met-$tRNA_{M*}$) transfer methionine residues only to internal positions of the polypeptide (Smith and Marker 1970, Takeishi, Sekiya, and Ukita 1970, Yoshida, Watanabe, and Morris 1970). In addition, experimental conditions can be devised under which synthesis of any peptide is dependent upon the presence of both Met-$tRNA_{F*}$ and a template RNA containing one of the procaryote initiator codons (Brown and Smith 1970).

Prior evidence was apparently not in accord with the suggestion that the polypeptide chains of hemoglobin are initiated from an N-terminal methionine residue in this manner. No N-terminal methionine residues had been found in short nascent globin chains isolated from either intact reticulocytes or derived cell-free systems (Gonano and Baglioni 1969, Rahamimoff and Arnstein 1969, Rich, Eikenberry, and Malkin 1966, Wilson and Dintzis 1969). Moreover, the mRNA isolated from reticulocyte polysomes can only serve as a template for *in vitro* synthesis of globin chains by a cell-free system derived from *E. coli* when supplemented specifically with N-acetyl-valyl-tRNA (Laycock and Hunt 1969). On the other hand, in cell-free systems derived from reticulocytes the synthesis of at least some α-chains may be initiated from 2-hydroxy-isovaleryl-tRNA, an analogue of valyl-tRNA (Rahamimoff and Arnstein 1969, Rich, Eikenberry, and Malkin 1966).

However, more recent analyses of the shortest nascent globin chains present in both intact reticulocytes and lysate systems have shown that both globin chains are initiated from N-terminal methionine residues (Housman et al. 1970, Jackson and Hunter 1970, Wilson and Dintzis 1970, Yoshida, Watanabe, and Morris 1970). This N-terminal residue is released from the growing peptide chain while the latter is quite short (Jackson and Hunter 1970, Yoshida, Watanabe, and Morris 1970). Similarly, the transient presence of N-terminal methionine residues has been reported in nascent chains of yeast isocytochrome *c* (Stewart et al. 1969), protamines of trout testis (Wigle and Dixon 1970), all of eight polypeptides specified by the genome of adenovirus-2 (Caffier et al. 1971), and of undetermined proteins of *Neurospora crassa* (Rho and DeBusk 1971). Further, N-acetylmethionine has been reported as the N-terminal residue of proteins from the thorax of the honeybee (*Apis mellifica*) (Polz and Dreil 1970). Nevertheless, it still remains to be established that *all* polypeptide chains are initiated with an N-terminal methionine residue.

Various N-acyl-amino acid residues, and particularly acetyl-amino acids, are present in the N-terminal positions of eucaryote proteins (e.g., Kreil and Kreil-Kiss 1967, Narita, Tsuchida, and Ogata 1968, Press, Piggot, and Porter 1966). Further, a variety of N-acetyl amino acids will serve as substrates in reactions of the aminoacyl-tRNA synthetase enzymes (e.g., Pearlman and Bloch 1963). The possibility that polypeptide chain synthesis in eucaryote cytoplasm might be initiated from N-acyl-aminoacyl-tRNAs has therefore been considered repeatedly. It now seems highly improbable that this is a common general mechanism of chain initiation in eucaryotes. In general, N-terminal acyl groups are commonly introduced into completed polypeptide chains and, more important, acyl-aminoacyl-tRNAs do not usually initiate peptide chain synthesis in cell-free extracts of eucaryotes at low magnesium concentrations (e.g., Mosteller, Culp, and Hardesty 1968[a], Siler and Moldave 1969). Similarly, the pyrrolidone-carboxylic acid N-terminal residues of certain immunoglobulin chains appear to be generated from N-terminal glutamate residues (Baglioni 1970), although pyrrolidone-carboxylyl-tRNA can be formed from glutamyl-tRNA (Moav and Harris 1967, Rush and Starr 1970).

However, the possibility that some peptide chains are initiated from N-acetyl-aminoacyl-tRNAs in eucaryote cells has yet to be excluded. For example, "nascent" chains of histones IV and IIb1 (see Chapter 7) synthesized in regenerating rat liver contain the N-terminal acetyl-serine residue of the completed peptide chain, and inhibition of protein synthesis by puromycin also blocks incorporation of acetate into the nascent chains (Liew, Haslett, and Allfrey 1970). These observations are entirely in accord with a mechanism of chain initiation from methionyl-$tRNA_{F*}$, followed by acetylation of the nascent chain after removal of the initial N-terminal methionine residue. However, N-acetyl-seryl-tRNA can be isolated from the regenerating liver, and present evidence is in accord with a mechanism of chain initiation from the acetyl-seryl-tRNA (Liew, Haslett, and Allfrey 1970). Similarly, the presence of N-acetyl-glycine in nascent peptides from hen oviduct minces, and particularly the formation of N-acetyl-glycyl-puromycin by such minces, are in accord with the suggestion that synthesis of ovalbumin may be initiated from N-acetyl-glycyl-tRNA (Narita, Tsuchida, and Ogata 1968, Narita et al. 1969).

These examples, together with that of initiation of hemoglobin α-chain synthesis from 2-hydroxy-isovaleryl-tRNA (Rahamimoff and Arnstein 1969, Rich, Eikenberry, and Malkin 1966), may possibly reflect a more general role of the eucaryote methionyl-$tRNA_{F*}$. Association, at an "entry site" of the ribosome, of this species of methionyl-tRNA with an initiator -ApUpG codon immediately preceding that for the N-terminal amino acid of the peptide chain would provide a specific mechanism for alignment of the mRNA "in register" against the acceptor and donor sites of the ribosome (Smith and Marcker 1970).[20] Transient incorporation of the methionine residue into the N-terminal position of nascent chains could then be an "optional" feature of the peptide chain-initiation process, but not an invariable essential requirement.

PROTEIN SYNTHESIS IN CELL ORGANELLES

Mitochondria and chloroplasts are semi-autonomous genetic entities which may play a major role in the processes of differentiation (e.g., S. Nass 1969). They are in many respects similar to saprophytic bacteria, from which they may indeed have evolved. Both types of organelle contain DNA and carry non-nuclear genes which are revealed in the phenomenon of *maternal inheritance*. Both types of organelle contain enzyme systems necessary to the replication of DNA, the synthesis of various species of RNA, and the synthesis of protein (S. Nass 1969, Spencer and Whitfield 1967). The mitochondria appear to have been more extensively studied than chloroplasts in these respects. However, no major difference between them has been reported to date and they will be considered as one class. Properties of the mitochondria have been extensively surveyed recently (M. M. K. Nass 1969, S. Nass 1969, Rabinowitz et al. 1969), and only two aspects particularly pertinent to the topic of regulation of protein synthesis will be considered here.

First, the components of the protein-synthesizing systems

[20] An essentially similar proposal has been developed in detail (Culp, McKeehan, and Hardesty 1969[a], 1969[b]) to account for the observation that initiation of polyphenylalanine synthesis in reticulocyte systems appears to be dependent upon the presence specifically of de-acylated phenylalanine-specific tRNA (Culp, Mosteller, and Hardesty 1968, Culp, McKeehan, and Hardesty 1969[a], Mosteller, Culp, and Hardesty 1968[b]). However, alternative interpretations of these data are possible.

of mitochondria and chloroplasts are quite distinct from those of the surrounding cytoplasm. Their ribosomes differ in size and composition (Borst and Grivell 1971, Fauman, Rabinowitz, and Getz 1969, Rawson and Stutz 1969, Rifkin, Wood, and Luck 1967, Stutz and Noll 1967). Complements of transfer RNAs, aminoacyl-tRNA synthetases, and elongation factors are distinct (Barnett, Brown, and Epler 1967, Buck and Nass 1969, Reger et al. 1970, Richter and Lipmann 1970). Even more striking, these organelles contain and initiate synthesis with N-formyl-methionyl-tRNA (e.g., Epler, Shugart, and Barnett 1970, Sala and Kuntzel 1970, Schwartz et al. 1967), whereas the cytoplasmic system does not. Second, the organelles differ also from the cytoplasm in showing a "bacterium-like" response to various inhibitors of protein synthesis. In particular, the organelles from most types of cell are sensitive to inhibition by the antibacterial agent chloramphenicol but insensitive to the antifungal agent cycloheximide, whereas the cytoplasmic system shows the converse responses (e.g., Lamb, Clark-Walker, and Linnane 1968, Goffeau and Brachet 1965, Loeb and Hubby 1968). These marked differences in properties provide a mechanism by which the synthesis of protein in the organelles may be regulated independently of the cytoplasm in response to an exogenous agent.

Despite the distinct character of the protein-synthesizing machinery of the two types of organelle, the latter are not completely independent from the synthetic machinery of the surrounding cytoplasm. A single circular molecule of mitochondrial DNA is not sufficiently large to code both for all the components of the protein-synthesizing machinery of the mitochondrion and all the proteins found in the mitochondrion. Indeed, one such molecule could code for only approximately 30 polypeptide chains of average molecular weight 20,000. Although some mitochondria may contain more than one (and possibly different) DNA molecule, it is clear that individual mitochondria do not contain sufficient genetic information to specify all the proteins found therein (M. M. K. Nass 1969). This conclusion is reinforced by the demonstration that approximately 10 percent of the mitochondrial genome specifies the distinctively mitochondrial ribosomal RNAs (Wood and Luck 1969, Zylber, Vesco, and Penman 1969) and a further portion specifies the transfer RNAs characteristic of the mitochondria (Nass and Buck 1969, Zylber and Penman 1969). Nevertheless, the organelle does contain genetic information specifying species of messenger RNA (Borst and Aaij 1969, Vesco and Penman 1969). The question is then posed of determining which proteins are coded for, and synthesized in, the organelle.

Early studies indicated that the bulk of the protein synthesized by the mitochondria was nonsoluble or structural protein (Roodyn, Reis, and Work 1961, Roodyn, Suttie, and Work 1962). More recent studies indicate that approximately 80 percent of the protein synthesized by the mitochondrion is structural protein of the inner membrane (Beattie, Basford, and Koritz 1967, Kadenbach 1967[a], Yang and Criddle 1969; but see Senior and MacLennan 1970). Protein of the outer membrane and many of the enzymically active constituents of the mitochondrion are made by the cytoplasmic system and subsequently migrate to the mitochondrion (Beattie, Basford, and Koritz 1967, Clark-Walker and Linnane 1967, Kadenbach 1967[b]). The proteins of the mitochondrial ribosomes and at least one of the aminoacyl-tRNA synthetases appear to be included in the list of the constituents synthesized outside the mitochondria (Davey, Yu, and Linnane 1969, Gross, McCoy, and Gilmore 1968, Küntzel 1969, Neupert et al. 1969). Some of the enzyme proteins which "migrate" into the mitochondria appear to undergo a modification of structure, or acquire prosthetic group, after entry into the mitochondrion (Gross, McCoy, and Gilmore 1968, Hayashi, Yoda, and Kikuchi 1969, Schiefer 1969).

The site of synthesis of the cytochromes has not yet been

conclusively established. Chloramphenicol and other "antibacterial" antibiotics completely inhibit accumulation of cytochromes *a*, a_3, *b*, and *c*, by yeast under conditions where growth is not affected (Clark-Walker and Linnane 1966). However, a resistant mutant strain with apparently normal protein-synthesizing machinery (for cytochrome-less mitochondria) still fails to accumulate any of these cytochromes (Davey, Yu, and Linnane 1969). The available data are in accord with the hypothesis that these cytochromes are synthesized in the cytoplasm, are transported into the mitochondrion, and then acquire a heme prosthetic group (Davey, Yu, and Linnane 1969, Kadenbach 1969, Schiefer 1969, Scott and Mitchell 1969). Interpretation is, however, complicated by the fact that the mutant studied by Davey and co-workers is a "cytoplasmic" mutant. Various interpretations are possible, including that of synthesis of protein in the cell cytoplasm on messenger RNAs synthesized in the mitochondrion (Attardi and Attardi 1967, Davey, Yu, and Linnane 1969).

Until recently the question of whether the cell nuclei actually synthesized any protein was of considerable debate. It now seems indisputable that eucaryote nuclei are indeed capable of independent "synthesis" of some protein, and that the major portion of the protein synthesis occurs in association with the nucleoli (e.g., Gallwitz and Mueller 1969, Stellwagen and Cole 1969[a], Zimmerman et al. 1969). However, the bulk of the nuclear basic histones and protamines which are believed to control gene expression appears to be synthesized in the cell cytoplasm (e.g., Kedes et al. 1969[a], Ling, Trevithick, and Dixon 1969, Robbins and Borun 1967). At the present time we know nothing of the role of proteins synthesized in the cell nucleus, and therefore synthesis of protein in the cell nucleus will not be discussed further.

CELL MEMBRANES AND LIPID FACTORS

Early studies of the role of cell membranes in the synthesis of protein were complicated by the observation that amino acid was rapidly incorporated into a lipid fraction (Hendler 1963, 1965, Hunter and Goodsall 1961, Hunter and James 1963). The primary products of the reaction are esters of phosphatidyl-glycerol (MacFarlane 1962), and aminoacyl-tRNAs can serve as primary donors of the amino acid residue (Gould et al. 1968, Nesbit and Lennarz 1968). This reaction is not believed to serve in the synthesis of protein (Hunter and James 1963).

Recent studies indicate that the polysomes which are bound to the cell membranes constitute a qualitatively distinct pool and are not freely interchangeable with those free in the cytoplasm of eucaryote cells (Aronson and Wilt 1969, Murty and Hallinan 1968). Membrane-bound polysomes appear to be engaged primarily in the synthesis of structural protein of the membrane itself and of proteins which are exported from the cell, whereas free polysomes appear to synthesize proteins of the cell sap (Ganoza and Williams 1969, Gaye and Denamur 1970, Redman 1969, Takagi, Tanaka, and Ogata 1970). The nascent chains of "export" proteins appear to elongate into the membrane (Sabatini and Blobel 1970), and the completed export proteins remain associated with vesicles of membrane before active secretion from the cell (Ganoza and Williams 1969, Peters 1962[a], 1962[b]).

The mechanism which serves to discriminate between the two classes of polysome remains to be elucidated. Fridlender and Wettstein (1970) have reported two differences in the complements of proteins in the large subunits of free and bound polysomes, respectively, of embryonic chick cells. It is not known whether this represents substitution of one protein by another, or conversion of one protein into the other. A similar difference between the large subunits of free and bound polysomes has also been reported for bacterial cells (Brown and Abrams 1970), and may prove to be a general phenomenon. Other possible factors which have been suggested include binding of the ribosomes to specific sites in the membrane by spermine (Khawaja and

Raina 1970), and association of the N-terminal portion of the nascent peptide chain with the membrane (Sabatini, Tashiro, and Palade 1966).

There is some suggestive evidence that association of polysomes with cell membranes is subject to regulation by steroid hormones (Cox and Mathias 1969, James, Rabin, and Williams 1969, Sunshine, Williams, and Rubin 1971).

A different role for lipid in the reactions of protein biosynthesis has been indicated by Hradec and Dušek (1968). These workers reported that activity of the aminoacyl-tRNA synthetase enzymes could be reversibly impaired by extraction of the lipid fraction. Activity was restored specifically by cholesterol or cholesterol esters. However, Hradec's group have reported both stimulatory and inhibitory effects of various lipid components of rat liver tissue (e.g., Hradec and Dušek 1968, Hradec and Štroufová 1960). It therefore seems possible that a variety of natural lipids may be capable of modifying the overall activity of a cell in synthesizing protein. Indeed, Moore and Umbreit (1965) have reported the activity in protein synthesis of all subcellular fractions derived from cells of *Streptococcus faecalis* to be directly proportional to the phospholipid content.

ACCESSORY FACTORS

The activity of some of the essential components of protein-synthesizing systems of both procaryotes and eucaryotes may be affected by accesory factors. Certain of these may under special circumstances be envisaged as exerting a regulatory function, and they will therefore be briefly enumerated without discussion.

Synthesis of polypeptide, both *in vivo* and by *in vitro* systems, is absolutely dependent upon an adequate concentration of magnesium ions. This requirement for magnesium can be partially satisfied, and the synthesis of polypeptide stimulated by, natural amines and polyamines (e.g., Takeda 1969[a], Tanner 1967). Extensive study has been made of the effects of these amines on synthesis of macromolecules, and of variations in cell content of amines under varying growth conditions or during differentiation. However, the possible role(s) of the natural amines as regulators of protein synthesis remain unclear. As the available data have been summarized recently (Cohen 1968, Cohen and Raina 1967, Herbst and Bachrach 1970), they will not be discussed further.

In addition to the polysome system for synthesis of protein, cells of *Escherichia coli* and *Alcaligenes faecalis* contain an enzyme system capable of synthesizing small specific oligopeptides. Extensive studies by Beljanski and his co-workers have shown that there are a minimum of four distinct polypeptide synthetase enzyme reactions. Each requires one specific ribonucleoside triphosphate and catalyzes incorporation only of a selected group of amino acids, yielding specific oligopeptides containing up to eight residues. The formation of peptide does not require ribosomes, or aminoacyl-tRNAs, but is markedly stimulated by messenger RNA(s) (Beljanski 1965, Beljanski, Beljanski, and Lovigny 1962). In addition to forming peptide, the enzyme system also attaches amino acid in ester linkage to specific base residues in the messenger RNA (Beljanski, Fischer-Ferraro, and Bourgarel 1968). A similar system has also been demonstrated in rat liver (Beljanski 1960, Zalta, Lachurie, and Osono 1960). To date, no direct effect of this system upon the polysome system has been shown, either as a source of peptide in an alternative mechanism of initiation of peptide synthesis or in affecting the availability of messenger RNAs.

Any consideration of mechanisms regulating the synthesis of protein in living cells must necessarily take into consideration the dynamic "turnover" of cellular constituents. Accumulation of protein is not synonymous with synthesis of protein. The detailed mechanism whereby protein is degraded during the processes of turnover in undamaged cells is perhaps now more of a mystery than that of biosynthesis.

Normal protein catabolism requires a source of energy and coenzyme A and is (at least in part) distinct from the autolytic degradation processes (Penn 1960, Simpson 1953, Steinberg, Vaughan, and Anfinsen 1956). A number of instances in which an increase in intracellular level of specific enzyme results from a decreased rate of degradation have now been established (Schimke 1966).

REGULATION OF PROTEIN SYNTHESIS

Of the various components of the protein-synthesizing systems described, most have already been directly implicated in one or more instance(s) of regulation of protein synthesis.

Regulation of the synthesis of specific proteins has been shown to be affected by:

(a) primarily, the level of specific messenger RNAs, determined by the balance between *transcription* of appropriate structural genes and degradation of messenger RNA,
(b) presence or absence of *translation inhibitor* proteins, and possibly specific initiation factors,
(c) availability of specific "minor species" of *transfer RNAs* and corresponding synthetase enzymes, and
(d) levels of various prosthetic groups, and interactions with other nonidentical subunits of a common protein.

Superimposed on the regulation of synthesis of specific proteins, synthesis of protein in general may be markedly affected by the levels of all other components of the systems.

In the ensuing chapters, I shall first describe some of the best-documented regulatory phenomena in bacterial cells, and then consider the mechanisms regulating these processes. These will then be compared and contrasted with regulatory processes elucidated in cells of higher forms.

There are evidently causes for these curious manifestations of the play of molecular forces. To indicate them in a precise manner would certainly be a difficult matter.—Louis Pasteur, 1861.

How it may be that from indistinct causes there may issue effects manifest and immediate. . . .—Leonardo da Vinci, ca. 1490.

CHAPTER 2: **NUCLEAR AND CHROMOSOME DIFFERENTIATION, GENE AMPLIFICATION, AND CELL CYCLES**

CONSTANCY OF THE EUCARYOTE GENOME AND CONCEPTS OF TOTIPOTENCY AND DIFFERENTIAL GENE ACTIVITY

It has been believed for many years (e.g., Sturtevant 1951) that, as a general rule, all the somatic cells of any individual multicellular organism contain identical genomes. There are a few well-documented cases in which embryonic development was clearly associated with a cytologically visible loss of chromosomes in the early stages of development. However, these have been recognized as obvious exceptions to the otherwise valid general rule that differentiation is not based upon selective or sequential loss of portions of the genome (e.g., Markert 1964). Indeed, differentiation demonstrably reflects intercellular variations in selection of portions of the total cell genome to be expressed, a phenomenon of *variable* or *differential gene activity* (e.g., Davidson 1968, Gurdon 1968[a]).

Initial stages in the diversification of cell types (i.e., of selection of portions of the genome expressed) begins in the early cleavage stages (e.g., Mintz 1967, Mintz and Palm 1969, Wolpert and Gingell 1970). This probably largely reflects localized variations in the distribution of organelles and other macromolecular components of the egg cytoplasm (e.g., Clement 1962, 1968, Davidson et al. 1965, Pucci-Minafra, Minafra, and Collier 1969). It may also partly reflect secondary minor regional differences in concentrations of critical metabolites. Indeed, Novick and Weiner (1957) have demonstrated with a model system that cells of identical genotype may display markedly different phenotypes in the presence of a critical threshold concentration of a regulatory substrate.

The body of evidence in support of the twin concepts of constancy of the genome and differential gene activity is truly monumental (see Davidson 1968, DuPraw 1968) and cannot be reviewed in detail here. Some of the more recent studies are particularly persuasive. For example, Laskey and Gurdon (1970) have raised fertile adult toads (*Xenopus laevis*) from functionally enucleate eggs into which they transplanted nuclei from mature adult tissue. Similarly, Steward and coworkers (Steward 1970, Steward et al. 1964) have raised mature flowering plants from single somatic cells. Moreover, no large differences have been found in the base sequences of DNA isolated from different tissues of one animal or different individuals of the same species (Britten and Kohne 1968, McCarthy and Hoyer 1964).

However, it seems important to note that the concept of constancy of the genome does imply considerably more than has actually been demonstrated by experiment. Indeed, it may prove important to our further understanding of the regulatory processes involved in differentiation to make explicit qualifications. The demonstration that nuclei of both developing and mature tissues can support normal development of enucleate eggs (e.g., Gurdon and Uehlinger 1966, King and Briggs 1956, Laskey and Gurdon 1970) indicates those nuclei to be *totipotent.* These specific nuclei must contain all the (essential) genes present in the original fertilized egg from which the donor organism developed. However, the demonstration that these particular nuclei are totipotent does not establish that all types of nuclei in all types of cell are equally totipotent. There is certainly a considerable body of evidence that the course of differentiation is not associated with *marked, irreversible* changes in the nuclear genome. However, the possibilities that particular developmental sequences entail (a) very small, but critical, specific losses or gains of part of the genome, or (b) *reversible* modification of the genome, cannot be considered to be formally excluded for most experimental systems (e.g., Schultz 1952). Indeed, we now know that one pronounced feature in some differentiating systems is a marked, *reversible* alteration of the genome which results from a process of "*gene amplification*" (see page 60). Further, it has recently been proposed that "commitment" of antibody forming cells to produce one particular antibody (see Chapter 11) might involve the irreversible loss of a small portion of the genome (Smithies 1970).

It may also prove useful to make an operational distinction between those processes which we know to lead to quantitative changes in the number of genes in the genome and possible unknown, potentially reversible qualitative modifications of the genome (e.g., Scarano and Augusti-Tocco 1967).

THE PHENOMENON OF DETERMINATION

The nuclei of individual cells undergo a series of "functional" or "physiological differentiations" which are regulated by the activities of the cell cytoplasm (Davidson 1968, Gurdon 1968[a], 1968[b]). At each stage of this series of *progressive determination* the future course of development becomes ever more specifically restricted or *determined,* and the cell progresses through a series of *determined states.* For some cell lines each of these determined states is transient or "unstable," and the cell develops continuously along its determined course to its final differentiated form. Other cells are destined to be the future germ cells and the course of their differentiation is brought to a temporary halt at an intermediate "stable" determined state until the onset of maturity. Still other cells are destined to become *stem* cells which, for example, both give rise to antibody-forming cells and propagate more stem cells throughout the life of the organism. In this type of determined state the *extent* of further differentiation has not been determined at the stage of temporary arrest of development, and it may be considered "metastable." A more extreme situation is found in the imaginal discs of some insects. The cells of these discs remain in an embryonic state throughout the larval period of development because of establishment of a "quasi-stable" heritable modification. Such cells can be propagated in the determined embryonic state for many generations by transplantation into the abdomen of an adult fly. During the course of serial transplantation some of the cells may undergo reversible changes to new quasi-stable, heritable determined states. This process of *transdetermination* has been reviewed by Hadorn (1965, 1966) and will not be further discussed. A similar quasi-stable determined state has been observed with a "multi-potent" cell line maintained in culture by Finch and Ephrussi (1967).

Thus the process of determination may progress through intermediate determined states showing a wide range of degrees of stability of determination. In at least some cases, the entire course of a sequence of progressive determinations occurs without irreversible modification of the genome (Gurdon and Uehlinger 1966, Steward 1970), and any possible reversible modifications are clearly readily reversible. However, it remains possible that these cases are no less exceptional than those in which substantial portions of the genome are irreversibly lost (Brown 1969, Schultz 1965, White 1951) or in which particular cells substantially increase the gene content by endomitosis and become polyploid (e.g., Painter 1940, Painter and Reindorp 1939).

It seems possible that there is also a wide range of degree

of stability of differentiation or reversibility in the determined states of fully differentiated cells. A general situation can be envisaged in which progressive determination is paralleled by reversible but progressive modification of the genome, such that the further differentiation has progressed the more difficult it becomes to reverse the modification.[1]

The frequency of successful demonstrations of totipotency of transplanted nuclei declines after the blastula stage of development of donor frogs (King and Briggs 1956, Subtelny 1965[a], 1965[b]). Similarly, the frequency of successful transplants is different for nuclei taken from different regions or tissues of the same embryo (Briggs, Signoret, and Humphrey 1964, Gurdon and Laskey 1970, Smith 1965). These observations may reflect differences in the frequencies with which chromosomes fail to separate during the first cycle of replication in transplanted nuclei taken from different sources. In turn, such differences may reflect other differences in the cell cycle between the various donor tissues (Briggs, Signoret, and Humphrey 1964, Gurdon and Laskey 1970). However, neither the data nor the latter interpretation appears to be at variance with the hypothesis that nuclei from different sources vary in the extent to which the genome has been reversibly modified, and in ease of reversion of such modifications.

Nevertheless, such a suggestion is very highly speculative and not supported by any particularly persuasive experimental evidence. Therefore, it will not be considered further.

HETEROCHROMATINIZATION AND DIFFERENTIAL GENE INACTIVATION

The initial stages of the processes leading to elimination of a portion of the genome appear (at least superficially) similar to the more common process of differentiation by

[1] It should be noted that demonstration of the totipotency of nuclei of adult tissues (Laskey and Gurdon 1970) was made with nuclei from cells growing rapidly in tissue culture.

heterochromatinization.[2] However, there is no compelling reason to regard elimination of genetic material other than as an exceptional process. Therefore, no further discussion will be made of this phenomenon, except to note that a regulated elimination of a specific portion of the cell genome constitutes a highly effective and very efficient means of controlling selection of the proteins synthesized by a cell.

The process of heterochromatinization, or *heteropyknosis,* leads to a physiological inactivation of portions of the genome and is effectively equivalent in function to a process of reversible gene elimination. Indeed, one of the general features of cell differentiation is that each step of the sequence of progressive determination requires a period of prior or simultaneous DNA synthesis (e.g., Ebert 1968, Fernandes, Castellani, and Kimura 1969, Nossal et al. 1967, Turkington 1968[a]), which has been attributed to a requirement to "clean the genes" (Ebert 1968, Gurdon 1969). Development of regions of heterochromatin selectively limits the spectrum of available genes which *may be differentially activated* and expressed within any given cell. Thus heterochromatic regions of a chromosome are believed to be "unconditionally repressed" and inactive, whereas regions of euchromatin may be either active or "conditionally repressed" and potentially active (see Chapter 7). The process is illustrated in its most extreme form by mature avian erythrocytes. In the case of mammalian erythrocytes the cell nucleus is actively expelled at the stage of the late erythroblast, or normoblast (Pinheiro, Leblond and Droz 1964, Skutelsky and Danon 1967). However, in birds and some other classes of animal the nucleus is retained, but the chromosomes become entirely heterochromatic. In the process the nuclei lose ability to support the synthesis of nucleic acids or protein and appear genetically inert (Cameron and

[2] Use of the terms heterochromatin and euchromatin is based upon cytological examination of the chromosomes after staining with dyes. Regions of heterochromatin stain intensely and are compact in structure.

Prescott 1964). This process of inactivation is at least partially reversible, for when the nucleus of a mature avian erythrocyte is introduced into the cytoplasm of other types of cell it may resume synthesis of nucleic acids (Harris et al. 1966, Johnson and Harris 1969) and control the synthesis of specific proteins coded only in the erythrocyte genome (Cook 1970, Harris and Cook 1969). Similarly, the micronuclei of ciliates appear to be largely, or totally, physiologically inert during vegetative growth and fission, yet they are "activated" to become the primary genetic determinant during conjugation (Gorovsky and Woodard 1969, Nanney 1964).

Physiological inactivation of an entire nucleus (Cameron and Prescott 1964) or of one complete set of parental chromosomes (Brown 1969) does, however, appear to represent an extreme situation. On the other hand, the inactivation of all, or virtually all, of an entire chromosome may be a more common event (e.g., Hsu, Schmid, and Stubblefield 1964, Lyon 1962, 1968, Mittwoch 1967, 1969). It has been repeatedly observed that in female mammalian cells one of the two X chromosomes is *heteropyknotic,* and in culture this particular chromosome appears to be "*late-replicating*" out of phase with its homologue and the autosomal chromosomes. Further, tissues of adult females which are heterozygous for sex-linked characters appear to be "genetic mosaics" in which only one member of each pair of alleles is expressed in each individual cell (e.g., Beutler, Yeh, and Fairbanks 1962, Davidson, Nitowsky, and Childs 1963).[3] It is generally believed that the gene allele which is not expressed in a particular cell is carried by the X chromosome which has become heteropyknotic and genetically inactive. Although this correlation has not been formally established, it is in accord with the common observation that *heterochromatic* regions of

[3] This situation is not observed in at least some types of viable interspecies hybrids, where there appears to be a selective inactivation of the X chromosome of paternal origin (Hamerton et al. 1969, Richardson, Czuppon, and Sharman 1971, Sharman 1971).

eucaryote chromosomes are less active than the other *euchromatic* regions in supporting the synthesis of ribonucleic acid. A similar late-replication and inactivation also occur for the male Y chromosome.

The fact that the entire chromosome appears to become inactivated in this phenomenon is not necessarily at variance with an increasing number of reports of developmental processes attributed to the action of genes located within the heterochromatic sex chromosome. Development of the heterochromatic condition does not appear to arise immediately after fertilization of the egg. Neither X chromosome is late-replicating during early cleavage of young hamsters and rabbits (Hill and Yunis 1967, Kinsey 1967, Lyon 1968). It therefore seems probable that the X chromosome which is destined to become heteropyknotic is expressed during the very early stages of development. Indeed, Morris (1968) has presented compelling evidence that at least one X chromosome is required for development of fetal mice to the stage of normal implantation, and a full complement of sex chromosomes is required for normal early development.

The mechanism responsible for determining which of the X chromosomes shall become inactivated is not known. In formulating what is known as the "Lyon hypothesis" to account for the cytological and genetic observations made for sex chromosomes, Lyon (1961) suggested that one X chromosome was inactivated in a random selection process. Most observations seem to be in accord with the postulated random inactivation of an X chromosome.[4]

Although the basis for inactivation of one of the sex chromosomes is unknown, it is recognized to be a finely regulated process. In individuals who inherit more than two sex chromosomes all but one of the X chromosomes is inactivated (e.g., Lyon 1962, Mittwoch 1967). However, in tissue cultures of tetraploid *syncarya,* formed by fusion of two different cell lines, two X chromosomes remain active and are ex-

[4] See footnote 3.

pressed (Silagi, Darlington, and Bruce 1969, Siniscalco et al. 1969). The process does not appear to involve any distinct or unique histone fraction (Comings 1967). Mukherjee and co-workers (1968) have reported that replication of the late-replicating chromosome is initiated and completed in synchrony with its homologue. They suggest that one X chromosome is replicated more rapidly than the other late in the period of chromosome replication and (as a corollary) more slowly during the early phase of replication. Possibly this reflects structural differences in the sites of initiation of replication of particular regions of the chromosome.

The most common situation is one in which only selected regions of particular chromosomes become heterochromatic and inactive. Individual tissues show variations in the pattern of distribution of heterochromatin and euchromatin at different stages of development, and the patterns vary for different tissues of the same organism (e.g., Beermann 1963, Clever 1964). These variations reflect the mechanisms regulating differential gene expression which are discussed in detail in Chapter 7 and will not be further discussed here. It is uncertain whether differential inactivation of only one member of a pair of autosomal genes in a process analogous to inactivation of entire sex chromosomes ever occurs. The possibility cannot be considered totally excluded, for it appears to be a common observation that antibody-forming cells heterozygous for "heavy-chain" alleles and for allotype alleles yield only antibody immunoglobulins containing one specific type of heavy chain and one allotype-specific amino acid sequence (Cebra, Colberg, and Dray 1966, Herzenberg, McDevitt, and Herzenberg 1968, Merchant and Brahmi 1970, Pernis et al. 1965). However, the mechanism and regulation of antibody synthesis show several other unusual features (some of which are considered in Chapter 11), and there are other possible interpretations of the data. For example, these might be further instances of "*preferential activation*" of only one allele of a pair. To date, this phenomenon has been clearly observed only with the few viable "extreme hybrid" progeny of selected interspecies crosses (e.g., between quail and domestic chickens); and never with intraspecies hybrids. In these cases only alleles of maternal origin are expressed in the early stages of development (e.g., Hitzeroth et al. 1968, Ohno et al. 1968, 1969, Wright and Subtelny 1971). Alternatively, it seems possible that synthesis of immunoglobulin peptide chains specified by only one member of a pair of alleles could reflect selective gene amplification (Krueger and McCarthy 1970, Little and Donahue 1970).

GENE AMPLIFICATION

A number of recent studies indicate that reversible modification of the cell genome by "*gene amplification*" is a major feature in the development of some types of cell. This process consists of the accumulation of multiple copies of one, or a few, specific gene(s) by repeated replication of selected portions of the genome. It is dramatically illustrated by the marked increase in the number of nucleoli which occurs in the late stages of oocyte maturation in fish and amphibia.

Nucleoli of various animal species have been the subject of extensive cytological and biochemical studies for many years (e.g., Brown 1967, Davidson 1968, DuPraw 1968). These intranuclear organelles contain circular molecules of DNA (Callan 1966, Miller 1966, Peacock 1965) and are very active sites of RNA synthesis (e.g., Miller and Beatty 1969[a], 1969[b], McMaster-Kaye and Taylor 1958, Sirlin, Kato, and Jacob 1961). A variety of experimental procedures have been exploited to establish that the nucleoli are the site of synthesis of the major RNAs of the cytoplasmic ribosomes, and of assembly of the ribosome subunits. These processes (see Chapter 8) involve the synthesis of a large precursor of the ribosomal RNAs (Edström and Daneholt 1967, Perry, Srinivasan, and Kelley 1964, Scherrer, Latham, and Darnell 1963). This is then methylated and cleaved before association of the two ribosomal RNAs with the corresponding ribosomal proteins (Greenberg and Penman 1966, Ozban, Tandler, and Sirlin 1964, Warner 1966, Warner and Soeiro 1967). Application of the technique of "DNA-RNA hybridization" has indicated that each somatic nucleolus contains more than 200 cistrons specifying the ribosomal precursor RNA in Hela cells (McConkey and Hopkins 1964), *Xenopus laevis* (Brown and Weber 1968[a]), *Drosophila melanogaster* (Ritossa and Spiegelman 1965), and chickens (Ritossa et al. 1966).

Nucleoli are derived from defined regions or loci of the chromosomes known as *nucleolar organizer* regions, and somatic cell nuclei typically contain one distinct nucleolus per haploid set of chromosomes. Analysis of the DNA of such cells by the technique of DNA-RNA hybridization shows the nucleolar organizer region to contain the same species of multiple cistrons as the homologous nucleolus (Chipchase and Birnstiel 1964, Ritossa et al. 1966). Genetic (Ritossa, Atwood, and Spiegelman 1966, Ritossa and Scala 1969, Schalet 1969) and physico-chemical evidence (Birnstiel et al. 1968, Brown and Weber 1968[a], 1968[b], Quagliarotti and Ritossa 1968) indicates these large numbers of apparently identical cistrons to be very closely linked, possibly alternating with genetic material of unknown function. Thus both the nucleolus and the nucleolar organizer region contain a large reiterated sequence of apparently identical cistrons. Further, studies made with various stocks of *Drosophila* and *Xenopus laevis* indicate these genes to be functionally redundant, for elimination of a substantial number does not lead to a change in phenotype or reduction in the ribosome content of the cell (Brown and Gurdon 1964, Kiefer 1969, Ritossa and Atwood 1966, Ritossa and Spiegelman 1965).

In contrast to the somatic cells, developing oocytes of fish and amphibia contain several hundred nucleoli per haploid genome; published numbers for *Xenopus laevis* range from 600 to 1600 nucleoli per oocyte (Brown and Dawid 1968,

Perkowska, MacGregor, and Birnstiel 1968). They form a cytologically distinct nuclear cap which contains the reiterated DNA sequence specifying ribosomal RNAs in an amount approximately three times that of the entire DNA content of the chromosomal genome. The localization of such a high content of one sequence of DNA bases permits ready demonstration of DNA-RNA hybridization in cytological preparations (John, Birnstiel, and Jones 1969, Pardue and Gall 1969). Thus during the course of oogenesis there is a very marked increase in the number of cistrons specifying ribosomal RNAs over and above the large number present in somatic cells.[5] Similar, but less extensive, amplification of the cistrons for ribosomal RNAs has also been reported for some invertebrates (Brown and Dawid 1968, Dawid and Brown 1970, Lima-de-Faria, Birnstiel, and Jaworska 1969) and may be inferred for at least some plants (Chen and Osborn 1970). However, similar amplification was not found in starfish oocytes (Vincent et al. 1969).

These examples of gene amplification are consequences of the maturation processes rather than causes. They lead to accumulation in the mature oocyte of a sufficient complement of ribosomes and polysomes to support subsequent embryonic development to the stage of the blastula (see Chapter 6). In contrast, a second well-documented example of the phenomenon very probably represents a case in which gene amplification regulates the course of subsequent development.

The giant multifiber *polytene* chromosomes of the salivary glands, and certain other tissues, of dipteran larvae contain enlarged regions of disperse structure known as chromo-

[5] These supernumerary cistrons differ from those of the nucleolar organizer in not containing 5-methyl·cytosine (Dawid, Brown, and Reeder 1970). Wallace, Morray, and Langridge (1971) have recently presented a persuasive argument that they are generated by reiterated replication of genes present in autonomous plasmid nucleoli. It is postulated that these cytoplasmic nucleoli are distributed exclusively to presumptive germ-line cells in the early stages of cleavage of the fertilized egg.

some "puffs" and Balbiani rings. Distributions of such puffs within the chromosomes vary in characteristic sequences which can be correlated with the course of embryonic development and may be modified in response to treatments which affect the latter (Beermann 1963, Clever 1964, Pavan 1965). Further, the puffs appear to be particularly intense sites of synthesis (Clever and Romball 1966, Pavan 1965, Rudkin and Woods 1959) of heterodisperse cRNAs (Daneholt et al. 1969[a], 1969[b], Edström and Daneholt 1967). In larvae of *Rhynchosciara angelae* and related sciarid flies the puffs are also sites of extensive synthesis of DNA or gene amplification (Bradshaw and Papaconstantinou 1970, Pavan 1965, Pavan and Da Cunha 1959). This occurs either immediately preceding, or concurrently with, the intense synthesis of RNA (see Pavan and Da Cunha 1959). However, the chromosomal puffs of other species of dipterans do not appear to be sites of gene amplification (e.g., Rudkin and Woods 1959).

At the present time it seems unlikely that the process of gene amplification serves as a general mechanism for controlling the course of cell differentiation. Fluctuations in levels of cistrons specifying ribosomal RNAs could be expected to affect overall rates of protein synthesis but not to determine selectively the species of proteins formed. On the other hand, selective gene amplification in chromosome puffs does not appear to be a general feature in all dipteran larvae. Furthermore, analyses of the DNA sequences isolated from different tissues of one animal, or different individuals of the same species, by the technique of DNA-DNA hybridization (Britten and Kohne 1968, McCarthy and Hoyer 1964) indicate the absence of similar extensive gene amplification in all tissues examined, except possibly for those engaged in synthesis of immunoglobulin chains (Krueger and McCarthy 1970, Little and Donahue 1970).

Nevertheless, the possibility that less marked processes of gene amplification may play a role in controlling cell differentiation cannot be totally dismissed on the basis of evidence currently available.[6] In both the well-documented types of example the segments of the genome which are amplified already contain large numbers of copies of a single species of gene and are said to exhibit *gene redundancy*. Amplification of these segments by reiterated replication generates temporary additional redundancy. Techniques which have proven adequate to demonstrate amplification in these cases are unlikely to be adequate for demonstration of any possible limited amplification of genes present in the genome as single, unique base sequences.

The possibility that such genes are amplified was particularly emphasized by a report of the presence in embryonic muscle, and other tissues, of cytoplasmic particles which Bell (1969) named *I-somes*. These particles contained DNA which served as template for synthesis of most of the rapidly labeled (presumed messenger) RNAs recovered in the cell polysomes. However, subsequent studies have established that these particles are artifacts (Fromson and Nemer 1970, Müller, Zahn, and Beyer 1970, Williamson 1970).

GENE REDUNDANCY

As noted above, the process of gene amplification generates a temporary gene redundancy. This consequence inevitably created some confusion between the two phenomena of gene amplification and gene redundancy. It is therefore appropriate to digress briefly to consider the phenomenon of gene redundancy.

One of the properties of a messenger RNA which may be used to provide either qualitative identification or a quantitative assay of concentration is that of forming an RNA-DNA hybrid with one of the strands of the DNA of the gene locus from which it is transcribed (Chapter 5). Application of procedures for formation of such DNA-RNA hybrids, *hy-*

[6] This reservation is especially important in the context of current discussion of the *possible* presence of RNA-dependent DNA polymerase activity in normal cells free of all virus.

bridization, permitted Yankofsky and Spiegelman (1962[a], 1962[b], 1963) to demonstrate that the RNAs of bacterial ribosomes are also transcribed in a similar manner from cistrons contained in the bacterial genome. Further, detailed analysis of the stoichiometry of the reaction indicated that *E. coli* contained approximately ten closely linked copies of the cistrons specifying each species of ribosomal RNA. Subsequently, Giacomoni and Spiegelman (1962) and Goodman and Rich (1962) used similar procedures to demonstrate the presence of some 40 to 50 cistrons specifying transfer RNAs. The most recent estimates for *B. subtilis* indicate some 40 cistrons for transfer RNAs, possibly as 2 operons (Bleyman et al. 1969, Smith et al. 1968), 3 to 4 cistrons for 5S ribosomal RNA, and 9 or 10 cistrons for each of the two major ribosomal RNAs (Smith et al. 1968). Thus even in bacteria there is some gene redundancy for the ribosomal RNA cistrons. Still greater redundancy for ribosomal RNA and transfer RNA cistrons is found in each haploid nuclear genome of *Neurospora crassa* and *Saccharomyces cerevisiae.* As noted above, there is extensive redundancy for the cistrons specifying the cytoplasmic ribosomal RNAs in eucaryote cells.

Early results obtained in application of the DNA-RNA hybridization assay procedure for presumed messenger RNA to cells of a variety of higher plants and animals were unexpected. In comparison with data obtained from microbial systems, both the rate and extent of formation of DNA-RNA hybrids were several orders of magnitude higher than expected from the greater degree of complexity of eucaryote genomes. Similarly, the extent and rate of reassociation of preparations of denatured DNAs were several orders of magnitude higher than predicted from data obtained with preparations of bacterial DNAs. Subsequent detailed analysis of the kinetics of the hybridization reaction and degree of "correct matching" of base pairs in the hybrids formed have led to evolution of our current belief that the genomes of metazoans contain many "families" of large numbers of reiterated identical, or very similar, base sequences.[7] Such "families" of redundant gene sequences are believed to account for approximately half of the cell DNA content in some species, and they range in apparent size from a few copies to several million copies of similar or identical base sequences. The remainder of the DNA is believed to consist of single copies of "unique" base sequences (e.g., see Britten and Kohne 1968).

The inherent precision of the analytical procedures used does not allow the conclusion that all of the redundant sequences of any one family are exactly identical. Indeed, our present concepts of the course of evolution deduced from comparisons of the amino acid sequences of both individual proteins and series of homologous proteins (e.g., Anfinsen 1959, Jukes 1966, King and Jukes 1969) would lead us to expect a number of similar base sequences to be present in the genome of any given organism. Further, concepts of the contribution of development of polyploidy in the evolutionary processes (e.g., Ohno, Wolf, and Atkin 1968, Stebbins 1966) and the demonstrable occurrence of gene duplications (e.g., Bridges 1936, Dixon 1966, Holmes and Markert 1969, Ingram 1963, Thompson, English, and Bleecker 1969) both lead to the same expectation. Nevertheless the extent of the reiteration of similar or identical base sequences appears far too large to be accounted for by these concepts alone. Thus we believe that the chromosomes of higher organisms have retained large numbers of (nearly) identical copies of a substantial proportion of the genes present in the genome.

It would be idle to speculate in detail upon all of the many unresolved problems posed by the presence of families of

[7] Until recently this belief rested entirely on these analyses and the analogy afforded by the demonstrable redundancy of the cistrons "coding" for precursor of the ribosomal RNAs. Small "satellite DNA" isolated from mouse cell nuclei appeared to consist of about a million reiterated identical, or nearly identical, copies of one short sequence of about 300 nucleotide base pairs (Waring and Britten 1966). Direct analysis indicates that this DNA probably consists of an even larger number of very similar, but not identical, shorter base sequences (Southern 1970).

highly redundant genes in the genomes of metazoan cells. However, one obvious question which must be considered is the relationship between the two phenomena of gene redundancy and gene amplication. A common feature of both phenomena is the requirement for some mechanism which ensures that all copies of a single gene maintain the identity of base sequence. In considering this requirement, Callan (1967) postulated a mechanism of gene amplification in which a series of "slave gene copies" formed by replication of a "master gene" were "matched" against the master gene, and "corrected" as necessary, during development of the "lampbrush" configuration of oocyte chromosomes (Figure 2.1). Whitehouse (1967[a]) further developed this proposal

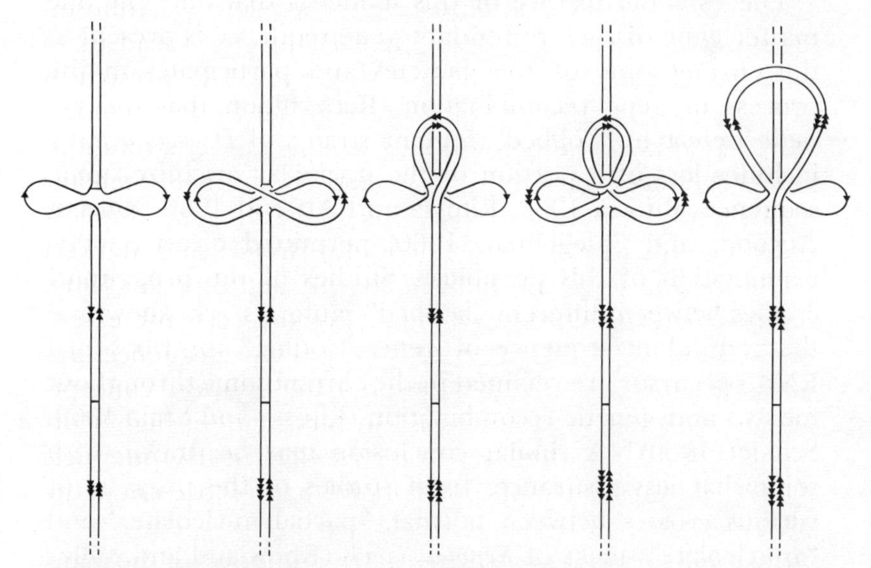

FIGURE 2.1. Model mechanism for matching of slave genes against the master gene

The opposite polarity of complementary polynucleotide chains is indicated by arrows pointing in opposite directions. The limits of repeated coding sequences are indicated by short lines at right angles to the polynucleotide chains. The first (master) sequence is marked with single arrows, the second (first slave) is marked by two, the third (second slave) is marked by three.

Reproduced from Callan (1967) with permission of the author and the Company of Biologists Limited.

to account for the phenomenon of gene redundancy in a "cycloid model" of chromosome structure and behavior. In this model all redundant gene copies are excised from the chromosome during maturation of gametes by recombination between the master gene and the last slave copy in the redundant sequence (Figure 2.2). Following fertilization and gene recombination, excised slave copies could either be reincorporated into the chromosome (Figure 2.2) and corrected as necessary by matching against the corresponding master gene, or replaced by reiterated replication of the master gene. Thus, in this model gene amplification and gene redundancy were considered to be different manifestations of the same phenomenon.

The essential feature of this model is that only the one master gene of each redundant gene sequence is present in the chromosome of the gametes and participates in the process of gene recombination. Recognition that the genetic defect in "bobbed" mutant strains of *D. melanogaster* includes loss of a portion of the nucleolar organizer gene sequence (Ritossa 1968, Ritossa and Atwood 1966, Ritossa, Atwood, and Spiegelman 1966) permitted direct genetic examination of this possibility. Studies of the progeny of crosses between different "bobbed" mutant stocks show that the redundant sequence of genes "coding" for ribosomal RNA precursor are retained in the chromosome throughout meiosis and genetic recombination (Ritossa and Scala 1969, Schalet 1969).[8] A similar conclusion may be drawn, with somewhat less assurance, from studies of the progeny of various crosses between normal, "partial nucleolate," and "anucleolate" stocks of *Xenopus laevis* (Knowland and Miller 1970, Miller and Knowland 1970). Thus redundancy of the

[8] Alternative interpretations based upon mutation in a hypothetical gene controlling the number of reiterated copies in the gene sequence cannot be totally dismissed, for some amplification of these genes ("magnification") does occur in some stocks (see also Atwood 1969, Henderson and Ritossa 1970, Tartof 1971). However, such interpretations appear to require formulation of very specific, multiple *ad hoc* proposals which seem implausible.

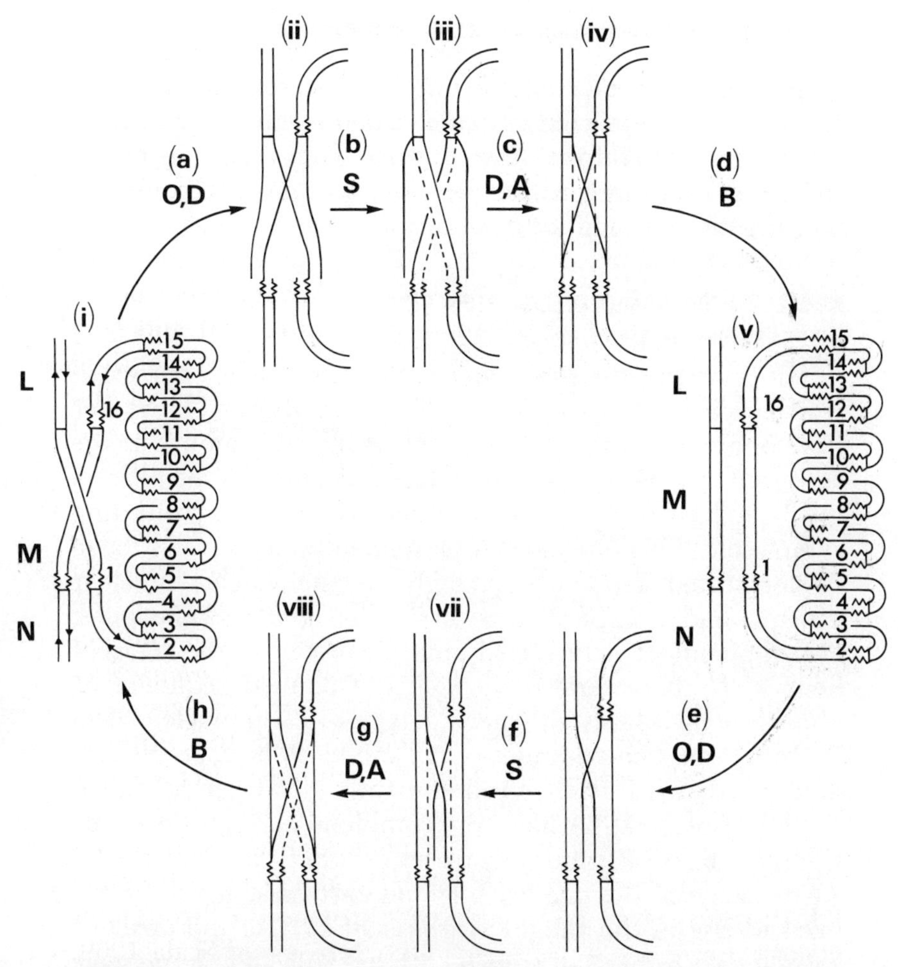

FIGURE 2.2. Proposed mechanism of excision and reincorporation of slave genes in the cycloid model of chromosome structure

The diagram illustrates the postulated behavior at meiosis of a gene, *M*, and 16 copies of it arranged in a consecutive linear sequence.

(a)–(d) Detachment of the 16 copies as a result of intrachromatid crossing-over between the first and last members of the series, leaving the master gene *M* in the chromatid in a position to undergo crossing-over with a homologous chromatid from the other parent.

(e)–(h) Reincorporation of the copies into the chromatid by crossing-over. The lines represent nucleotide chains, broken lines indicating newly synthesized chains. Wavy lines show the position of the operator of the gene. *A*, annealing; *B*, breakdown of unpaired chains; *D*, dissociation; *L*, *N*, neighboring genes to *M* and its copies; *M*, master copy of gene; *O*, breakage at operator; *S*, synthesis; 1–16 = 16 copies of gene *M*.

Reproduced from Whitehouse (1967[a]) with permission of the author and the Company of Biologists Limited.

genes for ribosomal RNAs is not maintained by gene amplification in the fertilized egg. The phenomenon of gene amplification is therefore quite distinct from that of gene redundancy, although it provides one obvious mechanism by which gene redundancy could arise.

NUCLEO-CYTOPLASMIC INTERACTION AND INTERSPECIES HETEROCARYA

The course of the program of differential gene expression which occurs within any given cell nucleus during embryonic development is ultimately controlled by the genome of the fertilized egg. However, immediate regulation of the process is effected by the physiological activity of the surrounding cytoplasm. This nucleo-cytoplasmic interaction has been demonstrated most convincingly in the elegant nuclear transplantation experiments of Gurdon and his colleagues. Development of a fertilized amphibian egg to the stage of neurula is characterized by a clearly defined sequence of phases of synthesis of specific types of ribonucleic acid (RNA). Prior to attainment of the midblastula stage the embryo synthesizes RNA predominantly of the type called "heterodisperse DNA-like," or "complementary," cRNA (see Chapters 6 and 8), and only traces of other classes of RNA.[9] At the stage of the late blastula and early gastrula there is a marked rise in the rate of synthesis of cRNAs and onset of substantial synthesis of transfer RNAs. Still later, at the stage of the late gastrula or early neurula, synthesis of the ribosomal RNAs becomes prominent (Brown and Littna 1966[a], 1966[b], Gurdon 1968[a]). Thus the different spectra of active and inactive genes at each of these developmental stages are reflected by the characteristic types of RNA which are synthesized. Conversely, analysis of the RNA being

[9] This type of RNA has formerly been designated dRNA ("DNA-like") or, less frequently, HnRNA ("heterodisperse-nuclear"). However, the IUPAC-IUB Commission on Biochemical Nomenclature now recommends use of the designation cRNA ("complementary RNA").

synthesized serves to identify the developmental status of the cell nucleus. In a series of experiments in which nuclei from one type of tissue were transplanted into functionally enucleated cytoplasm of oocytes or unfertilized eggs, Gurdon demonstrated that the injected nuclei rapidly acquire a new developmental status characteristic of the stage of development attained by the recipient cytoplasm (Gurdon 1968[a], 1968[b]). During the processes involved in this adjustment of developmental state of the nucleus there is a massive transfer of protein from the cell cytoplasm into the transplanted nucleus (Merriam 1969).

Similar dramatic changes in the developmental status of a cell nucleus occur in formation of interspecies hybrid heterocarya (Harris 1968, 1970). One of the most interesting cases is observed following introduction of nuclei of mature avian erythrocytes into the cytoplasm of Hela or mouse cells (Harris et al. 1966). The genetically inert erythrocyte nuclei increase markedly in volume and dry mass, the chromatin assumes a more disperse structure, and its DNA can more readily dissociate into a single-stranded configuration (Bolund, Ringertz, and Harris 1969, H. Harris 1967). These changes are believed to be the consequence of a substantial transfer of cytoplasmic protein into the erythrocyte nucleus, for very similar changes can be induced within intact avian erythrocytes by incubation in appropriate medium (Bolund, Darźynkiewicz, and Ringertz 1969, Ringertz and Bolund 1969). In heterocarya these changes are followed by resumption of synthesis of DNA and RNA in the erythrocyte nuclei (H. Harris 1967, Harris et al. 1966). Most of the RNA synthesized during the first 2 or 3 days after formation of the heterocaryon appears to be heterodisperse cRNA, and none of it is recovered from the cytoplasm. Thereafter, the nuclei develop functional nucleoli and synthesize all species of RNAs in proportions similar to those characteristic of Hela cell nuclei (Harris 1968, 1970, Sidebottom and Harris 1969). At this stage the heterocarya will synthesize specific proteins coded by the avian genome, at a rate which is independent of the number of avian nuclei in the cell (Cook 1970, Harris and Cook 1969).

That the response of the avian nucleus is determined by the cytoplasm of the host cell is shown by two additional types of heterocaryon. Formation of heterocarya in which avian erythrocyte nuclei are introduced into rabbit macrophages leads to activation of RNA synthesis by the avian nucleus but not resumption of synthesis of DNA. This cannot be attributed to an irreversible change in the rabbit macrophage nucleus which precludes further synthesis of DNA in the macrophage cell, for the rabbit macrophage nucleus is activated to resume DNA synthesis when inserted into a Hela cell (Harris et al. 1966). The nature of the materials which reactivate the avian nucleus is not yet known, but they presumably include the cytoplasmic components which regulate the course of established cell cycles (e.g., De Terra 1969, Mano 1970, Ord 1969). These include small, heat-stable materials other than the precursor substrates of the nucleic acids (Thompson and McCarthy 1968, 1970) and probably also specific proteins (Salas and Green 1971).

The composition of the cell cytoplasm is equally important in the less extensive nuclear differentiation or differential gene activity which occurs in the early stages of embryonic cleavage (e.g., Collier 1966, Davidson 1968, Fischberg and Blackler 1963). In none of the systems studied to date have the components responsible for this "*localization phenomenon*" been unequivocally established. Factors which may be responsible include unequal cell division in prior cleavages and uneven distribution within the developing oocyte of polysomes containing "maternal" messenger RNAs, or of mitochondria (S. Nass 1969).

INTERSPECIES SYNCARYA

Some understanding of the nature and mechanism of action of the primary factors controlling processes of chromosome elimination and differential gene activity in developing embryos can be expected from study of equivalent phenomena observed with "*interspecies somatic hybrids*," or syncarya, in tissue culture. Development of a syncaryon involves the fusion of nuclei of the two parental cell lines to yield a tetraploid cell containing all, or virtually all, the chromosomes of both original parental types. The fact that such syncarya are viable and can be propagated without an immediate mass loss of chromosomes indicates existence of a finely balanced mechanism regulating the nuclear cycle of the cells. As noted by Stern (1964), the mechanism prevents formation of a tetrapolar mitotic spindle (in at least some cells) and permits a normal mitotic cycle to occur. Further, many hybrids are formed from parental lines which differ markedly in growth rate, and the hybrid cell must possess a mechanism for coordinating the replication and distribution of the combined complement of chromosomes (Ephrussi and Weiss 1967).

Examination of the chromosome complement, or *karyotype*, of the progeny of the hybrid after propagation in tissue culture reveals a slow progressive diminution in chromosome number. The loss of chromosomes is not (totally) random. In hybrids derived from rat with mouse cells there is a preferential loss of chromosomes of rat origin, whereas mouse-hamster hybrids show preferential loss of murine chromosomes. Ephrussi and Weiss (1967) suggest that the primary factor in the coordinated control of such hybrids is the enclosure of the two genomes within a single nuclear membrane. Such a suggestion is entirely in accord with our belief that attachment of the chromosome to the nuclear membrane is a key step in replication of the chromosome (Comings and Kakefuda 1968, DuPraw 1968, Mizuno, Stoops, and Sinha 1971, Pawlowski and Berlowitz 1969). Thus the same mechanism may regulate the progressive loss of chromosomes during propagation of somatic hybrids as that responsible for chromosome elimination during normal embryonic development of certain animal species.

Differential gene activity has been shown to be substantially modified in a number of somatic hybrids. For example, syncarya obtained by fusion of pigmented hamster cells and unpigmented mouse cells are unpigmented and lack dopa oxidase or tyrosinase activity (Davidson, Ephrussi, and Yamamoto 1968). Similarly, the "inducibility" of tyrosine aminotransferase enzyme in rat hepatoma cells (see Chapter 6) is absent from hybrids formed by fusion with mouse fibroblasts (Schneider and Weiss 1971). Ribosomal RNA coded by human genes is lacking from human-mouse syncarya (Eliceiri and Green 1969). A similar suppression of ribosomal RNA formation is also indicated for hybrids formed between species of blackcurrant and gooseberry, which lack the "small chromosomes" containing nucleolar organizer characteristic of the blackcurrant (Keep 1962). The mechanism which *determines* the selection of genes differentially inactivated in such somatic cell hybrids may perform the same function during the course of normal differentiation.

CELL CYCLES

The various processes discussed above collectively comprise the overall phenomenon of differential gene activity. These processes serve to modify the complement of genes (and gene copies) available for expression in the synthesis of specific protein via the process of transcription (Chapter 8). Superimposed on these specific processes, almost all actively dividing cells show a cyclical variation in the total content of genes (active and inactive, nuclear and extranuclear) during the course of the "cell cycle." Further, the cell cycle itself undergoes differentiation during the course of embryonic development. Even in young embryos the length of the complete cycle varies markedly between different regions of the embryo (Flickinger, Freedman, and Stambrook 1967, Graham and Morgan 1966, Ozato 1969). As differentiation proceeds, the duration of the cycle varies both within the same tissue and between tissues (Fujita 1962,

Gamow and Prescott 1970, Graham and Morgan 1966, Löbbecke, Schultze, and Maurer 1969). These variations are themselves reflections of changes in the composition of the cell cytoplasm, as shown by studies of nuclear transplants and interspecies heterocarya discussed earlier in this chapter. Nevertheless, the processes of the "cell cycle" substantially affect the composition of the host cytoplasm, and marked changes in the pattern of total protein synthesized at different stages of the cycle have been demonstrated (Kolodny and Gross 1969).

The course of synthesis of specific proteins during the cell cycle often reflects the replication of the corresponding structural genes, with a discontinuous increase in actual or potential rate of synthesis occurring at a defined stage of the cycle (Braun and Behrens 1969, Mitchison 1969). In other cases there is no clear correspondence between the pattern of synthesis of a particular protein and replication of the coding gene (Klevecz 1969, Mitchison 1969). Interpretation of these various patterns is usually in terms of the twin phenomena of induced enzyme synthesis and enzyme repression (Goodwin 1966, 1969, Masters and Donachie 1966, Mitchison 1969) (Chapters 3 and 4), and will not be discussed further. In other cases the pattern observed probably reflects "compensation for gene dosage."

One particularly interesting example of an influence of phase of the cell cycle upon the course of formation of a specific protein has been reported by Martin, Tomkins, and Granner (1969). The synthetic steroid dexamethasone phosphate will induce formation of tyrosine aminotransferase enzyme by rat hepatoma cells in tissue culture if supplied during the late G1 and S phases of the cell cycle. However, the steroid fails to elicit this response when supplied during G2, M, and early G1 phases. This pattern of response appears to be determined by a mechanism which controls the synthesis of protein on preformed messenger RNA (Martin, Tomkins, and Granner 1969, Tomkins et al. 1969), which will be discussed in later chapters (see Chapters 6 and 9). A recent study indicates that cyclic fluctuations in the rates of antibody synthesis during the cell cycle are regulated by a similar mechanism (Buell and Fahey 1969).

Further discussion of the processes by which genes are either "inactivated" or expressed, and of the mechanisms regulating those processes, requires a prior definition of the phenomena in which the regulatory mechanisms are revealed. These are the content of the ensuing chapters.

A change must sometimes be made to things one is not accustomed to.—Hippocrates, ca. 400 B.C.

Seeing these wondrous dispensations of Nature, whereby these little creatures are created so that they may live and continue their kind, our thoughts must be all abashed. . . .—Antony van Leeuwenhoek, 1702.

CHAPTER 3: **INDUCED ENZYME SYNTHESIS**

ENZYMATIC ADAPTATION; EARLY STUDIES OF INDUCED ENZYME SYNTHESIS

The phenomenon of induced enzyme synthesis was first observed before the turn of the century. In what appears to be the first recorded study, Wortmann (1882) reported that cells of an unidentified bacterial species ("*Bacterium termio*") only produced enzyme(s) active on starch when grown in medium containing starch. On the other hand, he found that yeast cells readily produced invertase during growth in the absence of sucrose. In subsequent years additional reports were made of similar instances of an "enzymatic adaptation" in which a microorganism produced enzyme(s) active on a given substrate only during growth in the presence of that substrate.

A major advance in analysis of the nature of the phenomenon was reported in Dienert's (1900) now classic paper which laid the basic foundation for subsequent detailed analysis at a later date. This study included a demonstration that washed yeast cells developed increased activity of the "galactozymase" enzyme system (galactokinase, galactose-1-

phosphate uridylyl transferase, and epimerase) when incubated with galactose in phosphate buffer, without an exogenous source of nitrogen. In addition, Dienert reported five other major findings:

1. The composition of the growth medium markedly affected the level of galactozymase activity of the cells but had little effect on activity in fermenting glucose.
2. Only cultures growing in medium containing galactose or utilizable galactoside had high levels of galactozymase activity.
3. Cultures growing in medium containing both galactose and glucose had low levels of galactozymase.
4. When cells previously grown with galactose were transferred to medium without galactose the level of galactozymase activity fell.
5. And some strains of yeast formed high levels of galactozymase activity during growth without galactose or galactoside.

The evidence presented in this classic paper cannot be considered rigorous by contemporary standards. Nevertheless, the observations presaged subsequent rigorous demonstrations that the phenomenon of induced enzyme synthesis is distinct from the process of selection of mutant strains. Further, it served to define the experimental criteria by which the phenomenon was recognized pending the development of more sophisticated genetic criteria.

Unfortunately, Dienert's study seems to have made little impression at the time. Perhaps it was merely overshadowed by the rediscovery of Mendel's "laws of inheritance." Perhaps it was viewed as a "nice" laboratory example supporting Darwin's view of evolution, as at a later date some studies of selection of mutants were viewed as supporting Lysenko's (1948) belief in "directed changes in heredity." In any event, the stage was set for a protracted argument over the reality of the twin phenomena of selection of mutants and of induced enzyme synthesis. However, the curtain did not rise for another three decades, when Stephenson and Yudkin (1936) confirmed the findings of Dienert and showed that the increase in galactozymase activity took place without any net increase in cell number. Stephenson and her co-workers (Stephenson 1937, Stephenson and Stickland 1933) similarly resolved the induced synthesis of formic hydrogenlyase activity by cultures of *Escherichia coli* from cell growth. Added impetus to the study of induced enzyme synthesis was given by Karström's (1938) classification of enzymes of "*Betacoccus arabinosus*" into *adaptive* and *constitutive,* and by Yudkin's (1938) "mass action" hypothesis as a possible mechanism for the process.

In the ensuing three decades the phenomena of induced enzyme synthesis and related processes have been intensively studied and have provided the basis (or point of departure) for most of our current views of mechanisms regulating the synthesis of specific proteins. It is beyond the scope of this essay to review the history of this period or to consider all the wealth of relevant information available, contained in a series of excellent reviews by Epstein and Beckwith (1968), Jacob and Monod (1961[a]), Pollock (1959), Richmond (1968), and others. Rather, in the following pages I shall summarize the evidence obtained from the "model" systems for which rigorous evidence is available, which defines the phenomena of interest. Mechanisms by which the phenomena are regulated will then be considered in later chapters (see Chapters 5 and 9). It should, however, be emphasized at this juncture that, although our current views represent an internally consistent blend of evidence drawn from a variety of experimental systems, many of the essential points have been rigorously established for only a very limited number of model systems.

The inducible enzyme which has been the subject of the most intensive study is unquestionably the β-galactosidase enzyme of strains of *Escherichia coli,* first extensively exploited by Monod and his co-workers. Monod's early studies were necessarily limited to an analysis of the physiology of enzymatic adaptation (or induced enzyme synthesis) in growing cultures as evidenced by growth on, and development of the ability to ferment, a test carbon source. These early studies were severely hampered by lack of a sufficiently sensitive and specific assay for β-galactosidase enzyme activity and were seriously complicated by the effects of a second regulatory process which we now call "catabolite repression." Nevertheless, and indeed largely because of this effect of glucose—the "diauxie" (Monod 1942, 1944[a], 1944[b], 1945)—, Monod demonstrated that the process of enzymatic adaptation was a highly specific response to the particular test carbon source and that the rapidity of the response was markedly dependent upon the concentration of the test sugar (Monod 1943). The process required energy (Monod 1944[c]) and took place only under conditions permitting cell growth (Monod 1947). However, analysis of the kinetics of the increase of enzymic activity, and of the subsequent decline on subculture in the absence of specific substrate, made it highly improbable that the changes could be attributed to selection of mutant cells of high enzymic activity followed by a counterselection against them. A subsequent detailed analysis by Ryan (1952) fully established that this was not the mechanism. Meanwhile, other workers had demonstrated similar enzymatic adaptations taking place either before onset of cell multiplication, or without any net increase in cell numbers, for yeast "galactozymase" (Dienert 1900, Spiegelman, Lindegren, and Hedgecock 1944, Stephenson and Yudkin 1936) for "formic hydrogenlyase" of *E. coli* (Stephenson and Stickland 1933), "tetrathionase" of *Salmonella paratyphi B* (Knox and Pollock 1944), "nitratase" of *E. coli* (Pollock 1946), and other systems. Therefore the process of induced enzyme synthesis seemed independent of, even when coincident with, growth (Monod 1947).

The first direct demonstration that the ability to form induced enzyme in response to external substrate (i.e., the character of "*inducibility*" or *regulation*) is a genetic character was made for the "melibiozymase" and "galactozymase" enzyme complexes of strains of yeast (Lindegren 1945, Lindegren, Spiegelman, and Lindegren 1944, Spiegelman, Lindegren, and Lindegren 1945), using conventional genetic procedures. Shortly thereafter, Monod obtained indirect but strong circumstantial evidence (Monod 1949, Monod and Andureau 1946) that the process of induced β-galactosidase synthesis in *E. coli* is regulated genetically.

At this juncture the introduction of new experimental techniques and phenomena both raised and provided the means of answering a whole series of questions, the answers to which have led to the present position of β-galactosidase as the pre-eminent "model system" (see Beckwith and Zipser 1970).

INDUCED DE NOVO SYNTHESIS OF ENZYME PROTEIN

The introduction by Lederberg (1950) of *o*-nitrophenyl-β-galactoside as a chromogenic substrate of the enzyme provided a direct, specific, and highly sensitive assay of β-galactosidase activity. Prior to the use of this substrate it had not proved possible to demonstrate significant "basal" β-galactosidase activity in cultures of noninduced cells, that is, not exposed to external galactoside. The new assay showed that noninduced cells did contain very low but measurable levels of β-galactosidase. In a sense this qualified β-galactosidase as a candidate to be the model system, for whenever sufficiently sensitive assay procedures can be applied it has invariably been found that the *phenomenon of induced enzyme synthesis involves the increase in level of an enzymic activity which is present at a low level in noninduced cells of the same type.*

However, the demonstration of the presence of a "basal" level of enzyme in noninduced cells raised a number of interrelated problems. The first was whether the basal level of activity could be due to the presence of a small proportion of "*constitutive*" mutant cells which produced a high level of enzyme even in the absence of substrate. Although such mutants are present in cultures of noninduced cells (Ryan 1952), quantitative analysis by Holmes, Sheinin, and Crocker (1961) indicated that their numbers are insufficient to account for the basal level of enzyme. Thus the noninduced cells themselves produced a basal level of enzyme. This posed the dual problem of whether the basal and induced enzyme were related but different enzymes; and whether exogenous substrate could in some way affect the genetic material of the cell so that it produced an enzyme of altered structure and efficiency as a catalyst (i.e., a phenomenon of mass transitory mutation).

This question has, in fact, been examined in any detail only for the β-galactosidase enzyme of *E. coli* and the extracellular penicillinase, or β-lactamase I, enzyme produced by strains of *Bacillus cereus* and *Bacillus licheniformis.*[1] Evidence for all other systems is far less complete and usually rests upon the currently accepted concepts of the mechanisms controlling synthesis of inducible enzymes as uniquely demonstrated for the β-galactosidase system. Indeed, in the absence of the current body of internally consistent evidence information for other systems would be considered totally inadequate. For both β-galactosidase and the exopenicillinases a number of substrates and competitively inhibitory analogues are available (Jacob and Monod 1961[a],

[1] The strains of *B. cereus* used in the studies discussed in this chapter have been shown to produce two distinct species of extracellular β-lactamase enzyme active on penicillin. They differ in stability (e.g., Crompton et al. 1962, Kuwabara and Abraham 1969), size (Kuwabara 1970, Lloyd and Peacocke 1970), amino acid composition, and content of amino sugars (Kuwabara, Adams, and Abraham 1970).

The α- and bound β-penicillinases referred to here are the dominant species of β-lactamase I (Kuwabara and Abraham 1969), which is inactive on cephalosporin (e.g., Crompton et al. 1962). The minor species β-lactamase II is a cephalosporinase, which is unstable in the absence of zinc ions (Kuwabara 1970). Under the experimental conditions used in the studies to be discussed, β-lactamase II invariably contributed less than 5 percent, and most commonly less than 1 percent, of the total penicillinase activity (e.g., see Kuwabara 1970).

Monod and Cohn 1952, Pollock 1964). Both types of enzyme are antigenic (Cohn and Torriani 1952, Pollock 1956[a]). In the case of β-galactosidase the enzymic activity is profoundly affected by the presence of monovalent cations, including hydrions (Cohen-Bazire and Monod 1951, Cohn and Monod 1951). Thus a comparison can be made of the enzymic and antigenic properties of a β-galactosidase in a variety of ionic environments to define the identity of the molecular species of enzyme. When such a comparison was made, no differences were found between the basal and induced β-galactosidases, or the enzyme produced by a constitutive mutant. Similarly, Manson, Pollock, and Tridgell (1954) and Pollock (1956[a]) were unable to demonstrate any differences between the basal and induced penicillinases for *B. cereus* and *B. licheniformis;* nor were differences found for enzyme produced by a constitutive mutant (Kogut, Pollock, and Tridgell 1956).

Strictly speaking, a rigorous proof of the identity of the basal and induced levels would require a demonstration of identity of amino acid sequences. Current concepts of the mechanisms regulating the formation of inducible enzymes require such an identity of sequence and assign no role to exogenous substrate in modifying the primary structure of the enzyme (see Chapter 5). It is therefore unfortunate that the monumental endeavor of examining this crucial point has not been undertaken. Nevertheless, the available evidence for the β-galactosidase and penicillinase enzymes (and by circular argument all other inducible enzymes) indicates that the phenomenon of induced enzyme synthesis is one of *regulation of the quantity of enzyme formed.*

A measured level of enzymic activity is, however, not necessarily a true indication of the quantity of active enzyme present, and it was of considerable importance to determine whether observed increases in enzyme activity were indeed due to an increase in level of specific enzyme protein. The availability of highly specific antisera prepared against highly purified enzyme permitted direct examination of the possibility for both the β-galactosidase of *E. coli* and the penicillinases of *B. cereus* and *B. licheniformis.* In both cases the answer was clear-cut, but complex. Cohn and Torriani (1951, 1952) demonstrated that the increase in β-galactosidase activity was associated with a marked increase in the quantity of protein precipitable by highly specific antiserum. However, the precipitated antigen-antibody complex retained enzymic activity, and this feature of the reaction permitted demonstration of the presence of enzymically inactive cross-reacting material in noninduced cells. During the course of induced synthesis of β-galactosidase enzyme the level of cross-reacting material fell in a manner consistent with the hypothesis that the cross-reacting material was a precursor of the active enzyme (Cohn and Torriani 1953, Monod and Cohn 1952). Nevertheless, subsequent studies made with cells "prelabeled" with radioactive isotope indicated this hypothesis to be invalid (Hogness, Cohn, and Monod 1955, Rotman and Spiegelman 1954). Indeed, when subsequent genetic manipulation became possible, Cohn, Lennox, and Spiegelman (1960) demonstrated that partial hybrid species obtained by introducing the appropriate genes of *E. coli* into *Shigella dysenteriae* cells did not produce any detectable quantity of the cross-reacting protein, although they did produce β-galactosidase of the *E. coli* type in response to exogenous galactoside. The nature of the cross-reacting material present in noninduced cells of *E. coli* and the basis for the decline in level of this material during the induced synthesis of β-galactosidase still remain unexplained. However, the data rigorously established that the increased β-galactosidase activity is due to the *de novo* synthesis of new enzyme protein.

In the case of the penicillinase enzyme of *B. cereus* some of the enzymic activity was bound, and only a portion of the bound enzyme was neutralized by the specific antiserum (Pollock 1956[b]). This nonreactive γ-penicillinase fraction was isolated in a soluble form and found to have properties distinct from those of the exo-enzyme. However, using radioactively labeled amino acid as tracer, Pollock and Kramer (1958) demonstrated that this species of penicillinase enzyme was not a precursor of the exo-enzyme. The increased activity of extracellular α-penicillinase is due to the *de novo* synthesis of new enzyme protein. Nevertheless, subsequent studies by Citri and co-workers (Citri 1958, Citri and Garber 1958, Citri, Garber, and Sela 1960), and by Rudzik and Imsande (1970), have shown that the bound penicillinase unreactive with antiserum toward exo-penicillinase and the α-penicillinase are probably two different interconvertible configurations of the same molecule.[2] The precise relationships between the two conformations during the course of induced synthesis of penicillinase remain to be determined. Nevertheless, the foregoing studies with the β-galactosidase of *E. coli* and the penicillinases of *B. cereus* and *B. licheniformis* served to define the *process of induced enzyme synthesis* as one of *regulated de novo synthesis of new enzyme protein.* Similar demonstrations have subsequently been made for other enzyme systems, and this feature of the process is recognized as one experimental criterion which must be met to justify acceptance of an observed increase in enzyme activity as an instance of induced enzyme synthesis.

NATURE AND SPECIFICITY OF THE INDUCER

Introduction of *o*-nitrophenyl-β-galactoside as substrate for the assay of β-galactosidase activity entailed the explicit recognition that the enzyme did not have absolute specificity as a lactase. Indeed, subsequent study has shown that the enzyme is a β-galactoside galactohydrolase with galactotransferase activity (Aronson 1952, Wallenfels, and Malhotra 1961). The product of the action on lactose is in fact a complex mixture comprising free glucose and galactose, 5 disaccharides, 3 trisaccharides, and traces of tetrasaccharide

[2] Kuwabara and Abraham (1969) have, however, recently suggested that γ-penicillinase might consist of β-lactamase II lacking essential zinc ions.

(Burstein et al. 1965, Müller-Hill, Rickenberg, and Wallenfels 1964). Yudkin (1938) had postulated that a possible mechanism of induced enzyme synthesis was one in which substrate complexed with pre-existing free enzyme present in equilibrium with enzymically inactive precursor, thereby displacing the equilibrium in the direction of formation of more free enzyme. This "mass action" hypothesis required that galactosides with high affinity for the enzyme should readily induce synthesis of β-galactosidase. Conversely, galactosides without affinity for the enzyme should not induce its formation. Monod and his co-workers therefore compared the affinity of a series of β-galactoside substrates and competitive inhibitors for the enzyme with their efficiencies as inducers of β-galactosidase synthesis. The results which they obtained for the ML (Monod, Cohen-Bazire, and Cohn 1951) and, later, the K12 (Jacob and Monod 1961[a]) strains of *E. coli* are particularly instructive in a number of respects; they are summarized in Tables 3.1 and 3.2, respectively.

First, the data show that there is no correlation between the two properties of the series of β-galactosides for either strain. Melibiose (an α-galactoside) is a very effective inducer

TABLE 3.1. Induction of β-galactosidase by various galactosides for E. coli strain ML

Compound	Induction value * (a)	(b)
Isopropyl-β-thiogalactoside	.....	(100)
4 Glucosyl·1·β-galactoside (lactose)	104	36
6 Glucosyl·1·α-galactoside (melibiose)	100	35
Phenyl·1·β-galactoside	23	8
Galactose	18	6

* The induction value was defined (Jacob and Monod 1961[a]) as the specific activity developed by a wild-type strain under conditions of gratuity in presence of 10^{-3} *M* inducer, expressed as a percentage of the value obtained with isopropyl-thiogalactoside as inducer. Data for the latter inducer were not presented for this strain. Therefore values are expressed in column (a) with melibiose set as standard inducer. For convenience in comparison with Table 3.2, data are computed in column (b) for isopropyl-thiogalactoside as standard on the assumption that relative efficiencies for melibiose and isopropyl-thiogalactoside are the same as for *E. coli* strain K12. (Data from Monod and Cohn 1952.)

TABLE 3.2. Induction of β-galactosidase and thiogalactoside transacetylase by various galactosides for E. coli K12

Compound	Induction values*	
	β-galactosidase	β-thiogalactoside transacetylase
Isopropyl·β-thiogalactoside	100	100
Glucosyl·β-galactoside (lactose)	17	12
Glucosyl·α-galactoside (melibiose)	35	37
Phenyl·β-galactoside	15	11
Galactose	<0.1	<1

* See footnote to Table 3.1. (Data from Jacob and Monod 1961[a].)

of β-galactosidase synthesis in both strains, but it is neither a substrate of the enzyme nor bound to the enzyme to any detectable extent. Conversely, *o*-nitrophenyl-α-L-arabinoside (a substrate) and phenyl-β-thiogalactoside (a competitive inhibitor) complex with the enzyme but do not induce its synthesis; nor does the phenyl-thiogalactoside competitively inhibit the induced synthesis of enzyme. Thus the capacity of a galactoside to induce the synthesis of β-galactosidase is not dependent upon complexing of the inducer with either the active enzyme or any other molecule containing a receptor group of the same stereospecificity as that of the enzyme. The data in the tables also permit partial definition of the structural features required for effective induction.

Second, comparison of the values for β-galactosides which are effective inducers shows a very wide range in the levels of β-galactosidase activity attained under comparable experimental conditions. This posed the question of whether the exogenous galactosides are themselves the direct inducers of β-galactosidase synthesis, or whether they must first be metabolized by the cells to a derivative which is the true inducer. In other words, were some inducers inherently more effective than others, or were some better substrates than others for a second "endogenous inducer-forming" enzyme? If the former were the case, the data provide some information about the specificity of the receptor site with which the inducer interacts. In addition, comparison of the data for the two strains shows them to contain different receptor sites. Melibiose is twice as effective as lactose for the K12 strain but only equally effective for the ML strain. On the other hand, if exogenous galactoside were only effective through conversion to an endogenous inducer substance, the ability of a strain to respond to exogenous galactoside (the character of "inducibility") would be determined, at least in part, by the presence and activity of "inducer-forming" enzyme.

Fortunately, the answer to this question was largely rendered irrelevant by subsequent genetic studies, because the current answer is only partial and complex. Studies by Burstein et al. (1965) indicate that lactose is not *per se* an inducer of β-galactosidase formation and probably exerts its effect through conversion to allolactose (6-β-galactopyranosyl-glucose) and galactobiose (6-β-galactopyranosyl-galactose) by the action of the "basal" β-galactosidase. Similarly, penicillin may cause the accumulation in cells of *B. cereus* of peptidoglycan subunits of the cell wall, or "spore peptide," which are highly effective inducers of the synthesis of penicillinase (Ozer, Lowery, and Saz 1970). On the other hand, isopropyl-β-thiogalactoside appears to have the inherent activity of an inducer (Burstein et al. 1965, Zubay and Lederman, 1969). Genetic studies made for a number of inducible enzyme systems indicate a similar mechanism of regulation to that pertaining in the case of the β-galactosidase of *E. coli* (e.g., Adhya and Echols 1966, Buttin 1963[a]). However, the results of Burstein et al. (1965) require that caution be exercised in the interpretation of data for a system not yet amenable to direct genetic analysis, or in cases where the results of genetic analysis do not correspond with those found for the β-galactosidase system of *E. coli* (e.g., Englesberg, Squires, and Meronk 1969).

Third, galactose is not an inducer of β-galactosidase synthesis in *E. coli.* This is in keeping with the observation that free galactose is not a substrate of galactoside galactohydrolases (Pazur 1953, Wallenfels and Malhotra 1961) and could not directly give rise to endogenous inducer through the action of basal β-galactosidase. Although galactose was originally reported to have activity as an inducer of β-galactosidase synthesis by the ML strain, more recent studies of Müller-Hill, Rickenberg, and Wallenfels (1964) show that galactose is not an inducer of β-galactosidase formation in this strain. It remains to be determined whether the basal level of β-galactosidase is maintained by its own action on endogenously formed galactoside or reflects some other feature of the mechanism of induced enzyme synthesis, which will be considered at a later point.

Finally, the data of Table 3.2 also include some results for a second inducible enzyme formed in response to the presence of exogenous galactoside, thiogalactoside transacetylase. Comparison of the values for the two enzymes shows that the extent of the response (expressed as a fraction of the maximal response obtained with isopropyl-β-thiogalactoside as inducer) was quantitatively the same for both enzymes for all inducers. Thus the synthesis of the two enzymes was affected to the same extent and reflected a *coordinate regulation* of both syntheses. I shall return shortly to this feature of coordinate regulation.

Similar, though less extensive, studies have been made of the specificity of induction for a number of inducible enzyme systems. In some of these cases also nonsubstrate inducers have been reported; e.g., cephalosporin C for the α-penicillinase of *B. cereus* (Pollock 1957[a]), D-fucose for the "galactozymase" enzyme complex of *E. coli* (Buttin 1963[b], Kalckar, Kuruhashi, and Jordan 1959), and substrate analogues for the histidine degrading enzymes of *Aerobacter aerogenes* (Schlesinger and Magasanik 1965), and for the aliphatic amidase of *Pseudomonas aeruginosa* (Kelly and Clarke 1962).

CONDITIONS OF GRATUITY AND THE DIFFERENTIAL RATE OF ENZYME SYNTHESIS

The availability of such nonsubstrate inducers has proved extremely useful to subsequent investigations. In studies of the induced synthesis of an enzyme capable of metabolizing the exogenous inducer, analysis of the events regulating the process is complicated by the unknown effects of progressive changes in both the contribution of the metabolism of the inducer to the physiology of the cells and in the composition of the medium. Therefore, detailed analysis of the physiology of the process requires that the experimental conditions be chosen such that the inducer does not significantly contribute to any process other than regulation of the synthesis of inducible enzyme; that is to provide *conditions of gratuity* (Monod 1947, Monod and Cohn 1952). These conditions can be approximated by use of an inducer which cannot be metabolized by the enzyme being induced, in the presence of adequate concentrations of all essential nutrients. In a few cases this can be achieved with a nonmetabolizable analogue of naturally occurring inducers, and in other cases this situation can be effectively attained by the use of appropriate mutant strains of an organism selected for inability to metabolize substrate inducer.

The penicillinases of various strains of bacilli, and possibly some staphylococci, constitute a special case. Studies of the exo-enzyme of *B. cereus* by Pollock and his colleagues have shown that very brief exposure to benzyl-penicillin is sufficient to ensure continued synthesis of penicillinase enzyme at a high rate (Pollock 1950), which can persist for a number of cell generations (Pollock 1958). This is correlated with, and appears to be due to, firm binding of very small quantities of penicillin to specific receptors of the cell (Pollock and Perret 1951). The quantity of enzyme molecules formed is many times the number of penicillin molecules bound (Kogut, Pollock, and Tridgell 1956). In view of this "catalytic" effect of small quantities of "fixed" penicillin, the extreme rapidity of the "fixation reaction" and the nature of the reaction catalyzed, it seems reasonable to consider this enzyme to be formed under conditions of "*de facto* gratuity."

Penicillin is similarly firmly bound to cells of *Staphylococcus aureus* (Cooper, Rowley, and Dawson 1949), and this binding *may* be correlated with the induced synthesis of penicillinase by *certain* strains of this microorganism. However, there appears to be no simple general correlation between the ability of bacterial cells to bind penicillin to specific "receptor sites" and to produce penicillinase or β-lactamase enzymes active on substrate penicillins (Cooper and Rowley 1949, Pollock 1964, Rowley et al. 1950).

The experimental conditions of gratuity can be further refined by use of techniques of continuous culture (Monod 1950, Novick and Szilard 1950) to maintain constancy of composition of the environment. This is particularly desirable in the event that the organism is capable of metabolizing the inducer via a pathway not requiring activity of the inducible enzyme under study. For example, this refinement would be essential in detailed analysis of the induced synthesis of β-galactosidase by the K12 strain of *E. coli* in response to melibiose for, although this inducer is not a substrate of the enzyme, it can serve as the sole source of carbon for growth (Lester and Bonner 1952). However, the refinement of conditions of continuous culture is not necessary when the inducer cannot be metabolized by the cell to any detectable extent, provided allowance is made for possible effects of changes in environmental factors. This may be accomplished by use of the isometric plot introduced by Monod, Pappenheimer, and Cohen-Bazire (1952). A plot of the increase in enzyme activity (assumed to reflect enzyme content) against the increase in total cell protein (or to a first approximation cell mass) represents the proportion of the new protein which is the specific enzyme. The plot thus depicts the "*differential rate*" of synthesis of the enzyme. Any factor which *specifically* affects the synthesis of an inducible enzyme is reflected in a change in the differential rate of its formation, whereas factors which affect the synthesis of protein in general will not alter the differential rate. In the case of instances of induced enzyme synthesis with washed cell suspensions, or prior to onset of cell multiplication, the rate of enzyme synthesis clearly cannot be correlated with *net* increase in cell protein (e.g., Rickenberg and Lester 1955). However, in such cases there may be a considerable "turnover" of cell protein (Mandelstam 1958[a], 1958[b], Mandelstam and Halvorson 1960), and the increase of enzyme may be related to the increase of new protein by the use of "tracer" radioactive amino acid precursors.

KINETICS OF INDUCED ENZYME SYNTHESIS

Analysis of the kinetics of the early stages of induced enzyme synthesis requires either the formal demonstration or the explicit assumption that all cells are participating in the process. Precise analysis requires that conditions be such that all cells are participating with approximately equal rapidity. A formal demonstration that all (or virtually all) of the cells in the population participate in the initial stages of induced enzyme synthesis appears to have been made only for the β-galactosidase enzyme of *E. coli* and the penicillinase enzyme of *B. cereus.* Genetic analyses for numerous other systems make it extremely probable that the validity of the demonstration may be extended to all such systems, although this does require the (reasonable) explicit assumption that what has been shown to be the detailed mechanism regulating the induced synthesis of β-galactosidase in *E. coli* is also *precisely* the same mechanism in these systems.[3]

Benzer (1953) provided indirect but elegant evidence that all cells of a culture of *E. coli* synthesized β-galactosidase

[3] For example, demonstration that individual cells of a population *can* respond directly to inducer when tested for ability to give rise to a colony does not necessarily constitute a sufficiently rigorous proof that *all cells do* in fact *respond directly* in a large cell population. Indeed, in a highly artificial situation not all inducible cells of *E. coli* are induced to form β-galactosidase (Novick and Weiner 1957).

enzyme at approximately the same rate under conditions closely approximating those of gratuity. He also showed that the increase in rate of synthesis of the enzyme commenced very soon (within a matter of minutes) after addition of inducer. Further, on removal of inducer synthesis of enzyme ceased very rapidly and pre-existing enzyme was then partitioned equally among all cells of the culture during continued growth. Parenthetically, it may be noted that when the inducer served also as the source of carbon and energy for growth Benzer (1953) found substantial heterogeneity among the cells of the population. Subsequently, Rotman (1961, Rotman, Zderic, and Edelstein 1963) introduced the use of fluorogenic galactoside substrates to measure the enzymic activity of individual cells before and after induction, and demonstrated the participation of all cells in the process. Similarly, Collins (1964) demonstrated the synthesis of penicillinase enzyme to approximately the same extent by individual cells of *B. licheniformis.*

As noted above, under conditions of gratuity the increase in activity of inducible β-galactosidase enzyme commences very shortly after addition of inducer to a growing culture of *E. coli.* Indeed, by extrapolation of the isometric plot of increase in enzymic activity against increase in total cell protein (Monod, Pappenheimer, and Cohen-Bazire 1952) it was found that the rate of induced enzyme synthesis is maximal at about 3 minutes after addition of inducer at 37°C (Boezi and Cowie 1961, Pardee and Prestidge 1961). This time interval is independent of the inducer used and the concentration of inducer, although the subsequent differential rate of enzyme synthesis is influenced by both these factors. Conversely, enzyme synthesis continues for approximately 3 minutes after removal of the exogenous inducer, addition of a competitive inhibitor of the induction process, or of glucose. On the other hand, both the interval prior to the increase of enzymic activity and that of continued synthesis of enzyme after removal of inducer are temperature-dependent. Similar "preinduction" and "postinduced" periods of enzyme synthesis were also observed for other inducible enzymes of *E. coli* (Pardee and Prestidge 1961).

The subsequent development of special techniques for analyzing the events during the first few minutes after addition or removal of inducer (e.g., Kepes 1963, Nakada and Magasanik 1964) permitted a more precise delineation of the kinetics of enzyme formation during these periods. Thus, in studies made at 37°C, Kepes (1963) reported formation of some β-galactosidase enzyme after exposure of cells of *E. coli* to inducer for a period of only 20 seconds. The first detectable increase in enzyme activity was found within 1½ minutes. At the lower temperature of 30°C the first detectable increase in activity was observed some 3 to 4 minutes after addition of inducer (Nakada and Magasanik 1964). When the inducer was removed, enzyme synthesis continued for approximately 3 minutes at 37°C or 10 minutes at 30°C, in each case at a rate which declined exponentially (Kepes 1963, Nakada and Magasanik 1964). Analyses of this type have proved extremely useful to our knowledge of the mechanisms regulating the induced synthesis of enzymes, and they will be considered further in Chapter 5.

The kinetics of synthesis of most other inducible enzymes have not been studied as intensively as those of β-galactosidase. Leive and Kollin (1967) studied in parallel the synthesis of galactoside transacetylase and of β-galactosidase. They made the important observation that the time lag between addition of inducer and increase of enzyme activity was some 1½ to 2 minutes longer for the transacetylase than the β-galactosidase at 30°C. This difference reflects the relationship between the corresponding structural genes for the two enzymes, which will be discussed later. For present purposes, onset of the induced synthesis of the two enzymes can be considered to be simultaneous and within the 3-minute period derived by extrapolation of the isometric representation of the kinetics of enzyme formation (Monod, Pappenheimer, and Cohen-Bazire 1952, Pardee and Prestidge 1961).

ATYPICAL KINETICS OF INDUCED ENZYME SYNTHESIS AND PROTEIN PRECURSORS

Kinetics similar to those observed for β-galactosidase have been found for almost all inducible systems which have been studied under conditions of gratuity (e.g., Michaelis and Starlinger 1967). The kinetics are quite different even for β-galactosidase under other conditions (e.g., Moses and Calvin 1965). There are, however, two notable exceptions; namely, the penicillinase enzymes of bacilli and staphylococci, and histidine-degrading enzymes of *B. subtilis.*

Interaction between cells of *B. cereus* and inducer benzylpenicillin occurs very rapidly. The reaction is 50 percent completed within 1 minute at 35°C (Pollock 1952), and virtually all free unbound penicillin is destroyed within 5 minutes (Imsande 1970). An increase in the rate of penicillinase formation can be observed as early as 30 seconds after addition of penicillin to the cells (Imsande 1970). However, there is a "latent" period of 15 to 16 minutes before the maximal differential rate of enzyme synthesis is attained. Further, the differential rate of penicillinase formation then declines progressively for an additional 15 minutes, and thereafter the rate of continued penicillinase synthesis remains constant (Imsande 1970, Pollock 1951). This latter "linear" phase of enzyme formation and the "latent" period are qualitatively distinct. Effects of anaerobiosis (Pollock 1952) and ultraviolet irradiation (Pollock 1953, Torriani 1956) upon subsequent penicillinase formation differ for cells in the latent phase from those upon cells in the linear phase. On removal of all free inducer the induced synthesis of penicillinase continues for a number of hours (Pollock 1958)—the "Pollock effect." A similar Pollock effect has been reported for the penicillinase of *S. aureus* (Leitner et al. 1963). Events during the early phase of induced penicillinase formation in *S. aureus* are at least as obscure as for *B. cereus,* for the length of the lag phase preceding detectable increase in rate of synthesis has been reported between values as low as 3 to 4 minutes (Leitner et al. 1963) and longer than that observed with *B. cereus* (Richmond 1968). The Pollock effect can be attributed to the small quantities of penicillin irreversibly bound to the cells (Pollock and Perret 1951). The basis for the unusual pattern of induced synthesis of penicillinase is quite obscure, although hypothetical mechanisms can be formulated (Csányi, Jacobi, and Straub 1967, Imsande 1970, Pollock 1958). This pattern does not reflect the involvement of a "stable" protein precursor of the enzyme (Pollock and Kramer 1958).

In contrast, the kinetics of induced formation of histidase and N-formimino-L-glutamic acid formiminohydrolase (FGA hydrolase) by cells of *B. subtilis* do appear to reflect the formation of active enzymes via distinct protein precursors (Hartwell and Magasanik 1963, 1964, Kaminskas and Magasanik 1970). When histidine is added to a culture growing exponentially with glutamate as primary carbon source there is a lag period (at 37°C) of some 4 to 5 minutes before any detectable increase in level of histidase enzyme and the maximal rate of enzyme synthesis is attained at about 9 minutes. Addition of chloramphenicol to the culture at 5 minutes almost totally and immediately inhibits the incorporation of amino acid into protein, yet it has little or no effect on the increase of histidase activity for 2 to 3 minutes; indeed, the differential rate of increase in histidase activity increases roughly tenfold. By the use of radioactive leucine as a tracer, it can be demonstrated that formation of a nonenzymic protein precursor of histidase enzyme commences 2 minutes after the addition of inducer, and the protein formed during this period is converted into active enzyme after a further 3-minute interval. When exposure to inducing concentrations of histidine is terminated, formation of active histidase continues for a further 12 to 15 minutes. However, synthesis of the precursor follows the pattern for the β-galactosidase of *E. coli* (Hartwell and Magasanik 1963, 1964). Thus the difference between the kinetics of syntheses of β-galactosidase and of histidase is only superficial and due to the relatively slow conversion of precursor to active enzyme.

In more recent studies using slightly different conditions of culture, Kaminskas and Magasanik (1970) obtained similar evidence of conversion of enzymically inactive precursor of FGA hydrolase to active enzyme. They also obtained evidence that transcription of sufficient amounts of messenger RNAs to support subsequent formation of detectable quantities of both enzymes was already initiated (see Chapter 5) as early as 30 seconds after addition of histidine.

An even more complex situation than that for the histidase of *B. subtilis* appears to exist for the inducible cytochromes *c* of bakers' yeast—*Saccharomyces cerevisiae.* Cells grown under anaerobic conditions contain none of the cytochromes normally found in cells grown aerobically. When washed suspensions of such anaerobically grown cells are incubated aerobically, in the absence of an exogenous source of nitrogen, there is a lag phase, followed by rapid induced synthesis of cytochromes *a*, *b*, and *c* (Slonimski 1953). Detailed study of this system by Slonimski and his colleagues has shown that during this process two distinct molecular species of cytochrome *c* are formed which differ in primary structure, that is, amino acid sequence in the polypeptide chain (Slonimski et al. 1963). The two isocytochromes are not formed synchronously. Synthesis of the minor component, iso-2-cytochrome *c*, commences immediately on exposure to oxygen at a maximal rate, which subsequently falls. On the other hand, synthesis of the major component, iso-1-cytochrome *c*, commences slowly and, after a lag, becomes rapid at a stage coinciding roughly with the minimum rate of synthesis of iso-2-cytochrome *c* (Fukuhara 1966, Sels et al. 1965). The early phase of rapid formation of iso-2-cytochrome *c* appears to represent conversion of a protein precursor into iso-2-cytochrome *c* in a process not involving protein synthesis. On the other hand, the slow early synthesis *de novo* of iso-1-cytochrome *c* may be the consequence of an inhibition of its formation ("repression") by the precursor of the iso-2-cytochrome *c* (Fukuhara 1966, Fukuhara and Sels 1966, Slonimski et al. 1963).

CONCURRENT INDUCED ENZYME SYNTHESIS

It is often observed that addition of an inducer to a culture of microbial cells elicits the induced synthesis of more than one enzyme. The group of enzymes formed usually comprise a metabolic pathway. For example, the "galactozymase" complex of yeast and of *E. coli* formed in response to exposure to galactose consists of the enzymes galactokinase, galactose·1·phosphate uridylyl transferase, and UDP-glucose epimerase. Collectively they convert galactose to glucose·1·phosphate for further metabolism by the Embden-Meyerhof pathway. The induced synthesis of a group of enzymes constituting a metabolic pathway in response to one single inducer may reflect (a) the phenomenon of "*sequential induction,*" (b) "functional integration" of the corresponding structural genes into a single "*regulon*" or "*operon,*" or (c) a combination of both the former possibilities.

Less frequently, a single inducer may evoke formation of two enzymes which are not components of a single metabolic pathway, but catalyze similar reactions on related substrates. For example, with certain strains of yeast maltose will evoke the formation of a specific α-glucosidase which is active on maltose (a maltase) and simultaneously serve as the inducer of a second α-glucosidase which is virtually inactive with maltose as substrate (Hestrin and Lindegren 1951, Spiegelman, Sussman, and Taylor 1950). Similarly, melibiose (an α-galactoside) induces formation of both α- and β-galactosidases with the B strain of *E. coli* (Sheinin and Crocker 1961). With one outstanding possible exception, in all cases where the induced enzymes formed are not components of a single pathway, the inducer is a substrate of one of the enzymes and an unused substrate-analogue of the other enzyme. The concurrent synthesis of the group of enzymes in these cases is fortuitous, and detailed comparison of the processes involved in the concurrent synthesis of the various enzymes affords no additional insight into the mechanisms regulating the induced synthesis of enzymes.

Ironically, the one possible exception to this generalization is the case of β-galactosidase enzyme in strains of *E. coli*. Induced synthesis of this enzyme is associated with a parallel (coordinate) induced synthesis of a distinct β-thiogalactoside transacetylase and a protein facilitating the entry of β-galactosides into the cell, which Rickenberg et al. (1956) named β-galactoside "permease." The nature and precise function of β-galactoside permease, and of other related permeases and "exitases" (Cohen and Monod 1957, Kepes 1960, Rotman 1958, 1959), is far from clear. Recent studies with appropriate strains of mutant have shown that one component of the permease transport system is the so-called **M** protein of the cell membrane, which strongly binds β-galactosides and galactose (Fox, Carter, and Kennedy 1967). However, at least one other component is required for permease activity (Boos 1969), and unsaturated fatty acids are required for formation of an active transport system (Fox 1969). Nevertheless, despite uncertainty regarding the nature and mechanism of the transport system, it shows many properties of an enzyme system and for present purposes will be considered an enzyme. Insofar as both β-galactosidase and transacetylase activity require entry of substrate into the cell, the permease can be considered part of the same metabolic pathway as β-galactosidase. However, the relationship, if any, between the transacetylase reaction and those of β-galactosidase or galactoside permease is totally unknown (Zabin 1963, Zabin, Kepes, and Monod 1962).

SEQUENTIAL INDUCTION

There are two distinct processes by which a single inducer may evoke the induced synthesis of a series of enzymes comprising a complete metabolic pathway. Both processes occur in strains of pseudomonad during development of the ability to metabolize aromatic acids by the pathways shown in Figure 3.1.

The first is the process of *sequential induction*, which was discovered independently by Karlsson and Barker (1948), Stanier (1947), and Suda, Hayaishi, and Oda (1949). Whenever induced enzyme synthesis is elicited by an inducer which is a substrate of the enzyme formed, the latter will convert the inducer substrate into corresponding product

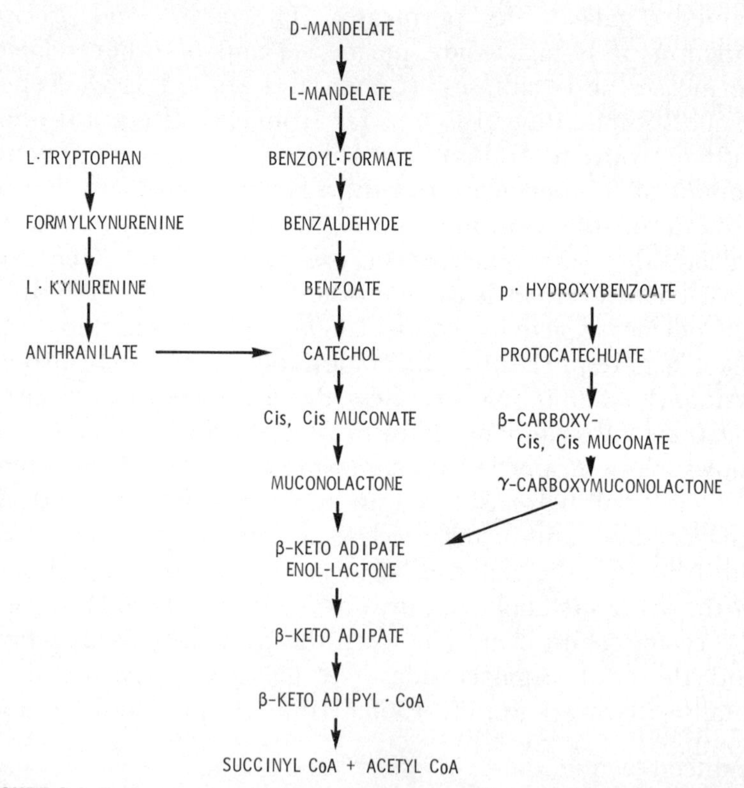

FIGURE 3.1. Pathways of metabolism of aromatic acids in strains of pseudomonad

After Cánovas, Ornston, and Stanier (1967) and Stanier (1954). There are separate NAD- and NADH-linked dehydrogenase enzymes for conversion of benzaldehyde to benzoate.

metabolite. Usually, although not invariably, this product will not serve as an inducer of the synthesis of the same enzyme. However, it *may* serve as both an *endogenous inducer* and a substrate of a second enzyme. Thus inducer substrate of the first reaction of a metabolic pathway may initiate a "cascade" of induced enzymes in sequence, and thereby *indirectly* evoke the induced synthesis of all enzymes of the pathway. Similarly, an exogenous supply of one of the intermediates of the pathway may elicit formation of all the enzymes acting upon it and metabolites of the later stages of the pathway, but not of enzymes acting upon precursors of that intermediate.

For example, cells of *Pseudomonas aeruginosa* (*fluorescens*), and other strains of pseudomonad, grown with D-mandelic acid as the carbon source contain all the enzymes needed to metabolize D-mandelic acid to intermediates of the tricarboxylic acid cycle (Hegeman 1966[a], Stanier 1954), whereas cells grown on glucose have negligible levels of all these enzymes. Similarly, cells grown on benzoic acid contain the enzymes necessary to oxidize benzoate through the later reactions of the pathway, but lack the enzymes for conversion of mandelate to benzoate. Further, a mutant strain unable to form the racemase enzyme which interconverts D- and L-mandelate (and hence unable to produce benzoate from the former) does not form the enzymes required to oxidize benzoate when exposed to D-mandelate. Thus the induced synthesis of benzoate oxidase in nonmutant cells growing on D-mandelate is dependent upon the prior induced synthesis of enzymes needed to convert D-mandelic acid to inducer benzoic acid (Stanier, Hegeman, and Ornston 1965). Similarly, a number of other enzymes participating in the interlocking pathways depicted in Figure 3.1 are also induced sequentially in various strains of pseudomonad (e.g., Hegeman 1966[a], Kemp and Hagerman 1968, Ornston 1966, Palleroni and Stanier 1964, Stanier, Hegeman, and Ornston 1965).

Such phenomena of sequential induction are of considerable potential value as a tool in the unraveling of complex metabolic pathways (Stanier 1947, 1950). However, analysis of these phenomena does not provide additional understanding of the underlying regulatory processes.

REGULONS

As noted above, when a mutant strain of pseudomonad which cannot form a functional mandelate racemase enzyme is exposed to D-mandelate it neither produces benzoic acid nor synthesizes the enzymes required for benzoate oxidation. Nevertheless, D-mandelate remains a "gratuitous" inducer of synthesis by the mutant of the four enzymes of the pathway from L-mandelate to benzoate (Figure 3.1). Therefore, formation of this group of enzymes is not controlled by a series of sequential inductions. On the contrary, all of the enzymes are formed (or not) in parallel as a response to one single inducer (Hegeman 1966[b]). Similarly, formation of the four enzymes is inhibited in parallel (i.e., repressed; see Chapter 4) by benzoate, catechol, or succinate (Mandelstam and Jacoby 1965).

Thus formation of all of this group of enzymes is coordinated by one single regulatory process, and they are said to be *coordinately regulated.* The structural genes corresponding to the various enzymes constitute one single *physiological unit of coordinate regulation,* responding to a single regulatory inducer. At the present time the precise relationship between these structural genes within the genome of the organism is not known. It seems desirable to use a descriptive generic term to describe such physiological units of regulation which carries no implication of a necessary fundamental arrangement of the structural genes in relation to each other, or of knowledge of the detailed mechanism of regulation. One useful term is the *regulon,* introduced by Maas and Clark (1964) to describe a coordinated physiological unit of regulation for which there is no architectural relationship between the structural genes of the unit. I propose to use the term regulon as a generic term describing all coordinated physiological units of inheritance; thereby, I use the term in a broader sense than Maas and Clark.

Three other such regulons are involved in the metabolic pathways given in Figure 3.1. The first is the group of enzymes which convert L-tryptophan to anthranilate, comprising tryptophan pyrrolase, kynurenine formamidase, and kynureninase. This regulon is of interest as one for which the product—L-kynurenine—serves as inducer for synthesis of all three enzymes, whereas substrate L-tryptophan is not an effective inducer. The second regulon is the group of enzymes which convert protocatechuic acid to β-ketoadipic acid, comprising carboxymuconate lactonizing enzyme, carboxymuconolactone decarboxylase, enol-lactone hydrolase, and β-ketoadipate CoA-transferase. This group of enzymes is also synthesized in response to the product β-ketoadipate. In this case three of the corresponding structural genes have been shown to be closely linked in the genome in *P. aeruginosa* (Kemp and Hegeman 1968), and this regulon may be of the type defined as an operon (see below). The third regulon probably consists of the enzymes which convert catechol to β-ketoadipate enol-lactone: namely, catechol oxygenase, muconate lactonizing enzyme, and muconolactone isomerase. With *P. aeruginosa,* synthesis of all three enzymes is induced by *cis-cis* muconic acid, and the corresponding structural genes are all closely linked in the genome (Kemp and Hegeman 1968). However, coordinate synthesis of all three enzymes was not formally demonstrated.

With *P. putida* the synthesis of catechol oxygenase was not coordinate with that of the other two enzymes (Ornston 1966). However, a recent genetic analysis indicates that the genes "coding" for all three of the enzymes are closely linked (Wheelis and Stanier 1970). It is therefore possible that the three enzymes represent two regulons in both strains of organism, or that the size of the regulon is different in both strains. At this time it would be premature to equate all or part of the regulon with an operon. The regulatory pattern of control of the enzymes of these convergent pathways is different for *Moraxella calcoacetica* from that of the two pseudomonads (Cánovas, Ornston, and Stanier 1967).

OPERONS

Many, but by no means all, of the regulons which have been examined in detail by fine genetic analysis are of the type named "*operons*" by Jacob et al. (1960). An operon is a "*genetic unit of coordinate expression*" or an "*integrated unit of coordinate regulation.*" The "typical" operon is defined by the properties of the *lac* operon of *E. coli* (Jacob and Monod 1961[a]), represented by the enzymes β-galactosidase, β-thiogalactoside transacetylase, and β-galactoside permease. This operon has been studied far more extensively than any other operon or regulon, and its properties have recently been reviewed in detail (Beckwith and Zipser 1970).

When noninduced, "wild-type" cells of *E. coli* are exposed to inducer galactoside, induced synthesis of all three enzymes occurs in parallel, or coordinately. The onset of induced synthesis of β-galactosidase precedes that of the transacetylase by a short interval, and on removal of inducer galactoside synthesis of transacetylase persists at a maximal rate for a brief period after that of β-galactosidase begins to decline (Alpers and Tomkins 1965, Leive and Kollin 1967). However, for many purposes it may be considered that onset of synthesis of the three enzymes and cessation of their synthesis occur virtually simultaneously.

The special experimental criteria which serve to define an operon rigorously require detailed genetic analysis and are therefore available only for experimental systems which are amenable to genetic analysis. However, the existence of two particular types of mutant organism may be considered presumptive evidence of a possible operon. The first of these types of mutant is one presumed to be due to a "point mutation" within a single gene (inferred from the nature of the mutagen used to induce the mutation) that results in loss of ability to regulate the synthesis of a group of related enzymes, the synthesis of which is coordinately regulated in the parental strain. Thus mutants which have simultaneously acquired the ability to synthesize β-galactosidase, β-thiogalactoside transacetylase, and β-galactoside permease constitutively (i.e., in the absence of exogenous galactoside) are *per se* presumptive evidence of the existence of the *lac* operon. Such mutants, in which a series of different characters are affected by a "single point" mutation, are said to be *pleiotropic.* Conversely, pleiotropic point mutants which are unable to synthesize any of the three enzymes in response to exogenous inducer serve as presumptive evidence of an operon. Another class of pleiotropic mutant—the so-called "polarity mutants"—provide even stronger presumptive evidence of an operon. Such mutants are characterized by the inability to synthesize one active enzyme, together with impaired but regulated production of one or more related enzymes. However, the fortuitous finding of such mutants does not constitute rigorous proof of the existence of an operon. For this purpose formal genetic analysis is required, and at this juncture it is appropriate to turn our consideration to the evolution of the body of genetic information now available concerning the β-galactosidase enzyme and *lac* operon of *E. coli.*

THE LACTOSE OPERON OF E. COLI

Genetic studies with the bacterium *E. coli* date from the reports of a process of gene recombination between multiply mutant strains by Lederberg and Tatum (1946) and Tatum and Lederberg (1947). These studies were extended to include utilization of lactose, a measure of β-galactosidase formation, as a genetic character (J. Lederberg 1947, E. M. Lederberg 1952), and the first chromosome maps included a site for what we now recognize as the *lac* operon. It is beyond the scope of this presentation to discuss the mechanisms of gene transmission and recombination in microorganisms, or to chronicle the history of the development of contemporary knowledge. These matters are covered in a number of excellent articles to which the reader is referred (Campbell 1969, Catcheside 1951, Hayes 1964, Jacob and Wollman 1961, Ravin 1965). The current status of knowledge of the *lac* operon in *E. coli* in particular has been recently reviewed (Beckwith and Zipser 1970, Epstein and Beckwith 1968). Only the major developments essential to an understanding of the nature of the operon and the mechanism by which synthesis of enzymes associated with it is regulated will be dealt with. For this purpose, it is most convenient to commence by identifying the various genetic elements known to be present in the *lac* operon. The elements and their organization within the operon (Beckwith 1967, Davies and Jacob 1968, Jacob and Monod 1961[a]) are indicated in Figure 3.2. Representative data indicating the effects of mutations on β-galactosidase and β-thiogalactosidase transacetylase in the various genes are given in Table 3.3.

FIGURE 3.2. The lactose operon of E. coli according to Ippen et al. (1968)

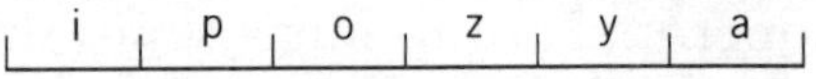

i	p	o	z	y	a

Regulator gene *i* lies outside the operon. Other genes or sites designated are promoter *p*, operator *o*, and structural genes *z*, *y*, and *a* coding, respectively, for β-galactosidase, β-galactoside permease, and β-thiogalactoside transacetylase.

The operon contains three structural genes, designated *z*, *y*, and *a*, which code for the polypeptide chains of β-galactosidase, β-galactoside permease, and β-thiogalactoside transacetylase, respectively. Lipid required for synthesis of active permease (Fox 1969) does not appear to be specified by any component of the operon. Abutting the structural gene *z* are, in sequence, a gene *o* known as the *operator*, a genetic element *p* known as the *promoter*, and beyond this the "inducibility" gene *i*, which is not part of the operon. In addition to these elements or genes, the operon must contain an entity corresponding to the *starter* of Davies and Jacob (1968) (which is the hypothetical "*promotor*" as postulated by Jacob, Ullmann, and Monod (1964)—see page 119) and may be a particular section of one of the defined genes.

TABLE 3.3. β-Galactosidase and β-thiogalactoside transacetylase levels of various types of E. coli mutant strains

	Noninduced cells †		Induced cells †	
Genotype *	β-Galactosidase ‡	Transacetylase	β-Galactosidase	Transacetylase
$i^+o^+z^+a^+$	<0.1	<1	100	100
$i^+o^+z^-a^+$	<0.1	<1	<0.1	100
$i^+o^+z^+a^-$	<0.1	<1	100	<1
$i^-o^+z^+a^+$	120	120	120	120
$i^+z^-a^+/Fi^-z^+a^+$	2 ‡	2	200	250
$i^-z^-a^+/Fi^+z^+a^+$	2	2	250	120
$i^-z^-a^-/Fi^-z^+a^+$	250	250	200	250
$i^so^+z^+a^+$	2	2	2	2
$i^so^+z^+a^+/Fi^+o^+z^+a^+$	2	2	2	2
$i^+o^cz^+a^+$	25	25	100	100
$i^+o^+z^-a^+/Fi^+o^cz^+a^+$	75	75	250	300
$i^+o^+z^+a^+/Fi^+o^cz^-a^+$	1	75	100	250
$i^so^+z^+a^+/Fi^+o^cz^+a^+$	150	150	150	150
$i^+o^oz^+a^+$	<0.1	<1	<0.1	<1
$i^+o^oz^+a^+/Fi^+o^+z^+a^+$	<0.1	<1	250	250
$i^+o^oz^+a^+/Fi^+o^cz^+a^+$	75	75	250	250
$i^+o^oz^+a^+/Fi^-o^+z^+a^+$	1	1	250	250
$i^+o^+p^-z^+a^+$	1	<1	2	<1
$i^+o^+p^-z^+a^+/Fi^+o^+p^+z^-a^+$	1	–	1.5	–
$i^+o^+p^-z^+a^+/Fi^+o^+p^+z^+a^+$	1	–	112	–

* Data from Jacob and Monod (1961[b]) before resolution of genes *y* and *a* used for purposes of illustration as of the designated genotype; and from Scaife and Beckwith (1966).

† Levels of activity are expressed as a percentage of the activity developed by the wild-type strain under conditions of gratuity with isopropyl-β-thiogalactoside.

‡ Some of the strains used in the compilation of Jacob and Monod were reported to contain a small number of constitutive mutants.

STRUCTURAL GENES OF THE lac OPERON

The three distinct structural genes were defined by studies of mutants in which a point mutation of the "mis-sense" type (see Jukes 1966) has occurred in one of them. In theory, point mutations could result in conversion of a codon specifying one given amino acid to a second codon specifying the same amino acid. Similarly, such a mutation could result in the synthesis of a β-galactosidase polypeptide chain in which one amino acid of the wild-type sequence has been replaced by a second amino acid without change in the physicochemical and catalytic properties of the molecule. Such "neutral mutations" have indeed been mimicked for the tryptophan synthetase enzyme **A** protein of *E. coli* in elegant experiments of Yanofsky's group (Yanofsky, Ito, and Horn 1966) and are believed to have occurred frequently throughout the course of evolution (see Arnheim and Taylor 1969, Jukes 1966, King and Jukes 1969). In practice, however, the mutants actually sought and studied have been those which have lost the ability to produce an enzymically active β-galactosidase protein, either in the presence or the absence of an inducer galactoside. Such mutants show induced synthesis of an enzymically inactive analogue of β-galactosidase (Perrin, Bussard, and Monod 1959), but normal induced synthesis of β-galactoside "permease" and β-thiogalactoside transacetylase, in the presence of inducer (Table 3.3, line 2). Similarly, mis-sense point mutations in the other two structural genes affect only the corresponding protein (Table 3.3, line 3: These data are from Jacob and Monod 1961[b] before genes *y* and *a* were resolved). The three structural genes are distinct entities and may be physically separated by the process of gene recombination. Nevertheless, formal genetic analysis shows them to be adjacent. Kinetic analysis of the induced synthesis of the three proteins shows the three structural genes to constitute a physiological unit of coordinate expression.

THE INDUCIBILITY GENE OF THE lac OPERON

Gene *i* was defined by the existence of constitutive mutants which produce maximal levels of β-galactosidase enzyme in the absence of inducer (Monod 1956) (Table 3.3, line 4). Formal genetic analysis shows the position of the locus to be as indicated in Figure 3.2. As indicated above, comparison of the data obtained for a series of galactosides with the ML and K12 strains of *E. coli* (Jacob and Monod 1961[a], Monod and Cohn 1952) shows some inducers to be relatively more efficient inducers with one strain than with the other. This may be considered presumptive evidence of the existence of a number of alleles of the *i* gene in different wild-type strains. Five different types of allele have been recovered in mutants obtained in the laboratory:

(a) i^- constitutive or unregulated.
(b) i^s "regulator noninducible," pleiotropic negative mutants which form only basal levels of the three enzymes even in the presence of high concentrations of inducer galactoside (Willson et al. 1964).
(c) $i^{s\,amber}$ secondary mutants containing a "nonsense" codon, which are constitutive in strains of *E. coli* lacking "*amber*-suppressor" gene alleles (Bourgeois, Cohn, and Orgel 1965).
(d) i^q "high repressor level" mutants which are fully inducible for all three enzymes and distinguished from the wild type inducible by genetic criteria (see below) (Müller-Hill, Crapo, and Gilbert 1968).
(e) i^{-d} constitutive mutants which are distinguished from those of class (a) by genetic criteria. (These mutants will not be further discussed.)

Each of the first four of these classes has proved particularly useful in the analysis of our contemporary concepts of the operon. However, it should be noted that only mutants of the first two classes were available at the time Jacob and

Monod (1961[a]) first presented their concepts of the operon.

The wild-type inducible or regulated strain is designated "*inducible-positive*" (i^+), and the constitutive or unregulated mutants of the first class are designated "*inducible-negative*" (i^-). This convention reflects the experimental observation that the allele "inducible" is dominant in action to that of the "constitutive" allele; regulated is dominant to unregulated. In early studies of the linkage relationships between genes z, y, and i, Pardee, Jacob, and Monod (1959) took advantage of the transitory partially diploid state of the zygote between the stages of gene transfer and gene recombination (Wollman, Jacob, and Hayes 1956) to assess dominance relationships between various alleles. Crosses were made in the absence of inducer galactoside between strains able to form active β-galactosidase and permease but inducible ($i^+z^+y^+$) and strains unable to form active β-galactosidase but producing permease constitutively ($i^-z^-y^+$). Thus the experimental conditions were such that neither parental strain could synthesize appreciable levels of β-galactosidase enzyme. When the cross was between an inducible male donor strain ($Hfr\ i^+z^+y^+$) and permease-constitutive female recipient strain ($F^-i^-z^-y^+$) rapid constitutive synthesis of β-galactosidase commenced within a few minutes of entry of donor genes into the recipient cytoplasm. This constitutive synthesis continued for up to 90 minutes, after which time further β-galactosidase formation required the presence of an inducer. On the other hand, in the reciprocal cross ($Hfr\ i^-z^-y^+$ by $F^-i^+z^+y^+$) no appreciable synthesis of β-galactosidase took place at any time in the absence of inducer. Thus, in the first type of cross unregulated synthesis of β-galactosidase in constitutive cytoplasm, determined by the female constitutive allele, gave way to regulated synthesis determined by the male inducible allele. In the second type of cross the constitutive allele of the male is never expressed in the inducible cytoplasm of the female. Pardee and colleagues therefore concluded that the inducible allele was dominant and controlled the synthesis of a cytoplasmic component which actively prevented unregulated β-galactosidase synthesis. In considering the possible nature of this cytoplasmic component, they presented a persuasive argument that this might be analogous to hypothetical *repressors* postulated (Vogel 1957[a], 1957[b]) as mediating a second type of regulatory process, *enzyme repression,* which will be considered in the next chapter. The function of the inducer of induced enzyme synthesis would then be one of *antagonizing the cytoplasmic repressor,* and the process of induced enzyme synthesis would be one of *derepression.* Synthesis of a basal level of enzyme in the absence of exogenous inducer would then be determined by the relative concentration of repressor and endogenously formed inducer (Fangman, Gross, and Novick 1967, Müller-Hill, Crapo, and Gilbert 1968).

At the time alternative interpretations could be advanced. However the discovery of an unusual *sex factor* which had incorporated an entire *lac* operon (Jacob and Adelberg 1959, Jacob et al. 1960) permitted the analysis of dominance relationships between alleles in stable partially diploid (*merogenote* or *heterogenote*) strains of *E. coli.* These strains contain one entire *lac* operon within the bacterial chromosome and a second complete *lac* operon within the cytoplasmic *Flac* factor or *episome* introduced into the cell by the process known as *sex-duction* (Hayes 1964, Jacob and Wollman 1961).[4] This latter process of sex-duction permits the geneticist to construct appropriate merogenote strains of *E. coli,* containing particularly useful pairs of different alleles of each of the genes of the *lac* operon and the regulator gene *i* in various combinations. The genetic structure of such strains is represented by indicating the chromosomal alleles in the conventional manner, and the alleles attached to the

[4] In such strains the basal level of enzyme in noninduced cells would be determined by the concentration of repressor molecules per operon (Buttin 1963[c], Germaine and Rogers 1970, Revel and Luria 1963).

sex factor are preceded by the symbol **F**. Thus a strain with wild-type alleles of the *z* and *i* genes in the chromosome and mutant alleles in the sex factor would be represented by the notation i^+z^+/Fi^-z^-. Two types of merogenote can be constructed. One contains the allele "inducible" linked to a *lac* operon containing the normal or wild-type allele for β-galactosidase, and the allele "constitutive" linked to an operon containing mutant allele(s) (i.e., i^+z^+/Fi^-z^-). In this arrangement or configuration the i^+ and z^+ alleles are said to be in the *cis* position, or much less frequently to be "*in coupling*." The alternative configuration has the allele "inducible" linked to an operon containing mutant alleles, and the allele "constitutive" linked to a normal operon (i.e., i^+z^-/Fi^-z^+). In this configuration the alleles i^+ and z^+ are said to be in the *trans* position, or "*in repulsion*."

Studies of merogenotes with the alleles z^+ and i^+ in the *trans* position readily showed the "inducible" allele i^+ to be dominant to "constitutive" allele i^- and to regulate expression of the z^+ allele *trans* with respect to it (Table 3.3, lines 5 and 6). Similar studies with merogenotes mutant in the structural genes served to confirm that there is only one structural gene, or *cistron* (Benzer 1957), for each of β-galactosidase, β-thiogalactoside transacetylase, and β-galactoside permease (Jacob and Monod 1961[a]). Further extension of the use of this so-called "*cis-trans* test of function" in selected merogenote strains led to the recognition both of additional alleles of the *i* gene and the operon itself.

Mutants of the "regulator noninducible" i^s type were initially isolated as strains unable to synthesize more than basal levels of β-galactosidase, β-thiogalactoside transacetylase, or β-galactoside permease, even in the presence of very high concentrations of galactoside inducer. Detailed genetic analysis showed that the mutation was located in the *i* gene locus and that i^s allele was dominant to the wild-type "inducible" allele (Willson et al. 1964: Table 3.3, lines 8 and 9). The existence of this type of mutant provided strong evidence in support of the concept of regulation of induced synthesis of β-galactosidase by direct antagonism of the action of a specific repressor molecule (Pardee, Jacob, and Monod 1959). The other classes of mutant alleles of gene *i* have served partly in providing circumstantial evidence of the nature of the repressor and, in the case of "high repressor level" i^q mutants, as a particularly useful source for the isolation of repressor (Müller-Hill, Crapo, and Gilbert 1968).

It should be noted that although regulator gene *i* is closely linked to the operator gene *o* it is not a component of the operon. This is shown by (a) independence of expression of the *i* gene from that of expression of the genes of the operon (Jacob and Monod 1961[a]), and (b) the formal demonstration (Gilbert and Müller-Hill 1966) that synthesis of repressor protein is not inducible.

THE REPRESSOR OF THE lac OPERON

The nature of the postulated repressor was not conclusively established until it was actually isolated and shown to have the predicted properties. With the wisdom of hindsight, we now know that the relevant evidence available at the time was highly ambiguous. It was originally suggested that the repressor was probably not a protein (Jacob and Monod 1961[a]). However, subsequent study of the effects of suppressor genes (Jacob, Sussman, and Monod 1962, Sarkar and Luria 1963) and of nonsense mutant alleles of the *i* gene (Bourgeois, Cohn, and Orgel 1965, Müller-Hill 1966), coupled with the failure to find any other gene specifying the nature of the repressor, led to reconsideration of the conclusion. Indeed, in presenting their concepts of the nature of *allosteric proteins* and of the control of cellular metabolism, Monod, Changeux, and Jacob (1963) suggested that the properties of the hypothetical repressor of induced enzyme synthesis could be readily interpreted in terms of an allosteric protein. Specifically, the repressor protein would possess an "active site" which (in noninduced cells) would combine with an appropriate segment of the *lac* operon and repress expression of the constituent structural genes by blocking transcription of the corresponding messenger RNA(s) (see Chapter 5). The repressor would also contain one or more (allosteric) specific site(s) with which inducer could combine and, in so combining, cause a reversible change in conformation of the repressor protein: an *allosteric transition.* In the altered conformation repressor would no longer combine with the *lac* operon, and the uncomplexed operon would then be expressed. Conversely, certain noninducer analogues of active inducers could be expected to compete with inducers for the allosteric site of the repressor to form complexes not displaced from the operator, thereby competitively inhibiting induced enzyme synthesis (Jayaraman, Müller-Hill, and Rickenberg 1966, Monod 1956). Further, the modified type of repressor molecule produced by a constitutive mutant would be expected to react with inducers to form complexes, some of which combine with the operator gene and thereby inhibit the constitutive synthesis of enzyme (Monod and Cohen-Bazire 1953[a]). Thus the regulator gene *i* would be the structural gene for repressor protein.

The elegantly simple concepts of the operon (including features yet to be discussed in the following pages) (Jacob and Monod 1961[a]) and of the repressor as an allosteric protein (Monod, Changeux, and Jacob 1963) were readily and widely accepted. Nevertheless, direct demonstration of the protein nature and function of the repressor had to wait until Gilbert and Müller-Hill (1966) partially purified the repressor of the *lac* operon. They then demonstrated (Gilbert and Müller-Hill 1967) that it is a protein with physicochemical properties essentially in accord with those predicted by Monod, Changeux, and Jacob (1963). Meanwhile, Zubay, Lederman, and De Vries (1967) demonstrated that the repressor protein had the predicted biological properties in a "coupled" cell-free system which was able to synthesize a portion of the β-galactosidase peptide chain *in vitro*. However, more recent studies of the process of transcription of the *lac* operon (see Chapter 5) indicate that at least one additional component is required for effective interaction between the repressor and operator gene (de Crombrugghe et al. 1971[b], Eron et al. 1971, Ohshima et al. 1970). Preliminary evidence indicates that a protein associated with the ribosomes (de Crombrugghe et al. 1971[b], Ohshima et al. 1970) is also required.

Isolated repressor protein is a relatively large molecule which appears to contain four subunit chains (Gilbert and Müller-Hill 1966, Riggs and Bourgeois 1968, Riggs, Suzuki, and Bourgeois 1970). It binds specifically and reversibly to DNA containing the wild-type *lac* operon but not to DNA lacking all genes of the operon (Gilbert and Müller-Hill 1967, Riggs et al. 1968; see however Lin and Riggs 1970). The extent of binding corresponds to one repressor molecule per *lac* operon, or more specifically, per wild-type operator gene (Riggs, Suzuki, and Bourgeois 1970). Complex formation with DNA containing so-called "operator constitutive"-mutant operons (see p. 117) is markedly reduced, to an extent predictable from the residual degree of increase in rate of β-galactosidase synthesis *in vivo* on addition of exogenous inducer (Gilbert and Müller-Hill 1967, Riggs et al. 1968). Interaction between repressor and operator gene occurs very rapidly, whereas dissociation of the complex is a relatively slow process (Riggs, Bourgeois, and Cohn 1970).

Both inducers of β-galactosidase synthesis and noninducer structural analogues readily complex with the repressor protein (Gilbert and Müller-Hill 1966); probably one molecule binding to each of the subunit chains (Riggs, Suzuki, and Bourgeois 1970). However, only the inducers cause dissociation of repressor-operator DNA complexes (Gilbert and Müller-Hill 1967, Riggs et al. 1968). Association of inducer with the complex facilitates dissociation of the latter, rather than prevents complex formation (Riggs, Newby, and Bourgeois 1970). It seems probable that this involves a change in conformation of the repressor protein; more specifically, reversal of an allosteric transition which occurs during complex formation with the DNA. However, this remains to be determined, as does the role of the additional factor(s) required (de Crombrugghe et al. 1971[b], Eron et al. 1971, Ohshima et al. 1970) for suppression of expression of the *lac* operon by repressor.

To date, the only other similar repressors which have been isolated and shown directly to be proteins with the predicted properties are those of two lysogenic bacteriophages (Pirrotta and Ptashne 1969, Ptashne 1967[a], 1967[b]), discussed in the next chapter.

The specificity of the allosteric sites of the repressor for complex formation with various inducers and inhibitory "antiinducers" (Gilbert and Müller-Hill 1966, Riggs, Newby, and Bourgeois 1970) is largely in agreement with that previously inferred from the efficiencies of various galactosides as inducers or inhibitors of induced β-galactosidase synthesis (Tables 3.1 and 3.2). Major differences are observed for lactose and galactose. However, lactose is not *per se* an inducer, but a substrate for production of endogenous inducer by intact cells (Burnstein et al. 1965). Conversely, although galactose is not an effective inducer of β-galactosidase synthesis by wild-type strains of *E. coli,* it is a good inducer for mutants lacking galactokinase and for "operator constitutive" mutants with high levels of galactoside permease (Llanes and McFall 1969[a], 1969[b]). Subject to the reservation that there may be similar complications, we may conclude that study of the induced synthesis of any other enzyme system in response to a series of substrates and analogues (Buttin 1963[a], Kalckar, Kurahashi, and Jordan 1959, Kelly and Clark 1962, Pollock 1957 [a]) gives a preliminary indication of the stereospecificity of corresponding postulated repressors.

THE OPERATOR GENE OF THE lac OPERON

Early analyses had served to define two of the properties of the reaction between the repressor and the cell genome. First, the reaction involved a highly specific site, for mutations in the *i* gene only affected the synthesis of proteins controlled by the *lac* operon. Second, the repressor had a pleiotrophic effect, regulating the time and extent of synthesis of the corresponding proteins coordinately. These two properties alone required the conclusion that the repressor acted directly on the genome and that the reaction involved a single highly specific site, which Jacob and Monod (1961[a]) termed the *operator*.[5]

One immediate consequence of this conclusion was the prediction that there should be two classes of mutant affected in the operator (Jacob and Monod 1961[a]). Mutants of the first type would be unable to react with repressor product of the *i* gene. Such mutants would synthesize β-galactosidase constitutively. Further, such "operator-constitutive" (o^c) mutants would show dominance of the mutant allele over the wild-type operator allele in the *cis-trans* test of function in partly diploid merogenotes. Mutants of the second class would be characterized by loss of ability to synthesize any of the proteins specified by the associated structural genes because of irreversible complexing of the repressor to the operator. These "operator-inactivated" (o^0) mutant alleles would be recessive in the *cis-trans* test of function to either wild-type or operator-constitutive mutant alleles. Mutants of both predicted types were isolated by Jacob et al. (1960) (Table 3.3, lines 1, 10, and 14). Genetic analysis indicated both to be affected in the same region of the genome,

[5] At the time alternative interpretations based upon an interaction with a cytoplasmic component of the "protein-synthesizing machinery" were not excluded (see for example Maas 1961, Stent 1964). These alternative interpretations have since been formally eliminated in the case of the *lac* operon. It seems highly probable that the results obtained for the latter can be considered generally valid. Nevertheless, in some instances (e.g., Engleberg, Squires, and Meronk 1969) additional hypotheses must be made to "bring the data into line" with requirements of the operon model. The note of caution recently sounded by Richmond (1968) is certainly in order.

at a locus very closely linked with the structural genes *z*, *y* (and subsequently *a*) and the regulator gene *i* (Jacob and Monod 1961[a]). Subsequent genetic studies have actually indicated that most, if not all, mutations of the o^c constitutive class arise by *deletion* of this portion of the genome, and that the o^0 noninducible class of mutants are not mutant in the operator locus itself (Beckwith 1964, Brenner and Beckwith 1965). However, they have at the same time served to confirm the existence of a "repressor-sensitive site" and to define more precisely the relationship of this operator to the other genes of the operon (Beckwith 1967, Davies and Jacob 1968, Ippen et al. 1968).

The physiological relationship between the operator and the structural genes of the *lac* operon was revealed by the application of the *cis-trans* test of function in merogenotes. Purely physiological studies had revealed that the structural genes comprised an integrated "unit of physiological expression," or regulon. Further, the existence of point mutations with *polar* (pleiotropic) effects upon the synthesis of all proteins coded by the structural genes of the operon clearly revealed some interdependence in the synthesis of these proteins. However, the nature of that interdependence could only be speculated upon. Application of the *cis-trans* test of function revealed the existence of a "*position effect*" (Jacob and Monod 1961[a]) similar to that observed by Sturtevant (1925) and Lewis (1951) for so-called pseudoalleles at partially defined gene loci of the fruit fly *Drosophila*. Thus, in diploids of the genetic constitution o^+z^+/Fo^cz^- the synthesis of β-galactosidase required the presence of exogenous inducer, whereas in diploids of the genetic constitution o^+z^-/Fo^cz^+ synthesis of the enzyme was semiconstitutive or almost totally constitutive (Table 3.3, lines 11 and 12). Even though strains of the two genotypes illustrated in lines 11 and 12 of Table 3.3 contained the same total complement of gene alleles, one strain was inducible and the other semiconstitutive. The genotypic difference between the two strains is in the arrangement of the various alleles. In the inducible strain the normal allele of gene *z* is *cis* to the "wild-type" operator *o*, and *trans* to the o^c mutant genome. In the semi-constitutive strain the normal allele of gene *z* is in the converse relationship to the operators. Therefore, we may conclude that the operator affects only the function of the structural genes which are *cis* with respect to it. In other words, each operator and the adjacent structural genes constitute a single entity which functions independently of any other homologous operon present in the same cell.

In formulating their concepts, Jacob and Monod (1961[a] noted the formal similarities between the considerable body of information which had been amassed for the *lac* operon and for situations observed in the study of *lysogenic bacteriophages* and predicted the existence of similar operons controlling the replication of such viruses. Further, they similarly indicated a number of systems in which the synthesis of inducible enzymes and repressible enzymes (next chapter) showed features entirely in accord with their operon hypothesis. Subsequent study has largely fulfilled the predictions made (Jacob and Monod 1961[a]), although in most cases other than that of two *lysogenic bacteriophages* alternative interpretations of the available data have not been rigorously excluded (e.g., Richmond 1968). These systems will not be considered in detail but will be referred to subsequently as appropriate—making the explicit assumption that these systems are indeed entirely analogous with the *lac* operon.

THE PROMOTER OF THE *lac* OPERON

In proposing their concepts of the operon, Jacob and Monod (1961[a]) also suggested a mechanism by which the operon could be expressed in the synthesis of β-galactosidase (see Chapter 5), which required that the operon contain a *site* from which the synthesis of messenger RNA commenced: the "*initiation point*." The demonstration that the original "operator constitutive" o^c mutants had deletions of the presumed operator region, and that "operator-inactivated" o^0 mutants contained point mutations in the *z* structural gene (Beckwith 1964, Brenner and Beckwith 1965), were not compatible with the hypothesis that the initiation point was located within the operator region defined by the deletions of the o^c mutants. Jacob, Ullmann, and Monod (1964) therefore examined an extensive series of o^c deletion mutants and found that they fell into two groups. The vast majority were like the original operator constitutive o^c mutants which had served to define the existence and position of the operator gene. These mutants contained an intact structural gene *z* and were either partly or fully constitutive for synthesis of β-galactosidase. A smaller group were mutants in which the deletion both extended into the structural gene *z* and eliminated the regulator gene *i*. These mutants did synthesize β-thiogalactoside transacetylase and β-galactoside permease constitutively, but at levels not exceeding 10 percent of the maximal levels of induced wild-type cells or "regulator-constitutive" i^- mutants. It was concluded that expression of the residual *a* and *y* structural genes of these mutants resulted from their fusion with another intact operon. This conclusion was formally validated in two specific cases. In one the residual genes were fused to an operon regulating biosynthesis of enzymes required for the synthesis of purines, and their expression was regulated coordinately with the purine operon (Jacob, Ullmann, and Monod 1965). In the other case they were fused to, and regulated by, an operon controlling tryptophan biosynthesis (Beckwith, Signer, and Epstein 1966, Hu 1969).

Comparison of the properties of these two classes of deletion mutant revealed the presence of a genetic element between the regulator gene *i* and the structural gene *z* with two interrelated properties. First, the element is normally essential for synthesis of any of the three proteins controlled by the operon. No operator-constitutive o^c deletion mutants were recovered in which the deletion extended into the structural gene *z* and did not also eliminate the regulator gene *i*. Thus, it appeared that the structural genes *y* and *a* could only be expressed when the region between the regulator gene *i* and the structural gene *z* was eliminated if they were fused to another operon containing a like element. Second, this region appeared to regulate the extent of expression of the genes of the operon. Levels of transacetylase and permease activities observed when genes *y* and *a* were fused to another operon were invariably very markedly less than when these genes were present in an intact *lac* operon. This genetic element was named the *promotor* (Jacob, Ullmann, and Monod 1964).

It was originally concluded from these analyses that the promotor is located between the operator gene *o* and the structural gene *z* (Beckwith 1967, Jacob, Ullmann, and Monod 1964). However, subsequent detailed genetic analysis has indicated that the element now known as the promoter is between the regulator gene *i* and the operator gene *o*, as shown in Figure 3.2 (Ippen et al. 1968, Miller et al. 1968).

Subsequent studies have concentrated on the second property of the postulated *promotor*, namely that of controlling the extent of expression of the structural genes of the operon. The experimental criteria which serve to define this function of the genetic unit *do not necessarily include that of serving as the initiation point* postulated by Jacob, Ullmann, and Monod (1964). It is therefore perhaps fortunate that the name of the unit has undergone a transition to *promoter*, while the hypothetical initiation point has been renamed the "*starter*" (Davies and Jacob 1968). The precise relationship of these two entities is of very considerable importance to an understanding of both the mechanism of action of the promoter and the mechanism by which the operon is expressed as an integrated unit (see Chapter 5). It has usually been assumed that the promoter is the site of initiation of transcription of messenger RNA, but this has not been formally demonstrated.[6]

Three classes of observation collectively define the existence of a promoter which controls the extent of expression of the structural genes of the *lac* operon. The first of these are the studies of "operator-constitutive" o^c deletion mutants discussed above (Beckwith 1964, Jacob, Ullmann, and Monod 1964, 1965). Second, Scaife and Beckwith (1966) isolated strains of *E. coli* which showed properties expected of mutants affected in the promoter region. Basal levels of the enzymes controlled by the operon were of the same order as those of the wild-type strain, and exposure to inducer elicited the induced synthesis of all three enzymes. However, the maximal levels of enzyme formed were markedly less than those obtained with the wild-type strain. Further, the mutation affected the expression of structural genes in the *cis* position but not of those of a second homologous *lac* operon. Thus the effects of these mutations were distinct from those known, or expected, of mutations in the operator *o* and regulator *i* genes. The site of these mutations was originally located between the operator *o* and structural gene *z* (Beckwith 1967, Scaife and Beckwith 1966). However, more detailed genetic analysis of the strains containing point mutations of this type places the mutant site, and hence promoter, between the operator *o* and regulator *i* genes (Arditti, Scaife, and Beckwith 1968, Ippen et al. 1968, Miller et al. 1968). Finally, in a study of regulator-constitutive i^- mutants isolated from an operator-inactive o^0 stock of

[6] Roberts (1969[a], 1969[b]) has demonstrated that the levels of specific mRNAs transcribed *in vitro* from the genome of bacteriophage λ can be correlated with mutations in the corresponding promoters. However, efficient regulation of gene transcription requires the presence of ribosomal protein(s) for which the site and mechanism of action are not known (see Chapter 5). Formal demonstration of the identity of initiation site and promoter should be possible by analyses of the average numbers of initiation sites per genome using techniques of Bautz and Bautz (1970). However, I shall accept the identity of the two sites in further discussion.

E. coli, Pardee and Beckwith (1962) found very marked variations in the levels of β-galactosidase enzyme formed by a series of different constitutive mutants.

STIMULATION OF EXPRESSION OF THE lac OPERON BY CYCLIC AMP

A number of recent studies have indicated that cyclic 3′5′-AMP (cAMP) plays a major role in the synthesis of inducible enzymes required for the utilization of a variety of carbon sources. Many of these studies have been concerned primarily with the effects of cAMP in relieving an inhibition of induced enzyme synthesis caused by the operation of a second regulatory phenomenon known as *catabolite repression.* These will be considered in relation to the phenomenon of catabolite repression in the following chapter. However, some studies have been made of the effects of cAMP on the phenomenon of induced enzyme synthesis *per se.*

Addition of an exogenous source of cAMP enhances the differential rate of β-galactosidase synthesis to an extent dependent upon the nature of the available source of energy (and in particular whether glucose is also present; see Chapter 4) with intact cells of *E. coli* (Ullmann and Monod 1968) and "spheroplast" preparations (Goldenbaum and Dobrogosz 1968, Pastan and Perlman 1968, Perlman, de Crombrugghe, and Pastan 1969). However, it does not affect the duration of the lag period before onset of rapid enzyme synthesis (de Crombrugghe et al. 1969, Pastan and Perlman 1969[a], Perlman and Pastan 1968). Further, a mutant strain with a marked deficiency of adenyl cyclase enzyme activity shows very markedly impaired induced synthesis of β-galactosidase enzyme and very poor (or no) growth on a number of media for which an essential prerequisite for growth is the induced synthesis of enzymes acting on the carbon source. Addition of cAMP markedly stimulates induced β-galactosidase synthesis and growth on the various test media (Pastan and Perlman 1970, Perlman and Pastan

1969).[7] Conversely, a mutant strain lacking enzyme degrading cAMP shows little effect of varying the carbon source upon β-galactosidase formation (Monard, Janeček, and Rickenberg 1969).

cAMP appears to stimulate transcription of messenger RNA from the *lac* operon (Jacquet and Kepes 1969, Varmus, Perlman, and Pastan 1970[a], 1970[b]), but the mechanism by which it does so is unknown. Chambers and Zubay (1969) observed a very marked stimulation by cAMP of the *in vitro* induced synthesis of β-galactosidase in a "coupled" system able both to transcribe messenger RNA and to synthesize protein. This stimulation involves formation of a complex between the cAMP and a specific cAMP-"binding protein" (Emmer et al. 1970, Pastan and Perlman 1970, Zubay, Schwartz, and Beckwith 1970). However, the function of this complex has not yet been determined (de Crombrugghe et al. 1971[b], Eron et al. 1971).

The site of action of cAMP appears to be the promoter of the operon. Either point mutation within this region, or deletion of most of the promoter, markedly reduces both sensitivity to stimulation by cAMP and effects of varying the carbon source in the culture (Pastan and Perlman 1968, Perlman, de Crombrugghe, and Pastan 1969, Silverstone et al. 1969). On the other hand, mutants affected in the operator *o* and regulator *i* genes show similar responses to those of the wild-type strain (Pastan and Perlman 1968, Perlman and Pastan 1968, Tyler and Magasanik 1969, Ullmann and Monod 1968).

THE GENERALITY OF PROMOTERS

Evidence for the existence in operons (or other genetic units) controlling synthesis of most other inducible (and re-

[7] Results with this mutant are complicated by the finding that it could grow on media containing galactose and fructose as sole carbon sources (for which induced enzyme synthesis would have been expected to be required), and that glucose metabolism was also impaired in the absence of an exogenous source of cAMP (Perlman and Pastan 1969).

pressible) enzymes of promoter regions determining the extent of enzyme formation is entirely circumstantial. Even in the case of the extensively studied *gal* operon (enzymes – galactokinase, galactose·1·phosphate uridylyl transferase and uridine diphosphogalactose·4·epimerase) of *E. coli* the evidence is incomplete. Genetic studies have provided mutants of the operator-constitutive o^c type (Buttin 1963[b], Fiethen and Starlinger 1970) and "leaky" pleiotropic mutants (Adhya and Shapiro 1969, Saedler and Starlinger 1967[a], 1967[b], Shapiro and Adhya 1969) similar to the promoter mutants of Scaife and Beckwith (1966). A series of regulator constitutive i^- mutants have different levels of the enzymes (Buttin 1963[a], Yarmolinsky, Jordan, and Weismeyer 1961). Furthermore, the expression of the genes of this operon is affected by the available supply of cAMP (Perlman, de Crombrugghe, and Pastan 1969, Tao and Schweiger 1970). Nevertheless, the adenyl cyclase-deficient mutant *E. coli* strain of Perlman and Pastan (1969) grows as well with galactose, albeit slowly, as it does with glucose as carbon source. Similarly, in the case of the L-arabinose (*ara*) operon (arabinose isomerase, ribulokinase, ribulose·5·phosphate epimerase) "operator-constitutive" o^c types of deletion mutant have been obtained (Englesberg et al. 1965, Englesberg et al. 1969), and substantial synthesis of the enzymes appears to be dependent upon the supply of cAMP (Perlman and Pastan 1969). However, this operon shows features of the regulatory process quite different from those for the *lac* operon (Englesberg, Squires, and Meronk 1969).

For other microbial inducible enzymes of particular interest the demonstration of a marked stimulation of induced enzyme synthesis by cAMP may be considered *presumptive* evidence of a related promoter. Such a demonstration has been made with *E. coli* for enzymes of the glycerol operon and unidentified enzymes required for fermentation of maltose and mannitol (de Crombrugghe et al. 1969, Perlman and Pastan 1969). However, caution must be exercised in the interpretation of such observations, for cAMP also stimulates the induced synthesis of tryptophanase by *E. coli* after transcription of further messenger RNA has been arrested (Pastan and Perlman 1969).

Other systems for which the available evidence is markedly more circumstantial, but nonetheless suggestive of a possible role for a promoter analogous to that of the *lac* operon of *E. coli,* are those of the penicillinase enzymes of *B. cereus, B. licheniformis,* and *S. aureus.* Pollock (1957[b]) reported a remarkably wide range of maximal levels of penicillinase enzyme formed by mutants derived from the **569** and **5** strains of *B. cereus.* In particular, for strains of the **5** series maximum levels spanned a 5000-fold range between the "micro-noninducible" strain **5** and the "magno-noninducible" derived **5/B** strain.[8] Similar series of mutant derivatives have been reported for *B. licheniformis* (Dubnau and Pollock 1965) and for *S. aureus* (Richmond 1967). Many of the mutant strains reported show similarities to operator-constitutive o^c and promoter mutants of the *lac* operon of *E. coli.* Unfortunately, neither *B. cereus* nor *B. licheniformis* is (yet) amenable to the necessary detailed genetic analysis to determine whether the observed points of similarity warrant the conclusion that a promoter is probably involved in regulation of penicillinase synthesis.[9]

In the case of *S. aureus* the necessary genetic analysis can be made. The results of extensive genetic analysis of the elements regulating the induced synthesis of penicillinase by this organism have recently been reviewed (Richmond 1968, 1969) and will not be presented in detail. To a large extent the accumulated data are in accord with the hypoth-

[8] The ever-increasing number of mutant strains of bacteria showing effects upon the production of one single enzyme exhibits very wide ranges in maximum level of enzyme formed, and in extent of response to exogenous inducer. Collins et al. (1965) have proposed a rational scheme of classification of mutant genotype based upon these two parameters. This classification is followed here.

[9] The recent demonstration of transduction with the **B569** strain of *B. cereus* (Yelton and Thorne 1970) will permit direct analysis of this possibility in the near future.

esis that penicillinase synthesis is controlled by an operon, located in an extrachromosomal plasmid (Novick and Richmond 1965) rather than in the bacterial chromosome. However, Smith and Richmond (cited in Richmond 1968) have failed to recover any operator-constitutive o^c deletion mutants of the type studied for the *lac* operon of *E. coli.* It remains to be determined whether the micro-noninducible mutants used by Richmond (1967) to define a regulatory site additional to the *i* locus are similar to the "promoter point-mutation" strains (e.g., strain **L8**) first discovered by Scaife and Beckwith (1966). Nevertheless, we must seriously consider with Richmond (1968) that regions analogous to the operator and promoter of the *lac* operon of *E. coli* may play no part in regulating the induced synthesis of penicillinase.

INDUCED ENZYME SYNTHESIS CONTROLLED BY OTHER OPERONS AND REGULONS

The conclusion that synthesis of inducible enzymes in general is *invariably* regulated by the complex interplay of inducer, repressor, operator, promoter, and cyclic AMP *in exactly the same manner* as for the *lac* operon of *E. coli* would certainly be premature. In a limited number of cases the cumulative body of evidence from studies of the genetics and the physiology of the enzymes concerned (including those of transcription of the corresponding messenger RNA; see Chapter 5) is entirely in accord with expectation from the current concepts of the nature and function of the *lac* operon of *E. coli.* This is particularly true for the enzymes of the galactose pathway (galactokinase, galactose·1·phosphate uridylyl transferase, and UDP-glucose epimerase) and corresponding *gal* operon of *E. coli* (Saedler et al. 1968). In a larger number of cases an entirely similar situation also appears to exist for *repressible* enzyme systems (Chapter 4). There are, however, a number of systems for which the responses observed for the *lac* operon of *E. coli* do not provide a sufficient explanation of the available data. Additional, or alternative, postulates must be made.

One of the aberrant systems to which reference has been made previously is that of the complex of "arabinose enzymes" of *E. coli* (pentose isomerase, ribulokinase, and ribulose·5·phosphate epimerase). Extensive studies of this system by Englesberg and his co-workers (Englesberg et al. 1965, Englesberg et al. 1969, Englesberg, Squires, and Meronk 1969) have established that synthesis of the enzymes is regulated coordinately, and that the corresponding structural genes are linked in an integrated genetic unit containing a gene formally equivalent to the operator of Jacob and Monod (1961[a]). Synthesis of this group of enzymes is controlled by the regulator gene *araC,* which is probably the structural gene specifying a "regulator protein" (Hogg and Englesberg 1969). However, regulation of the *ara* operon involves more, or something else, than the "negative" procedure of dissociation of a repressor from operator which we believe to occur in the activation of expression of the *lac* operon. In appropriate merogenote stocks the wild-type allele of the regulator *araC* causes a "*trans*-dominant *activation*" of expression of the operon, which indicates a "positive" process of regulation of function of the *ara* operon (Sheppard and Englesberg 1967). Englesberg and his colleagues (Englesberg, Squires, and Meronk 1969, Englesberg et al. 1969) suggest that in this system the postulated repressor "gene product" of the regulator *araC* can undergo a reversible change between two molecular configurations with different and distinct functions. In the noninduced cell it is present as a repressor in the configuration $\mathbf{P}_1$, and complexed with the operator gene of the *ara* operon. On adding inducer arabinose the repressor is displaced from the operator and undergoes a change in conformation to assume the configuration $\mathbf{P}_2$ of an "activator," in which state it can combine with a second ("initiator") site within the operon. This has properties of the promoter region of the *lac* operon and has been tentatively located between the operator gene and the first structural gene of the operon. An alternative interpretation which appears to be in accord with the available data would be that L-arabinose is not *per se* an inducer of the *ara* operon, and that the "product" of gene *araC* serves *both* as the repressor of the operon *and* as an enzyme which converts L-arabinose into "true" inducer.

There is indeed some independent evidence to warrant postulating a dual *direct* role of action of the "product" of a regulator gene from studies of the "arginine enzymes" of *S. cerevisiae* (Thuriaux et al. 1968). In this organism the amino acid arginine serves simultaneously to repress synthesis of enzymes of the arginine biosynthetic pathway and as an inducer of synthesis of the enzymes arginase and ornithine transaminase. The synthesis of both groups of enzymes is inversely regulated by a gene designated *ar-r*. Mutations in this gene locus cause a simultaneous loss of the ability to repress formation of enzymes of the biosynthetic pathway and loss of the ability to form the two inducible enzymes. Thus the product of this *ar-r* gene is required for repression of synthesis of one group of enzymes and for induced synthesis of enzymes of a second group. The mutant genotype is recessive to the wild type, and Thuriaux et al. (1968) consider this to indicate that the gene product of the *ar-r* gene is not *per se* a repressor of the enzymes of the degradative pathway. This formal demonstration of dual functions of the gene product of a regulator gene simultaneously in both a negative and a positive direction on two interrelated but different regulons can serve as a model to warrant the postulation of a similar dual function of the regulation of a single operon.[10] Nevertheless an interpretation of the data

[10] A more complex situation in which the product of a regulator gene simultaneously participates in repression of some enzymes of arginine biosynthesis and induced synthesis of ornithine δ-transaminase is known for certain mutant strains of *E. coli* (see Chapter 4). However, an additional structural gene is required for induced synthesis of ornithine δ-transaminase. There is no evidence of such a second structural gene in the case of the arabinose operon.

for the *ara* operon of *E. coli* along these lines does at the present time require the *ad hoc* assumption that a postulated repressor can serve two related but distinct functions within one operon. Similar reservations must also be made for other instances in which the regulation of synthesis of induced enzymes appears to be (Hatfield, Hofnung, and Schwartz 1969[a], 1969[b], Power 1967), or may be (Cove and Pateman 1969, Jones-Mortimer 1968[a], 1968[b]) subject to a positive control mechanism.

A second system which shows properties differing to some extent from those of the *lac* operon in *E. coli* is a regulon controlling the formation of the pyruvate dehydrogenase enzyme complex of *E. coli* (Henning and Herz 1964, Henning, Herz, and Szolyvay 1964, Henning, Szolyvay, and Herz 1964). In the case of the *lac* operon "extremely polar" mutations of the type originally thought to be operator-inactivated (o^0) are rare and restricted to one relatively small region of the structural gene *z*. Analogous mutations for the regulon (acetate locus) for the pyruvate dehydrogenase enzyme complex are relatively common and occur within a relatively large region of the structural gene for the decarboxylase component of the enzyme complex. Henning and his co-workers (1966) have therefore postulated that the polypeptide chain coded by one cistron of this regulon participates in induction of synthesis of the other polypeptides of the complex when it is not complexed in pyruvate dehydrogenase "aggregate."

Yet another system which shows aberrant behavior is the penicillinase enzyme of *B. cereus,* and probably that of *B. licheniformis.* As discussed previously, there is a substantial lag after addition of inducer penicillin before the maximum differential rate of penicillinase formation is attained (Imsande 1970, Pollock 1952), and enzyme synthesis continues at a linear rate after removal of free inducer (Imsande 1970, Pollock 1958, Pollock and Perret 1951). Imsande (1970) has interpreted these observations in terms of a nondiffusible, membrane-bound inducer, and an operon controlled by two regulator genes. One of the latter is postulated to be the structural gene for a repressor protein and the other the structural gene of an "inducer-binding site" of the cell membrane. It is further proposed that repressor spontaneously dissociates from the complex formed with the operator at an appreciable rate and is destroyed by interaction with binding site-inducer complex. At the present time this model is highly speculative but is now amenable to analysis.

In the case of the cytochromes *c* of *S. cerevisiae* synthesized when anaerobically grown cells are exposed to oxygen, the processes involved in the appearance of iso-1-cytochrome *c* differ from those required for formation of iso-2-cytochrome *c* (Sels et al. 1965, Slonimski et al. 1963). Detailed analysis of this system indicates that the polypeptide chain of the iso-2-cytochrome *c* is formed from a pre-existing precursor, and that the heme-free peptide chain may serve as the repressor of synthesis of iso-1-cytochrome *c* (Fukuhara 1966, Fukuhara and Sels 1966, Slonimski et al. 1963). The data available for this system can therefore be interpreted in terms of one or more operons similar to the *lac* operon of *E. coli.* However, it remains to be determined whether one operon of a higher degree of complexity than envisaged in the original concepts of Jacob and Monod (1961[a]), two separate operons, or additional regulatory processes are the essential minimum requirements needed to permit a satisfactory and complete theoretical explanation of the properties of this system.

In summary, the present evidence indicates that there are at least as many inducible enzyme systems for which the pattern of regulation of their synthesis differs somewhat from that of the *lac* operon and each other as there are systems for which the pattern is not demonstrably different. Similarly, there is no one invariant pattern of regulation of the synthesis of other enzymes of microorganisms (Chapter 4). It therefore seems possible that contemporary microorganisms have evolved, and are in the process of evolving (Cánovas, Ornston and Stanier 1967, Horowitz 1965, Pollock 1967), different but similar regulatory mechanisms. In this context the *lac* operon of *E. coli* can perhaps be considered to represent the extreme example in which the highest possible efficiency has been attained, both in architectural organization of essential genetic elements and in economy of number of component parts or steps needed to effect the regulation.

INDUCED ENZYME SYNTHESIS IN METAZOANS

The phenomenon of induced enzyme synthesis is probably not restricted to microorganisms, although most intensive studies have necessarily been undertaken with bacteria and yeasts for technical reasons. Obvious difficulties include performing the necessary detailed genetic analysis and resolving direct actions of inducer upon the responding cells from indirect actions via humoral factors. Recent studies (see Chapter 9) have served to emphasize the need to distinguish in higher organisms between an increase in rate of *de novo* synthesis of enzyme and an accumulation of enzyme synthesized *de novo* as a consequence of a reduced rate of degradation.

It is illustrative of the difficulties involved that one of the earliest cases reported, that of pancreatic lactase in the dog (Bayliss and Starling 1904[a], 1904[b]), was not formally shown to be incorrect until 1969 (Coughlan, Rajagopalan, and Handler 1969). Even today there is still considerable debate whether the nitrate reductase enzymes of a variety of plant cells are indeed inducible enzymes (e.g., Ferrari and Varner 1970, Kessler and Oesterheld 1970, Shen 1969).

What appeared to be the first clear-cut and conclusive demonstration of induced enzyme synthesis in a mammal was not reported until 1950. Knox and Mehler (1950, 1951) observed a rapid increase in liver tryptophan pyrrolase activity following administration of substrate L-tryptophan.

Increases in activity of some tenfold were observed at approximately 6 to 8 hours, and thereafter the level of enzymic activity rapidly declined to the normal basal level. The response was specific to the liver and to administration of the substrate, or the hormone hydrocortisone (see Chapters 6 and 9). However, subsequent studies have shown that the increased level of tryptophan pyrrolase activity found in this classic, and apparently definitive, study is not due to induced enzyme synthesis. Rather, there is an activation of pre-existing enzyme protein (Feigelson and Greengard 1961, Greengard and Feigelson 1961, Pitot and Cho 1961), coupled with a stabilization of the enzyme by substrate tryptophan against degradation (Schimke 1966, Schimke, Sweeney, and Berlin 1965). Responses of this system are further complicated by the effects of levels of various hormones in the circulation, age of the animal, and by ill-understood differences in the nature of the response when tryptophan is administered by different routes (Knox 1966, Knox and Greengard 1965, Labrie and Korner 1968). Similarly, the tissue levels of many other enzymes are markedly influenced by the hormonal status of the animal. Indeed, most of the reported cases of induced enzyme synthesis in tissues of higher plants and animals which have been intensively studied deal with responses to hormones rather than substrates (and will be considered later in Chapter 6). As a consequence most of the cases of induced enzyme synthesis in response to substrate reported for higher organisms cannot yet be considered rigorously established.

Most of the physiological problems inherent in studies with intact animals can be avoided by the use of preparations of perfused isolated organs, or better of homogeneous "established" lines of cells in tissue culture. These approaches have to date been largely used in analyses of responses to hormone treatments rather than substrate. However, Nebert and Gelboin (1968, 1970) have shown in embryonic hamster cells very rapid increases of microsomal aryl hydroxylase activity which appear to be due to an increased rate of *de novo* synthesis of enzyme in response to exogenous substrate, rather than a decreased rate of degradation of enzyme stabilized by substrate. Further, comparison of the responses of different generations of subculture of cells originally isolated directly from the tissues revealed changes in response to substrate coincident with change in genome, as reflected by changes in cell morphology.

It seems likely that further application of the techniques of analysis of cells in tissue culture will prove as fruitful in rigorously demonstrating the potential for induced enzyme synthesis in response to substrate by cells of higher forms as it has been for other types of regulatory phenomena first studied intensively with bacteria and yeasts. Further, the development of newer techniques for undertaking genetic analysis of such cells in culture will substantially contribute to an understanding of the mechanisms by which the controlling genes exert their action. The existence in higher forms of operons similar to those of *E. coli* or of other integrated groups of genes is currently a matter of controversy and speculation. Consideration of this topic will be deferred to Chapter 6.

It being a general rule in Nature's proceedings, that where she begins to display an excellency, if the subject be further searched into, it will manifest, that there is not less curiosity in those parts which our single eye cannot reach, than in those which are more obvious. — Robert Hooke, 1665.

In this task Nature has excelled in arranging everything so very suitably for the preservation and procreation of every kind of animal that anyone who surveys this matter carefully will discover nothing which is not wonderful, nothing not divine. — H. Fabricius, 1600.

CHAPTER 4: **ENZYME REPRESSION, CATABOLITE REPRESSION, TRANSIENT REPRESSION, AND IMMUNITY IN LYSOGENIC BACTERIA**

ENZYME REPRESSION AS THE COUNTERPART OF INDUCED ENZYME SYNTHESIS

To date, we know of only a few inducible enzymes which are essential components to anabolic, biosynthetic pathways. One of the first reported is the kynureninase enzyme of *Neurospora crassa* (Jakoby and Bonner 1953, Wainwright and Bonner 1959), which participates in both the pathway of synthesis of niacin and in the "tryptophan cycle." The nature of the induced response to L-tryptophan indicates that it serves in the economy of the cell primarily in the essentially catabolic process of converting L-kynurenine to anthranilic acid. Increase in the potential activity of a key enzyme of the niacin biosynthetic pathway appears to be a fortuitous consequence of the substrate specificity of kynureninase enzyme (Wainwright and Bonner 1959). Other reported cases have been listed in a review by Horowitz and Metzenberg (1965).

Studies of an apparent case of the induced synthesis of a key enzyme of a biosynthetic pathway, the acetylornithinase of *E. coli,* led to recognition of the phenomenon we now call *enzyme repression.* The biosynthetic pathway involved is one leading to the formation of ornithine, citrulline, and arginine from glutamic acid (Vogel 1953). Ornithine is formed via a series of acetylated intermediates of which the last one, N^{α}-acetylornithine, is deacylated by the enzyme acetylornithinase (Vogel 1953, Vogel and Bonner 1956). Examination was made of the growth patterns of an *auxotrophic* mutant strain of *E. coli* unable to synthesize N-acetylornithine. It was observed that cells grown with a supplement of acetylornithine grew equally readily on subculture into medium supplemented with either acetylornithine or arginine. In contrast, cells grown previously with arginine showed a pronounced lag in growth with acetylornithine, but not with arginine as the supplement (Vogel 1957[a], Vogel and Davis 1952). Moreover, when the mutant strain was inoculated into medium containing both acetylornithine and arginine the growth curve was diphasic. Early growth was associated with utilization virtually exclusively of the arginine, and later growth following exhaustion of the arginine was supported by acetylornithine (Vogel 1957[a], 1960). However, further analysis of this case of the apparent induced synthesis of acetylornithinase enzyme in response to acetylornithine showed that the underlying phenomenon was a release from inhibition of acetylornithinase formation by excess arginine. Rather than acetylornithine serving as an inducer of acetylornithinase synthesis, arginine was acting as an antagonist of enzyme formation. Vogel (1957[a], 1960) named this phenomenon of decreased rate of synthesis of acetylornithinase specifically in response to exogenous arginine "enzyme repression." The enzyme was said to be *repressible,* and arginine was defined as a "*repressor*" which *repressed* synthesis of acetylornithinase.

In formulating these concepts and terminology, Vogel (1957[a]) noted a number of reported cases in which growth of microbial cells in the presence of amino acids or pyrimidines, which they had the potential (genotypic) ability to synthesize for themselves, yielded cells with unexpectedly low actual (phenotypic) abilities to synthesize the corresponding metabolite. The first case discovered was that of the virtual absence of methionine synthase enzyme for cells of *E. coli* and *A. aerogenes* grown specifically in the presence of methionine (Cohn, Cohen, and Monod 1953, Wijesundra and Woods 1953). Similarly, tryptophan synthetase levels were markedly reduced in cells grown in the presence of tryptophan and also various structural analogues (Monod and Cohen-Bazire 1953[b]). Vogel (1957[a]) therefore suggested that the type of control observed for acetylornithinase might be a general phenomenon. He also explicitly noted that the term enzyme repression would embrace situations in which an added metabolite interfered with the process of induced enzyme synthesis.

Subsequent studies by a variety of workers have established that the regulatory process of enzyme repression is indeed a general phenomenon found not only with microbial cells but also involved in regulating such diverse processes as viral reproduction and the activities of cells of higher forms. The phenomenon, as defined by Vogel (1957[a]), is now recognized to be regulated by two entirely different mechanisms, which reflect the regulatory action of two distinct genetic elements. One of these corresponds to the situation observed for acetylornithinase, in which synthesis of an enzyme of a biosynthetic pathway is suppressed in the presence of the end product of the pathway, and the term "enzyme repression" or "*end-product repression*" is reserved to describe this phenomenon. The other type of situation corresponds with that envisaged by Vogel (1957[a]) in his explicit inclusion of cases in which the process of induced enzyme synthesis is suppressed by an added metabolite. Originally widely known as "the glucose effect," this type of suppression of enzyme formation is now known as *catabolite repression.*

In addition, as we have seen in Chapter 3, many instances

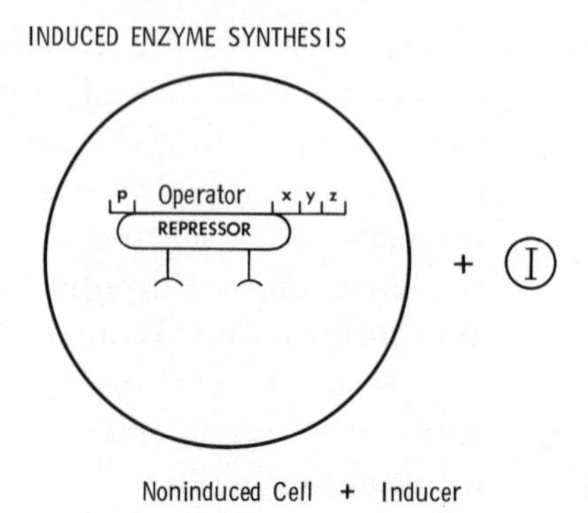

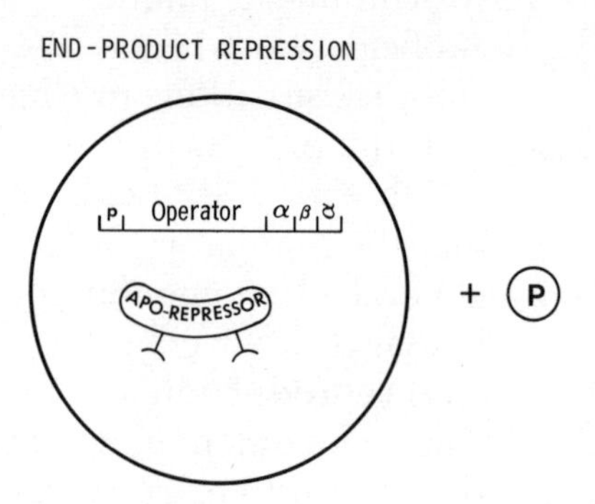

FIGURE 4.1. Schematic comparison of the processes of induced enzyme synthesis and end-product repression

of induced enzyme synthesis are now recognized to be cases of specific derepression of a repressible enzyme. Indeed, the similarities between the two processes figured prominently in the development of the concepts of the operon formulated by Jacob and Monod (1961[a]). They postulated that the two regulatory processes of induced enzyme synthesis and of end-product repression were formally related by a common underlying mechanism. The basic difference which was reflected in laboratory observations of the two phenomena was in the properties of the postulated repressor protein. In the case of induced synthesis of an enzyme the corresponding repressor protein was complexed with the appro-

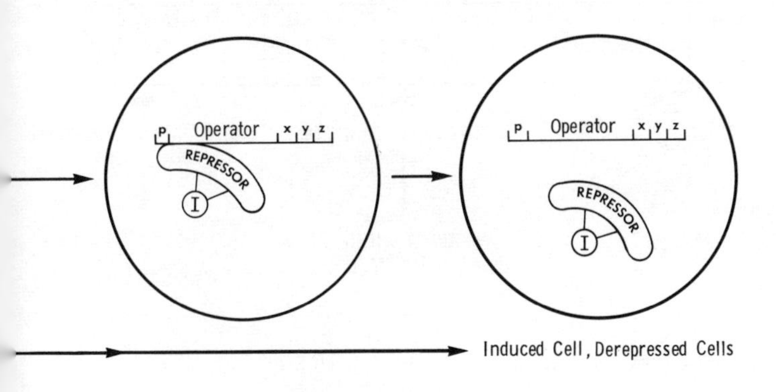

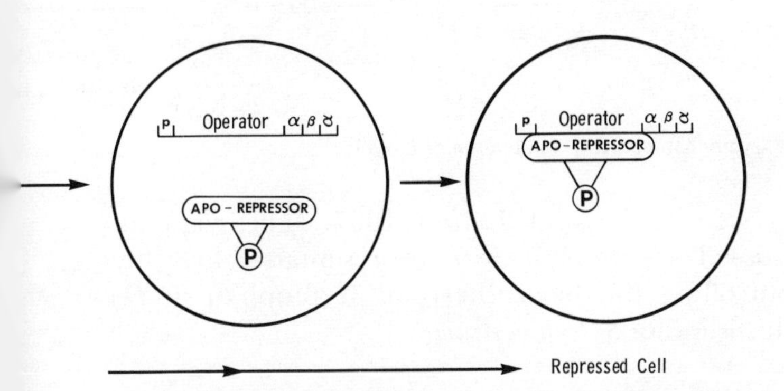

priate operator gene in the absence of inducer and displaced from the operator by complexing with inducer. The converse situation was postulated to occur for repressible enzymes (Figure 4.1). Here, the *co-repressor* end product complexed with an *apo-repressor* protein present in the cell cytoplasm, and the resulting complex, bound to the corresponding operator gene, hindered expression of the associated genes of the operon. It is therefore instructive to compare the properties of a "typical" repressible operon with the *lac* operon of *E. coli.* The arginine pathway and regulon (Maas and Clark 1964) cannot be used for this purpose as the component structural genes are not linked in a single

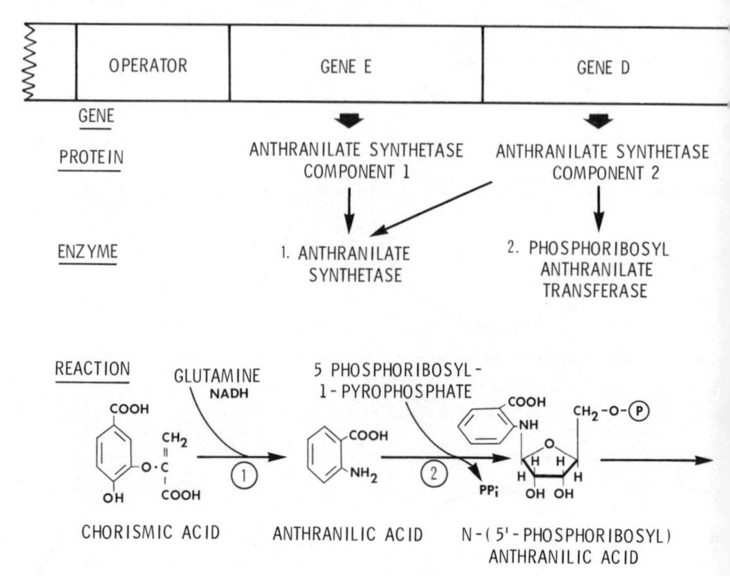

FIGURE 4.2. The tryptophan pathway and operon of E. coli

operon (e.g., see Taylor and Trotter 1967). This regulon will be discussed separately. More representative are the operons controlling the biosynthesis of tryptophan in *E. coli* and of histidine in *S. typhimurium.*

ENZYME REPRESSION AND THE TRYPTOPHAN OPERON OF E. COLI

The structure of the tryptophan operon of *E. coli* and the corresponding biosynthetic pathway are illustrated in Figure 4.2. This *trp* operon contains five structural genes in an order corresponding to the sequence of reactions in the pathway (Imamoto, Ito, and Yanofsky 1966, Yanofsky and Lennox 1959). As in the case of the *lac* operon, point mutations within one of the structural genes usually results in formation of an abnormal homologue of the normal corresponding protein. Indeed, Yanofsky and his colleagues have extensively exploited mutations within structural gene *A* in analyses of the genetic code and the nature of mutations (Yanofsky et al. 1964, Yanofsky, Ito, and Horn 1966), and in

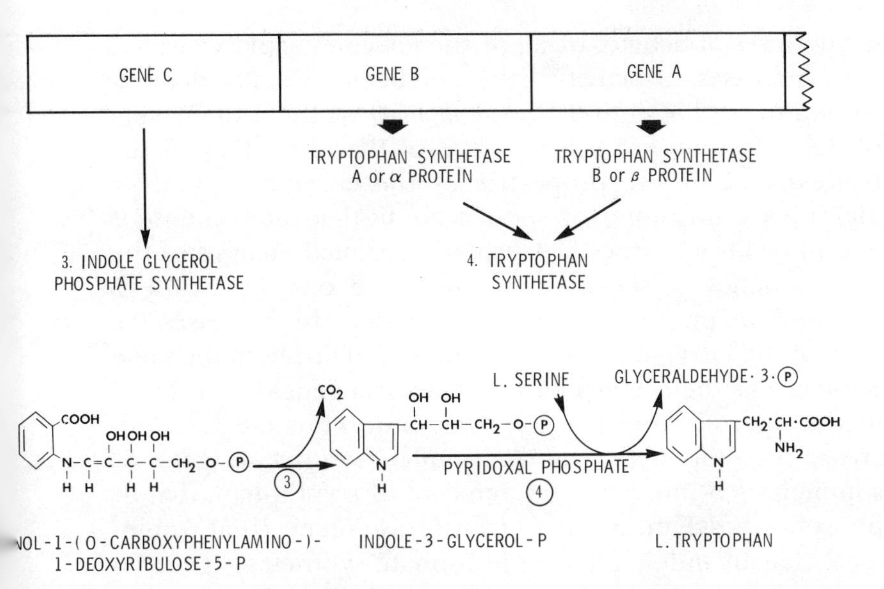

studies of the mechanism of action of regulator (Eisenstein and Yanofsky 1962, Lester and Yanofsky 1961), suppressor (Brody and Yanofsky 1965, Carbon, Berg, and Yanofsky 1966), and mutator (Yanofsky, Cox, and Horn 1965) genes. Point mutations within the region of a structural gene proximal (nearest) to the operator may have polar effects in reducing the levels formed of enzymes corresponding to the genes which are operator-distal (more distant) to it (Imamoto, Ito, and Yanofsky 1966, Yanofsky 1967). Small gene deletions extending into this same region may also have "reverse-polarity" effects upon the expression of the adjacent operator-proximal structural gene (Ito and Crawford 1965).

Regulation of expression of the genes is essentially coordinate (Ito and Crawford 1965, Lester and Yanofsky 1961). When cells of the wild-type strain are grown in medium containing a supplement of tryptophan the levels of each of the enzymes coded by genes of the operon are low. On removal of the exogenous tryptophan an increase in the level of activity of all of the enzymes rapidly ensues. The increases in activity have not been formally demonstrated to be due entirely to the *de novo* synthesis of enzyme protein, but the extensive studies of the physiology of the process and of the properties of the enzymes (Yanofsky 1967) leave no room for serious doubt that the response is one of *de novo* synthesis of protein. Detailed analysis of the early kinetics of the onset of increased enzyme synthesis (Ito and Imamoto 1968) has shown that the increases for each of the enzyme proteins commence in order, in the same sequence as the corresponding structural genes are ordered in the operon from the operator end. Thus (at 37°C) increase in component 1 of anthranilate synthetase (gene *E*) commences 3 minutes after removal of tryptophan, that of phosphoribosyl·anthranilate transferase (gene *D*) at 5 minutes, that of indole·glycerol·phosphate synthetase (gene *C*) at 7 minutes, and increase of the level of tryptophan synthetase *A* protein (gene *A*) at 9 minutes. Conversely, on reimposing conditions of repression the synthesis of the various proteins declines in sequence.

The system does show one anomaly, in that the formation of tryptophan synthetase A protein persists for more than two minutes after cessation of synthesis of indole·glycerol·phosphate synthetase. This is attributed to the synthesis by *E. coli* of a messenger RNA representing only the operator-distal portion of the operon, due to initiation of transcription at an internal site in the operon (Ito and Imamoto 1968). Indeed, regulation of synthesis of the enzymes specified by the *trp* operon is not strictly coordinate. In a comparison of the levels of enzyme activities of maximally repressed cells with those of unrepressed cells Morse and Yanofsky (1968) found that the enzymes fell into two distinct groups. Anthranilate synthetase and phosphoribosyl anthranilate transferase (genes *E* and *D*) comprised one group which was regulated coordinately. The remaining enzymes (genes *C*, *B*, and *A*) constituted a second group which was coordinately regulated, but to a different extent from the first group. A similar situation had previously been reported for the homologous *trp* operon of *S. typhimurium* by Bauerle and Margolin (1966, 1967). Present interpretations of the data are that both these *trp* operons contain two sites for the initiation of transcription of messenger RNA, that is, "promotors" as defined by Jacob, Ullmann, and Monod (1964) and renamed "starters" by Davies and Jacob (1968). Indeed, Morse and Yanofsky (1969[b]) have isolated nonrepressible mutant strains in which an operator-distal mutation in structural gene *E* has created a new point of initiation (promoter), and thereby simultaneously rendered the normal operator-repressor regulatory mechanism ineffective. These, and similar mutations in other operons, will be discussed further in Chapter 5. For present purposes it is sufficient to note that internal promoter sites may modify *the degree of coordination* of regulation of expression of the genes of the operon *without affecting the coordinate nature of the regulation.*

Regulation of the *trp* operon in response to the level of available tryptophan (Cohen and Jacob 1959, Yanofsky 1960) is effected *primarily* by a regulator gene designated *trp R*. In contrast to the regulator genes of the inducible enzymes previously considered (Chapter 3), gene *trp R* is totally unlinked to the operon it regulates (Cohen and Jacob 1959). The wild-type allele R^+ and the corresponding character of "*repressibility*," or regulation, is dominant in the *cis-trans* test of function to the allele R^- ("constitutive" and unregulated) (Morse and Yanofsky 1969[c]). Some of the constitutive *trp* R^- mutants studied by Morse and Yanofsky (1969[b]) are *amber* nonsense mutants, and the *trp R* locus is therefore the site of a structural gene coding for a specific protein, which is presumably the apo-repressor postulated by Jacob and Monod. Synthesis of the presumed repressor is apparently not regulated in response to the supply of tryptophan. The *trp* operon is also subject to regulation by a second unlinked gene designated *trp S*, which appears to be a structural gene for the enzyme tryptophanyl-tRNA synthetase (Doolittle and Yanofsky 1968, Hiraga et al. 1967, Ito, Hiraga, and Yura 1969, Kano, Matsushiro, and Shimura 1968), and it now appears probable that the co-repressor of the *trp* operon is tryptophanyl-tRNA.[1]

The operator of the *trp* operon of *E. coli* has been defined, and located adjacent to the structural gene *E*, by operator mutants equivalent to those originally used to define the operator of the *lac* operon. Thus Matsushiro and his co-workers (Matsushiro et al. 1962, 1965), and more recently Hiraga and colleagues (Hiraga et al. 1967, Hiraga 1969), have isolated mutants with the properties of operator-constitutive o^c deletion mutants. On the other hand, Sato and Matsushiro (1965) obtained a transducing bacteriophage containing an operator-inactive (o^0) type of deletion mutant, in which the genes of the *trp* operon appeared to be fused to one of the "immunity operons" of the phage genome and regulated coordinately with it. Existence of these two types of operator-deletion mutants also indicates the presence in the operator region of the operon of a promotor element as defined by Jacob, Ullmann, and Monod (1964). Insofar as distinct initiation sites or starters have been located in the interior of the operon, it seems probable that this promotor element is a complex of promoter as defined by Beckwith (1967) with a starter or initiation site.

The first reaction of the biosynthetic pathway controlled by the *trp* operon (Figure 4.2) is catalyzed by the enzyme anthranilate synthetase, which in wild-type strains of *E. coli* is susceptible to "*feedback inhibition*" by tryptophan. Mutations in the structural gene *E* which render the enzyme insensitive to feedback inhibition also reduce the extent to which

[1] Mosteller and Yanofsky (1971) have recently disputed this conclusion on the basis of results obtained in a comparison of the properties of various tryptophan analogues as substrates of tryptophanyl-tRNA synthetase and as co-repressors of the *trp* operon. However, the possibility that analogues not complexed with tRNA combined with nascent chains of anthranilate synthetase enzyme, and thereby affected translation of the cognate mRNA (see Chapter 9) was not eliminated.

tryptophan represses expression of the entire operon (Matsushiro et al. 1962, 1965, Moyed 1960, Somerville and Yanofsky 1965).[2] Until recently the basis of this pleiotropic effect was entirely a matter of speculation. However, there is now some presumptive evidence that expression of an operon is regulated in part by allosteric changes in conformation of "nascent" polypeptide chain "coded" by the first structural gene of the operon. Further discussion of this topic will therefore be deferred (see Chapter 9).

ENZYME REPRESSION AND THE HISTIDINE OPERON OF S. TYPHIMURIUM

Formation of the enzymes of the pathway for histidine biosynthesis in *S. typhimurium* is controlled by the *his* operon (Figure 4.3). Early analyses indicated that expression of this operon is regulated in the same manner as that established for the *trp* operon of *E. coli*. However, it now seems possible, and indeed probable, that the *primary* mechanisms controlling expression of these two operons are dissimilar.

The *his* operon contains nine structural genes specifying the enzymes of the pathway. These genes are arranged in sequence from the "operator" in reverse order to that of the corresponding reactions of the pathway, *except for the gene G coding for the enzyme catalyzing the initial reaction of the pathway* (Hartman, Rusgis, and Stahl 1965). The latter gene abuts a region which appears to contain operator plus promoter genes, defined by the properties (see Chapter 3) of presumed operator mutants (Ames, Hartman, and Jacob 1963, Ames, Martin, and Garry 1961, Hartman, Loper, and Sermon 1960, Roth, Antón, and Hartman 1966). However, for reasons which will be developed in the following paragraphs, it is now uncertain whether the *his* operon contains an operator gene of the type present in the *trp* or *lac* operons.

[2] It may be recalled (see Chapter 3, p. 118) that some of the presumed operator-inactivated o^0 mutants of the *lac* operon were shown to contain point mutations in the *z* structural gene (Beckwith 1964, Brenner and Beckwith 1965).

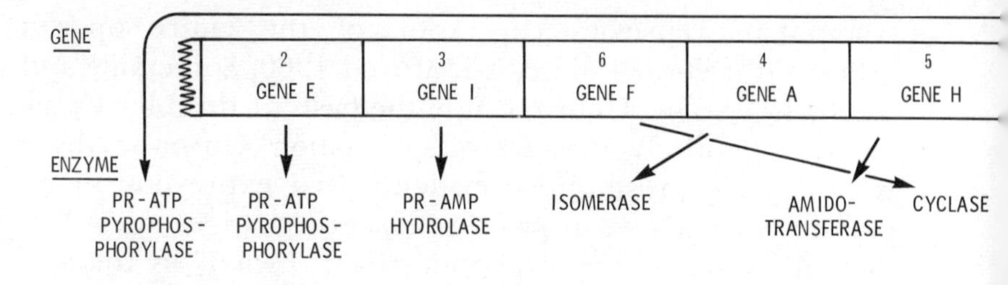

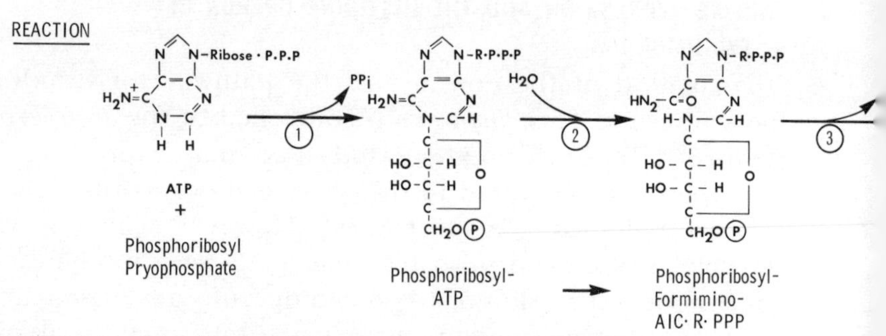

FIGURE 4.3. The histidine pathway and operon of S. typhimurium

Synthesis of all the enzymes controlled by the operon is coordinately regulated (Ames and Garry 1959, Ames et al. 1967) in response to the intracellular level of histidinyl-tRNA (Roth and Ames 1966, Roth, Antón, and Hartman 1966, Schlesinger and Magasanik 1964, Silbert, Fink, and Ames 1966). Further, point mutations in one of the genes may have polar effects upon the synthesis of all other enzymes for which the structural genes are "operator"-distal to the point of mutation (e.g., Ames and Hartman 1963, Ames et al. 1967).

Analysis of the mechanism by which histidine represses expression of the *his* operon is complicated by the additional role of histidine as a "feedback inhibitor" of the first enzyme of the pathway (Ames, Martin, and Garry 1961,

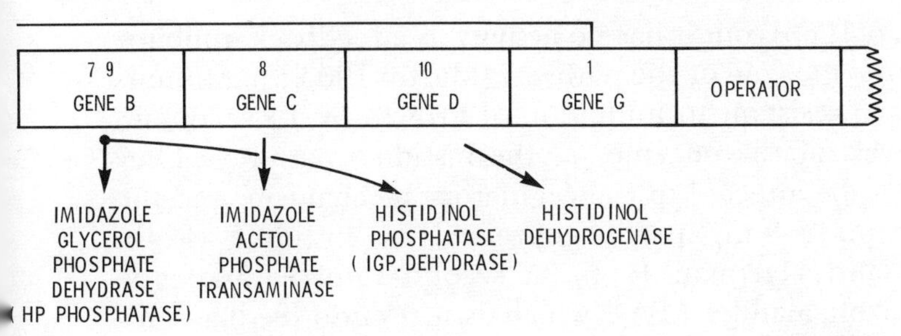

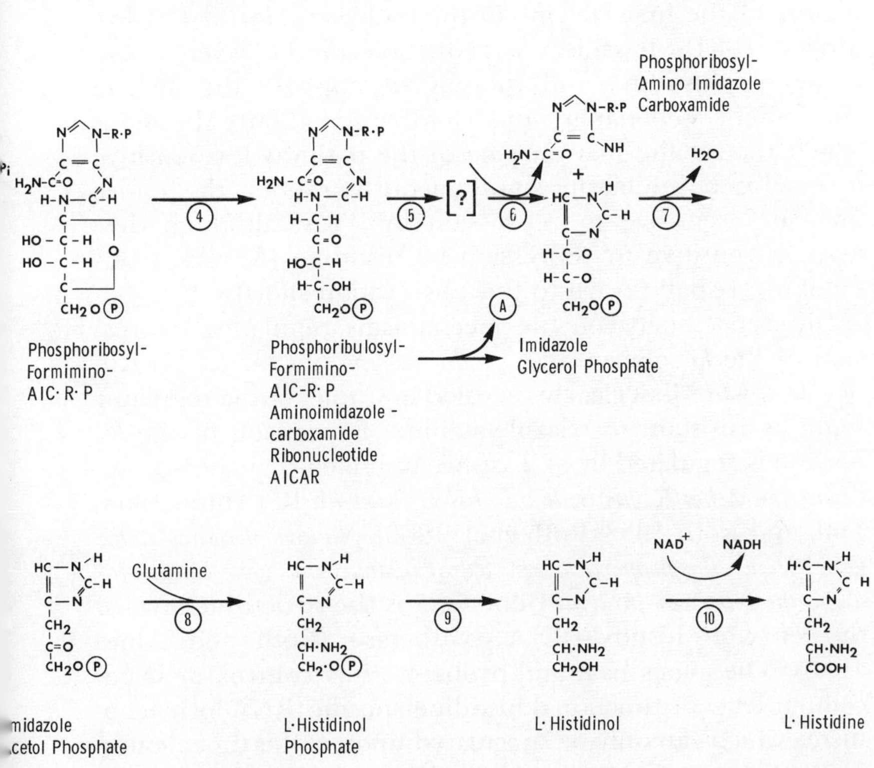

Martin 1963[a]). Therefore, many studies have been made with selected structural analogues of histidine. One particularly useful analogue is 1,2,4·triazole·3·alanine (TRA), which is incorporated into protein (and therefore into an aminoacyl-tRNA) and represses the *his* operon (Levin and

Hartman 1963), but it has no activity as a feedback inhibitor of the first enzyme of the pathway (Martin 1963[a]). Mutants which are resistant to inhibition of growth by TRA contain high levels of the enzymes of the histidine pathway. They appear to be affected in the regulatory mechanism, and the *his* operon is only partly repressed by histidine (Roth, Antón, and Hartman 1966). A second valuable analogue is 2·thiazole·alanine (TA), which is a pseudo-feedback inhibitor of the first enzyme of the pathway (Martin 1963[a], Moyed 1961). It causes a reduction in the level of endogenous histidine and thereby derepresses the operon (Berberich, Venetianer, and Goldberger 1966). Mutations which render the first enzyme of the pathway less sensitive to feedback inhibition simultaneously render the entire operon insensitive to repression by TRA, although they leave it sensitive to repression by histidine (Kovach et al. 1969[b]). I shall return to this observation shortly.

Differences between the mechanisms regulating expression of the *his* operon of *S. typhimurium* and the *trp* operon of *E. coli* were first clearly revealed in studies of the regulator mutants resistant to triazole·alanine. Expression of the *his* operon is regulated by at least five unlinked regulator genes, designated *his R, his S, his T, his U,* and *his W* (Antón 1968, Fink and Roth 1968, Roth et al. 1966). *None of these regulator genes shows the properties expected of a structural gene coding for a specific repressor protein.* Gene *his S* is the structural gene of the enzyme histidyl-tRNA synthetase (Roth and Ames 1966). The genes *his R* and probably *his W* control the intracellular level of functional histidine-specific tRNA by mechanisms which can only be speculated upon at this time (Antón 1968, Silbert, Fink, and Ames 1966). Functions of genes *his T* and *his U* are unknown. Thus analysis of the action of these regulator genes indicates that histidyl-tRNA is involved in regulation of the operon but fails to reveal the existence of a specific repressor protein.

The possibility that histidyl-tRNA synthetase enzyme is

an apo-repressor of the *his* operon (DeLorenzo and Ames 1970) cannot be eliminated. However, detailed analyses of the kinetics of repression of the operon indicate that the primary action of histidyl-tRNA is not that of complexing with the corresponding synthetase enzyme.

If cells of a "histidineless" mutant strain are grown in a minimal medium supplemented with a growth-limiting quantity of histidine the *his* operon is derepressed as the supply of essential amino acid is depleted. Growth and continued derepressed synthesis of enzymes specified by the *his* operon can be maintained in appropriate strains by addition of the histidine-precursor histidinol. When further histidine is added to such a culture synthesis of the enzymes of the histidine pathway ceases in sequence, in the same order as that of the corresponding structural genes from the "operator" (Goldberger and Berberich 1965, 1967). Thus (at 37°C) synthesis of phosphoribosyl-ATP pyrophosphorylase (gene *G*) ceases within 1 minute after addition of histidine, of imidazole phosphate transaminase (gene *C*) at 3 minutes, and of the isomerase (gene *A*) some 10 minutes after onset of repression of the operon. In contrast, when the analogue thiazole·alanine is added to the depleted culture before histidine, synthesis of all enzymes specified by the operon is arrested simultaneously (Kovach et al. 1969[a]).

It appeared probable that this marked difference in kinetics of repression reflected binding of the analogue to the "feedback-sensitive" site of the first enzyme of the biosynthetic pathway. In turn, this implied that the latter enzyme plays a role in regulating expression of the operon (Kovach et al 1969[a]). This conclusion has been confirmed by experiments using mutant strains in which sensitivity of the enzyme to feedback inhibition is altered (Kovach et al. 1969[b]). In a mutant hypersensitive to feedback inhibition prior addition of thiazole·alanine (TA) prevented repression of the operon by triazole·alanine (TRA), but not repression by histidine. Conversely, in feedback-resistant mutants the operon was not repressible by triazole·alanine, even in the absence of thiazole·alanine, but remained sensitive to repression by histidine. Thus the extent of repression of the operon by triazole·alanine appears to be markedly influenced by changes in conformation at the "feedback-sensitive" site of the first enzyme of the pathway, even though the analogue does not complex with this site.

The role(s) of the first enzyme, phosphoribosyl-ATP pyrophosphorylase, in regulating expression of the operon has not yet been established. Kovach et al. (1970) have recently demonstrated a selective binding of histidine-specific tRNA and histidyl-tRNA to the enzyme which does not involve the catalytic "active site." Thus it seems possible that the enzyme is an apo-repressor of the operon. However, there are indications that expression of the operon is probably regulated primarily by changes in conformation of ribosome-bound, nascent polypeptide chains.[3] In particular, the change in kinetics of repression from "sequential" to "simultaneous" (Kovach et al. 1969[a], 1969[b]) which results either from addition of the pseudo-feedback inhibitor thiazole·alanine, or from a mutational change at the feedback-sensitive site of the enzyme, is not an anticipated consequence of a change in structure of a repressor protein. On the other hand, inhibition of continued synthesis of enzymes specified by the operon by a conformational change in a nascent polypeptide chain could both rapidly and simultaneously repress expression of all genes of the operon (see Chapters 5 and 9).

REGULATION OF SYNTHESIS OF ENZYMES OF THE TRYPTOPHAN AND HISTIDINE PATHWAYS IN OTHER MICROORGANISMS

The organization and regulation of the *trp* operons of the closely related *E. coli* and *S. typhimurium* are very similar. It is therefore appropriate to note that this is not a character-

[3] Unusual features of the kinetics of derepression of the *his* operon (Goldberger and Berberich 1965, 1967, Marver, Berberich, and Goldberger 1966) are also in accord with this suggestion but may be interpreted in other ways.

istic arrangement for regulation of synthesis of enzymes of the tryptophan pathway in all microorganisms, or even all bacteria. Synthesis of the corresponding enzymes of *B. subtilis* is not coordinately regulated, and it seems probable that the variations in enzyme activities observed with various mutant strains reflect effects upon the assembly of a multifunctional enzyme aggregate (Whitt and Carlton 1968[a], 1968 [b]). The enzymes of the tryptophan pathway in *P. putida* present an even more complex situation. The six structural genes comprise three distinct linkage groups. Genes specifying anthranilate synthetase, phosphoribosyl·anthranilate transferase, and indole·glycerol·phosphate synthetase (genes *trp A*, *B*, and *D*, respectively) constitute one group of closely linked genes which is repressible by tryptophan, and may be an operon. The two peptide chains (**A** and **B**) of the tryptophan synthetase enzyme are specified by genes *trp E* and *F*, which are closely linked to each other and to a regulator gene controlling their inducibility by indole·glycerol·phosphate. Phosphoribosyl·anthranilate isomerase is specified by gene *trp C*, which is not linked to any of the other structural genes for enzymes of the pathway (Chakrabarty, Gunsalus, and Gunsalus 1968, Crawford and Gunsalus 1966, Gunsalus et al. 1968). A similar situation has been reported for the enzymes of the tryptophan biosynthetic pathway in *P. aeruginosa* (Fargie and Holloway 1965). Extensive surveys of the tryptophan pathways in a variety of microorganisms (DeMoss 1965, DeMoss and Wegman 1965, Hütter and DeMoss 1967, Wegman and Crawford 1968) have revealed a wide range of patterns of association of the enzymes and of the corresponding structural genes. Indeed, this diversity of organization may reflect attainment of different stages in the evolution of highly efficient regulons.

Similarly, the *his* operon of *S. typhimurium* is not the only pattern of organization of the genes of the histidine pathway. A similar type of operon does appear to occur in *E. coli* (Garrick-Silversmith and Hartman 1970, Goldschmidt et al.

1970) and in *S. aureus* (Kloos and Pattee 1965), but in *P. aeruginosa* the homologous structural genes comprise five separate linkage groups (Fargie and Holloway 1965, Mee and Lee 1967). The arrangement in *S. cerevisiae* (Fink 1966, Shaffer, Rytka, and Fink 1969) and *N. crassa* (Ahmed 1968, Ahmed, Case, and Giles 1964) is also different, although the precise nature of the organization is difficult to interpret. The enzymes corresponding to what may be a "truncated operon" (Horowitz and Metzenberg 1965) may be single multifunctional proteins (Minson and Creaser 1969). If more than one polypeptide is involved they are active only when associated together in a multifunctional enzyme aggregate which could be regulated as what Chovnick (Chovnick et al. 1969, Chovnick, Lefkowitz, and Fox 1956) has termed an "integrated physiological unit."

THE ARGININE REGULON OF E. COLI

In contrast to the *trp* operon, the structural genes for the arginine pathway in *E. coli* are not assembled in a single operon. Three of the genes appear to constitute a small operon, ordered in the sequence *arg C, arg B,* and *arg H,* and a further independent gene (*arg E*) is closely linked to them (Cunin et al. 1969). However, the remainder of the structural genes are widely separated in the bacterial chromosome (e.g., Taylor and Trotter 1967). Nevertheless, in most strains of *E. coli* (e.g., K, W, and C) synthesis of all eight enzymes is repressed in parallel by the end product arginine. Further, mutation in one single regulator gene *arg R* can render the entire group of enzymes nonrepressible (Gorini, Gundersen, and Burger 1961, Maas 1961, Vogel 1961, Vyas and Maas 1963). The one notable exception, wild-type *E. coli* strain B, differs from other wild-type strains precisely in the nature of the allele present at the *arg R* locus (Jacoby and Gorini 1967).

This pattern of response is very similar to that observed for an organized operon, and particularly an operon containing an internal initiation site which causes deviation from completely coordinate regulation of expression of its structural genes (e.g., Bauerle and Margolin 1967). The similarity was emphasized in the definition of the structural genes of the arginine pathway as constituting a *regulon* (Maas and Clark 1964). Indeed, Gorini, Maas, and Vogel all suggested (1961 references above) that each of the separate structural genes of the pathway might contain, or be linked to, a genetic region formally equivalent to an operator gene. Detailed examination of the possibility was, however, seriously complicated by the patterns of response to the level of arginine by the wild-type strain of *E. coli* B and certain mutants derived from other wild-type strains.

The enzymes of the arginine pathway are not repressible in cells of *E. coli* B, and the levels present in the cells are similar to those found in fully repressed cells of *E. coli* W or K12 (Ennis and Gorini 1961). In addition, the presence of exogenous arginine actually stimulates (i.e., induces) the synthesis of slightly higher levels of at least some of the enzymes of the pathway (Gorini and Gundersen 1961) in cells growing in medium containing glucose as the carbon source. The extent of this enhanced synthesis of enzymes of the arginine pathway is a function of the carbon source in the medium, and determined by a "modulator gene" designated R_x by Gorini and Gundersen. This particular regulator gene R_x is distinct from the primary regulator gene of the regulon (*arg R*) but closely linked to it. Furthermore, the regulation of enzyme synthesis attributed to the R_x gene is largely "masked" in recombinant strains containing the *arg* R^K allele derived from *E. coli* K12. To avoid confusion with the pattern of response observed with mutants derived from other wild-type strains of *E. coli,* it seems pertinent to note that the properties of the postulated gene R_x correspond with those of the promoter region of the *lac* operon of *E. coli* (Beckwith 1967).

The enzymes of the arginine pathway in *E. coli* W are repressible by exogenous arginine (Vogel 1960, 1961) and partly repressed by arginine produced endogenously (Gorini and Maas 1957). Mutant strains affected in the regulator gene *arg R* are fully derepressed, and the enzymes of the pathway are not subject to repression by exogenous arginine. Indeed, as in the case of *E. coli* B, in such mutants arginine induces the synthesis specifically of acetylornithine δ-transaminase. However, the mechanism and genes responsible for the specific induced synthesis of this enzyme are entirely different from those involved in *E. coli* B. The "arginine-inducible" transaminase enzyme is a distinct species of enzyme molecule from the normal "arginine-repressible" enzyme of the pathway and is specified by a distinct structural gene designated *arg M*. Formation of the arginine-inducible transaminase is, however, regulated by the same *arg R* regulator gene as that of the arginine-repressible homologue (Bacon and Vogel 1963, Forsyth, Carnevale, and Jones 1969, Vogel, Bacon, and Baich 1963, Vogel et al. 1967).

Synthesis of the inducible transaminase is "masked" in the wild-type *E. coli* W, and it is interesting to speculate upon the possible significance of this enzyme to the economy of the cell. One possibility suggested by the report of Fry and Lamborg (1967) is that the enzyme is in fact an N-formyl-L-methionine aminohydrolase enzyme required for removal of formyl groups from the N-terminus of nascent protein chains. The activity of the inducible enzyme as an ornithine transaminase may be a fortuitous consequence of its substrate specificity. In this case the dual role of the product of regulator gene *arg R* in simultaneously repressing one regulon and inducing synthesis of another enzyme will be entirely analogous to that observed in the case of the "arginine enzymes" of *S. cerevisiae* (Thuriaux et al. 1968).

Once the complications due to the R_x locus of *E. coli* B and the *arg M* locus of *E. coli* W were clearly recognized, the basic features of the regulation of the arginine regulon became evident. Regulation of the activities of a series of unlinked structural genes by a single unlinked repressor gene *arg R* required that the gene product of the regulator be diffusible. Formal analysis by Maas and his co-workers (Maas and Clark 1964, Maas et al. 1964) established that the wild-type regulatory allele *arg* R^+ is dominant to the *arg* R^- allele of unregulated, nonrepressible mutants. Furthermore, the regulatory gene of *E. coli* B (now designated *arg* R^B and including the region previously designated R_x) is an allele of the *arg R* gene of *E. coli* K12. Indeed, in an elegant series of experiments involving the formation of partial diploid strains containing an amber nonsense-mutant *arg R* allele and various types of amber-suppressor genes, Jacoby and Gorini (1969) demonstrated that substitution of one amino acid by another in the *arg R* gene-product protein can change the type of regulation from that observed with *E. coli* B to that found with *E. coli* K12.

Jacoby and Gorini (1969) also isolated mutants in which the formation specifically of ornithine transcarbamylase (gene *F*) was largely insensitive to repression by arginine. It may be recalled that it had been postulated that each of the structural genes of the *arg* regulon might contain, or be linked to, a gene segment corresponding to an operator gene. The properties of these mutants, and the site of mutation, were analogous to those expected of an operator-constitutive (o^c) mutant specifically of the gene *arg F*. Thus the isolation of these mutants appeared to provide direct experimental evidence in support of the hypothesis that the enzymes of the arginine pathway are controlled by a "dispersed operon," consisting of a series of discrete structural gene-operator-promoter complexes; all were apparently regulated by a common repressor (Jacoby and Gorini 1969, Karlström and Gorini 1969).

However, McLellan and Vogel (1970) have recently provided compelling evidence that the protein specified by the *arg R* gene does not act in the same manner as the repressor of the *lac* operon (see Chapter 3) or that inferred for the *trp* operon (see Chapter 5) of *E. coli*. Rather, it appears to promote degradation of preformed messenger RNA when excess arginine is present. Thus there is no evidence of regulation of expression of the *arg* regulon through interaction between a specific repressor protein and genetic elements equivalent to operator genes.

Participation of arginine-specific transfer RNAs ($tRNA^{Arg}$) in regulation of synthesis of enzymes of the arginine pathway was proposed by Vogel (1967, Vogel et al. 1967) in a speculative discussion of the possible mechanism of action of the regulator gene *arg R*. At the present time there is no direct evidence of a role of arginyl-tRNA in the regulatory process. It has now been formally demonstrated that the regulator gene *arg R* does not specify $tRNA^{Arg}$ (Jacoby and Gorini 1969) and that the processes of derepression do not lead to any demonstrable change in the cell complement of $tRNA_S^{Arg}$ (Leisinger and Vogel 1969). Further, mutations within the structural gene for arginyl-tRNA synthetase (gene *arg S*) (Hirschfield et al. 1968) do not affect the synthesis of the enzymes of the pathway in a manner comparable with the situation found for other pathways. However, what appears to be the protein gene product of the regulator gene *arg R* has been partially purified (Udaka 1970). Further, Hirvonen and Vogel (1970) have demonstrated that spheroplasts prepared from cells containing the *arg* R^- allele can be repressed in the presence of arginine by an extract of cells containing the *arg* R^+ allele. Thus possible interaction of arginyl-tRNA with the regulator protein and effects thereof upon synthesis of enzymes of the arginine pathway can now be examined directly.

THE ROLE OF AMINOACYL-tRNA IN ENZYME REPRESSION

Involvement of histidyl-tRNA in regulating expression of the *his* operon illustrates what may be a common feature for many operons, or regulons, controlling the pathways of biosynthesis of amino acids. As noted above, the regulator gene *trp S* controlling the *trp* operon of *E. coli* is the structural gene for the enzyme tryptophanyl-tRNA synthetase (Ito, Hiraga, and Yura 1969). Further, the suggestion that arginyl-tRNA plays an indirect role in regulation of the *arg* regulon is in accord with present knowledge of the mechanism of action of the *arg R* gene product (Leisinger, Vogel, and Vogel 1969, McLellan and Vogel 1970).

The pathway of synthesis of valine in *E. coli* can be regulated, at least in part, by the activity of valyl-tRNA synthetase (Eidlic and Neidhardt 1965, Williams and Freundlich 1969). Similarly, the threonine-isoleucine pathway of *E. coli* can be regulated by the activity of threonyl-tRNA synthetase (Nass, Poralla, and Zähner 1969) and of isoleucyl-tRNA synthetase (Dwyer and Umbarger 1968, Iaccarino and Berg 1971). However, results of a recent study by Hatfield and Burns (1970) raise the possibility that the role(s) of all three of these aminoacyl-tRNA synthetases in regulating enzyme synthesis may be indirect. Enzymes of the valine and threonine-isoleucine pathways are specified by the *ilv* operon. Hatfield and Burns have postulated that the apo-repressor of the operon is an inactive "immature," or nascent, species of the threonine deaminase which is converted to active "mature" enzyme through combination with any of the amino acids isoleucine, threonine, or valine. The immature species of enzyme can complex reversibly and specifically with leucyl-tRNA, and the primary mechanism of repression of the *ilv* operon may, therefore, be similar to that for the *his* operon of *S. typhimurium*. Other aminoacyl-tRNAs do not complex with the immature enzyme. However, in contrast to their effects when tested singly, a mixture of valine and isoleucine blocks "maturation" of the immature enzyme. Thus impaired activity of either of the corresponding aminoacyl-tRNA synthetases could lead to accumulation of the presumed apo-repressor.

Further, there are a number of reports in the literature of

repression of enzymes of a biosynthetic pathway which cannot be interpreted solely in terms of the action of a single species of aminoacyl-tRNA synthetase enzyme acting on all species of homologous specific tRNAs (Nazario 1967, Neidhardt 1966, Schlesinger and Nester 1969).

It would therefore be premature to conclude that the expression of all operons controlling biosynthetic pathways is in fact normally regulated by the activity of aminoacyl-tRNA synthetase enzymes.

REPRESSION OF ALKALINE PHOSPHATASE IN E. COLI

All the examples of end-product repression presented to this juncture have concerned pathways of biosynthesis of amino acids. With the exception of the regulator genes of the *his* operon in *S. typhimurium* of totally unknown function, all the regulator genes for such pathways identified to date can be classed into three major groups. The first specify a specific protein which (like their counterparts in the phenomenon of induced enzyme synthesis) *may* be a repressor protein capable of interacting directly with an operator gene (i.e., available evidence is compatible with such an hypothesis). Regulator genes of the second group affect the ability of the cell to synthesize the corresponding aminoacyl-tRNA, which may be the effective co-repressor. The *arg R* gene regulating expression of the *arg* regulon represents a third type, which may be a common type of regulator gene for coordinate expression of dispersed structural genes of related function.

It is, therefore, only fitting to consider a further system which has figured prominently in studies of the mechanism of gene action for which one of the regulator genes demonstrably does not fall into any of the foregoing categories. This is the alkaline phosphatase enzyme of *E. coli,* which was extensively studied by Garen and his co-workers in their analysis of the mechanism of action of *amber-* and *ochre-*suppressor genes (Garen 1968, Weigert et al. 1966).

Alkaline phosphatase protein is synthesized within the cell cytoplasm as enzymically inactive monomeric polypeptide chains. These migrate through the cell membrane and dimerize to active enzyme within the periplasmic space (Malamy and Horecker 1964, Schlesinger 1968, Torriani 1968). The association of monomer into dimer occurs readily *in vitro* in a nonenzymic reaction (Levinthal, Signer, and Fetherolf 1962), and the development of enzymic activity is customarily equated with synthesis of the polypeptide monomer. However, it remains to be formally determined whether *in vivo* the passage through the cell membrane, and even the dimerization process, is facilitated by the action of other enzymes, which would then play a regulatory role in the development of enzymic activity. This possibility does not seriously affect interpretation of the data which *are* available and will not be further considered.

The level of alkaline phosphatase activity formed is determined by the level of inorganic phosphate present in the growth medium. In media customarily used for growth of *E. coli* the level of activity is low, but high derepressed levels develop in medium of low content in inorganic phosphate (Garen 1960, Garen and Levinthal 1960). Three regulator genes have been shown to control the development of enzyme activity.[4] Of these, two are separate but closely linked cistrons at a gene locus originally designated *R2* (and more recently as *pho S*) and specifying a single regulatory protein (Garen and Otsuji 1964). It is therefore convenient (and customary) to continue to refer to these 2 cistrons as the single gene *R2*. This gene *R2* is located in the chromosome at some distance from the structural gene of the enzyme, whereas regulator gene *R1* is closely linked to the structural gene (Echols et al. 1961). Mutation in either regulator gene can lead to constitutive formation of alkaline phosphatase. The wild-type alleles of both genes are dominant in *cis-trans*

[4] Jones (1969) has presented evidence that formation of this enzyme may also be regulated by additional regulator genes.

tests of function, and some of the mutants at each locus recover the ability to regulate enzyme formation in the presence of *amber*-suppressor genes. Thus both regulator genes are structural genes for proteins (Garen and Echols 1962, Garen and Garen 1963, Torriani and Rothman 1961).

The protein specified by regulator gene *R2a* has been partially purified and does not have the properties of a repressor protein, or of a precursor thereof. Synthesis of the "*R2* protein" parallels that of alkaline phosphatase enzyme. Levels are high in derepressed cells of the wild-type strain and constitutive mutants affected in the regulator genes *R1* and *R2b* but low in repressed cells (Garen and Otsuji 1964). Proteins specified by the other two regulator genes have not yet been identified. However, there is strong circumstantial evidence that the gene product of regulator gene *R1* is also not a repressor protein. One class of mutant, $R1^c$, affected in this gene cannot produce alkaline phosphatase in medium of low phosphate content. The mutant allele is recessive to the wild-type allele in the *cis-trans* test of function, indicating that the gene product of this gene is required for formation of the enzyme as well as for repression of enzyme formation (Garen and Echols 1962). It is apparently not required for synthesis of the "*R2a* protein" (Garen and Otsuji 1964). Present evidence is in accord with a mechanism proposed by Garen and Echols (1962), in which gene *R1* controls the formation of an endogenous inducer of alkaline phosphatase formation, and the protein specified by gene *R2* converts the inducer to a repressor molecule in the presence of excess phosphate.

REPRESSION AND DIVERGENT BIOSYNTHETIC PATHWAYS

In Chapter 3 the complexity of regulation of synthesis of enzymes of converging catabolic pathways was illustrated. Even more complex patterns of regulation are observed in controlling the formation of the enzymes of diverging anabolic pathways. This complexity is enhanced by the effects of end products of the divergent paths in regulating the enzymic activity of the enzymes once formed by means of feedback (allosteric) inhibition. It is beyond the scope of this discussion to review fully these complex patterns, which are presented in excellent recent reviews by Datta (1969), Gibson and Pittard (1968), Kornberg (1966), Truffa-Bachi and Cohen (1968), and others. However, certain general features and the types of pattern can be illustrated by reference to three groups of divergent pathways.

One pattern is that observed for the "aspartic acid family" of amino acids in *E. coli,* for which the biosynthetic pathways from aspartic acid are illustrated in Figure 4.4. The first reaction of the common pathway is catalyzed by three different species of aspartokinase enzyme or *isozymes* (Cohen and Patte 1963, Patte, Le Bras, and Cohen 1967). End-product lysine both inhibits and represses formation of one of the aspartokinase enzymes; represses formation of diaminopimelate decarboxylase and partly of aspartic semialdehyde dehydrogenase; and also inhibits the activity of dihydrodipicolinic acid synthetase, the first enzyme of the branched pathway leading solely to lysine. The concerted effect of these actions of lysine serves to diminish the flow of intermediates into the synthesis of lysine, without impairing the capacity of the cells to produce the other end-product amino acids. Similarly, methionine noncoordinately represses formation of a second aspartokinase isozyme, one of two homoserine dehydrogenase isozymes, homoserine trans-succinylase, and the other enzymes of the pathway unique to methionine (Cohen and Patte 1963, Rowbury and Woods 1966). Maximal repression of the third aspartokinase isozyme, the second homoserine dehydrogenase, homoserine kinase, and the mutaphosphatase requires the presence of a combination of threonine and isoleucine (Cohen 1967).

This complex series of regulations of synthesis and activity of the various enzymes provides a magnificent illustration of the manner in which the economy of the cell is controlled.

In addition, it illustrates two fairly common features of the patterns of repression. The first is the "use" of a series of different isozymes to catalyze a reaction common to the synthesis of a series of end products, each being subject to repression specifically by one of the end products. This feature is not invariant, for various microbial cells can accomplish

FIGURE 4.4. Pathways of synthesis of amino acids of the "aspartic acid family" in E. coli

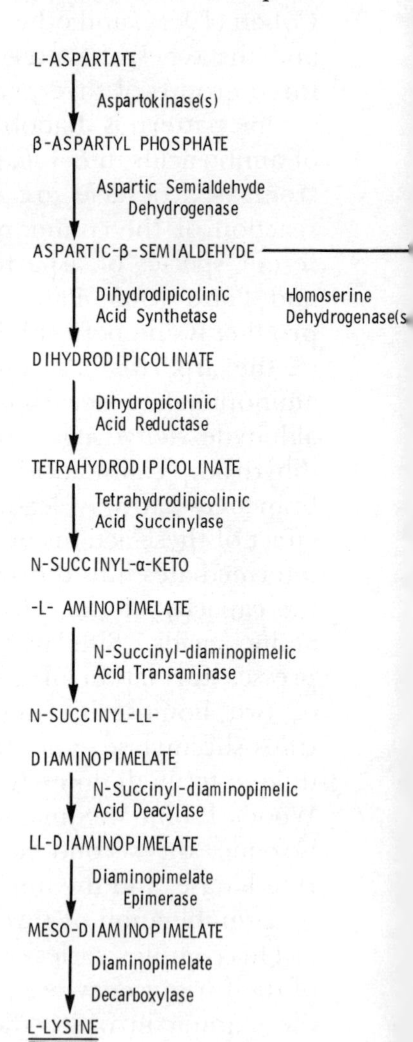

by feedback inhibition what other microorganisms achieve by enzyme repression. Various combinations of these two regulatory phenomena have been compared for the pathways of synthesis of the aromatic amino acids (Gibson and Pittard 1968, Nasser, Henderson, and Nester 1969). The second feature is the requirement for a combination of both threonine and isoleucine to repress the enzymes required for synthesis of both the amino acids. This is an example of the phenomenon termed "*multivalent repression*" by Freundlich, Burns, and Umbarger (1962, 1963).

This term was originally coined to describe the requirement for a combination of isoleucine, leucine, and valine to

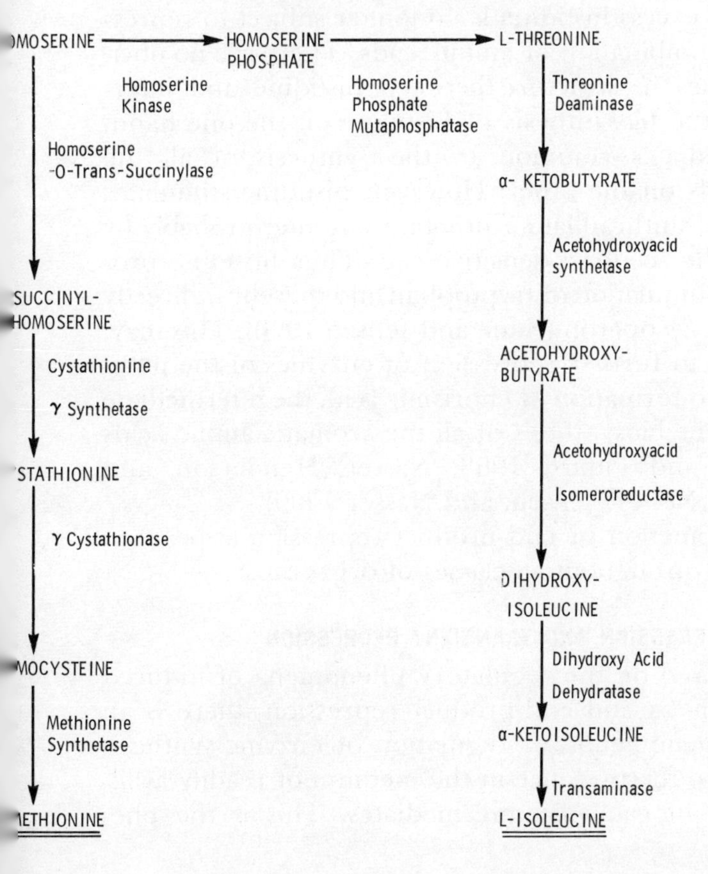

repress synthesis by *E. coli* or *S. typhimurium* of enzymes required for formation of isoleucine and valine, by pathways involving different intermediates. The "rationale" for this interlocked regulation becomes readily apparent when it is recognized that the same series of enzymes catalyzes the two homologous series of reactions (Leavitt and Umbarger 1961).

An unusual situation has recently been described by Chapman and Nester (1968) and Nester (1968) under the name "*cross-pathway regulation*" for cells of *B. subtilis.* Histidine *appears* to have a specific action in controlling the repression of enzymes required for synthesis of the aromatic amino acids. In particular, one class of mutants resistant to inhibition by excess histidine is no longer subject to repression by any combination of amino acids. There are no obvious similarities in structure between histidine and intermediates in the biosynthesis of histidine on the one hand, and intermediates common to the synthesis of all the aromatic acids on the other. However, histidine stimulates the activity of anthranilate synthetase enzyme; probably by binding to the feedback-sensitive site. Thus histidine promotes the accumulation of tryptophan and thereby indirectly represses the *trp* operon (Kane and Jensen 1970). This may, in turn, lead to further repression of enzymes of the pathway leading to formation of chorismic acid, the intermediate common to the biosynthesis of all the aromatic amino acids (see Gibson and Pittard 1968, Nasser, Henderson, and Nester 1969, Nester, Jensen, and Nasser 1969).

The phenomenon of end-product repression appears to occur widely throughout all classes of organisms.

CATABOLITE REPRESSION AND TRANSIENT REPRESSION

Superimposed on the regulatory phenomena of induced enzyme synthesis and end-product repression, there is an additional phenomenon of regulation of enzyme synthesis in response to the presence in the medium of readily available sources of catabolic intermediates. This is the phenomenon clearly described by Dienert in 1900, rediscovered by Monod (1942) in his studies of the "diauxie," and widely known until recently as "*the glucose effect.*"

For many years it had been recognized that the levels of various enzymes found in bacterial cells could be markedly affected by the addition of carbohydrate, and particularly glucose, to the medium. In some cases the effect of glucose or other carbohydrate was indirect and was due to changes in pH of the medium resulting from acids produced by incomplete oxidation of the sugar. However, in other cases glucose inhibited the formation of enzymes by some mechanism other than by serving as a source of acids (Gale 1943). Many of the enzymes present in markedly reduced levels during growth in medium supplemented with glucose are inducible enzymes. However, it is not a general rule that synthesis of *all* inducible enzymes is repressed in the presence of glucose. Indeed, in some cases where the inducible enzyme is not one acting on a carbohydrate substrate the provision of glucose, or an equivalent source of energy, is an absolute requirement for enzyme formation (e.g., Magasanik 1957, Pollock 1946). Nor is glucose the only carbon source able to repress enzyme formation (Mandelstam 1957, 1961, Monod 1942, Perlman, de Crombrugghe, and Pastan 1969). However, glucose is one of the most potent inhibitors of enzyme formation and the most commonly effective. As a consequence, this general phenomenon was named the "glucose effect" (Cohn 1956, Cohn and Monod 1953).

The basis of this glucose effect proved baffling for many years and was the source of considerable speculation. Subsequent events have shown the interpretation first formulated by Neidhardt and Magasanik (1956, 1957), and elaborated in detail by Magasanik (1961), to be the most nearly correct. The enzymes subject to end-product repression play key roles in a number of divergent anabolic pathways, which serve to balance most efficiently the flow of common metabolic intermediates and energy into production of all the materials required for metabolism and growth of the cell. Conversely, enzymes subject to the glucose effect participate in a variety of convergent catabolic pathways, controlling the flow of common intermediary metabolites and energy into the endogenous intracellular pools. Neidhardt and Magasanik therefore suggested that just as the divergent anabolic pathways were regulated by end-product repression, so too the convergent catabolic pathways were regulated by a similar end-product repression. In this case, the co-repressor would be some catabolite common to the metabolism of a variety of exogenous sources of carbon and energy. Hence, the glucose effect was renamed *catabolite repression* (Magasanik 1961).

As originally formulated, the concept of catabolite repression implied that a particular catabolite could serve as a co-repressor for the specific apo-repressor gene product(s) controlling the synthesis of one, or more, series of catabolic enzymes. Thus catabolite repression of formation of β-galactosidase formation in *E. coli* would involve complexing of the postulated co-repressor with the specific repressor protein of the *lac* operon. This feature of Magasanik's proposals appeared improbable at the outset and has since been demonstrated to be incorrect.

Nevertheless, the hypothesis in its original form immediately provided a ready (though incorrect) interpretation of a hitherto puzzling series of observations. For example, glucose and other carbon sources which are readily metabolized and good sources of energy are more potent (and more frequently effective) in causing catabolite repression than succinate and glycerol. Further, the extent to which an exogenous carbon source causes catabolite repression is dependent upon the physiological balance between rates of supply and use of key metabolites (Magasanik 1963). Thus repression of formation of β-galactosidase in *E. coli* by glucose is relieved under anaerobic conditions (Cohn and Horibata 1959, Okinaka and Dobrogosz 1967), when both the level of intermediates of the tricarboxylic acid cycle and the yield of energy per molecule of glucose consumed are reduced.[5] Moreover, enzyme formation in strains of bacteria which do not readily utilize glucose (at least without prior induced enzyme synthesis) is not repressed by glucose unless conditions are imposed which seriously impair growth (Mandelstam and Jacoby 1965, Neidhardt 1960). Conversely, succinate has been the carbon source of choice for study of induced synthesis of β-galactosidase by *E. coli* under conditions of gratuity (e.g., Monod, Pappenheimer, and Cohen-Bazire 1952). Yet it markedly represses synthesis of this enzyme by auxotrophic mutants which have been depleted of the supply of required amino acid (Mandelstam 1957, 1961), when the arrest of growth reduces the demand for intermediates of the tricarboxylic acid cycle as precursors of other amino acids.

However, a second series of studies with constitutive mutant strains of bacteria required modification to be made to the hypothesis as originally proposed. Indeed, prior studies had shown that constitutive formation of β-galactosidase by mutant strains of *E. coli* affected in the (*i*) regulator gene was still repressed by glucose (Brown and Monod 1961). Subsequent study of such mutants revealed the effects of glucose to be more complex. In a number of cases it was observed that glucose caused catabolite repression of β-galactosidase formation (Loomis and Magasanik 1964, Mandelstam 1962, Moses and Prevost 1966). On the other hand, with other combinations of bacterial strains and inducers glucose apparently either only slightly repressed β-galactosidase formation (Dénes 1961) or caused a marked temporary repression of enzyme synthesis followed by an apparent total relief from repression (Boezi and Cowie 1961, Moses and Prevost 1966).

[5] Enhancement of the repression by glucose of formation of enzymes of the tricarboxylic acid cycle (Gray, Wimpenny, and Mossman 1966) could have been attributed to a reduced supply of an inducer of induced enzyme synthesis.

This very marked but apparently temporary inhibition of β-galactosidase synthesis was termed *transient* repression (Boezi and Cowie 1961). Actually, analysis of the differential rate of β-galactosidase formation demonstrates that the period of marked *transient repression* is followed by a less severe *catabolite repression.* This is illustrated by curve D in Figure 4.5 for the case of induced synthesis of β-galactosidase. However, this analysis was not made in the earlier studies. Consequently, the process of "release" from transient repression appeared analogous to the related phenomenon of onset of induced enzyme formation under conditions of "diauxie" (Monod 1942), or to "escape" from inhibition by methyl-α-glucoside of induced enzyme synthesis (Johnson 1938, Wainwright 1953). Therefore, the results presented to this juncture could still be considered compatible with Magasanik's original concept of catabolite repression, subject to restrictions regarding the properties of "altered" gene products formed by constitutive mutants affected in the regulator gene *i*.

At this stage the situation was further complicated by the suggestion of McFall and Mandelstam (1963[a], 1963[b]) that the phenomenon of catabolite repression was merely a particular example of end-product repression. This suggestion was based upon a comparison of the effects of different carbon sources upon the synthesis of β-galactosidase, tryptophanase, and serine deaminase in a single strain of *E. coli.*[6] They found that both glucose and galactose repressed the induced formation of β-galactosidase, whereas pyruvate did not. On the other hand, induced synthesis of tryptophanase was reported to be markedly repressed by pyruvate but much less affected by glucose and galactose. This in turn led to a demonstration with cells of *P. aeruginosa* of what Man-

[6] The mechanism(s) by which formation of tryptophanase (Botsford and DeMoss 1971, Pastan and Perlman 1969[a]) and serine deaminase (McFall and Bloom 1971) are subject to catabolite repression differs from the (primary) mechanism of catabolite repression of synthesis of β-galactosidase and will not be further considered here.

delstam and co-workers (Mandelstam and Jacoby 1965, Stevenson and Mandelstam 1965) termed "*multisensitive end-product repression.*" It was shown that enzymes required for conversion of D-mandelic acid to benzoic acid were repressed when the culture was supplemented by benzoic acid, or derived metabolites, in addition to inducer mandelate.

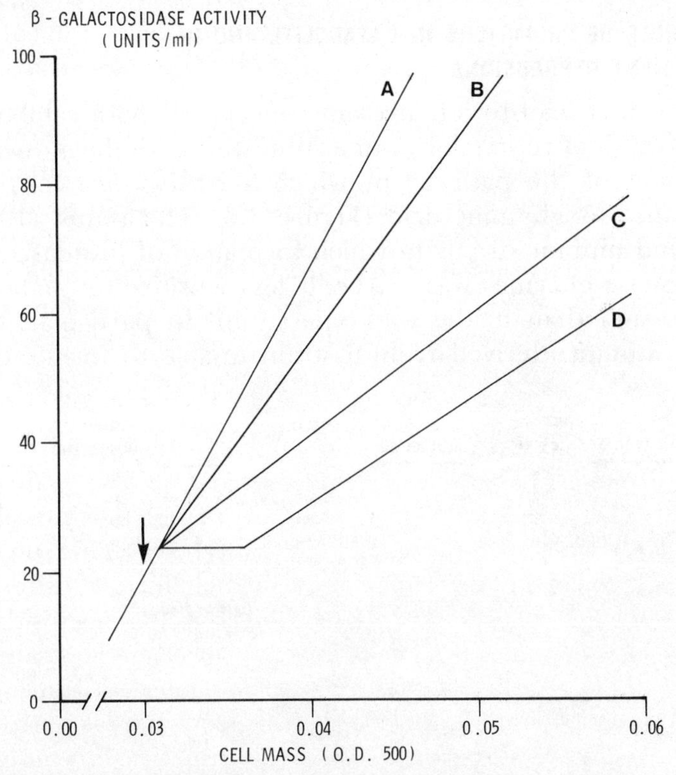

FIGURE 4.5. Reversal of transient repression and catabolite repression of β-galactosidase in E. coli 3000 by cyclic AMP (cAMP)

Inducer (isopropyl thiogalactoside, 10^{-3} M) was added at 0 minutes to a culture growing in synthetic medium containing 0.5 percent glycerol as carbon source. At 10 minutes, indicated by the arrow, the following additions were made:

A. None
B. 10^{-2} *M* glucose plus 5×10^{-3} *M* cAMP
C. 10^{-2} *M* glucose plus 10^{-3} *M* cAMP
D. 10^{-2} *M* glucose

Reproduced from Perlman, de Crombrugghe, and Pastan (1969) with permission of the authors and the editor of *Nature*.

Succinate and acetate also repressed formation of enzymes required for the metabolism of *p*·hydroxybenzoic acid (see Figure 3.1). However, a number of observations rendered the conclusion that catabolite repression and end-product repression were one and the same phenomenon rather unlikely (see Paigen, Williams, and McGinnis 1967).

THE ROLE OF PROMOTERS IN CATABOLITE AND IN TRANSIENT REPRESSIONS

The first lead to our present concepts of both catabolite and transient repressions came from studies of the inducible enzymes of the pathway by which *S. typhimurium* converts histidine to glutamic acid (Figure 4.6). Magasanik (1963) isolated mutant strains in which formation of histidase and urocanase enzymes was markedly less sensitive to catabolite repression than in the wild-type strain. In particular, one such mutant, derived from a strain unable to form either

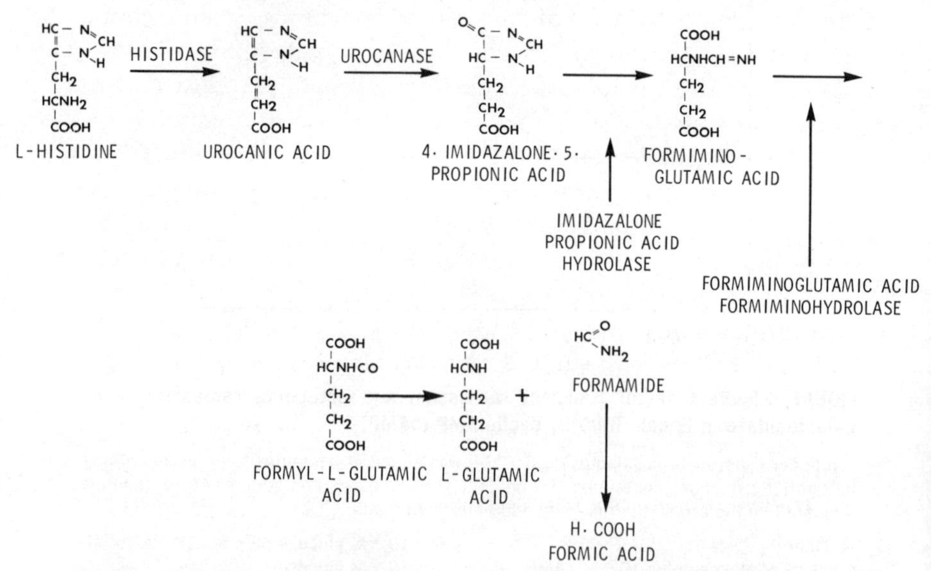

FIGURE 4.6. The pathway of metabolism of histidine to glutamic acid in S. typhimurium

enzyme, simultaneously acquired the ability to synthesize both. This reduced sensitivity to catabolite repression was specific to histidase and urocanase, and the causal mutation was located in a regulator gene closely linked to the corresponding structural genes.

Loomis and Magasanik (1964) subsequently studied the synthesis of β-galactosidase in transient merogenotes of *E. coli* (see Chapter 3). These were formed by introducing an inducible donor genome (i^+z^+) into cells unable to synthesize the specific repressor protein and β-galactosidase (i^-z^-), or even with the *i* regulator gene and most of the *lac* operon deleted from the recipient genome. In all cases, synthesis of β-galactosidase shortly after introduction of the donor genome was subject to catabolite repression. Thus the repressive catabolite was *almost* certainly not exerting its effects by combining with the product of regulator gene *i*. Very shortly thereafter McFall (1964[a], 1964[b]) reported finding pleiotropic mutants affected in the capacity to form D-serine deaminase. Synthesis of this enzyme was inducible in the parental strain and highly resistant to repression by glucose. In one of the mutants enzyme formation was totally constitutive and also subject to catabolite repression by glucose. Genetic analysis showed the mutation to be in a site closely linked to the structural gene of the enzyme and the mutant gene to be dominant for the structural gene *cis* with respect to it, but not for a homologue *trans* to it. Thus the site controlling sensitivity to catabolite repression appeared to be an operator gene, an alternative possibility previously considered briefly (Magasanik 1961). (It should be borne in mind that the concept of a separate promoter region was not yet clearly formulated.) Palmer and Moses (1967) subsequently reported that deletion of part of the operator of the *E. coli lac* operon specifically eliminated transient repression of synthesis of β-galactosidase by glucose. Although this study has since been challenged (Tyler and Magasanik 1969), it was soon followed by demonstrations that either mutations within the promoter region of the *lac* operon, or deletion of a portion of it, eliminated both transient repression and catabolite repression (Pastan and Perlman 1968, Perlman, de Crombrugghe, and Pastan 1969, Silverstone, Arditti, and Magasanik 1970, Silverstone et al. 1969).[7]

Identification of the site responsible for the sensitivity of synthesis of β-galactosidase to catabolite repression as the promoter region of the *lac* operon of *E. coli* indicates the underlying mechanism to be quite different from that of end-product repression. Neither the operator nor the regulator gene *i* (or its repressor protein gene product) are immediately involved in regulation via catabolite repression. Data for other systems are far less extensive, but two types of evidence indicate the conclusion to be general, at least for microorganisms. First, the studies by McFall of the D-serine deaminase of *E. coli* (1964[a], 1964[b]) indicate that a site controlling sensitivity to catabolite repression is located in a region of the genome showing properties of a promoter. Second, the rate of formation of inducible β-galactosidase in *E. coli,* and also of many other enzymes which are subject to catabolite repression, is markedly affected by the level of cyclic AMP.

ROLES OF CYCLIC AMP AND cAMP-"BINDING PROTEIN" IN CATABOLITE AND TRANSIENT REPRESSIONS

Cyclic AMP participates in many regulatory processes in animal tissues (Robison, Butcher, and Sutherland 1968). Makman and Sutherland (1965) demonstrated that this nucleotide is also present in cells of *E. coli* and released into the medium in the presence of glucose. At the time, func-

[7] Involvement of the promoter region in these phenomena has been the subject of considerable debate (see for example Moses and Sharp 1970[a], Moses and Yudkin 1968, Yudkin 1970). Two factors have complicated the analysis. One is the fact that cAMP (see next section) acts upon a number of processes involved in the overall phenomenon of "gene expression" (see Chapter 5). The other is that a number of intermediary metabolites formed from glucose may inhibit synthesis of β-galactosidase in mutant strains that are largely insensitive to repression by glucose. This observation will be considered separately (see p. 175).

tions of cAMP in bacterial cells were unknown, and the observation aroused little immediate interest. However, in recent years we have recognized that this nucleotide plays a major role in regulating the synthesis of enzymes which are subject to catabolite and transient repression in, at least, *E. coli* and related organisms (see Pastan and Perlman 1970).[8]

The extent of transient or catabolite repression of induced β-galactosidase formation appears to be correlated with the extent of decrease in intracellular concentration of cAMP (Aurbach, Perlman, and Pastan 1971, Monard, Janeček, and Rickenberg 1970). Mechanisms by which the cAMP concentration is lowered have not yet been elucidated. Transient repression appears to be correlated with the uptake and phosphorylation of the carbon source, or analogue, which causes repression, whereas catabolite repression requires further metabolism of the repressive carbon source. Neither form of repression is observed with mutant strains which lack either an appropriate permease enzyme required for uptake of the repressive compound, or the corresponding enzyme II of the PEP-phosphotransferase system required for its phosphorylation (Pastan and Perlman 1969[b], Tyler and Magasanik 1970).[9] On the other hand, the nonmetabolized analogue α·methyl·glucoside causes transient repression but not sustained catabolite repression (Cohn and Horibata 1959, Pastan and Perlman 1969[b], Wainwright 1953).

[8] Effects of cAMP on enzyme formation in gram-positive organisms (e.g., *B. cereus, B. licheniformis, B. subtilis*, or *S. aureus*) have not yet been formally examined in detail. However, it may be recalled that some of the early studies of the glucose effect were made with *B. subtilis* (e.g., Monod 1944[a]).

[9] There is an apparent disagreement between the results obtained by these two groups in studies with mutants deficient in other components of the PEP-phosphotransferase system. This probably reflects differences in the mechanisms of uptake of the sugars and analogues used.

The PEP-phosphotransferase system consists of three components. Enzyme I catalyzes the transfer of a phosphate group from phospho·enol·pyruvate to a small protein designated **Hfr**. An appropriate enzyme II, of a family of related species, then transfers the phosphate group from the latter to a sugar, or analogue, receptor. For a full description of the system see Kaback (1970).

A role of the nucleotide in both types of repression was indicated by studies of the effects of cAMP supplements upon the inhibition by glucose of induced β-galactosidase synthesis in *E. coli.* Perlman and Pastan (1968) found that low concentrations of cAMP both stimulated the induced synthesis of β-galactosidase and suppressed the transient repression of enzyme formation by high concentrations of glucose. Subsequent study showed that higher concentrations of cAMP prevented both transient and catabolite repression (Figure 4.5) and appeared to act at the promoter of the *lac* operon (Pastan and Perlman 1968, 1969[a], Perlman, de Crombrugghe, and Pastan 1969). Reversal of catabolite inhibition of β-galactosidase formation by cAMP has been confirmed directly (Goldenbaum and Dobrogosz 1968, Ullmann and Monod 1968), and indirectly by the finding that two mutants resistant to catabolite repression have defective cAMP-degrading enzyme (Monard, Janeček, and Rickenberg 1969). Similar studies have shown that cAMP also prevents repression by glucose of the synthesis of a number of other inducible enzymes in *E. coli,* and of β-galactosidase in other species of bacteria related to *E. coli* (de Crombrugghe et al. 1969). Moreover, a mutant strain of *E. coli* markedly deficient in activity of the adenyl cyclase enzyme which forms cAMP is unable to synthesize appreciable quantities of several inducible enzymes, even in the absence of glucose (Aurbach, Perlman, and Pastan 1971, Perlman and Pastan 1969, 1971).

There is a second type of pleiotropic mutant of *E. coli* in which the synthesis of various inducible enzymes appears to be permanently repressed, even in the presence of cAMP (Schwartz and Beckwith 1970). Extracts of such mutants fail to develop β-galactosidase activity *in vitro* under conditions where substantial increases in activity are obtained with equivalent extracts of other strains. The deficiency is due to a defect in a protein factor which appears to have a high affinity for cAMP. This protein has been named the "*catabo-*

lite gene activator protein" (*CAP*) by Zubay, Schwartz, and Beckwith (1970) and "*cyclic AMP receptor protein*" by Emmer et al. (1970).[10] The function of the complex of cAMP with this protein has not yet been determined. However, cAMP appears to stimulate transcription of mRNA from the *lac* operon (Jacquet and Kepes 1969, Varmus, Perlman, and Pastan 1970[a], 1970[b]).[11] Therefore, further discussion of this topic will be deferred to Chapter 5.

ROLE OF THE "CATABOLITE REPRESSION" GENE AND THE "EFFECTOR" OF CATABOLITE REPRESSION

Loomis and Magasanik (1967) reported isolation of mutant strains of *E. coli* resistant to catabolite repression by glucose as a consequence of mutation at a gene locus originally designated **CR** and recently renamed *cat* (Tyler et al. 1969). Formation of many enzymes by such mutants shows markedly reduced sensitivity to repression by glucose, but sensitivity to repression by glucose·6·phosphate and by a glucose-gluconate mixture remains high (Monard, Janeček, and Rickenberg 1970, Moses and Yudkin 1968, Rickenberg, Hsie, and Janeček 1968). Further, the "steady-state" intracellular levels of cAMP in mutant cells growing with glucose-gluconate mixture as carbon source are markedly less than that of cells growing with glucose as sole carbon source (Monard, Janeček, and Rickenberg 1970). Thus a defect in gene *cat* appears to prevent, or limit, conversion of glucose to a metabolite which effects the reduction in intracellular level of cAMP.

[10] Identity of the proteins partially purified by the two groups on the basis of different experimental criteria has not been formally established at the time of preparing this summary. Therefore, I shall use the generic term cAMP-"binding protein" in further reference to either protein.

[11] Glucose *may* repress both the transcription and translation of the mRNA coded for by the *lac* operon of *E. coli*. Yudkin (1969) has presented suggestive evidence of such dual effects, although the data are amenable to alternative interpretations. More recently, Contesse et al. (1970) have reported a premature termination of transcription of the operon in repressed cells which may reflect the consequences of an inhibition of the translation to which it is normally coupled (see Chapter 5).

The identity of this metabolite "effector" remains elusive. Goldenbaum and Dobrogosz (1968) observed that a concentration of cAMP which largely reversed catabolite repression of β-galactosidase synthesis by glucose had little effect in relieving cells repressed by gluconate, and even less effect on those repressed by glucose·6·phosphate. They therefore suggested that glucose·6·phosphate is the effector of catabolite repression. However, this possibility can now be eliminated. Levels of glucose·6·phosphate are as high in mutant cells resistant to catabolite repression by glucose as in those which are sensitive (Hsie et al. 1969), and mutants unable to produce substantial quantities of glucose·6·phosphate are sensitive to repression by glucose (Tyler and Magasanik 1970). Moreover, the properties of mutants deficient in adenyl cyclase activity (Perlman and Pastan 1969) are not in accord with the hypothesis that the effector is glucose·6·phosphate, either free or as a component of a complex.

Several metabolites of glucose can stimulate the activity of *E. coli* cAMP phosphodiesterase enzyme complex *in vitro* (Monard, Janeček, and Rickenberg 1970). On the other hand, formation of the enzyme is subject to catabolite repression (Aboud and Burger 1971). However, the role of this enzyme complex in regulating cAMP levels *in vivo* is highly uncertain (see Pastan and Perlman 1970). Conversely, pyruvate and ribose·5·phosphate inhibit activity of *E. coli* adenyl cyclase enzyme *in vitro* (Tao and Huberman 1970). However, the former is less effective than glucose in repression of induced synthesis of β-galactosidase (Aurbach, Perlman, and Pastan 1971, McFall and Mandelstam 1963[b]), and ribose·5·phosphate stimulates enzyme formation. Further, fructose·1·6·diphosphate and 5·phosphoribosyl·1·pyrophosphate are more effective than cAMP in relieving repression of β-galactosidase formation in a complex medium (Moses and Sharp 1970). Thus the possibilities that the effector is a metabolite formed only in trace amounts or that there is no single effector metabolite must be seriously considered.

CATABOLITE REPRESSION IN METAZOANS

The rigorous demonstration of possible catabolite repression in cells of higher forms is likely to prove a very formidable task. In addition to the obvious problems of resolving possible catabolite repression from end-product repression, it is necessary to distinguish between direct effects of a repressive catabolite and indirect effects of the nutritional status of the cells and of hormones. Further, cAMP is an essential cofactor of kinase enzymes acting on nuclear histones and possibly participates in regulation of the synthesis of specific proteins. At the present time, I am not aware of any compelling evidence of catabolite repression in cells of higher forms, although a few cases of possible catabolite repression have been reported (e.g., Adelman, Lo, and Weinhouse 1968, Jost, Hsie, and Rickenberg 1969, Wimhurst and Manchester 1970).

LYSOGENIC BACTERIOPHAGES AND IMMUNITY REPRESSORS

A further distinct variety of repression of protein synthesis is observed in bacteria which are described as *lysogenic*. Infection of *E. coli* with a *virulent* virus, or *bacteriophage,* initiates a series of events culminating in lysis of the host cell and release of newly replicated virus particles. The course of these events has been summarized recently by Cohen (1968) and will be further considered only with respect to regulation of the process of "gene transcription."

When cells of *E. coli* are infected with a temperate bacteriophage, such as *phage* λ, a proportion of the cells become lysogenized. The resulting lysogenic cells contain the entire genome of one infecting viral particle in a "latent state," the *prophage.* This prophage is incorporated into the chromosome of the host cell at a specific site, where it functions as an integral part of the bacterial chromosome. The prophage may be released (excised) from the chromosome either spontaneously or following various treatments collectively described as "induction." It then functions as an independent genetic entity and initiates the replication of infective viral particles. It is beyond the scope of this discussion to consider all aspects of this phenomenon, and interested readers are referred to the excellent treatments by Campbell (1969), Lwoff (1953, 1966), and Stent (1963). However, two features of this system are of particular significance in the present context and formed part of the base on which the concept of the operon was developed. First, the lysogenic cell contains a mechanism for suppressing the expression of the λ phage genome, both in the prophage and in any superinfecting λ phage which may arise by spontaneous induction of a neighboring cell. Second, this latter *immunity* is a highly specific phenomenon and protects the cell only against virulent infection by certain strains of λ phage.[12]

During vegetative reproduction of λ phage, in nonlysogenic or in induced cells, the various genes of the virus are expressed in a defined sequence (Campbell 1969, Jacob, Fuerst, and Wollman 1957) entirely analogous to that in cells infected with "T-even" phages (Cohen 1968). Discussion of the mechanisms regulating the course of this sequence of gene expression will be deferred to Chapter 5. The phage genes and genetic elements of particular interest in the present context are arranged in the sequence: __ *cIII*, **N**, *rex*, *cI*, *x*, *y*, *cII*, **O**, **P**, __. This is shown in greater detail in Figure 5.1.

Only two properties determined by phage genes have been detected in lysogenic cells (Bear and Skalka 1969). One is the property of immunity, determined by gene *cI* (Kaiser 1957). The other is the ability to prevent propagation (*exclusion*) of unrelated strains of virus in the λ-lysogenized cell, determined by a very closely linked gene *rex*.

The mechanism by which gene *rex* exercises this property of exclusion has not yet been established. One obvious possibility is that gene *rex* codes for a specific DNA methylase (or DNA demethylase) enzyme. The process of exclusion results

[12] The cell may also be resistant to infection by other virulent viruses because of a mutation leading to change in the site of virus attachment to the cell.

from the action of specific endonuclease "restriction enzymes" acting on DNAs not modified by the action of a host-specific DNA methylase enzyme (Arber 1965[a], 1965[b], Meselson and Yuan 1968, Roulland-Dussoix and Boyer 1969). Phage λ is particularly subject to "host modification" by DNA methylase enzymes (Arber 1965[a], 1965[b]). Although infection of *E. coli* B by phage λ has no effect on the *net* DNA methylase activity of the cells (Hausmann and Gold 1966), prior "induction" of formation of enzyme which destroys the methyl-donor (S-adenosylmethionine) eliminates both the phenomena of immunity and exclusion (Hirsch-Kauffmann and Sauerbier 1968). However, the function of gene *rex* can only be speculated upon at this time, and it will not be considered further.

Expression of all the λ genes specifying the replication of viral DNA and synthesis of viral protein is entirely dependent upon prior expression of viral genes **O** and **P**, and almost totally dependent upon prior expression of gene **N** (Ogawa and Tomizawa 1968). Genes **O** and **P** appear to be structural genes for an endonuclease enzyme (Shuster and Weissbach 1969), and gene **N** appears to be required both for closing the excised prophage genome as a circular molecule and for attachment to the cell membrane (Bøvre and Szybalski 1969, Hallick, Boyce, and Echols 1969).

These genes, together with genes *rex*, *cII*, and *cIII* and sites *x* and *y*, flank the gene *cI*. Gene *cI*, in turn, regulates expression of the genes **N**, **O**, and **P**. Jacob and Monod (1961[a]) noted that the genetic properties of this regulator gene were entirely analogous to those of the regulator *i* gene of the *lac* operon of *E. coli.* They therefore proposed that gene *cI* specified a repressor material which prevented λ phage replication by complexing with an operator gene similar to that postulated for the *lac* operon. Ptashne (1967[a], 1967[b]) subsequently isolated the protein specified by gene *cI* and demonstrated that it bound to preparations of λ phage DNA in a manner consistent with that expected of the hypothetical repressor. A number of studies of the physiology of cells containing the λ prophage had indicated that the regulator *cI* did in fact control two entirely independent sets of functions (e.g., Weisberg and Gallant 1966, 1967). Further analysis by Ptashne and Hopkins (1968) showed that the *cI* repressor protein did in fact complex with two sites within the λ genome; one corresponding to an operon containing gene **N** and *cIII*, and the other corresponding to the genes *x*, *y*, *cII*, **O** and **P** (Cohen and Hurwitz 1967, Kourilsky et al. 1968, Taylor, Hradecna, and Szybalski 1967).

Thus there are a number of features common to the mechanism controlling the onset of bacteriophage λ development in lysogenic cells and that regulating expression of the *lac* operon, considered in Chapter 3. The lysogenic state is maintained by repression of two key groups of genes through complex formation at specific sites in the genome with the repressor protein specified by the regulator gene *cI*. Similarly, the phenomenon of immunity is due to repression of the genome of the virulent superinfecting phage by the repressor protein specified by the regulator gene of the prophage genome *trans* to it in the cell. However, the mechanisms by which the prophage genome and the *lac* operon are derepressed differ. Moreover, detailed analysis of the course of derepression of the viral genome (see Chapter 5) indicates that the analogy with the *lac* operon may now be invalid.

Behold, I send my messenger before thy face which shall prepare thy way before thee. — Luke, "the beloved physician," First Century.

Molecular stimuli proceed from the nucleus into the cytoplasm; stimuli which, on the one hand, control the phenomena of assimilation in the cell and, on the other hand, give to the growth of the cytoplasm, which depends upon nutrition, a certain character peculiar to the species. — A. Weismann, 1889.

CHAPTER 5: **MESSENGER RNA AND GENE TRANSCRIPTION IN PROCARYOTES**

PREDICTION AND RECOGNITION OF MESSENGER RNA

A correlation between the rate of protein synthesis and the concentration of ribonucleic acid (RNA) in the cell cytoplasm was first recognized about 1940 by Brachet (1942) and by Caspersson (1941). In the following two decades the relationship was further refined as a correlation between the rate of protein synthesis and that of synthesis of RNA, either as an apparent net increase in quantity of RNA or as revealed in a process of "turnover." During the same period the experimental basis was laid for our concepts of the deoxyribonucleic acid (DNA) of structural genes as "coded genetic messages" specifying the primary amino acid sequence of proteins, and a number of tentative codes were proposed (see Crick 1966[a]). This period also marked the recognition that the vast majority of protein synthesis in most types of cell occurs in the cell cytoplasm in association with the ribosome particles (Chapter 1). Nevertheless, the nature of the relationships between protein synthesis and RNA metabolism and the role of RNA as the mediator of genetic control of the protein synthesized remained elusive. Indeed, material which we now recognize as *messenger RNA* (mRNA) was first reported in 1956 (Volkin and Astrachan 1956[a], 1956[b], Volkin, Astrachan, and Countryman 1958) to be present in cells of *E. coli* infected with bacteriophage.

The dramatic and explosive developments of the last decade were initiated by the elegant concepts of the operon and of messenger RNA as the mediator of gene expression formulated by Jacob and Monod (1961[a]). Certain of the properties of the postulated messenger RNA were dictated by the monumental body of evidence correlating the syntheses of RNA and protein and by existing concepts of the nature of the "genetic code." Other properties were specifically required to account for the special features of regulation of induced enzyme synthesis (Chapter 3). Thus the messenger RNA was postulated to be a "cytoplasmic transcript" of a corresponding structural gene and of a base composition reflecting that of the gene DNA. The total messenger RNA fraction of the cell would be very heterogeneous in size, *heterodisperse,* to reflect differences in complexity of arrangements of structural genes in organized operons. Under appropriate conditions, the messenger RNA would be associated with ribosome particles (at least temporarily) in the cell. Further, the messenger RNA would necessarily exhibit a very high rate of turnover, being very rapidly synthesized in response to addition of an inducer of induced enzyme synthesis and equally rapidly degraded on removal of the inducer.

Comparison of these predicted properties with those of the various known species of RNAs led Jacob and Monod (1961[a]) to make a tentative identification of the RNA reported by Volkin and Astrachan (1956[a], 1956[b]) as a bacteriophage-specific messenger RNA. Meanwhile, Spiegelman and his co-workers (Hall and Spiegelman 1961, Nomura, Hall, and Spiegelman 1960) had further characterized the bacteriophage-specific RNA formed in cells of *E. coli* infected with the phage T2 and tentatively concluded that it could serve as the specific template(s) for synthesis of phage protein. These conclusions were reinforced by the elegant demonstration that phage-specific mRNA was associated with pre-existing bacterial ribosomes (Brenner, Jacob, and Meselson 1961). Parallel studies revealed the presence of small quantities of material with properties predicted of messenger RNA in normal uninfected cells of *E. coli* and other bacterial species (Gros et al. 1961, Hayashi and Spiegelman 1961, Neidhardt and Fraenkel 1961). Thus prediction was dramatically substantiated essentially simultaneously, and the way was open for rapid progress toward our current rapidly evolving concepts of regulatory mechanisms.

Most of the early demonstrations of the existence of messenger RNAs were necessarily partial confirmations of most of the predicted properties postulated for such RNAs. The basic approach consisted of using a radioactive precursor to "label" rapidly synthesized ("pulse-labeled") RNA under conditions where a rapid regulatory response was anticipated: infection with phage, induced enzyme synthesis, or derepression of enzyme synthesis (Chapters 3 and 4). This pulse-labeled RNA was then isolated, in association with ribosomes and 30S subunits at high concentrations of Mg^{2+} and dissociated from the ribosomes at low concentrations of Mg^{2+}. The isolated fraction was then separated physically from "stable" ribosome and transfer RNAs and shown to be "DNA-like" or "complementary" RNA (cRNA) in base composition.

This basic approach was dictated largely by serious technical problems which had to be overcome before either the synthesis of messenger RNA under "steady-state" conditions of normal growth could be routinely assessed or the mRNA could be shown to direct the synthesis of specific protein in cell-free systems. Less than 5 percent of the total RNA of the cells is messenger RNA. Further, under normal conditions of exponential growth all classes of RNA are synthesized. Indeed, Midgley and McCarthy (Midgley 1962, Midgley and McCarthy 1962) demonstrated that the base-composition of the pulse-labeled RNA of cells of a number of bacterial species, when growing exponentially, corresponded to a mixture of one part of cRNA and two parts of ribosomal RNA, *rRNA*. Further, after a few minutes of "labeling" with radioactive precursor the base composition of the "labeled" RNA was indistinguishable from that of ribosomal RNA. It was therefore necessary both to suppress synthesis of ribosomal RNA and to examine only that RNA which was clearly distinct from ribosomal RNA in size and physicochemical properties. At the same time, the extreme susceptibility of mRNA to degradation by the very active nuclease enzymes of the bacteria examined presented a very serious source of potential experimental artifacts. The latter problem is greatly minimized by the use of current isolation procedures and by use of mutant strains of bacterium deficient in one or more of the nuclease enzymes of standard wild-type strains.

A basic procedure for distinguishing phage or bacterial messenger RNAs from all other classes of bacterial RNA was introduced and successfully exploited by Spiegelman and his co-workers. The procedure is based upon the property of complex formation between two strands of nucleic acid of complementary base sequence. In first applications of the technique of forming "hybrid" DNA-RNA complexes, Hall and Spiegelman (1961) made use of the marked differences in buoyant densities of DNA, RNA, and DNA-RNA complexes to isolate the last in cesium chloride density gradients. This extremely valuable technique of "DNA-RNA hybridization" has undergone many refinements in the ensuing years, largely to eliminate the inconvenience of the lengthy (and expensive) step of isolating the hybrid by density gradient centrifugation. Modified versions have made use of single-stranded, denatured-DNA trapped in columns of agar (Bolton and McCarthy 1962, 1964, McCarthy and Bolton 1964), adsorbed to columns of phosphocellulose (Bautz 1963[a], Bautz and Hall 1962) or to nitrocellulose membranes (Nygaard and Hall 1963) or powder (Riggsby

1969). The standard and widely used procedure is one refined by Gillespie and Spiegelman (1965).

The first applications of the technique of "DNA-RNA hybridization" were in qualitative demonstrations of the presence of presumed messenger RNAs in phage-infected cells and in normal cells under a variety of physiological conditions. Shortly thereafter it was rapidly exploited for a number of qualitative studies and quantitative assays. Some important applications which will not be discussed further include:

1. Assay of the fraction of the cell genome coding for the "stable" transfer and ribosomal RNAs (Giacomoni and Spiegelman 1962, Yankofsky and Spiegelman 1962[b]).
2. Isolation of a specific species of messenger RNA by cycles of hybridization with DNA derived, respectively, from wild-type and a deletion-mutant bacteriophage (Bautz and Hall 1962, Bautz and Reilly 1966).
3. Demonstration of the synthesis of messenger RNA "coded" by a cytoplasmic plasmid (Jenkins and Drabble 1969) and, more recently,
4. Isolation of the DNA coding-strand of the genes corresponding to a defined family of messenger RNAs (Jayaraman and Goldberg 1969).

This powerful technique of DNA-RNA hybridization has proved particularly important to our understanding both of the processes of synthesis of messenger RNA, *transcription,* and of regulation of the synthesis of protein at the level of transcription. Successful application of the technique demonstrated that the presumed messenger RNA was "DNA-like" in base sequence as well as base composition.[1] More detailed analysis also revealed that the mRNA was antiparallel to

[1] Actually, the very high degree of correspondence in base composition and sequence between mRNA and DNA template seen in these experiments was fortuitous, for only one of the two DNA strands is transcribed (see p. 190). However, at the time, this fortunate result was almost an essential requirement to establish identity of the presumed mRNA. Any other result would almost certainly have seriously retarded the subsequent advance in knowledge of control mechanisms.

the complementary DNA strand (Bautz and Heding 1964). Quantitative analysis also showed that the level of the specific mRNA fraction was raised during induced enzyme formation, both for enzymes of the *lac* operon and of the *gal* operon (Attardi 1963, Attardi et al. 1963, Hayashi et al. 1963).

During the same period rat liver nuclei (Weiss 1960) and various bacteria (Chamberlin and Berg 1962, Hurwitz et al. 1961, Weiss 1963) had been demonstrated to contain enzymes which synthesize an RNA in the presence of a DNA "template," and that the product was complementary in base composition and sequence to the primer DNA.[2] Improvements in the procedures used for isolation of the labile RNAs permitted demonstration of the predicted heterogeneity in size (e.g., Andoh, Natori, and Mizuno 1963, Monier et al. 1962, Sagik et al. 1962). In addition, incorporation of amino acid into polypeptide by cell-free systems of the type developed by Nirenberg and Matthaei (1961) was stimulated by partially purified preparations of presumed mRNAs (Bautz 1962, Monier et al. 1962), and even more marked stimulation was obtained *in vitro* by coupling systems for DNA-dependent synthesis of RNA and RNA-dependent synthesis of polypeptide (Ning and Stevens 1962, Wood and Berg 1962).

Thus, within a period of some two years, all but one of the essential properties predicted of the postulated messenger RNAs were partly substantiated for bacterial systems. The outstanding exception was the absence of any formal demonstration of the *in vitro* synthesis of a specific defined protein in response to preparations of the presumed messenger RNAs. This failure was usually attributed to development of impeding secondary structure in the isolated RNA during the process of isolation (e.g., Rich 1967, Willson and Gros 1964), for preparations of RNA obtained from the RNA-virus **f2** did promote synthesis of some specific viral protein

[2] These also were fortuitous results, for these early studies were made under conditions where both strands of the DNA template are transcribed equally readily.

under similar experimental conditions (Nathans et al. 1962). The stage was now set for examination in finer detail of the correspondence between properties expected of the hypothetical messenger RNA and of the materials identified in experimental systems and for critical appraisal of the regulatory function ascribed to this mediator of gene expression.

POLYCISTRONIC MESSENGER RNAs

At the time Jacob and Monod first presented their concepts of the operon and of messenger RNA they were unable to discriminate between two alternative mechanisms of expression of the structural genes of the operon. One possible basic mechanism consisted of the sequential transcription (from the operator region) of individual messenger RNA molecules corresponding to each of the structural genes. The alternative type of mechanism was one in which the structural genes of the operon, and possibly the operator itself, were transcribed as one large *continuous* messenger RNA molecule. Such a molecule would carry the "encoded genetic information" necessary to specify a series of polypeptide chains and would be a *"polycistronic" messenger RNA*. Resolution between the two basic types of mechanism was of considerable importance. The concept of polycistronic mRNAs immediately raised the need to consider the nature of the "punctuation marks" in translation of the genetic message, in addition to those mechanisms whereby the cell recognized the beginning and end of each discrete genetic message in the DNA of genome. Further, one important corollary of the concept of a polycistronic message is that all the peptide chains encoded within an individual operon should be synthesized in equimolar quantities, unless there are additional mechanisms controlling the translation and degradation of the messenger RNA. The existence of polar mutants affected in the *lac* operon, and in other presumed operons (Chapters 3 and 4), indicated that demonstration of a polycistronic mRNA was of itself *prima facie* evidence of such additional regulatory mechanisms.

Procedures used in the first conclusive demonstrations of materials with properties predicted of an mRNA were inadequate to make the necessary discrimination. The first isolation procedures used did not totally prevent degradation of the labile RNA, and as procedures were refined both the average size and the range of sizes of the isolated mRNA fractions increased (e.g., Bautz 1963[a], Gros et al. 1963, Sagik et al. 1962). Further, demonstration of the presence of a pulse-labeled cRNA of large size could not of itself be considered sufficient evidence of polycistronic mRNAs. This point has been especially emphasized by the demonstration within the nuclei of eucaryote cells of "giant" species of rapidly-labeled cRNA which do not serve as templates for the cytoplasmic synthesis of protein (e.g., Scherrer et al. 1966[b]). However, even at the time, it was recognized that equation of a large species of cRNA with a polycistronic genetic message required the adoption of several assumptions.

Results obtained in the first attempts to discriminate between the two possible mechanisms of transcription of the genes in a single operon appeared to provide clear identification of polycistronic mRNAs. The first such analysis was made for the histidine operon of *S. typhimurium* (see Chapter 4). Consideration of the extensive body of genetic information available regarding this operon led to estimates of the size of the operon and of a polycistronic mRNA which would correspond to all the structural genes of the entire operon. It was predicted that such a polycistronic mRNA would have a molecular weight of approximately 4×10^6 and would therefore sediment in a sucrose density gradient with an S value of approximately 38 [3] (Ames and Hartman 1963, Benzinger and Hartman 1962, Martin 1963[b]). Martin

[3] For convenience, it is customary to indicate the size of large polynucleotides by the sedimentation constant, or S value. In practice, the figure employed is usually an approximate value determined relative to some standard material examined under similar conditions, rather than an absolute value (S^{20}_{ω}) determined directly by rigorous procedures. This convention is followed here.

(1963[b]) sought evidence of such a polycistronic mRNA by examining the profile of mRNA preparations in a "dual-label" experiment. For this purpose he mixed the mRNA fraction of a mutant constitutive for the histidine operon which had been pulse-labeled with uridine-C^{14} and an mRNA preparation pulse-labeled with uridine-H^3 isolated from a deletion mutant lacking approximately three-quarters of the operon. Analysis of the profile showed the presence in the constitutive mutant mRNA of a 34S fraction missing from the mRNA fraction of the deletion mutant. Thus deletion of the major fraction of the operon was associated with a reduction in size of the 34S fraction of the mRNA. The correspondence between the size of this fraction and that predicted for the entire operon appeared to indicate that the histidine operon was probably represented by a polycistronic message. In a similar experiment Guttman and Novick (1963) obtained equivalent evidence indicating that the *lac* operon of *E. coli* was transcribed as a large 30S RNA. However, it now seems possible that the differences in profile observed in these studies can be attributed to unanticipated differences in the content of precursors of ribosomal RNAs in the two cell populations (Edwards et al. 1970).

Introduction of the technique of DNA-RNA hybridization (Hall and Spiegelman 1961) permitted a more direct analysis of the size of specific mRNA fractions. Thus hybridization of the mRNA fraction of *E. coli* with DNA isolated from defective bacteriophages containing genes of the bacterial *lac, gal,* or *trp* operons can be employed as assays for the presence of mRNAs corresponding to these operons. When coupled with a cycle of hybridization with DNA from the parental strain of bacteriophage not containing these bacterial genes, the procedure is highly specific. A similar, although less sensitive, method can be employed using the DNA of episomal particles containing appropriate structural genes. Application of these procedures has provided direct experimental evidence that the mRNAs for genes of the *lac* and *trp* operons of *E. coli*, the *his* operon of *S. typhimurium*, and several operons of various bacterial viruses are transcribed as polycistronic messenger RNAs (e.g., Iida et al. 1970, Imamoto 1968[b], Kourilsky et al. 1968, Venetianer, Berberich, and Goldberger 1968). Development of procedures for the isolation of pure preparations of portions of individual operons (Shapiro et al. 1969) should permit the assignment of even more precise values to the size of a number of bacterial and viral operons.

ASYMMETRY OF GENE TRANSCRIPTION

The prediction that messenger RNA would be DNA-like in base composition served as an essential criterion in the first demonstrations of the small quantities of such species of RNA within the total RNA isolated from virus-infected or normal cells (Gros et al. 1961, Hayashi and Spiegelman 1961, Nomura, Hall, and Spiegelman 1960). Nevertheless, it was recognized that such a correspondence in base composition was probably a fortuitous consequence of the limits of accuracy of assay of the base compositions of large polynucleotides. Our knowledge of the relationship between structural genes and the proteins they specified, as expressed by the "one gene–one enzyme" hypothesis (now the "one cistron–one polypeptide" hypothesis) (Beadle and Tatum 1941, Horowitz 1948), required that only one "translatable genetic message" be transcribed per gene. This posed the obvious first question of whether both strands of the DNA of the gene were transcribed into DNA-like RNAs, or whether only one species of RNA was specified by each two-stranded DNA. A question to which neither an analysis of overall base composition nor a simple demonstration of DNA-RNA hybrid formation could provide the answer.

Spiegelman and co-workers found that the mRNA isolated from cells of *E. coli* infected with the virus ϕX174 would hybridize with DNA of the two-stranded "*replicative form*" of the virus, but not with the single-stranded DNA of the infectious virus. They further demonstrated a similar *in vitro asymmetric transcription* of the circular DNA molecule of ϕX174 replicative form, and that rupture of the circular configuration led to transcription of both strands of the DNA molecule (Chandler et al. 1963, Hayashi, Hayashi, and Spiegelman 1963). Thus mRNA appeared to be transcribed *in vivo* exclusively from one coding strand of the DNA, which was specified by some structural feature of the DNA serving as the template for transcription. Shortly thereafter, similar asymmetry of transcription was demonstrated for the mRNAs specified by the phage ϕSP8 in cells of *B. subtilis,* and by the phage $\phi\alpha$ both in intact cells of *Bacillus megaterium* and in an *in vitro* system. In each case advantage was taken of marked differences between the two strands of the viral DNA in buoyant densities (Geiduschek, Tocchini-Valentini, and Sarnat 1964, Marmur and Greenspan 1963, Tocchini-Valentini et al. 1963). Asymmetry of transcription of the genome of phages T2 and T4 *in vivo* and *in vitro* was also inferred from the absence of self-complementary mRNAs and from quantitative assays of the fraction of the phage DNA participating in formation of DNA-RNA hybrids (Bautz 1963[b], Green 1964, Hall et al. 1963).

The unequivocal demonstrations of asymmetry of transcription of the genomes of the small bacterial viruses ϕX174, ϕSP8, and $\phi\alpha$ showed that only one strand of the entire viral genome was transcribed. However, this feature of transcription of all species of mRNAs exclusively from one coding strand of the genomes is not a general phenomenon also observed with larger viruses and bacteria. From purely genetic data, Beckwith, Signer, and Epstein (1966) were able to demonstrate that the DNA strand serving to "code" for transcription of the *lac* operon of *E. coli* was dependent upon the orientation of the operon in the chromosome. Further, following development of procedures for separating the two strands of the viral DNA (Hradecna and Szybalski 1967, Kubinski, Opara-Kubinska, and Szybalski 1966) it was demonstrated that both strands of bacteriophage λ serve as coding strands for transcription of different species of mRNAs *in vivo* and *in vitro* (e.g., Cohen and Hurwitz 1967, Taylor, Hradecna, and Szybalski 1967).

SEQUENTIAL TRANSCRIPTION AND DEVELOPMENT OF BACTERIOPHAGE λ IN LYSOGENIC CELLS

Very detailed analyses have been made of the course of gene transcription following "induction" of cells of *E. coli* containing the prophage genome of phage λ to yield mature λ-phage (see Gros et al. 1970, Kourilsky et al. 1969, Szybalski 1969, 1970, Szybalski et al. 1969).[4] These studies provide a dramatic illustration of the manner in which a highly complex developmental process may be controlled through the "*sequential transcription*" of different portions of the genome.

In an early study of the physiology of λ-phage development with a series of defective mutant strains, Jacob, Fuerst, and Wollman (1957) demonstrated that the course of development is characterized by a sequential expression of different portions of the phage genome. They were able to demonstrate a clear distinction between "early functions" and "late functions" of the sequence and to designate certain genes (as defined by the site of a mutation) as controlling early and late functions, respectively. Subsequent extensive genetic and biochemical analyses have amply confirmed and extended the concept of early and late functions in development of λ-phage and many other bacterial viruses. Indeed, for many purposes it is convenient to make a further distinction between "early" and "pre-early" functions, to designate the initial events in the developmental process. Enzymes catalyzing reactions of the early functions are termed "early enzymes" and the corresponding mRNAs

[4] The major qualitative transitions in the transcription program depicted schematically in Figures 5.1 and 5.2, and discussed in this chapter, are apparently supplemented by additional factors causing quantitative variations in the extent of transcription of portions of the genome expressed late in the developmental cycle.

as "early mRNAs." Corresponding terms are used with respect to the late and pre-early functions.[5]

The DNA of phage λ can be readily sheared into two half-molecules, which can be separated in density gradients because of marked differences in base composition. One half-molecule is enriched in adenine-thymine base pairs and carries genes controlling the early enzymes. The other is enriched in guanine-cytosine base pairs and carries genes controlling only late enzymes (Hogness and Simmons 1964, Kaiser 1962, Radding and Kaiser 1963). Use of such separated half-molecules of λ-phage DNA in the DNA-RNA hybridization procedures permits analysis of the stages of phage development at which different sections of the phage genome are transcribed. Application of such an analysis to the mRNAs produced by wild-type and by selected mutant strains of phage λ has proved a very powerful analytical tool.

In linear representations of the genome of phage λ (Figures 5.1 and 5.2) genes controlling early functions are located in the right half of the genetic map (e.g., Campbell 1969, Dove 1968), which corresponds to the "AT-rich" half-molecule. The "light" strand of DNA is oriented with 5′-terminal to the left and is designated the *l* strand; the "heavy" strand is oriented with the 5′-terminal to the right and is designated the *r* strand (see Wu and Kaiser 1967, Szybalski 1970).

Transcription of genes of the prophage genome present in uninduced cells is almost entirely from genes *cI* and *rex* of the "immunity region" (Kourilsky et al. 1969, Taylor, Hradecna, and Szybalski 1967). This transcription is from the light *l* strand of the DNA, and is said to be "to the left" or "leftward" (Szybalski 1970, Szybalski et al. 1969, Taylor, Hradecna, and Szybalski 1967). Leftward transcription of a

[5] Unfortunately, no standard system of nomenclature has been adopted. The alternate system most frequently used distinguishes between "immediate early" and "delayed early," or "intermediate early" functions. Other distinctions found in the literature include "very early" and "early"; class I and class II.

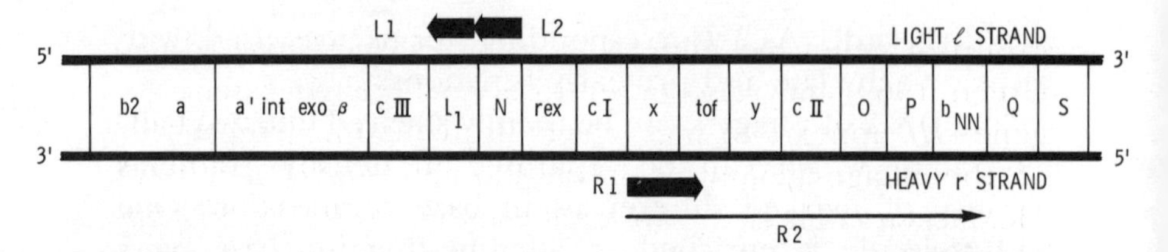

FIGURE 5.1. Strand selection and direction of transcription of pre-early genes of bacteriophage λ

Transcription of the pre-early genes of bacteriophage λ is represented by the arrows L1, L2, and R1. The narrow arrow R2 is intended to represent a less frequent continued elongation of the chain designated R1 due to unterminated transcription (see text). In this diagram the genome of the bacteriophage is not drawn to scale.

This diagram is based upon data of Cohen and Hurwitz (1968), Eisen et al. (1966), Gross et al. (1970), Hradecna and Szybalski (1969), Kourilsky et al. (1968), Kumar et al. (1969), Marcaud et al. (1971), Szybalski et al. (1969), and Taylor, Hradecna, and Szybalski (1967).

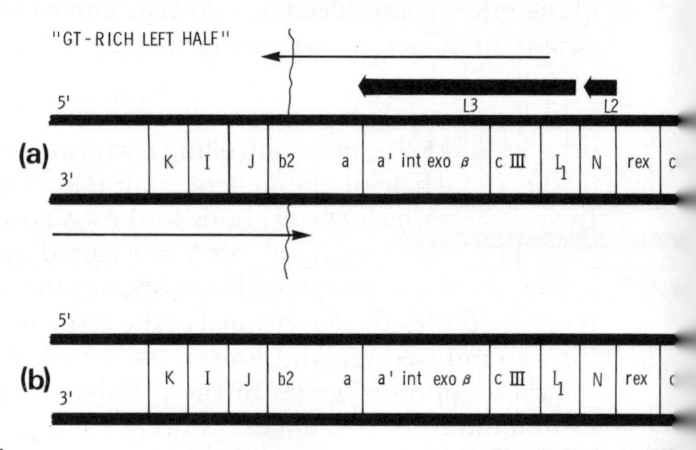

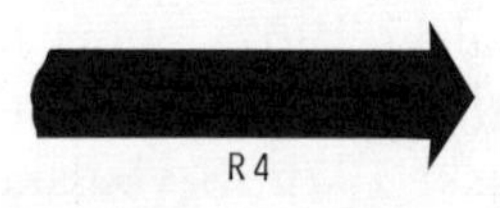

FIGURE 5.2. Strand direction and direction of transcription of early and late genes of bacteriophage λ during the developmental cycle

(a) Transcription early in the cycle. (b) Transcription late in the cycle. The width of the arrows denoting direction of transcription is intended to indicate relative frequencies of transcription.

This diagram is based on data of Cohen and Hurwitz (1968), Kourilsky et al. (1969), Kumar et al. (1969), Kumar and Szybalski (1969), Naono and Gros (1966), Nijkamp, Bøvre, and Szybalski (1970), Szybalski et al. (1969), and Taylor, Hradecna, and Szybalski (1967).

group of genes including **N** to *int* or *a'* (see Hayward and Green 1969) may also occur to a very limited extent, for gene *cIII* appears to play a role in maintaining repression of the phage genome (Malva, Razzino, and Calef 1969, McMacken et al. 1970). In addition, there may also be some restricted transcription from the *r* strand (Taylor, Hradecna, and Szybalski 1967), for gene *cII* may also participate in maintenance of the repression (McMacken et al. 1970).

When the cells are "induced" to produce mature virus, by inactivation of the phage repressor, transcription from gene *cI* (and presumably also *rex*) ceases rapidly (Heinemann and Spiegelman 1970, Kourilsky et al. 1969). The mechanism controlling this phenomenon is complex and appears to reflect some role of native repressor molecules in stimu-

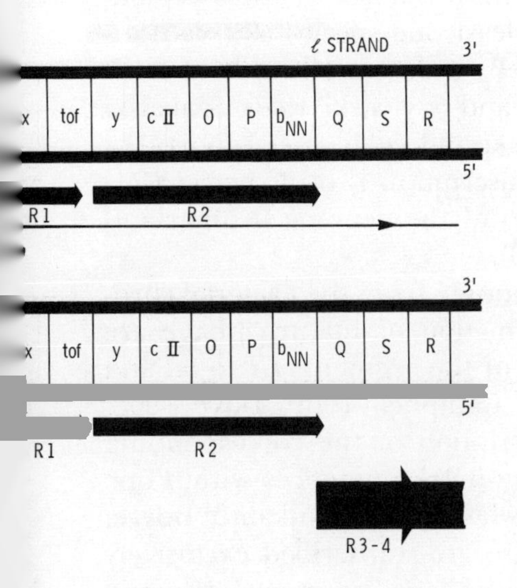

lating transcription of the cognate mRNA (Heinemann and Spiegelman 1970, Spiegelman 1971). In addition, it may involve an accessory role of gene *cIII* (Malva, Razzino, and Calef 1969) and the gene designated *tof* in Figure 5.1 (Eisen et al. 1970, Gros et al. 1970, Neubauer and Calef 1970, Oppenheim, Neubauer, and Calef 1970).

The process of induction also derepresses the operons controlled directly by the repressor of the "immunity region" and permits transcription of the pre-early mRNAs (Figure 5.1).[6] Two pre-early RNAs (L1 and L2) are transcribed "leftward" on the *l* strand from the pre-early gene **N** and the region designated l_1 (Kourilsky et al. 1968, 1969, Marcaud et al. 1971), probably as a single precursor molecule (Gros et al. 1970, Kourilsky et al. 1970). Some controversy exists regarding the number of pre-early RNAs transcribed "rightward" on the *r* strand from the promoter of the *x* region. There is a consensus that at least one species (R2) is transcribed from the genes *x* to **P**, inclusive (Kourilsky et al. 1968, 1969, Nijkamp, Bøvre, and Szybalski 1970, Szybalski et al. 1969). However, Szybalski and colleagues report that the bulk of the rightward transcription is of a species (R1) representing only the *x* region of the genome (Kumar and Szybalski 1970, Szybalski 1970).

Excision of the prophage genome from the bacterial chromosome and subsequent formation of mature phage are controlled by the early genes of the segment *cIII* to *int* and by gene **Q**, respectively (see Campbell 1969, Dove 1968, Szybalski et al. 1969). Transcription of the corresponding mRNAs (L3 and R2 of Figure 5.2) commences within one minute of induction (Kourilsky et al. 1969, Nijkamp, Bøvre, and Szybalski 1970). These also are transcribed exclusively from the AT-rich half of the prophage genome (Cohen and Hurwitz 1968, Kourilsky et al. 1968, Szybalski 1970, Taylor,

[6] Steinberg and Ptashne (1971) have recently made a formal demonstration that purified λ-repressor specifically inhibits the *in vitro* transcription of mRNAs from genes **N** and *tof*.

Hradecna, and Szybalski 1967) and represent about half of this section of the genome (Hradecna and Szybalski 1969).

Transcription of these early mRNAs is controlled by the "**N**-protein" specified by the pre-early gene **N**. There is very little, if any, transcription from the *l* strand of genes left of the region l_1 when gene **N** is either defective or deleted from the prophage genome, and the extent of all rightward transcription beyond gene **P** is markedly reduced (Kourilsky et al. 1968, Kumar et al. 1969, Nijkamp, Bøvre, and Szybalski 1970, Oda, Sakakibara, and Tomizawa 1969). **N**-protein probably functions as an "anti-termination" factor (see p. 229) which permits continued elongation of mRNAs by transcription of genes beyond the previous point of chain termination (Gros et al. 1970, Kourilsky et al. 1969, 1970, Roberts 1969[b], Szybalski 1970). Thus **N**-protein appears to mediate transcription from the promoter or gene **N** up to, and even beyond, the terminus of gene *int* (Gros et al. 1970, Szybalski 1970). Similarly, it permits transcription of gene **Q**, and also some limited transcription from genes of the regions **R-S** and **A-J** in the absence of functional product of gene **Q** normally required for their expression (Kourilsky et al. 1969, Nijkamp, Bøvre, and Szybalski 1970).

There is some controversy regarding effects of **N**-protein upon transcription from the pre-early genes **O** and **P**. A number of workers have reported marked stimulation (e.g., Brachet, Eisen, and Rambach 1970, Heinemann and Spiegelman 1970, Nijkamp, Bøvre, and Szybalski 1970). Others have observed little, or no, effect (e.g., Konrad 1970, Kourilsky et al. 1968). It therefore seems probable that these two genes are actually pseudo pre-early genes and that observed differences in the frequency of transcription as pre-early mRNAs reflect variations in the extent of fortuitous failure of the normal termination process in this region of the genome.

Transcription of the early genes of the region *cIII* to *int* is also regulated by a gene of the *x* region which has been variously designated *tof* (Eisen et al. 1966), *cro* (Eisen et al. 1970), and *zd* (Neubauer and Calef 1970).[7] The "gene-product" specified by *tof* severely limits transcription from the *cIII-int* region of the excised prophage genome and its progeny in the later stages of virus development. When the prophage genome is defective in gene *tof* there is uncontrolled transcription from the *cIII-int* region and excessive synthesis of the corresponding proteins (Eisen et al. 1966, 1970, Pero 1970). Conversely, rightward transcription of all mRNAs from the *r* strand of such mutant genomes is generally and markedly reduced (Nijkamp, Bøvre, and Szybalski 1970, Pero 1970). In contrast, *tof* exercises little, or no, control over transcription of pre-early gene **N** (Nijkamp, Bøvre, and Szybalski 1970), for which there appears to be a continuous requirement (Konrad 1970, Rabovsky and Konrad 1970).

The rate of mRNA synthesis rises markedly after replication of the prophage genome. Transcription is now predominantly rightward from the late genes **S-R** and **A-J** (Figure 5.2), although there is some continued synthesis of most pre-early and early mRNAs (e.g., Kourilsky et al. 1969, Lozeron and Szybalski 1969, Nijkamp, Bøvre, and Szybalski 1970, Oda, Sakakibara, and Tomizawa 1969, Takeda and Yura 1968). Several distinct species of late mRNAs have been recognized (Kourilsky et al. 1969, Oda, Sakakibara and Tomizawa 1969). However, present evidence indicates that transcription of all the late genes is initiated from a single promoter, located between genes **Q** and **S** (Herskowitz and Signer 1970, Kourilsky et al. 1970).[8] The gene-product of gene **Q** markedly stimulates transcription

[7] The genes designated by these various symbols have not been formally established as identical at the time of preparing this summary.

[8] Szybalski et al. (1970) have recently proposed that the "unit of transcription" defined by a single autonomous site for initiation of gene transcription, or promoter, be termed a *scripton*. It is envisaged that such a unit may contain several genetic elements which regulate the rates of transcription from different segments of the scripton. Therefore, scriptons are distinct from, but may contain, operons.

of the late mRNAs (Nijkamp, Bøvre, and Szybalski 1970, Oda, Sakakibara, and Tomizawa 1969) and is essential for normal production of mature virus (see Dove 1968). Present evidence (Naono and Tokuyama 1970) indicates that "Q-protein" may be a new species of RNA polymerase enzyme, a modified subunit thereof, or possibly a new specific "transcription factor" (see p. 225). Further discussion of its mechanism of action will be deferred.

Thus, the course of protein synthesis during development of bacteriophage λ is specified by the sequential transcription and expression of different portions of the viral genome.

SEQUENTIAL TRANSCRIPTION AND DEVELOPMENT OF BACTERIOPHAGES OF THE T-EVEN SERIES

Virus formation in cells of *E. coli* infected with bacteriophages of the "T-even" series is also characterized by sequential gene expression (see Cohen 1968). To date, it has not been possible to analyze the progression of sequential gene transcription in as fine detail as in the case of development of phage λ. The techniques which proved so useful in isolating clearly defined sections of the λ-genome have not yet yielded equally useful "marker" materials from the genome of phages of the T-even series (T2, T4, T6). At the present time, defined species of mRNA are usually recognized either by application of a cycle of DNA-RNA hybridizations with preparations of DNA containing and lacking, respectively, the corresponding structural gene (e.g., Bautz et al. 1966, Bautz and Reilly 1966) or by activity of the mRNA in coding for *in vitro* synthesis of the corresponding specific protein (e.g., Gesteland and Salser 1969, Salser, Gesteland, and Bolle 1967, Wilhelm and Haselkorn 1971, Young 1970).[9] Analyses of mRNA populations at different

[9] Coupling of either of these procedures with the genetic procedure for identification of the coding strand of the DNA developed by Jayaraman and Goldberg (1969) offers promise for a very detailed analysis of the course of sequential gene transcription in most bacteria-bacteriophage systems (Jayaraman and Goldberg 1970).

stages of the phage growth cycle are, therefore, usually of "families" of mRNA species revealed by a modified version of the DNA-RNA hybridization technique. For this purpose excess of a nonradioactive preparation of RNA is added to suppress binding of homologous molecules of radioactive RNA to the DNA through competition for the same complementary base sequences. The hybridization of radioactive molecules of a species of RNA common to the two preparations can be totally suppressed. Residual hybridization which cannot be eliminated by further excess of nonradioactive RNA is attributed to species of molecules absent from the "competing" RNA (see Gillespie 1968).

The physiology of the cycle of development of T-even phages is somewhat more complex than that of phage λ (see Cohen 1968, Luria 1970). Injection of the viral genome into the host cell causes abrupt and total cessation of protein synthesis specified by the host cell genome, and even that of RNA-containing viruses when they are introduced before infection with the T-even phage. This appears to result from synthesis of specific pre-early protein(s) which excludes all but the T-even virus-specific mRNAs from association with ribosomes (see Chapter 1) in initiation complexes (Kennell 1970, Klem, Hsu, and Weiss 1970), although translation of mRNAs already present in polysomes may also be impeded (Hattman and Hofschneider 1968). The rate of synthesis of bacterial mRNAs is immediately depressed, and all transcription of RNA from the bacterial genome ceases some 4 to 6 minutes after infection (Landy and Spiegelman 1968, Nomura et al. 1966). Indeed, within a few minutes of infection, the bacterial genome is degraded by endonuclease pre-early enzymes coded by the viral genome (Sadowski and Hurwitz 1969[a], 1969[b], Warner et al. 1970). The combined effect of these various complex processes is the creation of a highly efficient system for the synthesis specifically of viral components. Thus we are fortuitously provided with an experimental system in which all newly formed macromolecules can be identified with replication or expression solely of the phage genome. In fact, correlation of the sequential expression of viral genes with sequential transcription from different regions of the viral genome was first made with cells of *E. coli* infected with phages of the T-even series (Hall et al. 1963, Kano-Sueoka and Spiegelman 1962, Khesin and Shemyakin 1963).

During the first few minutes after infection pre-early and early mRNAs are transcribed almost exclusively from the *l* strand of the genome (Grau et al. 1969, Guha and Szybalski 1968). Onset of normal transcription of the early genes requires prior synthesis of pre-early protein (Grasso and Buchanan 1969, Salser, Bolle, and Epstein 1970). The function of this protein is currently a matter of controversy, for both the composition and "initiation site"-specificity of the host RNA polymerase enzyme are modified early in the course of viral development (see Skiff 1970, Travers 1971). However, present evidence from both *in vivo* and *in vitro* studies indicates that the early mRNAs are probably synthesized by unterminated transcription from the pre-early genes (e.g., Adesnick and Levinthal 1970, Brody and Geiduschek 1970, Milanesi et al. 1970, Salser, Bolle, and Epstein 1970).

As in the case of bacteriophage λ, transcription from at least some of the pre-early and early genes is restricted shortly after onset of replication of the viral genome (Adesnik and Levinthal 1970, Salser, Bolle, and Epstein 1970).

Transcription of late mRNAs occurs on both strands of the viral genome and is absolutely dependent upon prior synthesis of virus-specific protein (Guha and Szybalski 1968). Sustained synthesis at normal rates also requires continued replication of viral DNA and continuous expression of a gene designated 55 (e.g., Bolle et al. 1968, Bruner and Cape 1970, Pulitzer 1970, Pulitzer and Geiduschek 1970, Riva,

Cascino, and Geiduschek 1970[a]). The actual template for transcription of the late mRNAs is probably "nascent" progeny of the viral genome still containing "single-strand breaks" (see Hosoda and Mathews 1968, Lembach, Kuninaka, and Buchanan 1969, Riva, Cascino, and Geiduschek 1970[b]). The identity of the gene-product of gene 55 has not yet been established. Transcription of late mRNAs can be correlated with both modifications in structure and specificity of RNA-polymerase enzyme, and with synthesis of at least one new species of "transcription factor" (see Skiff 1970, Travers 1971). Further discussion of these changes will be deferred.

Analysis of the mechanisms regulating onset of transcription of late mRNAs by current procedures is also further complicated by the existence of a class of genes designated "quasi-late" or "pseudo-early," which includes that coding for the virus-specific lysozyme. The gene-products specified by these genes are late virus proteins, and normal expression of the genes requires transcription of late mRNAs. However, the pseudo-early genes of the parental infecting genome can be expressed to a limited extent when there is little, or no, replication of the viral DNA (Bruner and Cape 1970, Karam and Speyer 1970, Mark 1970). Analysis by the technique of DNA-RNA hybridization indicates that these genes are also normally transcribed to a limited extent as early mRNAs before the onset of DNA replication (e.g., Bautz et al. 1966, Salser, Bolle, and Epstein 1970). Thus the early mRNA fraction contains RNA which appears to be transcribed specifically from the structural gene for lysozyme (Bautz et al. 1966). However, it contains no material which can serve as template for *in vitro* synthesis of lysozyme in extracts prepared from cells late in the developmental cycle (Gesteland and Salser 1969). It therefore seems possible that pseudo-early mRNAs are transcribed from the noncoding strands of the corresponding genes. These pseudo-early mRNAs would then be "anti-messenger RNAs," which are believed to be generated by unterminated transcription from adjacent genes (Geiduschek and Grau 1970).

SEQUENTIAL TRANSCRIPTION AND DEVELOPMENT OF BACTERIOPHAGE T5

Buchanan and colleagues have exploited an unusual property of the bacteriophage T5 genome for studies of sequential gene expression. The DNA chromosome of this virus exhibits a point of "preferential breakage" when complexes with *E. coli* are sheared in a blender shortly after infection.[10] Bacterial cells recovered from such sheared complexes contain a fragment of the viral DNA that is known as the "*first step transfer* (*FST*) *fragment.*" This FST-fragment represents about 8 percent of the total virus genome (Lanni, McCorquodale, and Wilson 1964, McCorquodale and Lanni 1964). It contains genes which control degradation of the host genome, uptake of the remainder of the phage genome, and subsequent suppression of transcription from the FST-fragment (Lanni 1969, McCorquodale and Lanni 1970, Sirbasku and Buchanan 1970[a]).

Cells infected with the entire phage T5 genome show three distinct phases of transcription of mRNAs and synthesis of virus-specific proteins. In the first 4 to 5 minutes after infection a series of pre-early RNAs is transcribed exclusively from the FST-segment of the genome, and corresponding proteins of class I are synthesized. This phase of the developmental cycle is followed by one in which transcription of early mRNAs and synthesis of early class II protein commences, during which transcription of the FST portion of the genome steadily declines over a period of 5 to 6 minutes. Approximately 10 minutes after infection, transcription of mRNAs for late enzymes is initiated by a process dependent upon prior protein synthesis (and prob-

[10] This is presumably the site of one of the "single-strand breaks" in the T5 DNA (Abelson and Thomas 1966, Bujard 1969, Jacquemin-Sablon and Richardson 1970). However, this has not yet been established.

ably also replication of the viral genome). Thereafter the synthesis of late protein of class III rapidly ensues (Moyer and Buchanan 1969, McCorquodale and Buchanan 1968, Sirbasku and Buchanan 1970[a], 1970[b]).

SUCCESSIVE TRANSCRIPTION OF THE CISTRONS OF BACTERIAL OPERONS

The technique of DNA-RNA hybridization has also been used to demonstrate formally the successive transcription of structural genes of the *trp* operon of *E. coli.* For this purpose repressed cells were derepressed (see Chapter 4) in synchrony and the course of mRNA transcription followed. Preparations of isolated mRNA were hybridized with a series of preparations of DNA isolated from phages containing either the entire *trp* operon or lacking various combinations of the constituent genes. In this manner it was demonstrated that the structural genes of the operon are transcribed in sequence from the operator end of the operon. Transcription of the entire operon requires some 6 minutes at 37°C and is characterized by a progressive increase in size of the specific mRNA (Baker and Yanofsky 1968[a], 1968[b], Imamoto 1968[b], Imamoto, Morikawa, and Sato 1965).

Similar analysis by DNA-RNA hybridization has not been applied to analysis of the course of transcription of the *lac, gal,* and *his* operons of bacteria (Chapters 3 and 4). Appropriate mutant strains and experimental procedures are available for such analyses (Guha, Tabaczyński, and Szybalski 1968, Kumar and Szybalski 1969). However, pending such formal demonstration, successive transcription of these operons is inferred from the progressive increase in size of polysomes containing the corresponding mRNA (Schwartz, Craig, and Kennell 1970), and from kinetics of onset of increased synthesis of the various proteins for which the template has been shown to be polycistronic (Alpers and Tomkins 1965, 1966, Goldberger and Berberich 1965, Leive and Kollin 1967, Michaelis and Starlinger 1967, Venetianer, Berberich, and Goldberger 1968).

REGULATION OF PROTEIN SYNTHESIS AND CHANGES IN THE COMPLEMENT OF MESSENGER RNAs

Qualitative changes in the spectra of proteins synthesized in normal or virus-infected bacterial cells can be correlated with changes in the complement of mRNAs transcribed within the cells. Similarly, sequential expression of the entire genome of viruses acting on eucaryote cells can be inferred from the kinetics of synthesis of virus-specific proteins and of changes in the families of cRNAs or mRNAs found at different stages of the infective cycle (e.g., Aloni, Winocour, and Sachs 1968, Becker and Joklik 1964, Cheevers and Sheinin 1970, Kates and McAuslan 1967, 1968).

Nevertheless, a number of features observed in various studies are not readily compatible with the simple hypothesis that the *quantity* of specific protein formed is determined solely by the level of corresponding mRNA present in the cell. For example, the increase in rate of formation of specific protein in the phenomenon of induced enzyme synthesis (see Chapter 3) appears to be markedly greater than the increase in content of specific mRNA. The rate and extent of β-galactosidase formation by cells of *E. coli* are increased more than a thousandfold in the presence of methyl·β·thiogalactoside (Jacob and Monod 1961[a], Monod and Cohn 1952). However, the level of mRNA which hybridizes specifically with DNA enriched in genes of the *lac* operon was found to be increased a mere fivefold to fiftyfold (Attardi et al. 1963, Hayashi et al. 1963, Spiegelman and Hayashi 1963). A recent analysis, based on more specific hybridization procedures, indicates a much larger increase in level of specific mRNA in the range of sixtyfold to two hundredfold (Kumar and Szybalski 1969). Nevertheless, even this higher range of values falls far short of quantitatively accounting for the increase in β-galactosidase formation. Similarly, both early and recent estimates of the increases in levels of specific mRNAs in other cases of induced enzyme synthesis or enzyme derepression (see Chapter 4) are substantially less than the extent of increase in rate of enzyme formation (e.g., Attardi 1963, Attardi et al. 1963, Edlin et al. 1968, Stubbs and Hall 1968).

Thus, in none of these cases can the extent of the increase in rate of synthesis of specific protein be attributed *solely* to an enhanced rate of mRNA formation. Further, in at least the specific case of *E. coli* β-galactosidase, induced enzyme synthesis is not associated with an increase in rate of elongation of individual polypeptide chains. Rather, a larger number of nascent enzyme molecules are elongated simultaneously on each mRNA molecule (Lacroute and Stent 1968).

Corresponding discrepancies are also observed in the case of virus-infected cells. For example, approximately half of the protein made late in the infection of *E. coli* by phage T4 is coat protein, but no single species of mRNA has been found in such a large proportion (Adesnik and Levinthal 1970).

One disturbing feature of all early studies of natural mRNAs (i.e., those synthesized *in vivo*) was the inability to serve as a template for the *in vitro* synthesis of defined specific proteins. Indeed, presumed mRNAs of large size (as inferred from the sedimentation constant) were inactive as templates for *in vitro* incorporation of amino acid into any polypeptide (e.g., Willson and Gros 1964). Subsequent recognition of the many factors already known to be involved in the processes of protein synthesis (see Chapter 1) has substantiated the belief that early failures to demonstrate activity of isolated mRNAs as templates for *in vitro* synthesis of defined proteins were due to deficiencies of the experimental systems. *In vitro* synthesis of specific proteins has now been demonstrated for proteins specified by RNA-viruses (Cappechi 1966, Lodish 1968, Nathans 1965, Nathans et al. 1969, Zinder, Engelhardt, and Webster 1966) and by isolated mRNAs for phage-specific α-glucosyl·transferase (Young 1970) and lysozyme (Brawerman et al. 1969, Gesteland and Salser 1969, Salser, Gesteland, and Bolle 1967).

In addition, limited *in vitro* syntheses of *E. coli* tryptophan synthetase **A** protein (Yura, Marushige, and Imai 1963), phage-specific α-glucosyl·transferase (Gold and Schweiger 1969[a], 1969[b], Young 1970), and *E. coli* β-galactosidase (Chambers and Zubay 1969, de Crombrugghe et al. 1971) have all been demonstrated in coupled systems for transcription and translation.

Nevertheless, *in vitro* synthesis of a defined protein in a system supplemented by isolated preparations of a large bacterial polycistronic mRNA has not yet been obtained.

Thus various lines of evidence indicate that the *extent* of synthesis of a specific protein is not determined solely by the available level of corresponding mRNA. Additional mechanisms serve to determine the frequency of translation of that mRNA in synthesis of specific protein. These will be considered in Chapter 9.

COUPLED TRANSCRIPTION AND TRANSLATION

The rates of RNA synthesis observed in early studies with partly purified preparations of DNA-dependent RNA-polymerase enzyme were at least one order of magnitude too low to account for the rates of RNA synthesis observed *in vivo* (e.g., Bremer and Konrad 1964, Maitra and Hurwitz 1967). Recent studies indicate that much of this discrepancy was due to the absence of specific "transcription factors" required for transcription of various segments of the template DNA used in these studies (e.g., Skiff 1970, Travers 1971), which are considered later in the chapter. Nevertheless, the *rate of chain elongation* is also one order of magnitude lower *in vitro* than in the intact cell. For example, at 37°C the rate of elongation of nascent RNA molecules has been determined to be 2.5 to 5 nucleotides per second *in vitro* (Bremer and Konrad 1964, Hayward and Green 1970) and calculated to range from 28 to 50 nucleotides per second *in vivo* (Bremer and Yuan 1968[a], 1968[b], Manor, Goodman, and Stent 1969). Effects of the newly discovered transcription factors upon rates of chain elongation have yet to be examined. However, it seems probable that the discrepancy between *in vivo* and *in vitro* rates reflects some feature of the structural organization of the cell. Treatment of intact cells of *E. coli* to render them permeable to substrates of the RNA-polymerase enzyme reduces the specific enzyme activity to that obtained *in vitro* with an equivalent quantity of *crude* cell extract (Gros et al. 1967).[11]

Consideration of the evidence then available led Stent (1964, 1966, 1967) to propose that ribosomes were attached to the nascent mRNA chain and that synthesis of the corresponding protein was initiated on the incomplete mRNA molecule. Movement of ribosomes along this mRNA during the translocation step following formation of each peptide bond (see Chapter 1) would serve to "peel" nascent mRNA from the DNA template on which it was being transcribed. This "peeling" would, in turn, serve to make additional sections of the DNA template accessible for transcriptions. Thus the transcription process would be coupled to, and regulated by, the processes of protein synthesis.

In formulating this proposal, Stent noted an important corollary. The hypothesis required that transcription and translation of the mRNA molecule be initiated from the same end of the mRNA molecule. In turn, this required a specific orientation of the coding strand of the DNA of the genome relative to both the mRNA and to the corresponding proteins. This prediction has been borne out by a wide variety of studies and particularly by data obtained from the *trp* operon of *E. coli.* Yanofsky and his colleagues have made extensive analyses of the amino acid substitutions found in the tryptophan synthetase **A** protein components produced by a series of mutant strains. Guest and Yanofsky (1966) correlated these substitutions with both codon assignments

[11] Possible loss of essential transcription factors (see p. 225) cannot be formally excluded by evidence currently available. However, the size and properties of factors identified to date indicate such a loss to be improbable.

of the genetic code and location of the causal mutations within the structural gene for this protein. This permitted them to deduce the relative orientations of the coding strand of the DNA and directions of transcription and translation of the corresponding mRNA. As predicted by Stent (1964) both transcription and translation commence from the operator-proximal end of the structural gene. More recently, both transcription of the entire *trp* operon and synthesis of the corresponding enzymes have been shown to commence from the operator-terminal end of the operon (Baker and Yanofsky 1968[a], Imamoto 1968[a], Ito and Imamoto 1968, Morse, Baker, and Yanofsky 1968).

The coupling of translation and transcription envisaged by Stent's proposal requires the existence of complexes in which ribosomes are attached to DNA through nascent mRNA chains. Such complexes (Figure 5.3) have been isolated both from intact cells and from *in vitro* systems containing components required for coupled transcription of mRNA and synthesis of protein (Byrne et al. 1964, Eiserling et al. 1964, Miller, Hamkalo, and Thomas 1970, Revel et al. 1968). In these complexes the ribosomes are attached via the 30S subunit to short mRNA fibers (Figure 5.3). Complexes are formed only in the presence of the initiation factors re-

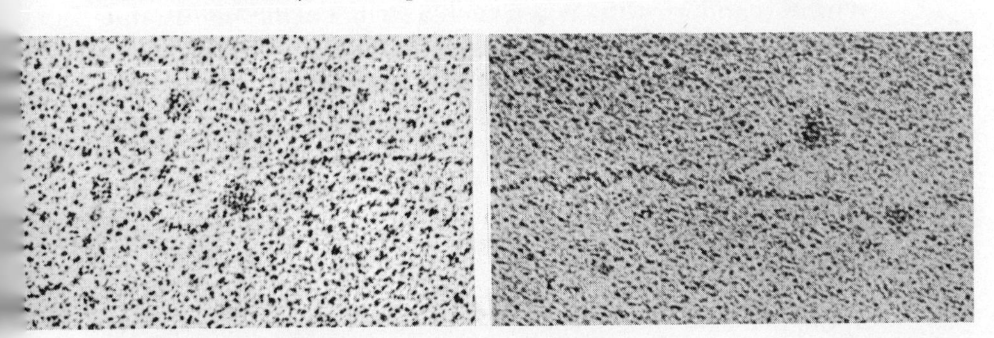

FIGURE 5.3. DNA-mRNA-ribosome complexes of E. coli

Left, complex isolated from intact cells. Right, complex formed in vitro.

Electron micrographs made after rotatory shadowing (×135,000), courtesy of Dr. M. Herzberg.

quired for protein synthesis, and they contain nascent polypeptide (Byrne et al. 1964, Eiserling et al. 1964, Revel et al. 1968). Similar complexes have also been isolated from *in vitro* systems for RNA transcription derived from *E. coli* and supplemented with homologous ribosomes, under conditions where little or no peptide synthesis occurs (Das, Goldstein, and Lowney 1967, Shin and Moldave 1966).

Addition of ribosomes plus initiation factors to *in vitro* systems for transcription of RNA stimulates the rate of RNA synthesis up to twofold. In addition, it permits transcription of mRNA sequences of large molecular weight and causes displacement of transcribed mRNA molecules from the DNA template (Jones, Dieckmann, and Berg 1968, Revel and Gros 1967, Shin and Moldave 1966).

As originally formulated, Stent's proposal envisaged the *in vivo* transcription of mRNA to be *obligately coupled* with translation of the mRNA in synthesis of protein (Stent 1964, 1967). However, a number of studies have indicated that coupling with translation is not an essential requirement for the *in vivo* transcription of mRNA. The majority of these demonstrations have been made with so-called *stringent* strains of *E. coli* (see Chapter 9), which also lack the ability to synthesize one, or more, amino acid(s) essential for protein synthesis and growth. When such a stringent *auxotrophic* mutant is deprived of all endogenous sources of essential amino acid(s) it shows no further *net* synthesis of protein or accumulation of RNA (Edlin and Broda 1968). However, there is continued synthesis of protein and of mRNA in a process of turnover, and induced enzyme synthesis (Chapter 3) may be observed with such "starved" cells (Mandelstam 1960, Nakada and Magasanik 1964). Quantitative assays have been made of the effect of starvation for essential amino acid(s) upon the rate and level of formation of specific mRNAs. Thus Morris and Kjeldgaard (1968) demonstrated that the rate of transcription of mRNA of β-galactosidase enzyme (and presumably the entire *lac* operon) was not impaired under conditions where the ability to translate the mRNA in formation of active enzyme was reduced to 5 percent of normal "unstarved" values. Further, in "stringent" strains unable to synthesize one of the enzymes required for formation of tryptophan the level of "*trp* operon mRNA" may be even higher than in unstarved derepressed cells (Edlin et al. 1968, Lavallé and De Hauwer 1968). A similar preferential transcription of the "*his* operon mRNA" has also been reported for histidine-starved cells of *S. typhimurium* (Venetianer 1968). In other studies transcription of mRNA independently of its translation has been demonstrated through the use of chloramphenicol to inhibit synthesis of all types of protein, or the use of bacteriophage containing nonsense mutations to preclude synthesis of specific proteins (Gesteland, Salser, and Bolle 1967, Nakada and Magasanik 1964).

Nevertheless, several lines of evidence indicate that the processes of transcription and translation are coupled under normal physiological conditions.[12] The rate of elongation of nascent peptide chains *in vivo* corresponds very closely with that for the nascent mRNA (Lacroute and Stent 1968, Rose, Mosteller, and Yanofsky 1970, Wilhelm and Haselkorn 1970). Free polyribosomes, unattached to the cell DNA, are rare (Miller, Hamkalo, and Thomas 1970). Analysis of the kinetics of onset of formation of the enzymes controlled by the *lac* and *trp* operons of *E. coli* (see Chapters 3 and 4) indicates that translation of the operator-proximal section of polycistronic mRNAs commences before transcription of the entire operon is completed (Alpers and Tomkins 1966, Imamoto and Ito 1968, Leive and Kollin 1967, Morse, Baker, and Yanofsky 1968). Moreover, inhibition of translation of mRNA transcribed from the operator-proximal genes of an operon *can* prevent accumulation of mRNA corresponding

[12] Perhaps the most compelling evidence is that provided indirectly by the observations that *in vitro* transcription from the *lac* operon of *E. coli* is unregulated in the absence of ribosomal proteins (de Crombrugghe et al. 1971[b], Eron et al. 1971, Ohshima et al. 1970).

to the operator-distal genes of the operon (Morse 1971). This is probably due to nuclease digestion of the exposed section of completed mRNA molecules which are not protected by complexing with ribosomes (Contesse, Crépin, and Gross 1970, Morse 1971, Morse and Yanofsky 1969[a], Summers 1971, Varmus, Perlman, and Pastan 1971).[13] Further studies of Morse et al. (1969) indicate that all the ribosomes which associate with a given molecule of polycistronic mRNA do so and initiate protein synthesis before transcription of the entire mRNA chain is completed. The complete cycle of transcription plus translation is envisaged as one in which a series of ribosomes attach to the end of a nascent mRNA molecule and move along the elongating mRNA in a cluster, with the exposed 5′-terminal of the mRNA chain degraded shortly after the traverse of the last ribosome of the cluster (Morse et al. 1969, Morse and Yanofsky 1969[a]). A similar integrated cycle is also inferred (Wilson and Hogness 1969) for synthesis of enzymes controlled by the *gal* operon of *E. coli* (see Chapter 3) from an analysis of effects of growth conditions upon the kinetics of enzyme formation.

Thus the rate and level of synthesis of a specific protein may be partly controlled by the number of ribosomes which are integrated into the polyribosome complex formed on each growing nascent mRNA chain. Present evidence is in accord with this suggestion (Lacroute and Stent 1968, Rose, Mosteller, and Yanofsky 1970).[14]

DEGRADATION OF MESSENGER RNA AND RIBONUCLEASE V

Jacob and Monod (1961[a]) defined susceptibility to rapid metabolic degradation as one of the essential properties of all mRNAs necessary to account for the rapid cessation of

[13] The kinetics of mRNA accumulation in the absence of chloramphenicol by strains containing the suppressor allele *suA* (Morse 1971) do not appear to be compatible with the suggestion that this strain is deficient in the endonuclease postulated to attack nascent mRNA not complexed with ribosomes (see also Varmus, Perlman, and Pastan 1971).

[14] Baker and Yanofsky (1970) have recently revised the estimate of the average number of ribosomes attached to each nascent mRNA chain and have also concluded that degradation of the mRNA molecules is probably initiated at random.

induced enzyme synthesis on removal of specific inducer. Subsequent studies have amply confirmed that "metabolic instability" is a general property of most species of bacterial mRNAs. Indeed, with cells growing aerobically at 30 to 37°C the average "half-life" of all species of mRNAs, as measured by function or hybridization with DNA, is of the order of 1 to 3 minutes (e.g., Fan, Higa, and Levinthal 1964, Kepes 1963, Schwartz, Craig, and Kennell 1970). However, a few situations are known in which mRNAs appear to be stable (e.g., Del Valle and Aronson 1962, Summers 1970), probably because of inactivation or loss of "degrading enzyme(s)."

Identification of the enzyme(s) responsible for this rapid degradation is of considerable importance, both to further understanding of the regulation of protein synthesis and to development of cell-free systems for *in vitro* studies. Nevertheless, until very recently, two factors combined to delay provisional identification of these enzymes. One is the presence in *E. coli* and other bacterial species examined of a variety of enzymes capable of degrading ribonucleic acids. Certain of these enzymes could be eliminated from consideration as those primarily responsible for the continuous degradation of mRNAs by the isolation of strains lacking the enzyme (Cammack and Wade 1965, Gesteland 1966), by analysis of the degradation products formed (e.g., Andoh, Natori, and Mizuno 1963, Castles and Singer 1969, Sekiguchi and Cohen 1963, Singer and Tolbert 1965), or by determination of the direction of degradation of large polycistronic mRNAs (Morikawa and Imamoto 1969, Morse et al. 1969). Nevertheless, these approaches did not identify the enzyme(s) responsible for the normal degradation of mRNA. Nor could this type of approach eliminate the possibility that particular regulatory mechanisms involved the action of other nuclease enzymes (Yamasaki and Arima 1969).

The analysis was further complicated by prevailing interpretations of the important observation that the rate of mRNA degradation is markedly reduced under anaerobic conditions and in the presence of inhibitors of the translation process (e.g., Fan, Higa, and Levinthal 1964, Lembach and Buchanan 1970, Summers 1970, Woese et al. 1963). Under these conditions the reduced rate or arrest of protein synthesis prolongs the period of association of mRNA with the ribosomes. That portion of the mRNA molecule associated with the ribosome particle is considerably more resistant to the action of added nuclease enzymes than uncomplexed regions of the molecule (Steitz 1969, Takanami and Zubay 1964). It therefore seemed probable that persistent association of untranslated mRNA with ribosomes would retard degradation of the mRNA (e.g., Castles and Singer 1969). However, recent studies indicate that the persistence of mRNA under conditions of impaired protein synthesis is due to a more direct inhibition of one enzyme which is responsible for mRNA degradation.

Kuwano et al. (1969) discovered in extracts of *E. coli* an exo-nuclease enzyme, RNase V, with novel properties which indicate that it degrades the mRNA while bound to the ribosomes. Enzymic activity is totally dependent upon all of the components required for elongation and translocation of "nascent" peptide chains, except that de-acylated tRNA will serve in lieu of aminoacyl-tRNA. Further, activity is inhibited by those inhibitors of protein synthesis which retard *in vivo* degradation of mRNA, and by cAMP (Kuwano, Apirion, and Schlessinger 1970[a], Kuwano et al. 1969, Kuwano and Schlessinger 1970, Kuwano, Schlessinger, and Apirion 1970[a]). In addition, the enzyme appears to act specifically on mRNAs and artificial template-RNAs, and degrades them in the direction observed for mRNA degradation *in vivo* (Kuwano et al. 1969, Kuwano, Schlessinger, and Apirion 1970[b]). However, degradation does not proceed to completion, and breakdown of the resulting "core" may require the activity of other nuclease enzymes.[15]

[15] Present evidence does not exclude the possibility that RNase V action is initiated by a prior endonucleolytic cleavage of the molecule and the "core polynucleotide" product represents an untranslatable base sequence.

INACTIVATION OF mRNA AND THE POSTULATED TICKET REGION

Early analyses indicated that each molecule of mRNA supported the translation of an average of some 6 to 20 polypeptide chains before being inactivated and degraded (e.g., Levinthal, Keynan, and Higa 1962, Woese et al. 1963). A large body of circumstantial evidence appeared to indicate that this did not simply reflect the probability of random interaction between "degrading enzyme" and the mRNA molecule before attachment of the latter to a protective ribosome. In particular, the rate of mRNA degradation (i.e., half-life) under normal aerobic conditions is independent of the composition of the growth medium and initial level of mRNA (e.g., Lavallé and De Hauwer 1968). It has therefore been tacitly understood that the extent to which a given mRNA molecule may serve as template for the synthesis of specific protein is determined by a specific control mechanism. Indeed, Morse and colleagues (Morse et al. 1969, Morse and Yanofsky 1969[a]) have concluded that incorporation of the last ribosome into a polyribosome is either regulated by, or in turn regulates, degradation of mRNA by exonuclease action from the 5′-terminal of the molecule.[16]

A number of workers have recently challenged the generality of this conclusion. Schwartz, Craig, and Kennell (1970) observed that the β-galactosidase activity of cells of *E. coli* continued to rise for several minutes after total cessation of transcription from the *lac* operon. More important, approximately one-third of the increase in activity occurred after the predicted time of inactivation of all molecules of the corresponding mRNA. They have therefore concluded that initiation of inactivation and degradation of individual mRNA molecules occur at random. However, the kinetics of increase in activity of the transacetylase enzyme coded by the operator-distal gene of the same operon (see Chapter 3) corresponded very closely with those predicted from the

[16] See footnote 14.

proposals of Morse and colleagues. It therefore seems unlikely that the sustained increase in β-galactosidase activity is due to continued translation of complete polypeptide chains from residual mRNA.

The accumulation of large and discrete species of mRNAs in the later stages of infection of *E. coli* with phages T4 and T5 is not in accord with the proposals of Morse and colleagues (Adesnik and Levinthal 1970, Sirbasku and Buchanan 1970[a]). However, late mRNA isolated from cells infected with phage T4 is susceptible to degradation by RNase V *in vitro* (Kuwano, Apirion, and Schlessinger 1970[b]). Thus accumulation of the RNAs *in vivo* may reflect a disruption of the normal control mechanisms of the cell which is observed earlier and in more dramatic form when *E. coli* is infected with the phage T7. In this case all species of mRNA, including those of the host cell, appear to be relatively stable and the cells appear to be deficient in "mRNA-degrading" activity (cf. Summers 1970, 1971).

Thus it seems premature to reject the hypothesis that termination of incorporation of ribosomes into a polyribosome is *normally* correlated with onset of inactivation of the constituent mRNA in uninfected cells.

One particularly interesting possibility has been suggested by Sussman (1970, Sussman and Sussman 1969) in what may be described as the "commuter pass model." They postulate that the template sequence of bases in an mRNA molecule is preceded by a regulatory sequence of *untranslated* bases, or "tickets." It is envisaged that each time a ribosome attaches to, and moves along, the mRNA a small portion of the regulatory base sequence is removed by transient exonuclease action. This process of ribosome attachment and associated shortening of the mRNA continues until all tickets are detached and the first "available" initiator codon removed. Thereafter, no further ribosomes can be incorporated into the polyribosome and inactivation of the template sequence commences.

At the present time this suggestion is entirely speculative. It has not yet been examined directly by experiment, and there is a paucity of relevant circumstantial evidence. There are lengthy sequences of untranslatable bases at the 5′-termini of the genomes of RNA-containing bacteriophages (e.g., Adams and Cory 1970, Billeter et al. 1969, Ling 1971). However, this class of RNAs appears resistant to degradation by RNase V *in vitro* (Kuwano, Apirion, and Schlessinger 1970[b]) and relatively stable *in vivo* (Doi and Spiegelman 1963, Jaenish, Jacob, and Hofshneider 1970). Recent studies of the *lac* operon of *E. coli* by Beckwith and co-workers indicate that the corresponding polycistronic mRNA also may contain an initial sequence of many untranslatable bases. The deft isolation of a portion of this operon as a pure chemical permits the assignment of promoter and operator regions of the operon to a DNA segment containing approximately 400 base pairs (Shapiro et al. 1969). Genetic analysis has located the operator region between the promoter and first structural gene of the operon (Ippen et al. 1968). Thus initiation of transcription from the promoter requires that a considerable portion of this non-coding segment of the operon be transcribed into a nontranslatable RNA sequence. However, the 5′-terminal section of mRNA transcribed from the *lac* operon has not yet been examined for the presence of an untranslatable sequence. On the other hand, it has been demonstrated that nascent pre-early mRNA transcribed *in vitro* from the genome of phage T4 contains a 5′-terminal sequence of 100 to 200 bases which lacks sites for efficient binding of ribosomes into initiation complexes (see Chapter 1) and is presumably not translated into protein (Revel, Herzberg, and Greenshpan 1969). This segment does appear to contain sites at which the ribosomes can complex without formation of normal initiation complexes, but it has not been determined whether the segment is susceptible to RNase V action during polyribosome formation under conditions which permit translocation to occur.

POLARITY AND INTERNAL PROMOTERS

One expectation implicit in the original concept of polycistronic mRNAs is that all of the individual cistronic portions will be translated in the synthesis of protein equally readily. Thus there would be absolutely coordinate synthesis (see Chapters 3 and 4) of equal numbers of each of the polypeptide chains specified by the polycistronic mRNA. This expectation has been borne out by data available so far for the *galactose* (Wilson and Hogness 1969) and *arabinose* (Lee 1971) operons of *E. coli*, and largely so for the *trp* operon of *E. coli* (Morse, Baker, and Yanofsky 1968). However, the prediction of strictly coordinate synthesis of a series of proteins does not appear to be generally valid. Early detailed analysis for the enzymes of the *histidine* operon of *S. typhimurium* appeared to provide the most persuasive evidence of accord with prediction (Whitfield, Smith, and Martin 1964). However, a more recent analysis indicates that the first cistron of the mRNA is translated three times as frequently as all the others (Whitfield et al. 1970). Similarly, for the *lac* operon of *E. coli* the molar ratio of β-galactosidase to transacetylase formation is of the order of 5 (Brown, Brown, and Zabin 1967). The mechanism responsible for disruption of coordination of enzyme synthesis in these particular cases is not yet known.

However, the coordinate translation of all the segments of a polycistronic mRNA is known to be subject to modification in a variety of ways. Indeed, the combined effects of several factors serve to ensure the noncoordinate synthesis of proteins specified by the genome of RNA-containing viruses. These include inhibition by viral coat protein of the synthesis of the protein specified by the viral genome (Eggen and Nathans 1969, Robertson, Webster, and Zinder 1968, Spahr, Farber, and Gesteland 1969), possibly "delayed" replication of a portion of the viral genome (Robertson and Zinder 1969), plus effects determined by the structure of the ribosomes and by associated initiation factors, or other proteins (Klem, Hsu, and Weiss 1970, Lodish 1969). These mechanisms will be considered in Chapter 9.

A more frequent cause of very marked deviation from the expected coordinate synthesis of a group of proteins is the presence in the operon of a *polar* mutation, of which there are three major types. One is a nonsense mutation in the operator-proximal end of a cistron. Such a mutation not only leads to inability to synthesize the corresponding protein but may also reduce the extent of expression of all genes of the operon which are operator-distal to the mutation (e.g., Imamoto, Ito, and Yanofsky 1966, Newton et al. 1965). This latter effect is the phenomenon of *polarity*. Polarity may also arise from the presence of an incorrect base sequence generated by insertion or removal of a few nucleotide bases in a "frame-shift" mutation (e.g., Ames and Whitfield 1966, Malamy 1966, Tsugita et al. 1969), or the insertion of a large segment of unrelated DNA (Jordan, Saedler, and Starlinger 1968, Shapiro 1969). In these cases, all polypeptide specified from the origin of the error to the first nonsense, terminator codon in the sequence is abnormal, and the extent of expression of distal genes of the operon again may be reduced.

The extent to which expression of genes operator-distal to a polar mutation is depressed is termed the degree of *polarity*. It ranges from relatively slight impairment (weakly polar) to virtually total suppression (extremely polar). The degree of polarity appears to be determined largely by the length of the base sequence between the polar mutation and the *operator-distal* border of the affected cistron, but independent of the position of that cistron within the operon (Michels and Reznikoff 1971, Newton 1969).

Various types of mechanism have been proposed to account for the phenomenon of polarity (e.g., Morse and Yanofsky 1969[a], Zipser 1969). These will not be reviewed in detail, for the available data provide a highly confusing and often contradictory picture. However, one group of observations is particularly pertinent to identification of the enzyme(s) responsible for rapid degradation of mRNAs in normal bacterial cells. When isolated from a strain of *E. coli* containing a strongly polar mutation in either the *lac* or *trp* operon the corresponding mRNA is of smaller average size than that from wild-type strains. It consists of a mixture of molecules of normal length and molecules lacking the base sequences operator-distal to the point of the polar nonsense codon (Contesse, Crépin, and Gros 1970, Contesse, Naono, and Gros 1966, Imamoto and Yanofsky 1967[a], and 1967[b]). Further, the average rate of inactivation of the operator-distal segment of the mRNA molecules of normal length (as determined by the half-life) is the same as in wild-type strains (Carter and Newton 1969, Hermann and Pardee 1965). This *appears* to be strong presumptive evidence that the phenomenon of polarity is due to premature termination of transcription at the site of the polar mutation and was originally so interpreted. However, subsequent analyses have revealed that the entire operon is transcribed and the operator-distal portion of the shorter species of molecules is selectively degraded (Baker and Yanofsky 1968[b], Contesse, Crépin, and Gros 1970, Morse and Yanofsky 1969[b]). This selective degradation cannot be attributed to the known activity of RNase V.[17]

A second feature implicit in the original concept of transcription of a polycistronic mRNA from an operon is that each cistron is transcribed in sequence from the promoter, or starter, element under the control of the operator gene. Thus, under steady-state conditions, all cistrons of the operon would be represented in the total mRNA comple-

[17] In contrast, the phenomenon of catabolite repression appears to be associated with abortive transcription of mRNA chains which are terminated prematurely (Contesse et al. 1970).

Parenthetically, it may be noted that there is no selective degradation of portions of mRNAs containing nonsense codons which are transcribed in *E. coli* from the genomes of mutant T-even bacteriophages (Bautz 1966, Gesteland, Salser, and Bolle 1967). This observation reinforces the conclusion that the infected cells are deficient in "mRNA-degrading" enzymes (see p. 216).

ment (nascent, complete and partly degraded) by the same number of transcripts. However, in at least certain cases, an excess of transcripts of the operator-distal cistrons appears to be synthesized in a process not closely regulated by the operator gene (Bauerle and Margolin 1966, Imamoto and Ito 1968, Morse and Yanofsky 1968). Similarly, deletion, nonsense or frameshift mutations in one structural gene of an operon *may* be associated with unregulated synthesis of enzymes specified by other genes of the operon. Usually, the rate of this unregulated synthesis is considerably less than the maximal rate observed for fully derepressed or induced wild-type operons (Englesberg et al. 1969, Jordan and Saedler 1967, Morse and Yanofsky 1969[b]). These various phenomena all appear to be due to the presence in the operon of a base sequence which can function as an internal or substitute promoter.

FUNCTIONS OF PROMOTERS AND ROLE OF CYCLIC AMP

Direct analyses of the levels of specific mRNAs present in intact cells indicate that the promoter controls the extent of transcription from an operon (Contesse, Crépin, and Gros 1969, 1970, Varmus, Perlman, and Pastan 1970[a], 1970[b]). Further, both direct and indirect analyses indicate that the action of a cAMP-binding protein complex (de Crombrugghe et al. 1971[b], Emmer et al. 1970, Zubay, Schwartz, and Beckwith 1970) at the promoter region (see Chapters 3 and 4) enhances the frequency of initiation of new mRNA chains (Jacquet and Kepes 1969, Perlman, de Crombrugghe, and Pastan 1969, Perlman and Pastan 1968). However, the underlying mechanisms of these effects are not known. The fact that cAMP stimulates β-galactosidase formation in constitutive mutant strains (Perlman and Pastan 1968, Ullmann and Monod 1968) indicates that the nucleotide-protein complex does not act primarily on the interactions between repressor, operator, and inducer molecules. Augmentation by cAMP of the amount of enzyme formed in the translation of a fixed number of mRNA chains (Aboud and Burger 1970, Pastan and Perlman 1969[a], Yudkin 1969) indicates an effect upon a reaction involving the ribosomes. Indeed, the data obtained in studies with intact cells appear compatible with the hypothesis that cAMP facilitates association of ribosomes with an untranslated segment of the mRNA corresponding to the promoter (see p. 124), and thereby stimulates coupled transcription of the operon (see p. 207).

Analysis of the roles of the promoter element and cAMP-binding protein complex in cell-free systems is currently in the preliminary stages. Results obtained with coupled systems in which the *in vitro* synthesis of β-galactosidase enzyme is dependent upon concurrent transcription of mRNA from the *lac* operon of *E. coli* parallel those obtained with intact cells. Induced synthesis of the enzyme is markedly stimulated by cAMP, and the quantity of mRNA corresponding to the *lac* operon which can be recovered is also increased (Chambers and Zubay 1969, de Crombrugghe et al. 1970, 1971[b], Zubay, Schwartz, and Beckwith 1970). Conversely, even in the presence of cAMP, there is little formation of β-galactosidase or transcription from the *lac* operon when the components of the system are derived from a mutant strain which is believed to synthesize a defective binding protein (Emmer et al. 1970, Zubay, Schwartz, and Beckwith 1970).

On the other hand, analyses of the roles of promoter element and cAMP-binding protein complex upon the transcription of RNA from the *lac* operon by highly purified preparations of RNA polymerase enzyme reveal that a number of additional components may participate in regulating the process. In most studies the extent of transcription from the operon has been found to be almost totally unregulated by either the specific repressor of the operon or by the cAMP-binding protein complex (Eron et al. 1970, Ohshima et al. 1970, Pastan and Perlman 1970).[18] Indeed, the only clear

[18] In contrast, inhibition by purified λ-repressor of the transcription of mRNAs from genes **N** and *tof* of the λ-phage genome by purified *E. coli* RNA-polymerase requires no additional components (Steinberg and Ptashne 1971).

effect of the latter was an increase in the proportion of transcripts synthesized on the correct coding strand (de Crombrugghe et al. 1971[b], Eron et al. 1970). It is not known whether this reflects an interaction with the enzyme in which the cAMP-binding protein functions as an accessory transcription factor (see p. 225), or some interaction with the nascent mRNA strands in a partial reaction of the processes which normally couple transcription to translation. At least one component of the ribosomes which is removed by extracting the initiation factors for peptide chain synthesis is required for regulation of transcription from the *lac* operon (de Crombrugghe et al. 1971[b], Ohshima et al. 1970). Moreover, the stimulation of transcription by cAMP-binding protein complex was further enhanced in the presence of 5′-diphosphoguanosine-2′,3′-diphosphate (de Crombrugghe et al. 1971[b]). However, de Crombrugghe et al. (1971[a]) have very recently reported regulation of the *in vitro* transcription of the *lac* operon in the absence of ribosomal protein, and largely unaffected by the guanosine tetraphosphate. At present it is unclear what factors may account for the contrasting results obtained by this group of workers.

Thus, at the very superficial level at which comparisons can be made at this time, the data obtained from experiments with cell-free systems appear largely in accord with conclusions made from studies with intact cells.

Effects of cAMP on the processes of translation due to formation of a complex with the elongation factor **G** and inhibitions of RNase V and GTPase activity are probably independent of the effects exerted at the promoter region of the operon (Kuwano and Schlessinger 1970).

SITES OF INITIATION AND TERMINATION OF TRANSCRIPTION

As discussed earlier in the chapter, the transcription process is asymmetric, and the course of viral development is characterized by the sequential transcription of different families of mRNA molecules. Considerable effort has been devoted to analysis of the factors determining the coding strand of the template DNA and the points of initiation of transcription.

Analysis of the course of chain elongation, both *in vivo* and *in vitro,* shows that transcription is initiated from the 5′-terminus and continues by attachment of incoming nucleotide to the 3′-terminal of the growing chain (Bremer et al. 1965, Goldstein, Kirschbaum, and Roman 1965, Maitra and Hurwitz 1965). Initiation commences from a pyrimidine residue of the coding DNA strand, for the 5′-terminal base in the transcribed RNA is (virtually) invariably a purine residue (Bremer et al. 1965, Jorgensen, Buch, and Nierlich 1969, Sugiura, Okamoto, and Takanami 1969). Indeed, all transcription of ribosomal RNAs and almost all the transcription of transfer RNAs are from the pyrimidine-rich heavy DNA strand in, at least some, bacteria (Margulies, Remeza, and Rudner 1970) and also eucaryote mitochondria (Aloni and Attardi 1971[a], 1971[b], Nass and Buck 1970). Similarly, as considered earlier in the chapter, the bulk of the mRNA synthesized during development of bacteriophage λ is transcribed from the heavy DNA strand. Moreover, a more general correlation has been demonstrated between the number of clusters of pyrimidine residues in a strand of viral DNA and the extent to which the strand serves as template for transcription of mRNA during development of that virus (Kubinski, Opara-Kubinska, and Szybalski 1966, Mushynski and Spencer 1970[a], 1970[b], Szybalski et al. 1969). Thus one of the factors involved in selection of the coding strand and initiation site of transcription appears to be recognition of specific clusters of pyrimidine residues.

Features which define points of termination of transcription have not yet been analyzed.[19]

[19] The observation that almost all of the 3′-terminal residues of RNAs transcribed *in vitro* from the genomes of phages T4 and T7 are uridine residues (Maitra 1970) raises the possibility that clusters of purine residues may define points of termination of transcription.

RNA POLYMERASE AND SIGMA, PSI, AND RHO FACTORS

We believe that all species of RNAs present in a bacterial cell are transcribed by a single RNA polymerase enzyme (see Geiduschek and Haselkorn 1969, Richardson 1969).

The enzyme of *E. coli* has been highly purified. It is a large, complex molecule which contains a series of different types of polypeptide subunit. The subunit composition of the *holoenzyme* of normal cells is depicted as $\alpha_2\beta\beta'\sigma(\omega_{1\text{-}2})$, with corresponding molecular weights ($\pm$10 percent) of 41,000, 145,000, 150,000, 86,000, and about 12,000, respectively (Berg and Chamberlin 1970, Burgess 1969, Travers and Burgess 1969, Zillig et al. 1970[a]).[20] It is not clear whether the small polypeptide ω is a true component of the enzyme or a tightly bound contaminant, for it has no known function in the activity of the enzyme (Burgess 1969).

The holoenzyme can synthesize RNA upon a variety of template DNAs *in vitro,* including calf thymus DNA and the genome of bacteriophage T4. Transcription from bacteriophage DNAs is asymmetric, and the enzyme selectively synthesizes the viral pre-early mRNAs (Bautz, Bautz, and Dunn 1969, Brody and Geiduschek 1970, Sugiura, Okamoto, and Takanami 1970, Summers and Siegel 1969, Travers 1969). However, preparations of high activity with calf thymus DNA recovered from columns of phosphocellulose have very low activity in transcribing RNA from the genome of bacteriophage T4 (Burgess et al. 1969). Moreover, the RNA which is synthesized with viral genomes as the template appears to be initiated at random from both strands of the DNA. These alterations of template and initiation site specificities are due to resolution of the sigma subunit (σ) from the *core enzyme,* or *PC-enzyme,* and reversed by addition of the subunit (Bautz, Bautz, and Dunn 1969, Burgess et al. 1969, Goff and Minkley 1970, Sugiura, Okamoto, and Takanami 1970).

[20] There is some variation in values reported in the literature, usually within the range of $\pm$10 percent. Those given are from Berg and Chamberlin (1970).

It appears probable that residual activities of the core enzyme can be attributed to initiation from localized regions of denaturation which are fortuitously present in the template DNAs (Travers 1971, Travers and Burgess 1969, Vogt 1969). However, the possibility of some limited inherent initiation site specificity in the core enzyme cannot be eliminated (see Goff and Minkley 1970, Matsukage, Murakami, and Kameyama 1969).

The sigma subunit has no inherent activity as an RNA polymerase and has no effect upon either the rate of elongation or the release of "nascent" mRNA chains (Burgess et al. 1969, Darlix et al. 1969). Rather, it appears to act as a cofactor of the core enzyme in stimulating the initiation of transcription from a specific category of promoters (Bautz and Bautz 1970, Travers and Burgess 1969). Further, the initiation-site specificity of the holoenzyme of *E. coli* (Travers 1969, 1970[a]), and also that of *B. subtilis* (Kerjan and Szulmajster 1969, Whiteley and Hemphill 1970), may be modified when the cognate σ subunit is replaced by that from another source. Conversely, the specificity shown by the σ subunit may be altered by replacement of the core enzyme with that from another species of bacterium (compare Bautz, Bautz, and Dunn 1969, Losick, Shorenstein, Sonenshein 1970, Travers 1969, Whiteley and Hemphill 1970). Thus the σ subunit serves as a cofactor of the core enzyme in recognition of the promoters from which initiation of transcription is stimulated.

The mechanism of action of the σ subunit, or factor, is not yet known (see Travers 1971). Both holoenzyme and core enzyme can rapidly form (probably nonspecific) reversible complexes with bacteriophage DNAs (DiMauro et al. 1969, Richardson 1966[a]). However, complexes formed by binding of the holoenzyme at promoter sites are relatively stable (Bautz and Bautz 1970, Hinkle and Chamberlin 1970, Zillig et al. 1970[b]). Nevertheless, σ factor alone does not bind to bacteriophage DNAs (Darlix et al. 1969). Indeed, the σ factor is released from the enzyme-DNA complex shortly after initiation of RNA synthesis *in vitro* (Dunn and Bautz 1969, Krakow, Daly, and Karstadt 1969, Travers and Burgess 1969), and probably also *in vitro* (Khesin 1970, Pettijohn, Stonington, and Kossman 1970). The precise stage of the processes of initiation of RNA synthesis at which the factor is released is, however, not yet established. Present evidence is in accord with the suggestion that the σ subunit dissociates from the holoenzyme immediately after formation of an enzyme-promoter complex, with an attendant change in conformation of the latter (Travers 1971).

In contrast to the genomes of bacteriophages which can develop in *E. coli,* the DNA of the host bacterium is a very poor template for *in vitro* transcription of RNA by highly purified preparations of the holoenzyme of *E. coli.* Indeed, no synthesis of ribosomal RNAs can be demonstrated, even though some 40 percent of the RNA transcribed *in vivo* is of these species. An additional transcription factor, named *psi* (ψ_r), is required specifically for initiation of *in vitro* synthesis of ribosomal RNAs, and probably also transfer RNAs (e.g., Primakoff and Berg 1970). In the presence of this factor up to 30 percent of the RNA transcribed from *E. coli* DNA *in vitro* by homologous holoenzyme is rRNA. Unlike the σ subunit, ψ_r factor has no effect upon the core enzyme but rather modifies the promoter specificity of the complete holoenzyme. The mechanism of this effect is not known, but it is believed to be due to association with the enzyme rather than the DNA (Travers, Kamen, and Schleif 1970). Activity of ψ_r is inhibited by the guanosine tetraphosphate 5′·diphosphoguanosine (2′·3′) diphosphate (ppGpp) (Travers, Kamen, and Cashel 1970), which accumulates when certain auxotrophic mutant strains of *E. coli* are depleted of essential amino acids (see Chapter 9). It has therefore been suggested that the bacterial cell may contain a series of psi factors which serve as secondary determinants of promoter specificity, and the activities of which are regulated by allosteric transitions in conformation in response to specific effector metabolites (Travers 1971, Travers, Kamen, and Schleif 1970).

Specifically, it has been postulated that the cAMP-binding protein (CAP) which acts at the promoter region of the *lac* operon (see Chapters 3 and 4) may be a psi factor. However, effects of diphosphoguanosine·diphosphate of the stimulation of transcription by cAMP-binding protein complex (de Crombrugghe et al. 1971[a], [b]) are not in accord with this proposal.

Davison, Pilarski, and Echols (1969) have isolated a further type of transcription factor, named **M**, which stimulates *in vitro* transcription by both core enzyme and holoenzyme. This factor is markedly more active with viral DNAs as template than bacterial DNA and is probably a ribosomal protein (Davison, Pilarski, and Echols 1969, Echols, Pilarski, and Cheng 1968). The mechanism of action of the factor is not known, and it seems possible that factor **M** is identical with the ribosomal protein required for regulation of transcription from the *lac* operon of *E. coli* (Ohshima et al. 1970).

Thus, at the present time, it remains possible that requirement for a psi factor is a feature unique to the transcription of the cistrons for rRNAs, which are disproportionately transcribed *in vivo.* Conversely, regulation of the transcription of other cistrons of the bacterial genome may be through coupling with translation in a reaction involving specific ribosomal initiation factors, or by *labile* modifications of the σ subunit.[21]

Termination of transcription of the cistrons for ribosomal RNAs appears to require no accessory factors (Pettijohn, Stonington, and Kossman 1970). However, normal termi-

[21] Phosphorylation of the σ subunit by protein kinase enzyme markedly enhances the stimulation of transcription by core enzyme (Martelo et al. 1970). It is not yet known whether this reaction is of physiological significance. cAMP-binding protein appears not to be a protein kinase (Pastan and Perlman 1970). However, a protein kinase is a component of the ribosomal proteins of at least one type of cell (Kabat 1971).

nation and release of completed mRNA chains from the RNA polymerase-DNA complex require a specific termination factor, designated *rho*, ρ (Roberts 1969[b]). In the absence of factor ρ, purified preparations of RNA polymerase appear to terminate transcription of mRNAs at random, yielding a product both highly heterogeneous in size and bound to the enzyme (Bremer and Konrad 1964, Goff and Minkley 1970, Richardson 1966[b], Roberts 1969[b]).[22] Addition of ribosomes stimulates transcription and causes some release of mRNA from the template DNA. However, the process is associated with formation of larger species of mRNA, without attendant reduction of the heterogeneity in size (Shin and Moldave 1966). Factor ρ causes termination of transcription at defined positions in the template DNA, thereby reducing the heterogeneity and average size of the RNA, and release of the transcript RNA from the enzyme (Davis and Hyman 1970, Goff and Minkley 1970, Roberts 1969[b]). The mechanism of action remains to be determined.

Factor ρ has no direct effect on the processes of initiation of transcription. However, it may have an indirect effect, in the sense that failure to terminate synthesis at a defined "termination site" will permit continued transcription into an adjacent segment of the genome. As considered earlier in the chapter, such unterminated transcription appears to be responsible for initiation of synthesis of bacteriophage early mRNAs. This does not appear to reflect failure of the normal termination process. It appears, rather, to be specifically regulated, for there is some indirect evidence that the gene product of bacteriophage λ gene **N** interacts with one of the RNA polymerase subunits (Pironio and Ghysen 1970).

[22] Chain termination at specific sites, with release of the transcribed RNA, can occur under conditions of very high ionic strength. However, even in these cases, the RNAs formed are usually considerably larger than any species transcribed *in vivo* (e.g., Richardson 1970).

REGULATION OF SEQUENTIAL GENE TRANSCRIPTION BY MODIFICATION OF RNA POLYMERASE CONSTITUTION AND OF TRANSCRIPTION FACTORS

Until recently the sequential transcription of different families of mRNAs during viral development was widely believed to reflect a series of cycles of derepression and repression. Indeed, the extensively studied λ-phage repressor (see Chapter 4) provided precedent, and the report of a virus-coded repressor of pre-early mRNA synthesis in cells of *B. subtilis* infected with phage SPO1 (Wilson and Geiduschek 1969) provided confirmation. However, the demonstrations of selective transcription of phage pre-early mRNAs from highly purified viral DNAs *in vitro* rendered invalid interpretations based solely on the phenomenon of repression.

The observation that in cells of *E. coli* infected with bacteriophage T4 the RNA polymerase activity falls dramatically and then rises again (Skold and Buchanan 1964) raised the possibility that sequential gene transcription was regulated through replacement of the host enzyme by a new "phage-coded" species. Recent studies indicate that this is, in fact, the probable regulatory mechanism in cells infected with the different bacteriophage T7. It has been established that the viral pre-early gene **1** "codes" for a distinct new species of RNA polymerase, which consists of a single polypeptide chain (Chamberlin, McGrath, and Waskell 1970, Gelfand and Hayashi 1970, Summers and Siegel 1970). The mechanism by which transcription of the pre-early genes is arrested has not yet been established, but it is probably that of inactivation of the bacterial RNA polymerase enzyme through loss of the σ subunit.

In contrast, sequential transcription from the genome of bacteriophage T4 is not regulated by synthesis of an entirely new species of RNA polymerase enzyme. Rather, the promoter specificity of the bacterial enzyme is altered by both modifications of the constituent subunits of the core enzyme and synthesis of new transcription factor(s). Various types of evidence indicate that the pre-existing α, β, and β' subunits of the core enzyme are conserved thoughout the period of viral development (DiMauro et al. 1969, Goff and Weber 1971, Haselkorn, Vogel, and Brown 1969, Stevens 1970). On the other hand, all but the β subunit of the holoenzyme have been shown to be modified at some stage during the course of sequential gene transcription.

The first effect of infection with this virus appears to be rapid inactivation of pre-existing σ factor. Synthesis of new protein is not required, and the process is essentially completed within one minute (see Seifert 1970, Seifert et al. 1969). Thereafter, σ factor of the bacterial holoenzyme cannot be recovered from the infected cells (Bautz, Bautz, and Dunn 1969, Seifert et al. 1969). In turn, this is rapidly followed by modification of the α subunit of the core enzyme through attachment of a 5′-adenylate residue, in a reaction which is half completed within 2 minutes of infection (Goff and Weber 1970, Seifert et al. 1969, Walter, Seifert, and Zillig 1968). This process does require synthesis of a new virus-coded protein. The ω peptide of the host holoenzyme is also replaced by a newly synthesized species (Stevens 1970), which is presumably coded by a viral pre-early gene.

These modifications lower the affinity of the core enzyme for σ factor (Travers 1969) but have not yet been shown to have any effect upon the course of gene transcription. Accurate and efficient transcription of RNAs from the viral genome remains dependent upon the presence of a σ factor (Bautz, Bautz, and Dunn 1969, Travers 1969, 1970[a]). Thus it seems probable that all pre-early and early mRNAs transcribed in the first few minutes after infection are initiated within the first minute. Further, the promoter specificities of holoenzymes formed with different σ factors are indistinguishable from those of corresponding holoenzymes containing the unmodified core enzyme (Travers 1969, 1970[a]). It seems possible that the modifications serve to modulate the action of the termination factor ρ and permit unterminated transcription from the pre-early genes through the early genes, but this remains to be determined.

Formation of a new species of transcription factor commences within 5 minutes after infection. This factor specifies transcription of the viral early mRNAs by either the host or modified core enzyme (Travers 1969, 1970[a]).[23] Properties of the factor (Seifert 1970, Travers 1969) indicate that it is a highly labile species of σ subunit (σ^{T4}) coded by a pre-early gene, although the possibility that it is a complex of host σ unit with a new species of ρ factor cannot be entirely eliminated (Travers 1971).

Transcription of the viral late mRNAs also appears to be selectively stimulated by a specific factor (Snyder and Geiduschek 1968, Travers 1970[b]). Data presently available indicate that it may be a further species of σ subunit, but this remains to be determined. Additional modifications of the enzyme β and β' subunits also occur at about the time of onset of transcription of the late mRNAs (Khesin 1970, Travers 1970[b], Zillig et al. 1970[b]). The significance of this modification is currently unknown.

Studies of the RNA polymerases of other virus-infected bacterial cells have been far less extensive. Present evidence indicates that changes similar to those observed for the bacteriophage T4 also occur during development of bacteriophage λ (Gariglio, Roodman, and Green 1969, Gros et al. 1970, Naono and Tokuyama 1970) and the virus SPO1 of *B. subtilis* (Geiduschek and Sklar 1969).

The process of sporulation in *B. subtilis* also appears to be associated with loss of the σ subunit, and cleavage of the β' subunit, of the RNA polymerase present in vegetative cells (Kerjan and Szulmajster 1969, Losick, Shorenstein, and Sonenshein 1970).

[23] Present data are inadequate to establish that the change in selection of families of mRNAs transcribed is actually due to prior synthesis of T4-sigma factor(s). The first transition occurs before new T4-sigma factor can be demonstrated to be present by techniques currently available.

DICHOTOMY OF CONTROL MECHANISMS

Five different control mechanisms have already been demonstrated to regulate transcription of mRNAs required for synthesis of specific proteins in procaryote cells.

1. Control of the interaction between repressor protein and operator gene by inducer or co-repressor.
2. Control of initiation of transcription at the promoter region of the operon by the action of cAMP–binding protein complex.
3. Control of selection of promoter elements from which transcription is initiated by modification of the σ factor, and other subunit, complement of the RNA polymerase enzyme.
4. Control of the length of the segment of genome transcribed by "unterminated transcription," possibly through modification of subunits of the RNA polymerase enzyme.
5. Control of transcription dependent upon replication of a viral genome.

Of these five, only the first two have been convincingly demonstrated to regulate the activities of normal (including lysogenized) vegetative cells. Conversely, only the last three have been convincingly demonstrated in the course of a developmental process which has been irrevocably initiated. It remains to be determined whether this reflects a real difference between phenomena of transitory, reversible regulatory processes and those of sequential, irreversible developmental processes.

Shall we, however, say that this is the universal process of nature for all animals and for men?—L. Spallanzani, 1785.

. . . for an unwearied and laborious patience has discovered the intermediate degrees by which the situation, figure and symmetry, are insensibly reformed. —Albrecht von Haller, 1764.

CHAPTER 6: GENE TRANSCRIPTION AS A FACTOR IN REGULATORY MECHANISMS OF METAZOANS

EARLY STUDIES OF SIMILARITIES BETWEEN MECHANISMS OF GENE EXPRESSION IN METAZOANS AND IN BACTERIA

The elegant and detailed concepts of the operon as a functionally integrated gene complex and of messenger RNAs (mRNAs) as mediators of gene expression (Jacob and Monod 1961[a]) were originally formulated to provide an interpretation of data obtained primarily from extensive studies of the *lac* operon of *E. coli* (Chapters 3 and 5). It was clear that additional proposals would be required to take account of the complexity of the organized chromosomes and discrete nuclei of metazoans. Further, as formulated, certain of the properties attributed to the postulated mRNAs were not in accord with observations made with metazoan cells, as, for example, the continued synthesis of hemoglobin in maturing reticulocytes after loss of the cell nucleus. Nevertheless, it was immediately apparent that, with appropriate modification in detail, the basic concepts of operons and of messenger RNAs could provide a unified interpretation of a vast body of hitherto only tenuously related data obtained with metazoan systems.

Prior studies, both *in vivo* and *in vitro,* had shown the most intense site of protein synthesis to be the *microsome* fraction, a complex of membrane-bound ribosomes with fragments of membrane of the endoplasmic reticulum. Relatively low rates of either synthesis or accumulation of new protein were observed with the cell nuclei (Allfrey, Daly, and Mirsky 1953, Keller, Zamecnik, and Loftfield 1954, Siekevitz 1952, Zamecnik and Keller 1954). Further, both completion of nascent hemoglobin chains and limited *de novo* synthesis of entire hemoglobin chains *in vitro* had already been demonstrated with cell-free systems derived from reticulocytes (Bishop, Leahy, and Schweet 1970, Kruh et al. 1961, Lamfrom 1961, Schweet, Lamfrom, and Allen 1958).

On the other hand, the cell nuclei were known to be the location of the bulk of the genome. They had been shown to be the major site of RNA synthesis, and migration of RNA synthesized in the nucleus out into the cell cytoplasm had been repeatedly demonstrated (Goldstein and Micou 1959, Goldstein and Plaut 1955, Ts'o and Sato 1959, Woods 1959, Zalokar 1959). Further, two distinct classes of nuclear RNA had been distinguished on the basis of physicochemical properties, base composition, intranuclear location and kinetics of incorporation of radioactive precursor (Allfrey and Mirsky 1957, Logan 1957, Sibatani et al. 1962, Vincent 1957). Indeed, repeated attempts had been made to identify one of the nuclear RNAs as precursor of the cytoplasmic RNA. Meanwhile, autoradiographic analyses of the course of RNA synthesis had indicated the nucleolus to be the source of most (if not all) cytoplasmic RNA and had provided suggestive evidence that RNA formed in association with chromatin was exported to the cytoplasm via the nucleolus (McMaster-Kaye and Taylor 1958, Rho and Bonner 1961, Sirlin and Jacob 1960, Sirlin, Kato, and Jones 1961, Woods 1959). Further, the DNA-dependent RNA polymerase enzyme had been partially purified from rat liver nuclei (Weiss 1960) and from nuclei of other mammalian tissues (Goldberg 1961, Krakow and Canellakis 1961).

These various observations were clearly entirely in accord with the concept of mRNAs transcribed on nuclear genes and transported, via the nucleolus, to the ribosomes of the endoplasmic reticulum. This immediate apparent accord between experimental observation and theoretical requirement stimulated the deliberate search for more conclusive evidence of a role of the postulated mRNAs in metazoan cells. Initial studies were necessarily restricted to formal demonstrations that the most rapidly synthesized fraction of nuclear RNA was that associated with the chromatin and DNA-like, or complementary (cRNA), in base composition (Allfrey and Mirsky 1962, Woods 1959). However, with development of the technique of DNA-RNA hybridization it was formally demonstrated that cytoplasmic and nuclear RNAs contained at least some common base sequences (Hoyer, McCarthy, and Bolton 1963). Other studies led to the further demonstrations that: (a) nuclear RNA contained material active as a template for *in vitro* incorporation of amino acid into polypeptide (Barondes, Dingman, and Sporn 1962, Brawerman, Gold, and Eisenstadt 1963, Kenney and Kull 1963); (b) isolated chromatin could serve as an *in vitro* template for synthesis of an RNA which would stimulate *in vitro* incorporation of amino acid into a defined protein (Bonner, Huang, and Gilden 1963, Bonner, Huang, and Maheshwari 1961); and (c) that protein was synthesized on polyribosome structures containing a "thread" of single-stranded RNA of the correct size to serve as a specific mRNA (Gierer 1963, Warner, Rich, and Hall 1962, Wettstein, Staehelin, and Noll 1963). Thus, at this juncture, a process of transcription of cytoplasmic mRNAs upon the nuclear chromatin entirely analogous to that postulated in the case of bacteria appeared to be largely established.

Technical limitations and other restrictions had precluded rigorous genetic analysis of metazoan systems in as fine detail as had been possible with bacteria. In general, the majority of the mRNAs isolated from metazoan polysomes are too small to be deemed polycistronic mRNAs (e.g., Howell, Loeb, and Tomkins 1964, Penman et al. 1963, Warner, Knopf, and Rich 1963). However, there was some suggestive evidence of the possible existence of operons in the fruit fly *Drosophila melanogaster* (Lewis 1951, 1963, Sturtevant 1925), and the possibility of a post-transcriptional scission of polycistronic mRNAs cannot yet be formally eliminated (e.g., Scherrer and Marcaud 1968). Further, various schema of interdependent operons could be formulated (Jacob and Monod 1963, Monod and Jacob 1961) to account for the sequential program of differential gene activity characteristic of embryonic development (Chapter 2).

Despite the lack of rigorous genetic evidence of mechanisms regulating protein synthesis at the level of gene transcription, there was a compensatory massive body of cytological and biochemical data. In contrast to the situation in procaryotes, the chromosomes of eucaryotes typically contain substantial quantities of histone, or equivalent basic protein. This histone is not distributed at random throughout the chromosome. Rather, it is precisely organized to delineate specific sites or sections of the chromosome as regions of *heterochromatin* and *euchromatin,* respectively, readily distinguished from each other by conventional staining procedures. The sequential program of differential activation of different portions of the genome which is characteristic of the course of embryonic development is paralleled by progressive changes in the distribution of regions of heterochromatin throughout the chromosomes (see Chapter 2). Elucidation of the regulatory mechanism controlling these two coordinated sequences of events has been slow, and even now is far from complete. Nevertheless, a role of the histone component of heterochromatin as an inhibitor (repressor) of gene transcription was a "latent" proposal implicit in the suggestion of Stedman and Stedman (1950) that histones and protamines served to regulate gene function. Formulation of precise concepts of the role of messenger RNAs provided the spur necessary to convert the Stedmans' latent proposal into an explicit working hypothesis to be examined by a variety of approaches. Development of *in vitro* systems for the assay of DNA-dependent RNA polymerase activity permitted the demonstration that isolated chromatin would serve as a template for the synthesis of RNA, whereas the nucleohistone fraction of native chromatin or reconstituted complexes of DNA with histone were virtually inactive as templates (e.g., Barr and Butler 1963, Bonner and Huang 1963, Bonner, Huang, and Maheshwari 1961). This, in turn, was followed by the formal demonstration that both heterochromatin regions of the chromosomes *in situ* and isolated preparations of "compact" chromatin are relatively poor templates for transcription of mRNA, whereas euchromatin and isolated "diffuse" chromatin contain sites of intense RNA synthesis (e.g., Frenster, Allfrey, and Mirsky 1963, Fujita and Takamoto 1963, Kroeger 1963, Littau et al. 1964).[1]

Additional evidence of the existence in metazoan cells of regulatory mechanisms probably acting at the level of gene transcription was also afforded by early studies of the effects of hormones upon the synthesis of RNA. Thus, for example, an increased rate of incorporation of precursor into RNA is one of the early responses to injection of 17 β-estradiol into ovariectomized rats, and this response precedes the hormone-induced increase rate of incorporation of amino acid into protein (e.g., Hamilton 1963, 1964). Similarly, administration of hydrocortisone to adrenalectomized rats stimulated incorporation of precursor into the RNA of liver nuclei (Kenney and Kull 1963). The probability that the action of hormones was at the level of increased gene transcription, rather than decreased degradation of mRNA, was

[1] In most studies identity of isolated "compact" chromatin with heterochromatin, and of "diffuse" chromatin with euchromatin, has not been established. However, the "constitutive heterochromatin" and "euchromatin" preparations isolated by Yasmineh and Yunis (1971) appear to correspond to the fractions defined by cytological procedures.

enhanced by observations made of the Balbiani rings and puffs of the chromosomes of dipteran salivary glands. Formation of certain of these structures can be induced specifically in response to the hormone ecdysone, and they are sites of intense RNA synthesis. Both induction of the morphological change in structure of the chromosome and gene expression in the synthesis of RNA within the structure are inhibited by the antibiotic actinomycin (Clever 1964, Clever and Rombull 1966).

Thus, within three years of publication of the detailed concepts of the nature of mRNAs and their roles as mediators of gene expression in procaryotes (Jacob and Monod 1961[a]), the presence of similar mRNAs, with largely similar properties, also appeared to be essentially established in eucaryotes. The volume of critical evidence was certainly less for eucaryotes than for procaryotes, and there were obvious differences in detail, notably in the apparently greater stability of eucaryote mRNAs *in vivo* and in the regulation of transcription by histones, hormones, and embryonic inducers in metazoan cells.

At this juncture, the available evidence seemed to indicate that the fundamental processes of gene expression, and regulation thereof, were basically the same in eucaryotes and procaryotes. It was obvious that lack of genetic information about the eucaryote systems in the fine detail available for many specific bacterial operons and bacteriophages, coupled with inherent technical difficulties, precluded certain types of analysis of the nature and functions of mRNA in eucaryotes which were possible with bacterial systems. However, the apparent basic similarities between eucaryote and procaryote systems seemed to warrant the assumption of identical basic processes. Thus the stage was set for analysis of mechanisms regulating gene transcription in eucaryote cells and in mixed cell populations organized in defined tissues.

Actually, with the wisdom of hindsight, we can now recognize that there already was a diverse assortment of experimental data available to indicate that the basic processes of gene expression are considerably more complex in eucaryotes than in bacteria. However, in the first flush of enthusiasm these were overlooked and had no immediate impact on the course of subsequent developments. Therefore, for convenience, further reference to these observations will be deferred to appropriate later sections of this chapter.

DISCRIMINATION BETWEEN COMPLEMENTARY RNAs AND MESSENGER RNAs

As in studies with procaryotes, various experimental criteria have been used to make provisional identification either of the presence of an mRNA in preparations of RNA isolated from metazoan cells or of a regulatory mechanism involving transcription of new species of mRNA. Unfortunately, application of certain of these criteria to eucaryote cells presents technical problems not encountered with bacterial cells. Even isolation of the bulk of the rapidly-labeled RNA fraction of eucaryote cells in an undegraded condition requires special procedures (e.g., Arion, Mantieva, and Georgiev 1967, Attardi et al. 1966, Georgiev et al. 1963, Gross, Malkin, and Hubbard 1965, Joel and Hagerman 1969, Kidson and Kirby 1964) and may well not have been achieved in many of the reported studies. Identification of mRNAs and quantitative assay of mRNA levels in eucaryote cells is, therefore, technically more difficult than in the case of bacteria.

From the outset, tentative identification as mRNAs of materials isolated from eucaryote cells, or subfractions, on the basis of overall nucleotide composition required a very broad interpretation of the term "DNA-like." In general, the actual degree of correspondence between the base composition of pulse-labeled RNA isolated after a brief exposure to radioactive precursor and the base composition of the corresponding DNA was so slight as to be of itself quite unconvincing. Rather, the material differed substantially from both ribosomal and transfer RNAs in being more DNA-like, in containing higher molar proportions of adenine and uracil, and (at a later date) in the absence of minor bases. This discrepancy did not initially present any serious obstacles to provisional identification of such pulse-labeled RNAs as mRNAs. On the one hand, a vast body of data accumulated over several decades required the conclusion that only a very small fraction of the genome of a eucaryote cell was actively expressed at any given stage in the life cycle of any particular cell, or development of any given tissue. Of that fraction only one coding strand could be expressed in the formation of specific gene product.[2] On the other hand, at the time we knew only of mRNAs and the two classes of ribosomal and transfer RNAs, from which the pulse-labeled RNA clearly differed. Nevertheless, it was generally recognized that actual identification of pulse-labeled RNA with messenger RNA on the basis of a difference in base composition from those of metabolically stable ribosome and transfer RNAs was not warranted. Accordingly, a tacit agreement gradually evolved to define materials recognized by this criterion as DNA-like (dRNAs) or more recently complementary RNAs (cRNAs), and that additional criteria of mRNA *function* are mandatory for the identification of an mRNA. Subsequent developments to be discussed below have more than justified the cautious insistence upon this explicit restraint in not making premature identification of pulse-labeled RNAs as mRNAs.

[2] The total absence of relevant data did permit some early consideration of the possibility that in eucaryotes one strand of DNA served as coding strand for the defined gene product, and the other as coding strand for intranuclear synthesis of corresponding histone. The possibility of the second strand serving directly as template for intranuclear protein synthesis has been proposed (Naora 1968), but would probably require identical species of proteins to be synthesized in both nucleus and cytoplasm. Present evidence is not compatible with concurrent use of both strands of a segment of DNA for transcription of *functional mRNAs*. The possibility of transcription of both strands of a nuclear gene seems highly unlikely but is not formally eliminated.

Development of procedures for forming DNA-RNA hybrids (Chapter 5) permitted the formal demonstration that such RNAs are indeed DNA-like in base sequence and can hybridize with selected portions of the DNA of the genome (e.g., Denis 1966[a], 1966[b], McCarthy and Hoyer 1964). Further, preparations of DNA-like RNA isolated from cell nuclei showed some *in vitro* activity as templates for incorporation of amino acid into polypeptide (Brawerman, Gold, and Eisenstadt 1963, DiGirolamo, Henshaw, and Hiatt 1964, Kenney and Kull 1963). Conversely, mRNAs were isolated from cytoplasmic polysomes and shown to have template activity (DiGirolamo, Henshaw, and Hiatt 1964, Henshaw, Revel, and Hiatt 1965, Munro, Jackson, and Korner 1964, Staehelin et al. 1964). Indeed, cRNA has been shown to form polysomes containing only pre-existing ribosomes in the anucleolate mutant line of *Xenopus laevis* (Gurdon and Ford 1967). Thus there was ample basis for concluding that at least a portion of the cRNA fraction of eucaryote cells is indeed messenger RNA.

Nevertheless, the practice of defining such preparations as cRNAs is maintained. A firm operational distinction is made between:

(a) the synthesis and function of cRNA as a measure of gene expression, and
(b) the synthesis or template function of mRNAs as defined by one or both of two other criteria.

One criterion used to make the latter definition is that a preparation of cRNA be isolated from cytoplasmic polysomes (e.g., Kedes and Gross 1969[a], 1969[b], Munro, Jackson, and Korner 1964, Staehelin et al. 1964). The second is that the RNA have demonstrable activity as a template for the *in vitro* incorporation of amino acid into polypeptide (e.g., Barondes, Dingman, and Sporn 1962, Brawerman, Gold, and Eisenstadt 1963, Brentani and Brentani 1969) or a defined protein (e.g., Heywood and Nwagwu 1968, Laycock and Hunt 1969, Lockard and Lingrel 1969). In some instances both of these two criteria have been satisfied (e.g., DiGirolamo, Henshaw, and Hiatt 1964, Henshaw, Revel, and Hiatt 1965). Indeed in the specific case of the globin chains of rabbit hemoglobin, the mRNA has been shown to be present in polysomes actively engaged in their synthesis (Gierer 1963, Warner, Knopf, and Rich 1963), to be of approximately the correct length in electron micrographs to be a template for one globin chain (Scherrer and Marcaud 1968, Warner, Rich, and Hall 1962),[3] to be pulse-labeled (Marks et al. 1962, Scherrer and Marcaud 1968), and to serve as template for the *in vitro* synthesis of rabbit globin chains by a system derived from *E. coli* (Laycock and Hunt 1969).

Thus the conceptual difference between cRNAs and mRNAs is that the latter are those cRNAs which have demonstrable additional properties of messenger RNAs, whereas cRNAs in general may contain species of RNA with other physiological functions. This is, in fact, a fundamental distinction of very considerable importance, to which I shall return in consideration of our available knowledge of the process and regulation of gene transcription of cRNAs (see Chapter 8).

At the present time we lack both the fundamental knowledge and necessary experimental procedures to study the transcription in eucaryotes of either specific mRNAs, or of cRNAs in general, by the techniques which have been so successfully exploited with procaryote cells (Chapter 5) and for the ribosomal RNAs (rRNAs) of eucaryotes (Chapter 8). Some initial success has been achieved in analysis of the course of transcription of the genome of virulent viruses (Becker and Joklik 1964, Hudson, Goldstein, and Weil 1970, Kates and McAuslan 1967, Oda and Joklik 1967, Wagner

[3] A recent analysis by Lim and Canellakis (1970) has indicated that this species of mRNA may contain an additional untranslatable terminal base sequence which is very rich in adenine residues, and may correspond to a segment of "tickets" (see Chapter 5).

and Roizman 1969) and of the viral genome incorporated into "transformed cells" (e.g., Fujinaga and Green 1970, Fujinaga, Pina, and Green 1969). It has also been shown that almost all of the RNA synthesized in rat liver mitochondria is transcribed from one DNA strand (Aaij et al. 1970, Aloni and Attardi 1971[a], Nass and Buck 1970). Some study has also been made of the transcription of so-called satellite DNAs (e.g., Harel et al. 1968). It is technically possible to correlate fractions of a cRNA population from mouse cells with subsets of the mouse chromosomes (Maio and Schildkraut 1969), and transcription in the testis of *Drosophila hydei* of a cRNA specific to the Y chromosome has been demonstrated (Hennig 1968). However, in all these cases the degree of resolution of the cRNAs which can be achieved is considerably less than that required to identify the transcripts of individual genes.

Fortunately, a useful indirect method of demonstrating the presence in eucaryote cells of specific mRNAs is often applicable. This procedure is based upon the ability of actinomycin D or other antibiotics to complex with DNA and inhibit the synthesis of all species of RNA (Franklin 1963, Goldberg, Rabinowitz, and Reich 1962, Shatkin 1962). This antibiotic also has a number of additional cytotoxic and other "side effects," including enhancement of the degradation of pre-existing polysomes (e.g., Acs, Reich, and Valanja 1963, Armentrout, Schmickel, and Simmons 1965, Honig and Rabinovitz 1966, Revel, Hiatt, and Revel 1964), and considerable caution must be exercised in the interpretation of observed effects of the antibiotic.[4] However, the synthesis of a defined specific protein by cells in which the transcription of all species of cRNA has been demonstrably blocked

[4] In general, these effects appear to be responses to higher concentrations of actinomycin than those required to inhibit the process of transcription, particularly in the case of precursors of ribosomal RNAs (e.g., Perry and Kelley 1970). Nevertheless, concentrations of the antibiotic sufficient to inhibit all transcription of RNA may have important "side effects" (e.g., Kiefer, Entelis, and Infante 1969, see page 249).

constitutes definitive proof of the presence in the cell of the corresponding mRNA at the time of total arrest of transcription. Application of this indirect type of probe for the presence of a functional mRNA coding for the synthesis of a defined protein has proved very useful in the preliminary analysis of mechanisms of hormone action (e.g., Hamilton 1964, Tomkins et al. 1965) or mechanisms regulating the course of embryonic development (e.g., Brachet and Denis 1963, Gross, Malkin, and Moyer 1964), and particularly valuable in cases where the supply of experimental material is extremely limiting (e.g., Wainwright and Wainwright 1966, 1967[c], Wilt 1965[a], Chapter 9).

In the following pages the term mRNA will be reserved for materials isolated from polyribosomes or shown to be present by the formation of corresponding protein under conditions where transcription of new cRNAs is suppressed. The term cRNA will be used to describe all other species of DNA-like RNAs, even where there has been an experimental demonstration of the presence of material serving as template for *in vitro* incorporation of amino acid into protein. This restriction is required by the observation that unmethylated precursors of ribosomal RNA can serve as such templates *in vitro* (Nakada 1965) and also avoids any implication that such preparations consist solely of mRNA.

SEQUENTIAL GENE TRANSCRIPTION DURING OOGENESIS AND EMBRYONIC DEVELOPMENT IN XENOPUS LAEVIS

At the present time we know virtually nothing about mechanisms controlling selection of the coding strand of gene DNA in eucaryote cells. However, inability to perform sophisticated analyses of the type applied with procaryotes has not precluded the demonstration of precisely regulated programs of sequential transcription in the course of embryonic differentiation and other developmental sequences. Undoubtedly, the two most intensively studied experimental materials are the stages of the maturing oocyte through development to the neurula embryo of *Xenopus laevis* and of various species of sea urchin (Davidson 1968, Gross 1967, 1968, Gurdon 1968[a]).

The smallest of the immature oocytes of *Xenopus laevis* contain few ribosomes and little ribosomal RNA (Thomas 1970). Early in the course of maturation the oocyte develops several hundred discrete nucleoli, each containing a reiterated sequence of very many copies of the cistrons specifying ribosomal RNAs (rRNAs).[5] This is the phenomenon of gene amplification (see Chapter 2). The large number of nucleolar copies of these cistrons collectively contains a quantity of DNA three times that of the entire chromosomal genome. Transcription of these cistrons into precursor molecules which are subsequently processed into mature ribosomal RNAs is quantitatively the most prominent feature of the program of gene transcription in the maturing oocyte. Coordinate transcription of these cistrons and those for the 5S rRNAs accounts for more than 98 percent of all gene transcription at the lampbrush stage of the maturing oocyte of *Xenopus laevis* (Brown and Littna 1964[b], Davidson, Allfrey, and Mirsky 1964, Davidson et al. 1966). The quantity of ribosomal RNA synthesized and assembled into a cytoplasmic store of ribosomes during this stage of maturation is sufficient to support normal development to the stage of gastrulation, and very little further synthesis of rRNA occurs until that stage is attained (Brown and Gurdon 1964, Brown and Littna 1964[a], 1966[a], Emerson and Humphreys 1971).

Nevertheless, a small proportion of nonribosomal RNAs is transcribed at this stage of maturation. They include transfer RNAs (tRNAs) and a heterogeneous, or "*heterodis-*

[5] These copies do differ from the corresponding chromosomal cistrons in that approximately 4.5 percent of the bases present in the latter are 5-methyl-cytosine (Dawid, Brown, and Reeder 1970). Absence of the substituent group from the DNA of the supernumerary nucleoli may be an important factor in determining their fate following fertilization of the mature oocyte (see Wallace, Morray, and Langridge 1971).

perse," cRNA fraction which ranges in size from 4S to greater than 30S, and represents some 3 percent of the total genome (Brown and Littna 1964[b], Crippa, Davidson, and Mirsky 1967, Davidson et al. 1966). Some of the transcribed cRNAs are transcripts of reiterated base sequences, but at least some 10^4 unique (i.e., nonreiterated) genes are transcribed (Davidson and Hough 1969, 1971). Approximately two thirds of this maternal cRNA is relatively stable and persists through fertilization and development to the midblastula stage (Crippa, Davidson, and Mirsky 1967, Crippa and Gross 1969, Davidson et al. 1966, Davidson and Hough 1971). Indeed, even under conditions where all further transcription is blocked, the fertilized egg will develop *apparently* normally to the late blastula or early gastrula stage (Brachet and Denis 1963). Most of these maternal mRNAs are, in fact, probably not available for translation in the unfertilized egg, as has been demonstrated for other species (Smith, Ecker, and Subtelny 1966, Spirin 1966), nor can many of them be found in functional polysomes in the early cleavage stages (Crippa and Gross 1969). The proteins specified by the maternal mRNAs have not been identified, although the identity of at least some can be inferred from the fact that they will support cleavage and early development. However, they do not appear to include mRNAs for enzymes of the category termed "housekeeping enzymes," for similar species of cRNA were not found in tissues of high metabolic activity taken from mature individuals of the same species (Crippa, Davidson, and Mirsky 1967) or in postgastrula stage embryos (Denis 1966[b]).

As the oocyte attains maturity the loops of the lampbrush chromosomes disappear, and the chromosomes lose their characteristic structure. At the same time all transcription of RNA ceases (Brown and Dawid 1968). Ovulation occurs in response to a rise in the circulating level of pituitary hormone. The nucleus ruptures, and the chromosomes, which hitherto have been arrested in the prophase stage of the first meiotic division, complete the first meiotic division and enter the metaphase stage of the second meiotic division. During this process the large number of excess nucleoli disappear and the germinal vesicle ruptures. A small quantity of heterodisperse RNA is transcribed both during the course of ovulation and by the unfertilized egg. A portion of this RNA can be recovered from the ribosome fraction (Brown and Littna 1964[a], 1966[a]) or shown to be active *in vitro* as a template for the incorporation of amino acid into protein (Slater and Spiegelman 1966[a]). Maturation of the oocyte is associated with the accumulation of large numbers of mitochondria and yolk platelets. Indeed, approximately two-thirds of the DNA content of the mature oocyte is attributed to the mitochondria plus yolk platelets (Dawid 1965, 1966, Hanocq-Quertier et al. 1968). It has been suggested that the transcription which occurs during ovulation and in the unfertilized egg is of the mitochondrial DNA (Gurdon 1968[a]).

Fertilization of the egg is followed by an immediate increase in the rate of synthesis of RNA. By analogy with results obtained with sea urchins (Chamberlain 1971, Craig 1970, Selvig, Gross, and Hunter 1970), it seems probable that much of this RNA is transcribed from the DNA of the mitochondria (see also Caston 1971). The RNA synthesized is almost exclusively heterodisperse cRNA, and at least some of the smaller species are present in the ribosome fraction of the cleaving embryo (Brown and Littna 1964[a], 1966[a], Gurdon and Woodland 1969).[6] Translation of these mRNAs is not essential for apparently normal development to the stage of the hatching tadpole, for suppressal of their transcription by actinomycin does not preclude development of apparently normal tadpoles (Brachet and Denis 1963).

Nevertheless, results obtained from comparable studies with sea urchin embryos indicate that at least some of the mRNAs transcribed during the early cleavages are, in fact,

[6] It seems probable that a very small proportion of the RNA transcribed is ribosomal RNA (Emerson and Humphreys 1971).

normally translated in protein synthesis. Suppression of the transcription of cRNA by actinomycin has little early effect upon the rate of total protein synthesis and does not prevent apparently normal development of a swimming blastula (Gross and Cousineau 1964, Gross, Malkin, and Moyer 1964). However, as for *Xenopus laevis,* during this period of normal development heterodisperse cRNA is transcribed from both the nuclear and mitochondrial components of the genome (Comb 1965, Craig 1970, Gross, Kraemer, and Malkin 1965, Hartmann and Comb 1969). In addition, rRNA is transcribed at an average rate per genome approximately equal to that found at later stages of development, but accounts for only about 6 percent of the 28S RNA transcribed during this period (Emerson and Humphreys 1971). The cRNAs include mRNAs active as template for the *in vitro* incorporation of amino acid into protein; they may be isolated from the ribosome fraction (Crippa and Gross 1969, Gurdon and Ford 1967, Kedes and Gross 1969[a], Nemer and Infante 1965, Rinaldi and Monroy 1969, Slater and Spiegelman 1968). In particular, the new cRNA fraction contains at least 3 distinct mRNAs found in small polysomes which are actively engaged in the synthesis of "histone" during the course of development to the stage of the blastula.[7] These and other new mRNAs normally found in the polysome fraction at this time are not transcribed in the presence of actinomycin, and the corresponding proteins are not synthesized (Crane and Villee 1971, Kedes and Gross 1969[a], Kedes et al. 1969[a], Nemer and Lindsay 1969). Indeed, despite the apparently normal development of a swimming blastula in the presence of actinomycin, there is a substantial frequency of abnormal mitosis which has been attributed to a deficiency of histone (Kiefer, Entelis, and Infante 1969).

[7] The basic proteins of chromatin isolated from embryos prior to the blastula stage are not the histones characteristic of the tissues of mature animals (e.g., Hnilica and Johnson 1971).

As the *Xenopus laevis* embryo develops from the midblastula to the late blastula stage there is a dramatic decline in the content of "maternal cRNAs" and a very marked rise in the rate *per embryo* of transcription of large cRNAs (Bachvarova et al. 1966, Brown and Gurdon 1966, Brown and Littna 1966[a], Crippa, Davidson, and Mirsky 1967, Davidson, Crippa, and Mirsky 1968). This cRNA is metabolically unstable, and during continued incubation appears to give rise to a small quantity of stable cRNA of smaller size (Brown and Gurdon 1966). These cRNAs represent entirely new species not present in the complement of cRNAs present at early stages of development (Davidson, Crippa, and Mirsky 1968).[8] A further change in the selection of cRNAs transcribed also occurs at the midgastrula stage (Davidson, Crippa, and Mirsky 1968, Denis 1966[b]), and about this time different regions of the embryo transcribe different populations of cRNAs. The dramatic rise in the rate of cRNA transcription at the midblastula stage of development is very shortly followed by onset of rapid transcription of tRNA (Bachvarova et al. 1966, Brown and Littna 1966[b], Woodland and Gurdon 1968). In turn, this is followed at the midgastrula stage by onset of rapid transcription of the rRNA cistrons of the ectoderm-mesoderm cells of the animal half of the gastrula (Brown and Littna 1964[a], 1966[a], Flickinger 1969, Woodland and Gurdon 1968) and after a delay of several hours also in the endodermal cells (Woodland and Gurdon 1968).[9] Thereafter, each distinct stage of development is characterized by the presence of a distinctive complement of cRNAs (Denis 1966[b]).[10]

[8] It should be noted that the comparisons of cRNA complements made in these studies were done under conditions which would only reveal differences in transcription of highly reiterated base sequences.

[9] It has not yet been formally demonstrated that rRNA (or tRNA) is transcribed prior to this stage. However, available data are not at variance with the hypothesis that the average rate of transcription of rRNA *per genome* is constant throughout development to beyond the stage of the gastrula (Emerson and Humphreys 1971).

[10] It should be noted that the comparisons of cRNA complements made in these studies were done under conditions which would only reveal differences in transcription of highly reiterated base sequences.

Control of the course of this finely regulated program of sequential transcription up to the stage of the gastrula resides in the genome, both nuclear and cytoplasmic, of the haploid oocyte and derivative diploid zygote. It is subject to substantial modification by the status of the surrounding cytoplasm (Gurdon 1968[b], Woodland and Gurdon 1969; see Chapter 2). Indeed, the mature oocyte produces at least one factor which specifically inhibits transcription of ribosomal RNAs by young immature oocytes (Crippa 1970, Shiokawa and Yamana 1967, 1969). Nevertheless, the status of the cell cytoplasm itself reflects the course of prior expression of the genome. The gross features of the program (at least) are still observed with dispersed cell suspensions and partial reaggregates of cleaving embryos and blastodiscs (Landesman and Gross 1968, Shiokawa and Yamana 1967). With the development of clear morphogenetic movements, the expression of the genome of any given cell becomes subject to the influence of "*organizers*" or "*inducers*" produced by neighboring cells (e.g., Ebert 1965, Tiedmann 1966).

Similar sequential patterns of transcription of different classes of RNA have been established for sea urchin embryos (Davidson 1968, Denny and Reback 1970, Gross 1968, Slater and Spiegelman 1970) and for loach (*Misgurnus fossilis*) embryos (Aitkhozin, Belitsina and Spirin 1964, Rachkus et al. 1969[a], Spirin 1969[a]). The pattern of synthesis in avian and mammalian embryos has not yet been determined in detail. However, it clearly differs from that of the amphibians and sea urchins in important respects. Transfer RNA appears to be transcribed as early as the 2-cell stage, and both it and ribosomal RNA are transcribed at the 4-cell stage. Indeed, the embryo cannot form a nucleolus or develop beyond the 4-cell stage when transcription is blocked by actinomycin (Mintz 1964, Woodland and Graham 1969).

ACTINOMYCIN AS A PROBE IN ANALYSIS OF SEQUENTIAL GENE TRANSCRIPTION IN METAZOANS; A PARTIAL PROGRAM OF TRANSCRIPTION IN DICTYOSTELIUM DISCOIDEUM

The small size of the polysomes on which histones and related proteins are synthesized (Borun, Scharff, and Robbins 1967, Kedes and Gross 1969[b], Ling, Trevithick, and Dixon 1969), coupled with the distinctive composition of these proteins, permits a direct analysis of the transcription of the corresponding mRNAs (Kedes and Gross 1969[b]). However, in general, the techniques which have served so well in establishing the patterns of transcription of the various classes of RNA in eucaryote cells cannot be applied to analysis of the course of transcription of specific mRNAs. Nor can the course of transcription of specific mRNAs be inferred with confidence from the kinetics of appearance of a specific protein, for we now know of many situations in which there is a substantial delay between transcription of a particular mRNA and translation of that RNA in synthesis of the corresponding protein. Indeed, it is a characteristic feature of those unfertilized eggs which have been studied in detail that the pre-existing "maternal" mRNAs which support development to the stage of the blastula are present in "inactive polysomes." These inactive polysomes are particularly resistant to dissociation at low concentration of Mg^{2+} and are said to be "sticky" (Maggio et al. 1968, Vittorelli, Caffarelli-Mormino, and Monroy 1969). The inability of these polysomes to support translation of the contained mRNA appears to be due primarily to association with additional protein (Harris 1967, Mano 1970, Mano and Nagano 1966, Monroy, Maggio, and Rinaldi 1965, Piatigorsky 1968, Stavy and Gross 1967), but may also be partly due to an inadequate supply of essential transfer factors (Castañeda 1969, Stavy and Gross 1969[a]). These inactive polysomes are converted to active polysomes and serve in the synthesis of protein on fertilization. Additional examples of such a "translational control" of protein synthesis will be considered in Chapter 9.

At the present time the only general approach to study of the course of transcription of specific mRNAs in metazoans is based upon the use of actinomycin to block transcription of new cRNAs. Thus acquisition of the ability to synthesize at least a detectable quantity of a defined protein in the presence of actinomycin may be taken to define provisionally the stage of embryonic development at which transcription of the corresponding mRNA first becomes significant. Development of such a "resistance to actinomycin-inhibition" has been a particularly useful probe in the analysis of the course of transcription of specific mRNAs, even in such model systems as sporulating bacteria or the fruiting bodies of slime molds which are already amenable to a limited measure of genetic analysis. The case of the fruiting body of *Dictyostelium discoideum* is particularly instructive (Sussman 1966, Sussman and Sussman 1969). Under favorable conditions this organism grows as discrete amoebae. However, on depletion of nutrient certain of the cells, possibly as a consequence of aneuploidy (Ashworth and Sackin 1969), become "initiator cells" and secrete a chemotactic agent *acrasin,* which is probably cyclic 3′·5′ AMP (Bonner 1970, Chassey, Love, and Krichevsky 1969, Konijn et al. 1967). The initiator cells serve as foci toward which the other cells "stream" to form aggregates, which develop into migrating slugs, or pseudoplasmodia. The latter finally lose their motility and undergo a sequence of morphogenetic movements and differentiation to yield a mature fruiting body (Figure 6.1).[11] Three morphologically and biochemically distinct, differentiated regions can be recognized in the fruiting body: a mass of spores, a cellulose-ensheathed stalk, and a basal disc. Differentiation of this structure involves intense carbohydrate metabolism and the synthesis of at least three distinct polysaccharides (Sussman and Sussman 1969).

The kinetics of development of activities of a number of

[11] Migration of the aggregate is not an essential prerequisite to differentiation of a fruiting body (Newell and Sussman 1970).

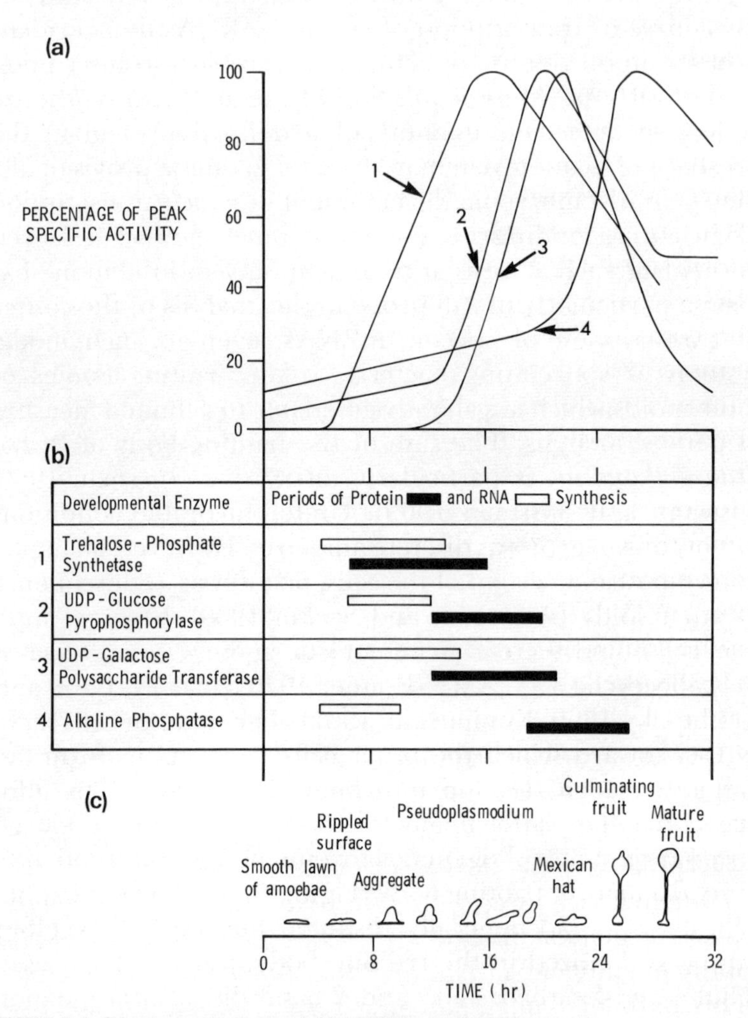

FIGURE 6.1. Development cycle of Dictyostelium discoideum

(a) Course of variations in levels of activity of selected enzymes after transfer of washed amoebae to minimal medium. (b) Periods of synthesis of new species of these enzymes and of transcription of the corresponding mRNAs. (c) Course of morphogenesis.

Based on data from Loomis (1969) and Roth, Ashworth, and Sussman (1968).

enzymes of carbohydrate metabolism have been analyzed (e.g., Coston and Loomis 1969, Loomis 1969, Roth, Ashworth, and Sussman 1968, Sussman and Sussman 1969), and illustrative examples are given in Figure 6.1. In all cases, the observed increases in enzymic activity appear to be due to the *de novo* synthesis of protein rather than activation of a pre-existing, enzymically inactive species of protein. Analysis has been made of the effects of suppression of transcription of RNA upon the course of synthesis of most of these defined enzymically active proteins, and it has permitted the compilation of "timetables" describing the sequence of transcription of the corresponding mRNAs (e.g., Figure 6.1). Mutant strains have been isolated in which the time required for completion of the program of sequential gene transcription and for development of the fruiting body are altered. However, the sequence of transcription is unchanged and remains correlated with the course of morphogenesis (Loomis 1970). Such timetables of sequential gene transcription are of particular interest in providing definitive evidence of the control of synthesis of a number of defined proteins at a level other than of transcription.[12] Indeed, in the specific case of UDPG-pyrophosphorylase enzyme the transcription process is essentially completed before any detectable translation of the specific mRNA occurs.

REGULATION OF GENE TRANSCRIPTION BY HORMONES

Our first real insight into mechanisms regulating the processes of gene expression in procaryotes came from studies of the phenomenon of induced enzyme synthesis (Chapter 3). This particular phenomenon has not yet been conclusively

[12] The term "translational control" is customarily used to describe situations in which the synthesis of a specific protein is limited by some other factor than transcription of appropriate mRNA. However, evidence to be discussed later indicates that additional regulatory mechanisms may be interposed between the nuclear transcription of a cRNA and the appearance of specific mRNA in the cytoplasmic polysomes. Therefore, I prefer to reserve the term translational control for mechanisms regulating translation of mRNA present in polysomes.

demonstrated to occur in cells in the organized tissues of metazoans. However, a related phenomenon of synthesis of specific enzyme(s) as a response of exposure of a "target" organ to certain hormones, both *in vivo* and *in vitro,* is common. It was, therefore, hoped that substantial insight into the regulatory mechanisms of metazoan tissues would be gained from studies of the mechanism of action of hormones. This hope has not been fulfilled, and it now seems more probable that elucidation of the mechanisms regulating transcription and "processing" of cRNA will permit interpretation of the mode of action of some hormones. Nevertheless, as a formal analogy to the phenomenon of induced enzyme synthesis in microorganisms, the phenomenon of hormonally induced enzyme synthesis has been the subject of intense study, and a wealth of data concerning effects of hormones upon various aspects of the processes of gene expression has been amassed. I propose to consider only three hormones or synthetic substitutes for illustrative purposes: namely, 17β-estradiol (1,3,5-estratriene-3, 17β-diol) hydrocortisone, and cyclic 3′·5′ AMP (cAMP).

REGULATION OF GENE TRANSCRIPTION IN THE UTERUS BY ESTRADIOL

The estrogenic hormones control both development of the secondary female sex characters and the large number of physiological changes which collectively comprise the proliferative phase of the estrous cycle. In addition, they may cause development of tumors in a variety of mammalian tissues (Jensen 1966). At one time it was considered possible that the various changes might reflect action of the hormones on the activity of one, or a few, key enzyme(s). Indeed, estradiol does have a direct effect upon the activity of a transhydrogenase enzyme of "target" tissues (Villee, Hagerman, and Joel 1960). However, it became clear that the hormones necessarily had effects upon the activity or synthesis of a considerable number of proteins and might involve the activation of large numbers of genes. Attention was therefore diverted from a search for effects of estrogens upon the activity or synthesis of specific enzymes to an examination of their effects upon the processes of protein synthesis and gene transcription.

Thus McCorquodale and Mueller (1958) reported that a single administration of estradiol to ovariectomized rats caused a marked and sustained rise in the level of activity of a series of aminoacyl-tRNA synthetase enzymes. Similarly, chick oviduct homogenates prepared 24 hours after administration of a single dose of estradiol showed higher activity than equivalent control homogenates for *in vitro* incorporation of amino acid into polypeptide (Kalman and Opsahl 1961). Most subsequent analyses of the effects of estradiol on gene transcription and protein synthesis in the uterus (Figure 6.2) have been made with tissue from ovariectomized rats. Similar results have been obtained *in vivo* and *in vitro,* with both mature and immature females, although there may prove to be some differences in detail.

Immediately following exposure to estradiol there are transient decreases in the rate of uptake of radioactive amino acids into the uterine cells, and of their incorporation into the proteins of all cell fractions. They are followed by a return to the initial rates within 2 to 4 hours, and thereafter by a general stimulation of total protein synthesis. This attains a maximum rate per genome similar to that in the uterus of a normal mature female in diestrus after some 24 to 36 hours of exposure to the estrogen (Hamilton 1968, Means and Hamilton 1966, Noteboom and Gorski 1963). The increase in rate of total protein synthesis is associated with, and probably preceded by, an increased rate of synthesis of total RNA, and there is an early detectable rise in the *apparent* total nuclear RNA-polymerase activity (Gorski, Noteboom, and Nicolette 1965, Hamilton 1968, Hamilton, Widnell, and Tata 1965, Notides and Gorski 1966). However, analysis of the kinetics of incorporation of RNA precursors in the early phases of response to the hormone is complicated by a parallel stimu-

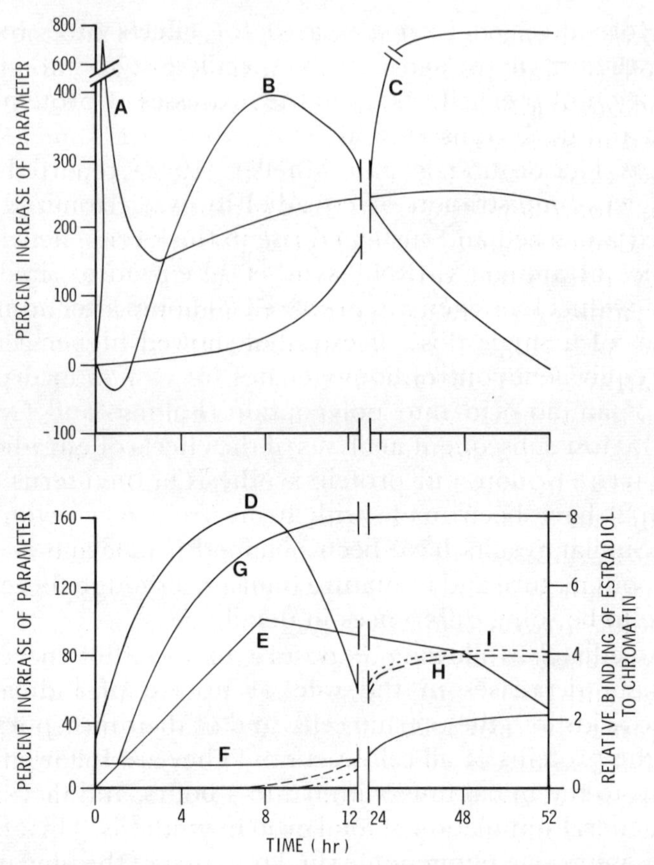

FIGURE 6.2. Uterine responses following injection of 17 β-estradiol to an ovariectomized mature female rat

A. Specific activity of nuclear RNA rapidly labeled by ^{3}H-uridine administered at zero time (c.p.m./mg).
B. Specific activity of nuclear protein labeled by ^{3}H-methionine administered 10 minutes before sacrifice (c.p.m./mg).
C. Cytoplasmic level of ribonucleoprotein.
D. Binding of ^{3}H-estradiol to chromatic (c.p.m./genome).
E. *In vitro* template activity of isolated chromatin.
F. Chromatin-associated RNA.
G. Activity of RNA-polymerase stimulated by Mg^{2+} (nucleolar).
H. Activity of RNA-polymerase stimulated by Mn^{2+} at high ionic strength.
I. Nuclear protein per genome.

Zero time values are from control animals not injected with hormone.
From data of Hamilton (1968) and colleagues.

lation of the rates of uptake of the precursors into the uterine cells (Billing, Barbiroli, and Smellie 1969, Hamilton 1968, Hamilton, Widnell, and Tata 1965). Most of the RNA transcribed in these early phases appears to be ribosomal RNA, but all classes of RNA are transcribed (Gorski and Nelson 1965, Moore and Hamilton 1964).[13] Indeed, recent studies have demonstrated early changes in the complement of cRNAs transcribed in response to exposure to estradiol, and the cytoplasm of stimulated cells contains species of cRNA found only in the nuclei of control cells (Church and McCarthy 1970). Moreover, transcription of demonstrable quantities of mRNA(s) required for synthesis of a new species of "estrogen-induced" protein occurs before any detectable increase in the rate of incorporation of precursor into RNA (DeAngelo and Gorski 1970).

The earliest response to estradiol recorded to date is the observation that within 15 seconds of administration of estradiol to ovariectomized rats there was a doubling in concentration of cAMP specifically in the cells of the uterus (Szego and Davis 1967). This "second messenger" hormone appears to be responsible for the stimulated rates of uptake of RNA precursors and for observed increases in the rates of uptake of amino acids (Noall and Allen 1961, Sharma and Talwar 1970).

Within 2 minutes of administration of estradiol there is a demonstrable binding of some of the hormone to uterine chromatin. The level of hormone bound per genome rises progressively to a maximum at about 8 hours and then steadily declines (Hamilton 1968, Teng and Hamilton 1968). Nuclei isolated from hormone-treated animals synthesize

[13] One serious technical problem in the analysis of the transcription processes in metazoan cells which has not been fully solved is the devising of methods for the complete recovery of all species of RNA in undegraded form. Joel and Hagerman (1969) have recently reported development of an improved isolation procedure, with which they obtained no indication of a selective stimulation of the synthesis of ribosomal RNA. Their conclusion does not, however, substantially affect the overall picture of the response of the uterus to estradiol as currently understood.

RNA *in vitro* more rapidly than those from control animals (Gorski 1964, Hamilton, Teng, and Means 1968, Hamilton, Widnell, and Tata 1968). A demonstrable increase in the activity of uterine chromatin as a template for *in vitro* transcription of RNA occurs within 15 minutes of exposure to estradiol. This increase rises to a maximum level at 8 to 12 hours and also then progressively falls (Barker and Warren 1966, Teng and Hamilton 1968). During this same period of time there is a progressive doubling in the "apparent" activity (i.e., with endogenous chromatin as template) of the RNA polymerase enzyme of the nucleolus, which is primarily involved in transcription of precursors of the ribosomal RNAs (Hamilton, Teng, and Means 1968, Hamilton, Widnell, and Tata 1968; see Roeder and Rutter 1969).

Increased template activity of the chromatin is correlated with displacement of histone III by nonhistone protein (Barker 1971, Hamilton and Teng 1969, Hamilton, Teng, and Means 1968, Teng and Hamilton 1969). This includes a species of acidic protein which appears to be specific to the uterus and synthesized at a markedly enhanced rate in response to exposure to estradiol (Barker 1971, Smith et al. 1970, Teng and Hamilton 1970[a]).[14] There is also an increase in the RNA content of chromatin in the first 15 minutes after injection of estradiol, but no further increase in the following 8 to 12 hours. This displacement of histone by acidic protein, and possibly also the increase in content of RNA, appears to be an important feature of the mechanism controlling selection of the genes to be transcribed into cRNAs, which is considered in detail in the following chapter. The genes activated during this early phase of response to estrogen are presumably predominantly the cistrons specifying ribosomal RNAs contained in the enlarged and more numerous nucleoli of the uterine cells of hormone-treated animals. However, they include the cistron(s) coding

[14] Data currently available do not eliminate the alternative hypothesis that estradiol stabilizes this protein.

for an "estrogen-induced" protein of unknown function. Synthesis of this protein is demonstrable some 30 to 40 minutes after exposure to estradiol, both *in vivo* and in organ culture *in vitro*. The rate of synthesis is maximal after 1 to 2 hours and then declines (Barnea and Gorski 1970, DeAngelo and Gorski 1970, Mayol and Thayer 1970). In addition, early transcription of new species of nuclear cRNAs is revealed by DNA-RNA hybridization assays (Church and McCarthy 1970). Newly transcribed RNA is first demonstrable in the cytoplasmic polysomes some 80 minutes after administration of estradiol, and the rate of accumulation in the cytoplasm rises dramatically after 2 to 4 hours (Hamilton, Widnell, and Tata 1968).

A second phase of response to estradiol is initiated some 12 hours after administration of hormone by a mechanism quite distinct from that regulating the first phase. At this time there is a further rise in the content of both RNA and nonhistone protein and a slight increase in the activity of one (or more) nuclear RNA polymerase enzyme not primarily active in transcribing ribosomal RNA cistrons. These changes are associated with onset of decreases in the extent of estradiol bound to the chromatin and in the template activity of the chromatin in the transcription of RNA, and with an increase in the content of histone in the chromatin (Hamilton 1968, Hamilton and Teng 1969, Hamilton, Teng, and Means 1968, Teng and Hamilton 1969). This series of changes presumably reflects changes in the species of mRNAs transcribed and transported into the cytoplasmic polysomes, for the cell content of ribosomes increases over a further period of 12 hours at an average rate roughly equal to that observed at the time of transition to the second phase (Figure 6.2).

This sequence of responses to estradiol by the primary "target organ" of the ovariectomized *mature* female probably reflects the events of the estrous cycle. Slightly different or additional responses might be expected, and have been reported, for the immature uterus. Different and less marked responses to estradiol are observed in other target organs such as the liver (e.g., Hamilton 1968). These and the long-term responses of male tissues to continuous exposure to estrogens will not be discussed.

HORMONE RECEPTOR PROTEINS

Studies of the interaction between estradiol and the uterus have proven singularly useful in providing a partial answer to one major problem in our understanding of the various modes of hormone action: namely, the nature of the mechanism which determines the specificity of response of the target organ(s) (Gorski, Shyamala, and Toft 1969). At physiological concentrations, estradiol is taken up by the uterus against a concentration gradient, and the bulk of the hormone is concentrated in the nuclei of the uterine endometrium and myometrium, in association with the chromatin (Jensen 1966, Jensen, De Sombre, and Jungblut 1967, Maurer and Chalkley 1967, Noteboom and Gorski 1965). This response appears to be due to the presence in the uterine cells of a highly specific "receptor protein" (Jensen, De Sombre, and Jungblut 1967, Toft and Gorski 1966, Shyamala and Gorski 1967).

The concentration of receptor protein per cell genome changes markedly under a variety of physiological conditions, but remains low and constant following ovariectomy unless estradiol is administered to the animal (Feherty et al. 1970).

The receptor protein is present in the cytosol predominantly in the form of a readily dissociable multimeric complex (S value 8 to 9.5) of a smaller monomer (S value 4 to 5S) (De Sombre, Puca, and Jensen 1969, Gorski, Shyamala, and Toft 1969, Korenman and Rao 1968). The protein isolated from the nucleus appears to be an acidic protein entirely in the form of the 5S monomer which is reversibly bound at sites which are largely specific to uterine chromatin

(Maurer and Chalkley 1967, Puca and Bresciani 1968, Steggles et al. 1971). Present evidence indicates that estradiol is bound noncovalently to the 9S multimer receptor protein in the cytoplasm and that the protein-hormone complex migrates into the nucleus in the monomeric form, there to complex with the chromatin (Arnaud et al. 1971, Gorski, Shyamala, and Toft 1969, Jensen et al. 1968, Maurer and Chalkley 1967, Shyamala and Gorski 1969). Such a migration of receptor protein has not yet, however, been formally established, nor do we know whether additional factors are involved in the specific concentration of estradiol against a concentration gradient. The relationship between this "receptor protein" and the "repressor protein" reported by Talwar et al. (1964) also remains to be formally examined.

A similar receptor protein appears to be present in the anterior pituitary, a second target organ of estradiol (Leavitt, Friend, and Robinson 1969). Related "hormone-binding" proteins have also been recently reported in target cells for various steroid hormones (Gardner and Tomkins 1969, Morey and Litwack 1969, O'Malley, Sherman, and Toft 1970, Swaneck, Chu, and Edelman 1970) and for plant hormones (Hardin, O'Brien, and Cherry 1970, Matthysse 1970, Matthysse and Abrams 1970).

REGULATION OF HEPATIC TYROSINE AMINOTRANSFERASE LEVELS BY GLUCOCORTICOID HORMONES

The mode of action of hydrocortisone and related glucocorticoid hormones has been the subject of very extensive investigation. This group of hormones acts primarily on the liver as target organ, and many studies have been made of their effects upon the processes of protein synthesis and gene transcription in liver tissue. Qualitative effects similar to those of estradiol upon the uterus have been reported, although in less detail (e.g., Brossard and Nicole 1969, Dahmus and Bonner 1965, Jacob, Sajdel, and Munro 1969, Kenney, Wicks, and Greenman 1965, Lukács and Sekeris

1967, Schmid, Gallwitz, and Sekeris 1967, Tomkins et al. 1965). The lesser detail available for this group of hormones is due in small measure to the use of animals under a wide variety of physiological conditions, or at various stages of liver regeneration or of embryonic development. However, the lack of information comparable to that obtained in studies of effects of estradiol upon the uterus is due primarily to the fact that effects of glucocorticoid hormones may be observed upon the synthesis of one specific hepatic enzyme protein.

Administration of hydrocortisone to either normal or adrenalectomized rats leads to a rapid and marked, but transient, rise in the level of hepatic tyrosine aminotransferase activity (Lin and Knox 1957). The increase in enzyme activity involves the *de novo* synthesis of enzyme protein and is associated with early effects of the hormone in stimulating transcription of both nuclear and mitochondrial RNAs (Garren, Howell, and Tomkins 1964, Kenney 1962, Kenney and Kull 1963, Mansour and Nass 1970). Until recently, effects of hydrocortisone upon the level of this specific hepatic enzyme had been considered to reflect the effects of the hormone upon the processes of protein synthesis and gene transcription in general.

However, study of the response to glucocorticoid hormone by hepatoma cells in tissue culture indicates that the hormone selectively induces an increased rate of enzyme formation. In particular, the increase in activity takes place without any detectable change in the rate of RNA synthesis (Gelehrter and Tomkins 1967, Kenney et al. 1968). The increase in enzyme activity is due to an enhanced rate of *de novo* synthesis of enzyme protein rather than a decreased rate of degradation of active enzyme (Granner et al. 1968, Granner, Thompson, and Tomkins 1970, Reel, Lee, and Kenney 1970, Tomkins et al. 1969). Prior transcription of specific mRNA is required (Granner, Thompson, and Tomkins 1970, Peterkofsky and Tompkins 1968). Nevertheless, the primary action of the steroid hormone in stimulating formation of tyrosine aminotransferase appears to be upon the stability and translation of the mRNA, rather than directly upon the process of gene transcription (Auricchio, Martin, and Tomkins 1969, Tomkins 1968, Tomkins et al. 1969).[15]

Thus, despite demonstrable effects of hydrocortisone and analogues upon the process of gene transcription in the liver, these hormones may regulate the synthesis of some specific proteins by mechanisms acting other than at the level of gene transcription. It should, however, be emphasized that this conclusion does not imply that effects upon gene transcription are necessarily subsequent to prior effects at the level of translational control. Early effects of glucocorticoid hormone upon the adhesiveness, electrophoretic mobility, and surface antigens of the hepatoma cells require both gene transcription and protein synthesis (Ballard and Tomkins 1969, 1970), and significant rises in nuclear RNA polymerase activity within one minute of exposure of liver nuclei to hormone have been reported (Lukács and Sekeris 1967). Rather, the point to be stressed is that the hormone may have effects upon the synthesis of

[15] This conclusion is based in part upon the observation that addition of a high concentration of actinomycin to preinduced cells in inducer-free medium containing serum has no effect upon the rate of degradation of preformed tyrosine aminotransferase, and promotes a further increase ("super-induction") in enzyme level (Auricchio, Martin, and Tomkins 1969, Martin, Tomkins, and Bresler 1969). In contrast, the phenomenon of super-induction is not manifest with cells in medium lacking serum, and actinomycin does inhibit degradation of preformed enzyme (Auricchio, Martin, and Tomkins 1969, Lee, Reel, and Kenney 1970). It has been suggested (Lee, Reel, and Kenney 1970) that super-induction in the presence of serum is due to stimulation of translation by residual insulin (see Gelehrter and Tomkins 1970, Lee, Reel, and Kenney 1970). However, the effect of actinomycin in the absence of serum appears to be due to inhibition of an "enhanced degradation of enzyme" (Auricchio, Martin, and Tomkins 1969). Moreover, the stimulatory factor in serum appears not to be insulin (Gelehrter and Tomkins 1969, 1970). In addition, the response of the cells to glucocorticoid hormone varies during the cell cycle in a manner which cannot be attributed to changes in the supply of insulin (Martin and Tomkins 1970, Martin, Tomkins, and Bresler 1969).

protein distinct from effects it may have upon transcription. The precise site of action of glucocorticoid hormone in regulating synthesis of tyrosine aminotransferase activity is not known.

Tomkins and colleagues have suggested that it serves to antagonize a "repressor of translation" and stabilize the specific mRNA against degradation (Martin and Tomkins 1970, Martin, Tomkins, and Bresler 1969, Tomkins et al. 1969). Present evidence is in accord with the suggestion, but direct examination of the possibility with cell-free systems will be required for elimination of alternative interpretations.

ROLES OF CYCLIC AMP IN REGULATION OF GENE EXPRESSION IN METAZOANS

Cyclic 3′·5′ AMP is a hormone active on a wide spectrum of tissues and serves as a "second messenger" mediating the action of at least 14 other hormones and affecting the activity of a number of preformed enzymes (Robison, Butcher, and Sutherland 1968). As noted above, it stimulates the uptake of RNA precursors by the uterus (Sharma and Talwar 1970). Under appropriate conditions, it may also induce formation of tyrosine aminotransferase enzyme in fetal rat liver by a mechanism quite distinct from that of hydrocortisone (Greengard 1969[a], 1969[b], Holt and Oliver 1969, Wicks 1968, 1969). In addition cAMP has been shown to induce formation of phosphopyruvate carboxylase enzyme in neonatal rat liver (Yeung and Oliver 1968), and of serine dehydratase in mature rat liver (Jost et al. 1970). The mechanism by which cAMP induces these increases of enzyme levels is not yet known. However, cAMP is known to markedly stimulate the activity of a class of enzymes of particular relevance to the process of gene transcription in metazoan cells. These are the widely distributed histone and protamine kinase enzymes (Gill and Garren 1970, Gutierrez and Hnilica 1967, Jergil and Dixon 1970, Kuo and Greengard 1969,

Langan 1968[a], Miyamoto, Kuo, and Greengard 1969, Walsh, Perkins, and Krebs 1968, Yamamura et al. 1970). Phosphorylation of histone or protamine and of nuclear acidic protein is one of the processes observed in a number of regulatory phenomena to be considered later in Chapter 7.

Phenomena analogous to that of end-product repression, and probably also that of catabolite repression, in procaryotes (Chapter 4) are fairly common in tissues of metazoans. The underlying mechanism has not yet been extensively investigated and may be distinct from that in procaryotes. This appears to be the case in the inhibition by glucose of increase in liver serine dehydratase activity in response to high dietary levels of amino acid. Suppression of the increase in enzyme activity is due partly to inhibition of enzyme synthesis and partly to enhancement of the rate of degradation of preformed enzyme. The primary site of inhibition of enzyme synthesis appears to be at the stage of release of completed peptide chain from the polysome complex and associated reactions of the ribosome cycle (Chapter 1), thereby "trapping" the specific mRNA in a complex unable to support further rounds of protein synthesis. These various effects of glucose can, however, be reversed by administration of cAMP, and the hormone alone stimulates synthesis of the enzyme (Jost, Hsie, and Rickenberg 1969, Jost, Khairallah, and Pitot 1968, Pitot and Jost 1968).

HISTONES AS HOMOLOGUES OF THE REPRESSOR PROTEINS OF PROCARYOTES

A role of histones as regulators of gene function was first proposed in 1950 by the Stedmans (Stedman and Stedman 1950). With the development of concepts of messenger RNA, and growing understanding of the phenomena of induced enzyme synthesis and enzyme repression in procaryotes (Chapters 3 and 4), this proposal became refined into the concept of histones serving as the repressors of gene transcription in metazoans. Early demonstrations that isolated histones markedly inhibited the activity of DNA-dependent RNA polymerase activity provided some measure of support for the proposal. The evidence fell far short of being compelling but served to stimulate extensive studies of the categories, distribution, and chemistry of the histones, and of their possible role as specific repressors of gene transcription. This has resulted in the accumulation of a vast body of experimental data concerning the histones, some of which is contradictory. Nevertheless, certain general conclusions may be drawn (e.g., Bonner et al. 1968[b], Bonner and Ts'o 1964, Hnilica 1967). These will be asserted here, and points of particular interest will be considered in more detail in Chapter 7.

1. Histones show little specificity in their interaction with DNA and are of themselves unable to recognize specific (operator-like) base sequences in the manner proposed for the repressors of procaryotes. Selection of the spectrum of genes to be expressed appears to be determined in part by acidic proteins, and in part also by specific chromatin-associated RNAs.
2. Histones do, however, function as "adjunct repressors" of the transcription of genes not selected for expression.
3. The activity of histones as repressors of gene transcription is markedly affected by enzymic modification of the histone, and by the presence of acidic phosphoproteins.
4. In addition to activities of the histones upon the processes of gene transcription, and in their interaction with DNA, the histones play a major structural role in determining the architecture of the chromosome.

CONCURRENT GENE EXPRESSION IN EUCARYOTES

A number of genetic and regulatory phenomena have been observed with eucaryote cells or organisms which resemble phenomena known to be controlled in procaryotes by the integrated gene complexes defined as operons (Chapters 3 and 4). However, no unequivocal demonstration has been made of the presence of such an operon [16] in the genome of any eucaryote. Thus genetic and biochemical studies of three of the enzymic activities required for synthesis of histidine by *Neurospora crassa* indicated the possible existence of a small operon controlling synthesis of three distinct proteins of a multifunctional enzyme aggregate (Ahmed 1968, Ahmed, Case, and Giles 1964), analogous to the enzyme aggregate controlling synthesis of tryptophan in the same organism (DeMoss and Wegman 1965). However, more recent studies have indicated that all three enzymic activities are catalyzed by a single multifunctional enzyme (Minson and Creaser 1969) analogous to the tryptophan synthetase enzyme of *N. crassa* (Bonner 1964). It is possible that such multifunctional enzymes are specified by structural genes which have evolved by fusion of cistrons originally present in an operon (Bonner 1964), but they are clearly distinct from the operons of contemporary procaryotes. Similarly, analysis of four of the enzymic activities required for synthesis of histidine by *Saccharomyces cerevisiae* indicates the possible presence of an operon controlling synthesis of a series of distinct components of an enzyme aggregate. Yet the properties of particular mutant strains are not readily interpretated in terms of such an operon (Fink 1966, Shaffer, Rytka, and Fink 1969).

Equally complex situations have been observed for a number of "compound loci," or "genetic units of function" (Chovnick et al. 1969), of *Drosophila melanogaster*. In certain of these cases present evidence indicates that the units of function are not operons but rather represent cistrons coding for components of enzyme aggregates (e.g., Chovnick et al. 1969, Welshons 1965). In other cases the majority

[16] The term "operon" is used here in the sense originally defined by Jacob and Monrod (1961[a]). This term has since been adopted (e.g., Georgiev 1969[a], Scherrer and Marcaud 1968) to describe related, but different, postulated gene complexes of metazoan genomes (see Chapter 8).

of the available evidence can be readily interpreted in terms of an operon, but the exceptional properties of certain experimental stocks preclude elimination of alternative interpretations (e.g., Lewis 1963).

A variety of situations are known in which tissue levels of two or more enzymes vary concurrently. One particularly good illustrative example is provided by the enzymes of the pathways of gluconeogenesis and glycolysis in rat liver (Weber et al. 1966). The enzymes may be classified into three groups as follows:

1. *Key gluconeogenic enzymes* catalyzing unidirectional reactions which limit the rate of synthesis of glucose; comprising glucose·6·phosphatase, fructose·1·6·diphosphatase, phosphoenol·pyruvate carboxykinase, and pyruvate carboxylase.
2. *Key glycolytic enzymes* catalyzing rate-limiting unidirectional steps of glycolysis; comprising glucokinase, phosphofructokinase, and pyruvate kinase.
3. *Bifunctional enzymes* present in excess and catalyzing reversible reactions.

Activities of enzymes of group 1 vary roughly in parallel in response to changes in hormonal status of the animals or changes in dietary regime.[17] Similarly, activities of enzymes of group 2 also vary roughly in parallel to the same changes, but in the converse directions. Analysis of the effects of metabolic inhibitors strongly indicates that the observed changes in enzymic activities reflect changes in the levels of enzyme proteins. Weber et al. (1966) have proposed that the various structural genes for the enzymes of each of the three classes comprise distinct "*functional genic units.*"

[17] Adenylate kinase (ATP-AMP phosphotransferase) activity also varies in parallel with those of the enzymes in group 1 (Adelman, Lo, and Weinhouse 1970). It is unclear whether this reflects a role of this enzyme in controlling intracellular concentrations of nucleotides which can regulate activities of enzymes in both groups 1 and 2, or is merely a coincidence.

These postulated gene complexes are considered to be equivalent in function to the operons of bacteria. However, there is no evidence to indicate that the structural genes are actually adjacent, linked components in the genome. Indeed, Britten and Davidson (1969) have presented a model in which totally unlinked structural genes can be controlled in a coordinate manner as functional genic units or "gene batteries" (see Chapter 7).

EPITOME; PARALLELS AND DIVERGENCES

In summary, for each of the major classes of mechanism regulating gene transcription in procaryotes there is an analogous counterpart in eucaryote cells. Indeed, the correspondence can be further extended by noting the analogies between the prophages of temperate bacterial viruses (Chapter 4) and the incorporated segment of viral genome in "transformed" metazoan cells (e.g., Dulbecco 1969, Fujinaga and Green 1970). Nevertheless, in general, the homologous mechanisms of metazoans and procaryotes differ substantially in detail. Appropriately, the protistan eucaryotes show features in common with, and also differing from, both procaryotes and metazoans.

Now some parts are the special instrument of an action; others have been so constituted that it seems as if the action could not take place without them; others have been laid down in order that the action may be better performed; and still others have been created for the protection and preservation of the whole.—H. Fabricius, 1621.

In the study of nature, as a philosopher has observed, the principals are certain general effects from primary causes, from which spring countless secondary effects; the art of linking the first to the second is to walk sightless along a highway from which a thousand by-ways lead astray.—X. Bichat, 1805.

CHAPTER 7: **REGULATION OF GENE TRANSCRIPTION IN METAZOANS**

The early proposal of the Stedmans (1950) that the histones regulate gene expression in metazoans was largely overlooked until the formulation by Jacob and Monod (1961[a]) of specific concepts of mechanisms regulating gene expression in procaryotes. In the ensuing years, the properties of the histones and their roles in the regulation of gene expression have been the subject of intensive study. Early experiments indicated that all classes of histone can inhibit RNA polymerase activity *in vitro,* restrict transcription from chromatin, and markedly affect the course of embryonic development (e.g., see Bonner 1965, Hnilica 1967). However, it seemed very highly probable that the number of individual genes, or groups of genes, regulated independently of all others in the genome of a given cell was much greater than the numer of distinct molecular species of histone. Therefore, it seemed equally highly improbable that histones regulated gene expression in eucaryote cells in the same manner as that envisaged for the very specific repressor proteins of bacterial cells and bacteriophages (see Chapters 3 and 4). Subsequent studies, to be discussed in the following pages, show that selection of the restricted portion of the genome of a eucaryote cell to be expressed in gene transcription is not determined by the histones. Rather, the histones appear to reinforce the selection by repressing transcription from the unselected portion of the genome.

A number of difficulties complicate the interpretation of the vast body of available data concerning the regulation of gene transcription in metazoans. One is the problem of distinguishing between the various interdependent functions of histones as regulators of both gene transcription and chromosome replication, and as structural components of the chromosomes. Equally complex is the problem of discriminating between processes of gene activation through modification of histone already present in chromatin and chromosome replication involving the incorporation of newly synthesized histone. Therefore, I propose to discuss a series of distinct topics which currently appear particularly pertinent to our understanding of mechanisms regulating the synthesis of specific proteins.

HISTONE NOMENCLATURE

It is beyond the scope of this exercise to consider in detail the chemistry of the histones and nucleohistones. They have been thoroughly presented in several recent reviews (e.g., Bloch 1969, Bonner et al. 1968[a], Busch 1968, Dixon and Smith 1968, Georgiev 1969[a], Hnilica 1967, Stellwagen and Cole 1969[a]). However, there is considerable confusion in the nomenclature used to describe the various classes of histone, and it is necessary to consider briefly the isolation procedures upon which the schemes of nomenclature are based.

The histones may be broadly classified into three major groups on the basis of the amino acid composition. These are, respectively, the lysine-rich, slightly lysine-rich, and arginine-rich histones. Johns et al. (1960, 1961) used columns of carboxymethylcellulose (CMC) to partly resolve these three classes of histone, which were designated in their order of elution from the column as f1, f2, and f3, respectively. However, separation of the fractions was incomplete, and this group of workers turned its attention to procedures based upon differences in the solubilities of the major classes of histone in acidic solutions (Hnilica, Johns, and Butler 1962). This led to development of methods for separation of isolated histone into the major classes and also a number of discrete subfractions. These were designated according to the nomenclature based upon behavior on CMC columns (see Johns 1964, Hnilica 1967).

Other workers (Rasmussen, Murray, and Luck 1962, Satake, Rasmussen, and Luck 1960) used columns of the resin Amberlite CG-50 and obtained good separation of histone preparations into the three major classes and some partial resolution into subfractions. The principal components were designated in order of elution from the column as Ia, Ib, IIa, IIb, and III plus IV. Further separation of the last mixture into two fractions was obtained by use of a similar column of resin under different operating conditions.

The initial separation obtained on columns of Amberlite CG-50 was adequate to provide the first step in isolation of homogeneous preparations of purified histones (Fambrough and Bonner 1968). Arginine-rich histones were further resolved by gel electrophoresis (Fambrough, Fujimura, and Bonner 1968) into distinct species of "alanine-rich" histone III and "glycine-rich" histone IV. The two fractions obtained by Rasmussen, Murray, and Luck (1962) do not correspond to these distinct molecular species of histone, but equivalent fractions are isolated by other fractionation procedures.

Over the past decade a number of other isolation procedures have been developed. Of these, the most selective are those employing sequential extractions of isolated nuclei or chromatin with solutions of progressively decreasing pH value (Murray 1966, Seligy and Neelin 1970, Vidali and

Neelin 1968). Histone fractions isolated by these various procedures have, with one exception, been most commonly designated according to Bonner's modification (e.g., Bonner et al. 1968[a]) of the nomenclature introduced by Rasmussen, Murray, and Luck (1962). The exception is histone fraction IIa, which has the amino acid composition of an arginine-rich histone, and may be generated from the latter type of histone by reduction (Murray 1966, Vidali and Neelin 1968). I shall use this nomenclature here.

The nomenclature introduced by Johns et al. (1960) is still used extensively. Of other systems in the literature, the only one in current usage is that of Busch and his colleagues (Busch 1968, Starbuck et al. 1968). The various systems of nomenclature in current use are compared in Table 7.1. Histone fraction V is a class of "serine-rich" and arginine-rich histone found only in erythrocytes and other cells of the erythropoietic series of birds (Neelin et al. 1964, Neelin 1968) and other vertebrates in which the mature erythrocyte retains a nucleus (Edwards and Hnilica 1968, Nelson and Yunis 1969, see also Vendrely and Picaud 1968). Some additional resolution of the lysine-rich fractions has been reported in recent studies. However, it now seems improbable

TABLE 7.1. Classification of histones

System of nomenclature *			Type
1	2	3	
Ia, b, c	f1	VLR	Lysine-rich or very lysine-rich
IIb1	f2a2		Slightly lysine-rich or intermediate lysine-rich
IIb2	f2b		
IIa	---	---	
III	f3	60% AL	Alanine-rich, arginine-rich
IV	f2a1	GAR	Glycine-rich, arginine-rich
V	f2c		Arginine-lysine and serine-rich avian erythrocyte-specific

* Systems given are based on: (1) Bonner et al. (1968[a]) with corrected designation for histone IIa; (2) Hnilica (1967); (3) Busch (1968), Starbuck et al. (1968).

that the total number of species of histone present in the somatic cells of any higher organism exceeds 10 to 12 (e.g., Bonner et al. 1968[b], Georgiev 1969[a], Panyim and Chalkley 1969, Stellwagen and Cole 1969[a]).

ELEMENTARY ASPECTS OF THE PROPERTIES OF HISTONES AS STRUCTURAL COMPONENTS OF CHROMATIN

Interpretations of the detailed structures of the various configurations assumed *in situ* by eucaryote chromosomes throughout the course of the cell cycle are still a matter of considerable controversy far beyond the scope of my discussion. Similarly, the precise structure of preparations of isolated chromatin is not known. Indeed, Clark and Felsenfeld (1971) have recently demonstrated that: (a) in saline solutions of high ionic strength some of the protein of isolated chromatin is bound to DNA in a labile equilibrium; and (b) the proportion of the chromatin DNA accessible to nuclease digestion is markedly increased in solutions of ionic strength substantially lower than that of a physiological saline. Present evidence is in accord with the hypotheses that at high ionic strength protein is bound to DNA at the same sites as *in vivo,* and that reduction of the ionic strength below that of a physiological saline leads to a reversible dispersion of condensed chromatin. However, many of the tentative conclusions made in the following sections are based upon evidence which will have to be re-examined in the light of the observations of Clark and Felsenfeld. Therefore, I shall also not consider the structure of chromatin in detail. Nevertheless, some of the properties of histones as structural components of chromatin are of considerable importance to our current views, and probably also to future understanding, of their roles in regulation of gene transcription.

Our present concepts of the basic structure of chromatin are derived largely from two major classes of study. One has been direct examination of the structure of chromosomes *in situ* in cells or isolated nuclei. The other has been analysis of the properties of "*native*" chromatin isolated, either directly from whole cells or from preparations of cell nuclei, using solutions of ionic strength approximating that of a physiological saline. Virtually all the DNA of such chromatins *appears* to be encased in an almost continuous sheath of protein, according to extensive data from a variety of cytological (e.g., DuPraw 1968, Miller 1965), physicochemical (e.g., Butler 1966, Hnilica 1967, Zubay and Doty 1959), and enzymic studies (e.g., Itzhaki 1971, Murray 1969). This protein sheath contains a number of components additional to the histones. The distribution of the various classes of histones along the DNA of the chromatin appears to be heterogeneous, but nonrandom. Each class of histone appears to be present in all fragments of calf thymus chromatin, but there may be small regions of homogeneity or selective enrichment in one class of histone (Murray 1969, Ohlenbusch et al. 1967).[1] In addition, there appears to be no substantial difference in the proportions of the various classes of histone present in the "active" (or derepressed) and "inactive" (or repressed) fractions of chromatin preparation, or between nucleoli and the remainder of the nucleus, when isolated by use of solutions of ionic strengths up to that of a physiological saline (Comings 1967, Frenster 1965[a], Grogan, Desjardins, and Busch 1966).[2]

One characteristic feature of the chromatin fibers of eucaryote chromosomes is their highly complex conformation. This is usually attributed to several rounds of supercoiling

[1] For reasons which will be developed in the following sections, it appears probable that the histones play roles in regulating gene transcription in metazoan cells at two levels of repression. One of these is relatively reversible, or conditional, and the other is relatively irreversible, or unconditional (see p. 282). It might therefore be anticipated that differences between various types of cell from the same organism in the complement of unconditionally repressed genes would be reflected in the distributions of histones in the corresponding chromatins. In the extreme case of cells of very highly specialized function this may be manifest as an apparent "clustering" of histones (e.g., Paoletti and Huang 1969, Wilhelm, Champagne, and Daune 1970).

[2] Variations have been observed in the proportions of the various classes of histone in chromatin isolated from different tissues of the same organism (e.g., Fambrough, Fujimura, and Bonner 1968, Stellwagen and Cole 1969[a]). This type of variation is briefly considered on page 299.

of a basic nucleoprotein fiber and additional complexing with protein (e.g., see DuPraw 1968). Studies of the interaction between purified DNA and isolated histone fractions indicate that at least the first round of supercoiling is due primarily to histones of the arginine-rich classes and possibly also histone IIb2 (e.g., Ohba 1966, Richards and Pardon 1970).[3] Conversely, extraction of histones I and V from preparations of chicken erythrocyte chromatin does not disrupt the inherent supercoiling, whereas further removal of the arginine-rich histones III and IV eliminates the supercoiling (Murray et al. 1970). Further, the arginine-rich histones are highly active in causing retraction of the loops of newt oocyte lampbrush chromosomes (Izawa, Allfrey, and Mirsky 1963). In contrast, the lysine-rich histones can more readily "cross-link" and stabilize the two-stranded structure of DNA than other classes of histone (e.g., Hnilica 1967, Olins 1969, Sluyser and Snellen-Jurgens 1970). This class of histones appear to be the component primarily responsible for the compact, condensed structure of inactive chromatin (e.g., Littau et al. 1965, Mirsky et al. 1968).

It seems probable that erythrocyte-specific histone V causes both supercoiling and cross-linking of the chromatin. The properties of this histone as a repressor of gene transcription (see p. 290) are similar to those of typical arginine-rich histones (Seligy and Neelin 1970, 1971). On the other hand, prior extraction of histone V from intact chicken erythrocytes leads to the isolation of chromatin with a less compact structure than that of chromatin isolated from "saline-washed" nuclei (Brasch, Setterfield, and Neelin 1971). Moreover, the content of lysine-rich histone I remains constant as the chromatin condenses during maturation,

[3] It is of obvious significance that, despite the extensive evolutionary divergence between contemporary peas and cows, the histone fractions III and IV obtained from these two organisms are extremely similar. The histones IV from pea seedling and calf thymus differ in amino acid sequence at only two positions in the 102 residues of the chain, and in extent of methylation of 3 to 5 lysine residues (DeLange et al. 1969). Complete amino acid sequences have not yet been presented for histones III from both sources, but present evidence indicates a similar near-identity of sequences (Fambrough and Bonner 1968).

while the content of histone V increases markedly (Mazen and Champagne 1969). Therefore, it seems possible that the latter histone functions both as a "super-repressor" of transcription from the erythrocyte genome and also in promoting condensation of the chromatin (Seligy and Neelin 1970).

TEMPLATE ACTIVITY OF ISOLATED CHROMATIN

Isolated chromatin will serve as a template for the *in vitro* transcription of RNA by either endogenous or added RNA polymerase enzyme. It is inherently very markedly less effective as such a template than an equivalent quantity of deproteinized DNA, and only a very restricted portion of the DNA present in the chromatin is actually transcribed (Bonner et al. 1968[b], Bonner and Huang 1963, Paul and Gilmour 1966, Seligy and Miyagi 1969, Seligy and Neelin 1970). The restricted portion which is transcribed varies for different tissues of the same organism and is larger for metabolically active tissues than inactive tissues (Bonner et al. 1968[b], Bonner, Huang, and Gilden 1963, Paul and Gilmour 1968, Smith, Church, and McCarthy 1969). Further, *in vitro* template activity of chromatin isolated from "target" organs increases after administration of hormone (Dahmus and Bonner 1965, Hamilton 1968).

More important, the species of RNAs transcribed *in vitro* from isolated chromatin are not readily distinguishable from those transcribed *in situ* in the cells from which the chromatin was isolated (Bonner, Huang, and Gilden 1963, Paul and Gilmour 1968, Smith, Church, and McCarthy 1969, Tan and Miyagi 1970). Thus the structural features of the chromosomes which define the template specificity *in situ* in the living cells *appear* to be preserved in the isolated chromatin.[4]

[4] Comparisons of this type made to date have been limited to examinations of the species of RNAs transcribed from the highly redundant DNA base sequences (see Chapter 2). Further, the analyses have been made with preparations of RNA transcribed *in vitro* by exogenous preparations of bacterial RNA polymerase enzymes. Therefore, it would be premature to conclude that the template specificity of isolated chromatin is *identical* with that exhibited *in situ*. Nevertheless, it is clear that the procedures employed in the isolation of native chromatin do not totally eliminate all of the inherent template specificity of the chromatin *in situ* within the intact cell.

Indeed, studies with isolated chicken erythrocyte nuclei have shown that condensed chromatin may be reversibly dispersed by reduction of the ionic strength of the medium without major associated changes in the apparent activity as a template for RNA synthesis (Brasch, Seligy, and Setterfield 1971).

As noted in the preceding section, there appears to be no major difference in the proportions of the various classes of histone in active and inactive fractions of isolated chromatin. Thus it seems improbable that regulation of a change in a program of gene transcription can be attributed to gross changes in the *relative* proportions of the histones at the sites of gene transcription. However, transcription of RNA from a given gene locus may not require the total displacement of histone (or other protein) from the vicinity of that gene. Rather, a decrease in the affinity of the DNA for the protein due to chemical modification of the latter, in one of the processes to be discussed later in the chapter, could render the locus accessible to RNA polymerase enzyme. It is therefore pertinent to note that histone present at the sites of active gene transcription *may* be more susceptible to digestion by trypsin than that at other regions of the genome (see Allfrey et al. 1966, Robert and Kroeger 1965).

HISTONES AS INHIBITORS OF GENE TRANSCRIPTION FROM CHROMATIN

Early comparisons of compact, repressed chromatin and diffuse, unrepressed chromatin isolated from calf thymus (Frenster, Allfrey, and Mirsky 1963, Littau et al. 1964) indicated that they probably differed in histone content (Frenster 1965[a]). More recently, Bonner (1967) and Marushige (see Bonner et al. 1968[b]) have reported the content of total histone in chromatins active as template to be substantially less than that of homologous repressed chromatins. Similarly, Hamilton and Teng (1969) have found that variations in template activity of chromatin isolated from uteri exposed to estradiol are correlated with converse changes in the content of total histone.

These observations are reinforced by demonstrations that *selective* removal of histones from isolated nuclei or chromatins results in marked increases in activity as template for *in vitro* synthesis of RNA (e.g., Allfrey, Littau, and Mirsky 1963, Paul and Gilmour 1968, Seligy and Neelin 1970, Spelsberg and Hnilica 1971[a]). However, even total removal of the histone to yield "dehistoned" chromatin does not increase the template activity to that of completely deproteinized DNA (Figure 7.1), indicating that some other protein component of the chromatin also serves to repress gene transcription (Paul and Gilmour 1968, Seligy and Neelin 1970, 1971, Spelsberg and Hnilica 1971[b], Spelsberg, Hnilica, and Ansevin 1971).

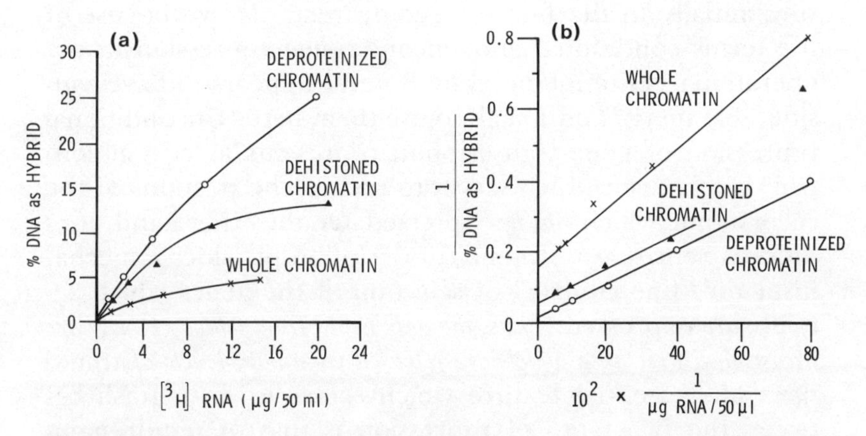

FIGURE 7.1. Repression of template activity by the proteins of calf thymus chromatin

RNA synthesized *in vitro* from native chromatin, dehistoned chromatin, and protein-free DNA was analyzed by the technique of DNA-RNA hybridization (see Chapter 5), using 5 μg aliquots of calf thymus DNA. (a) Percentage of the DNA hybridized with RNA as a function of the amount of RNA transcript present. (b) Double-reciprocal plot of the same data, which permits extrapolation of the data to the "saturation" level of hybridization at infinite concentration of transcript RNA.

Data of Paul and Gilmour (1968). Reproduced from the *Journal of Molecular Biology* with permission of the authors and Academic Press, holders of the copyright.

CONDITIONAL AND UNCONDITIONAL REPRESSION

Regulation of gene transcription in metazoan cells by two distinct mechanisms of repression was first explicitly postulated by Bonner (1967). It was proposed that some segments of the genome were unconditionally repressed in the formation of complexes with histones, especially those of classes I and II. These segments of the genome could be derepressed only by displacement of the histone during the processes of gene replication. However, other segments of the genome could be derepressed in the absence of concomitant gene replication and were conditionally repressed. Bonner further suggested that the repressor molecules responsible for conditional repression might be the nonhistone chromatin acidic protein.

Our concepts of mechanisms regulating gene transcription from the chromatin of metazoan cells have already changed substantially in the few succeeding years. However, use of the terms conditional and unconditional repression in the operational sense intended by Bonner appears to have considerable merit, and I shall so use them here. Unconditional repression of gene transcription represents a "coarse control" mechanism which serves to restrict the portions of the cell genome which *may be* expressed. On the other hand, conditional repression of gene transcription provides a mechanism for "fine control" of selection of the genes which actually *are* expressed. *Thus, the only genes from which cRNAs are transcribed are those which are released from both levels of repression.* The essential feature which serves to distinguish between the two types of repression is that of requirement for *DNA replication in release from repression.*[5] Thus, *only* conditional repression may be relieved in the absence of concomitant gene replication, whereas *both* types of repression may be relieved during the course of gene replication. This

[5] Some additional qualification is required in the case of mature sperm, for some genes present in the paternal genome may be activated without DNA replication in the early stages of embryonic cleavage (see Chapter 2).

feature also permits some measure of further operational distinction between processes essential to derepression of the genome and those involved in related phenomena, such as the total replacement of histone by protamine in maturing fish spermatids, which is considered in detail in Chapter 10.

NONHISTONE PROTEIN AND SELECTION OF THE SPECTRUM OF GENES TRANSCRIBED

The nonhistone protein fraction of chromatin includes a complex mixture of acidic proteins which is specific to the tissue of origin and, at least to some extent, also the species of organism from which the chromatin is isolated (Dastugue et al. 1970, Dravid and Burdman 1968, Elgin and Bonner 1970, Kleinsmith, Heidema, and Carroll 1970, Teng, Teng, and Allfrey 1970, 1971[a]). It also contains a number of enzymic components (e.g., O'Connor 1969, Wang 1968, Weiss 1960) and is rich in RNA (Hnilica 1967). It may also include specific hormone-binding proteins (see Chapter 6).

Derepressed, active chromatin has a higher content of total acidic protein and of phosphoprotein than repressed, inactive chromatin (Frenster 1965[a], Marushige, Brutlag, and Bonner 1968). Further, the levels of total acid protein and phosphoprotein in the chromatin appear to show much wider variation between tissues than do those of histones. In general, chromatins isolated from more active tissues have the higher contents of acidic protein (Bonner et al. 1968[b], Burdick and Himes 1969, Dingman and Sporn 1964, Kruh, Tichonicky, and Wajcman 1969, Marushige and Dixon 1969).[6] More recently, it has been reported that synthesis of particular species of chromatin acidic protein is selectively stimulated in cells of "target" organs in response to treat-

[6] Acidic phosphoprotein and basic histone readily form complexes *in vitro* (Langan 1967, Spelsberg and Hnilica 1969). Indeed, Johns and Forrester (1969) have reported the nonspecific adsorption to DNA-histone of acidic protein from both cytoplasm and nuclear sap. Caution must therefore be exercised in attributing significance to changes in the total content of nonhistone protein of isolated chromatin.

ment with hormones (Shelton and Allfrey 1970, Smith et al. 1970, Teng and Hamilton 1970[b]). In addition, synthesis of chromatin acidic protein is selectively maintained at a high rate during mitosis, in contrast to that of other nuclear proteins (Stein and Baserga 1970).

The possibility that the chromatin acidic protein might serve to regulate gene transcription in eucaryotes has been considered for a number of years. For example, Frenster (1965[a]–[d]) developed proposals in which selected gene loci were *activated* through displacement of histone by nuclear acidic protein, or other polyanions. In contrast, Bonner (1967) suggested that the chromatin acidic proteins might serve as repressors in conditional repression of gene transcription. However, the role(s) of these proteins as activators or repressors of gene transcription in the highly organized and dynamic structures of native chromatins still remains to be established. Indeed, the data currently available present a confused and ambiguous picture. As noted above, dehistoned chromatin does not have the full template activity of its constituent DNA. Thus at least some of the residual protein *can* inhibit transcription of RNA. On the other hand, nuclear acidic protein has been shown to complex with histone and thereby reduce the extent of inhibition by the latter of *in vitro* synthesis of RNA (Langan 1967, Wang 1966). Recent studies by Spelsberg and Hnilica (1969) show that the acidic nuclear protein will only prevent, and not reverse, inhibition of *in vitro* RNA synthesis due to complexing of isolated histone with purified DNA. In contrast, a more recent study of Kamiyama and Wang (1971) indicates that the nonhistone proteins of rat liver chromatin (added in large excess) *can* reverse inhibition of transcription from selected genes present in homologous condensed chromatin. Indeed, addition of the entire nonhistone fraction of the chromatin renders the template specificity of the condensed chromatin equivalent to that of the corresponding dispersed chromatin. However, the physiological significance of these effects is currently unknown.

On the other hand, studies of the spectra of RNAs transcribed from reconstituted nucleoproteins reveal a highly specific role of at least some of the components of the acidic protein fraction as "*selector protein*" which specifies the portion of the genome to be transcribed. A preparation of chromatin can be reversibly dissociated into its three major component fractions – DNA, histones, and nonhistone protein – by centrifugation in gradients of cesium chloride (Bonner et al. 1968[a], Huang and Bonner 1965). The reconstituted chromatin which is formed when all three components are allowed to re-associate under controlled conditions appears to be essentially equivalent to the original native chromatin as a template for the *in vitro* synthesis of RNA (Figure 7.2). In contrast, complexes formed in the absence of the nonhistone protein fraction have little, or no, activity as a template for RNA synthesis, and they show no resemblance to native chromatin in the sequences of RNA which are transcribed (Gilmour and Paul 1969, Paul and Gilmour 1968). The restricted specificity of the reconstituted chromatin appears to be due entirely to the nonhistone protein fraction, for the histone component can be replaced by the corresponding fraction from other tissues of the same animal (Gilmour and Paul 1970, Spelsberg and Hnilica 1970, Spelsberg, Hnilica, and Ansevin 1971).[7]

CHROMATIN-ASSOCIATED RNA AS A SELECTOR OF GENES TRANSCRIBED

One of the most controversial problems still to be resolved is that of the nature and functions of RNAs present in preparations of isolated chromatins.

On the one hand, extensive studies by Bonner and his colleagues indicate that one characteristic feature of the nonhistone protein fraction of chromatin is a high content (up to

[7] Yoshida and Shimura (1970) have recently reported differences in effects of preparations of total histone isolated from the posterior and middle silk glands, respectively, of silkworm larvae upon transcription of RNA from chromatins isolated from the same two sources. However, no differences were observed between the two preparations in experiments with reconstituted DNA-histone complexes.

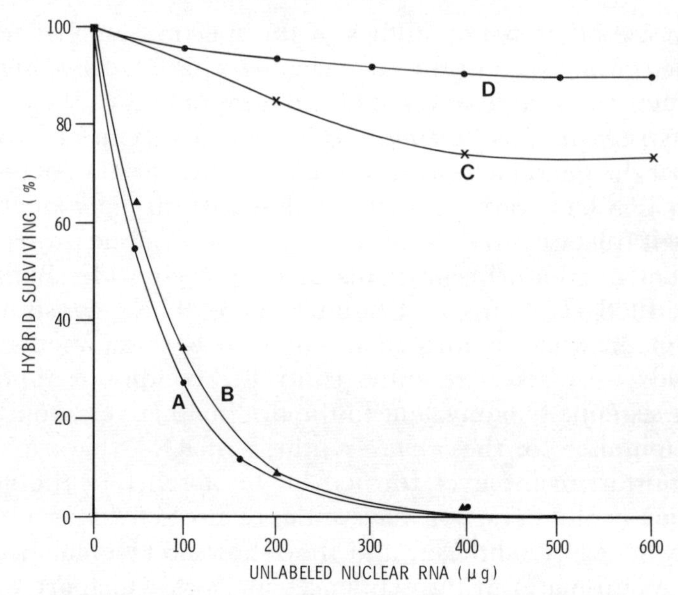

FIGURE 7.2. Specification of the portion of the chick embryo genome transcribed by the chromatin-associated RNA

Radioactive RNAs were synthesized *in vitro* from the test templates. Formation of hybrids between 50 μg of chick DNA and 25 μg of radioactive RNA was examined in the presence of increasing quantities of competing nonradioactive RNA isolated from chick embryo nuclei.

Templates were: (A) native chromatin; (B) reconstituted chromatin; (C) chromatin reconstituted after degradation of the chromatin associated RNA of the acidic protein fraction; (D.) Random re-aggregate of the components of chromatin.

Data of Huang and Huang (1969). Reproduced from the *Journal of Molecular Biology* with permission of the authors and Academic Press, holders of the copyright.

40 percent) of a distinctive type of RNA (e.g., Bonner et al. 1968[b]). This RNA is covalently linked to the nonhistone protein, but the nature of the bond has not yet been determined. A series of such "chromatin-associated RNAs" (caRNAs) has been isolated from various tissues (Bekhor, Kung, and Bonner 1969, Bonner and Widholm 1967, Huang and Bonner 1965, Huang and Huang 1969, Jacobson and Bonner 1968, Shih and Bonner 1969, Sivolap and Bonner 1971). These caRNAs are small (40 to 60 nucleotides) and are rich (5 to 10 moles per 100 moles of nucleotide

base) in dihydropyrimidine residues (Jacobson and Bonner 1968, Shih and Bonner 1969). In most of the species examined the dihydropyrimidine is dihydrouracil, which has been reported to be in the open-ring configuration as β-ureidopropionic acid (Huang and Bonner 1965). However, the caRNA of a rat ascites tumor contains dihydroribothymidine in place of uracil.[8]

Further, the caRNAs *appear* to be transcribed from highly reiterated segments of the genome (Sivolap and Bonner 1971). Preparations isolated from any given tissue will readily form hybrids with approximately 5 percent of the base sequences in homologous DNA, a level comparable with that observed for the RNA transcribed from the chromatin. Moreover, preparations of caRNAs isolated from different tissues of the same organisms contain different populations of caRNAs (Bekhor, Bonner, and Dahmus 1969, Bonner and Widholm 1967, Dahmus and McConnel 1969).

In contrast, a number of other workers have failed to find any evidence of the presence of RNAs of this distinctive type in chromatins isolated from various sources (Benjamin et al. 1966, Commerford and Delihas 1966, De Filippes 1970), even when using conditions as nearly identical as possible to those used by Bonner (Heyden and Zachau 1971).[9] Further, the experimental conditions adopted by Bonner for the formation of DNA-RNA hybrids do not require a high degree of specificity in the matching of base sequences, and do not ensure formation only of highly specific hybrids (Mc-

[8] It seems of considerable importance to determine whether this difference is a characteristic feature of all tumors. On the one hand, it could reflect the increased content and altered complement of "transfer RNA methylase" enzymes of tumor tissues (Borek 1969), and this information would assist in resolution of the current dispute concerning the intracellular site of methylation of transfer RNA (e.g., Muramatsu and Fujisawa 1968, Sirlin and Loening 1968). On the other hand, introduction of methyl groups affects the configuration and biological activity of RNAs (e.g., Capra and Peterkofsky 1968, Pillinger, Hay and Borek 1969, Shugart, Novelli, and Stulberg 1968).

[9] A number of workers have reported the presence or isolation of various novel species of nuclear RNAs. However, the significance and functions of these RNAs are totally unknown and cannot profitably be speculated upon at this time.

Conaughy, Laird, and McCarthy 1969). It has therefore been concluded that the bulk of the material of the size of Bonner's caRNAs consists of partially degraded transfer RNA (De Filippes 1970, Heyden and Zachau 1971).

Nevertheless, reconstitution of apparently native chromatin from the dissociated components has been observed only under conditions which permit formation of specific (as well as less-specific) hybrids between caRNAs and homologous DNAs. Similar complexes formed after destruction of the caRNA component have approximately the same level of activity as templates for *in vitro* synthesis of polyribonucleotide chains as have the corresponding reconstituted chromatins (Figure 7.2). However, the spectrum of RNAs transcribed from the complex without caRNA is totally different from that obtained with the native and reconstituted chromatins (Bekhor, Kung, and Bonner 1969, Huang and Bonner 1969). Thus the template specificity of a chromatin *appears* to be determined in part by some component of the caRNA fraction which is both attached to chromatin selector protein and complexed with the homologous section of the DNA.

However, it seems highly improbable that selection of the spectrum of genes transcribed from the chromatin is determined *solely,* or even primarily, by the content of caRNAs. Administration of estradiol to an ovariectomized rat leads to a rapid and progressive increase in the template activity of uterine chromatin for *in vitro* synthesis of RNA, which reaches a maximum after 8 to 12 hours and then declines. However, after a small initial increase which is complete within 15 minutes, the (total) RNA content of the chromatin remains almost constant for a period of 8 hours, and then rises again as the template activity declines. Similar, but less marked, changes are observed in the template activity of the liver chromatin. In contrast, the RNA content of the liver chromatin remains unchanged for 4 hours and then increases (Teng and Hamilton 1968). Moreover, the tissue and species specificities exhibited by the nonhistone proteins of chromatins are presumptive evidence that these proteins play more specific roles than merely serving as apo-proteins of caRNA-protein complexes.

PHOSPHORYLATION OF CHROMATIN ACIDIC PROTEIN

The proteins of the cell nucleus include a phosphoprotein fraction which: (a) serves as a substrate for specific phosphoprotein kinase enzymes, (b) possesses some inherent phosphoprotein kinase activity, and (c) complexes with histone and reduces its activity as an inhibitor of *in vitro* RNA synthesis (Kleinsmith and Allfrey 1969[a], Langan 1967, 1968[b]). Further, there is an intense metabolic turnover of the phosphate groups of nuclear phosphoproteins (Kleinsmith and Allfrey 1969[b], Kleinsmith, Allfrey, and Mirsky 1966[a]), and specifically in chromatin (Benjamin and Goodman 1969).

The role of this turnover of phosphate groups has not yet been determined. However, two lines of recent evidence indicate that it may play a role in one, or more, process(es) involved in regulation of the spectrum of genes transcribed into RNA. When lymphocytes are stimulated to proliferate by phytohemagglutinin there is a marked increase in the rate of phosphorylation of nuclear phosphoprotein before the increase in rate of gene transcription or the later onset of gene replication (Kleinsmith, Allfrey, and Mirsky 1966[b]). Further, both the complement and the pattern of phosphorylation of chromatin phosphoproteins differ markedly for different tissues of the same animal (Teng, Teng, and Allfrey 1970, 1971[a], 1971[b]).

HISTONES AS NONSPECIFIC ADJUNCT REPRESSORS OF GENE TRANSCRIPTION

Data considered in preceding sections of this chapter show that gene transcription from the DNA of eucaryote cells may be repressed, in different manners, by two distinct components of the homologous chromatin. Transcription from dehistoned chromatin is repressed by the nonhistone protein fraction which specifies the spectrum of genes actually transcribed from the intact chromatin. At the present time we do not know whether this fraction does, in fact, repress or activate gene transcription within the complex assembly comprising intact chromatin. However, we do know that the acidic protein fraction is highly specific in its action, and, at least to some extent, it may be considered as equivalent in function to the specific repressor proteins of procaryotes.

In contrast, the basic histone fraction is the most effective inhibitor of the template activity of protein-free DNA, and it quantitatively accounts for the bulk of the repression of gene transcription from intact chromatin. This fraction appears to be unspecific in its action. Repression of gene transcription from intact chromatin is presumably due to complex formation between the histone and DNA not "shielded" by the acidic protein fraction. Nevertheless, the histone protein does not appear to be distributed randomly along the chromatin fiber, but rather in some form of organized pattern. This probably reflects both differences in affinities of the various types of histone for complex formation with acidic protein, and changes in affinities for mutual interaction when associated with acidic protein. Thus, again at least to some extent, some, or all, types of histone may be regarded as corepressors of repressor proteins, or adjunct repressors.

Results obtained in analyses of the roles of the various classes of histone as inhibitors of gene transcription from isolated chromatins depend upon the degree of specificity of the processes used to effect stepwise removal of histone. Selective removal of *only* the lysine-rich histones of class I has no effect upon the *in vitro* template activity of chromatin isolated from avian erythrocytes. The activity increases markedly on further removal of the characteristic serine- and arginine-rich histone of class V. A small additional increase in template activity is obtained on extracting the remaining slightly lysine-rich histone IIb and arginine-rich histones III and IV (Seligy and Neelin 1970). Similarly, there is little increase in template activity after selective removal of the histone I fraction from chromatins of other tissues. In contrast, large increments in activity are obtained on removing either slightly lysine-rich histone IIb or the arginine-rich histones III and IV (Bonner et al. 1968[b], Seligy and Neelin 1971, Spelsberg and Hnilica 1971[a], 1971[b]).

On the other hand, it has been reported that the greatest increase in template activity is obtained on removing the lysine-rich histones I by procedures which also extract some nonhistone protein (Georgiev 1968, Georgiev, Ananieva, and Kozlov 1966). Indeed, the extracted chromatin appears to be almost equivalent to protein-free DNA as a template for *in vitro* synthesis of RNA, and it has been suggested that the lysine-rich histones I are repressors of transcription from the highly reiterated base sequences of the DNA (Georgiev 1968).[10] However, removal of histone from the chromatin by these procedures also extracts a large proportion of the nonhistone protein (Georgiev, Ananieva, and Kozlov 1966). It is therefore not clear to what extent the increases in template activity observed in these studies are due specifically to removal of lysine-rich histone or solely to removal of nonhistone protein, respectively. Appraisal of the roles of these two components must be made from other types of experiment.

Nevertheless, the very marked differences in responses which can be attributed to extraction of some of the non-

[10] Koslov and Georgiev (1970) have recently further suggested that repression of transcription by histones I is not a secondary consequence of effects upon the structure of chromatin. Rather, it is suggested that the histone blocks movement of the RNA polymerase along the DNA template and leads to synthesis of shorter RNA chains. However, these suggestions are based in part upon incomplete data and in part upon conclusions which are at variance with the authors' own published data.

histone protein serves to re-emphasize that transcription of the genome of a metazoan may be regulated by two types of mechanisms of repression.

HISTONE MODIFICATION AS A FACTOR IN REGULATION OF CHANGES IN SPECTRA OF GENES TRANSCRIBED

The strength of the binding of histone to DNA, and the inhibition of *in vitro* RNA polymerase activity, can both be substantially altered by chemical or enzymic modification of the histone molecule. Modifications with these consequences *in vitro* which have been shown to occur in intact cells include acetylation, methylation, and phosphorylation. The extent of such modifications to all classes of histones has been shown to vary during the course of biological responses involving a change in the selection of genes transcribed. A variety of patterns of histone modification have been reported in different regulatory processes, and the data show no unique invariant sequence of histone modification. For example, suspensions of nondividing human peripheral lymphocytes in tissue culture medium show a high rate of acetylation of arginine-rich histones and a very low rate of acetylation of lysine-rich histones. However, when stimulated to proliferate by phytohemagglutinin they show an immediate marked rise in the rates of acetylation of both types of histone (Pogo, Allfrey, and Mirsky 1966[b]).[11] In contrast, immediately following partial hepatectomy the remaining liver tissue shows a rapid transient rise in the rate of acetylation of arginine-rich histone, whereas there is little acetylation of lysine-rich histone for a period of 16 hours (Pogo et al. 1968, Pogo, Pogo, and Allfrey 1969).

These marked variations in patterns of histone modification are probably largely, if not entirely, a reflection of the

[11] Monjardino and MacGillivray (1970) have recently reported that histone acetylation is not stimulated by highly purified preparations of phytohemagglutinin. However, effects of these preparations upon the synthesis of RNA were also unimpressive.

diverse interdependent roles of the histones noted earlier in the chapter. Some of the observed modifications occur only during the *de novo* synthesis of histone which normally accompanies gene replication (e.g., Marzluff and McCarty 1970, Shepherd, Noland, and Hardin 1971). Such modifications would be expected as a consequence, but not a cause, of release from unconditional repression, superimposed upon any other histone modifications controlling the derepression. This appears to be particularly pertinent in the case of methylation reactions. Arginine-rich and slightly lysine-rich histones can be readily methylated *in vitro* by enzymes isolated from various sources (e.g., Kaye and Sheratzky 1969, Patterson and Davies 1969), and also in isolated nuclei (Allfrey, Faulkner, and Mirsky 1964). However, there is relatively little methylation of these histones *in vivo* (Gershey et al. 1969, Orenstein and Marsh 1968). Further, the methylation of these histones which does occur in association with a change in program of gene transcription usually takes place after onset of DNA replication and is probably due to the attendant *de novo* synthesis of histone (Tidwell, Allfrey, and Mirsky 1968).

Analysis of the roles of histone acetylation and phosphorylation is considerably more complex. Both processes occur in nondividing cells in the absence of demonstrable changes in the spectrum of genes transcribed and at elevated rates in the early stages of changes in programs of gene transcription (e.g., Allfrey et al. 1966, Pogo et al. 1968). On the other hand, both processes also occur during the terminal stages of differentiation of nondividing specialized cells in which the entire genome becomes unconditionally repressed (e.g., Adams, Vidali, and Neelin 1970, Dixon et al. 1969, Sung and Dixon 1970). Sung and Dixon (1970) have presented a persuasive argument that one function of such modification reactions is to facilitate the incorporation of histone, or protamine, into a nuclear protein with an organized structure by preventing formation of random nucleoprotein aggregates. Data currently available (see Chapter 10) are entirely in accord with this suggestion but, more important for present purposes, also strongly indicate that these reactions *also* facilitate dissociation of the bonds between histone already present in chromatin and the DNA.

It therefore seems probable that certain of the histone acetylation and phosphorylation reactions are directly and causally involved in processes regulating changes in programs of gene transcription. This appears to be particularly the case in phenomena of release from conditional repression. Indeed, some general features appear to be emerging from the accumulated data.

ACETYLATION OF ARGININE-RICH HISTONES AND RELEASE FROM CONDITIONAL REPRESSION

In all tissues examined to date, a high rate of gene transcription has been correlated with rapid acetylation of the arginine-rich histones. Thus high rates of acetylation of this class of histones has been observed with isolated nuclei from calf thymus, HeLa cells, rat brain, and rat liver (Allfrey, Faulkner, and Mirsky 1964, Bondy, Roberts, and Morelos 1970, Gallwitz 1970, Gallwitz and Sekeris 1969), with both nondividing and phytohemagglutinin-"transformed" lymphocytes (Darźynkiewicz, Bolund, and Ringertz 1969, Handmaker and Graef 1970, Ono et al. 1969, Pogo, Allfrey, and Mirsky 1966[b]), granulocytes (Pogo, Allfrey, and Mirsky 1967), regenerating liver (Pogo et al. 1968), and polycythemic spleen (Takaku et al. 1969). Conversely, little or no acetylation of the arginine-rich histone fraction occurs in mature avian erythrocytes (Allfrey 1968) and in phytohemagglutinin-"suppressed" granulocytes (Pogo, Allfrey, and Mirsky 1967).

This correlation between rates of gene transcription and of histone acetylation has been extended to the level of the chromatin. Acetylation of the histone fraction in the diffuse active chromatin of calf thymus nuclei occurs markedly more rapidly than acetylation of the compact repressed chromatin, and the extent of acetylation of histone isolated from the active chromatin is some 70 percent greater than that of histone in the inactive chromatin (Allfrey 1968, Allfrey et al. 1966).[12] Virtually all of the acetylation which occurs with isolated nuclei *in vitro* is into ε-N-acetyl·lysine residues of histones III and IV (Gershey, Vidali, and Allfrey 1968, Vidali, Gershey, and Allfrey 1968). In a related study of lymphocytes briefly treated with phytohemagglutinin, Darźynkiewicz, Bolund, and Ringertz (1969) found that only those cells which exhibit both an increased ability to bind acridine orange to DNA (a measure of the degree of "opening up" of chromatin structure) and active acetylation of histone showed significant rates of gene transcription. Thus, at least in part, *acetylation of histones III and IV is correlated with transcription from genes in the active chromatin already derepressed by some other mechanism.* Indeed, the difference between active and inactive chromatin does not lie in the extent to which the histone complement can be acetylated, but in the absence from inactive chromatin of some component required for generation of acetylcoenzyme A (Pogo et al. 1968). The nature of the mechanism by which the structure of the chromatin is opened up and selected genes are derepressed has not yet been established. However, it appears probable (see p. 297) that it involves phosphorylation of the lysine-rich histones I (Cross and Ord 1970, Handmaker and Graef 1970).

A second general feature is observed in phenomena in which there is a marked, dramatic rise in the rate of gene transcription due to release from conditional repression. Namely, the rise in rate of gene transcription is preceded by a marked rise in the rate of acetylation of arginine-rich histones and a decrease in the rate of "turnover" of the acetyl

[12] A recent report that there is no preferential acetylation of histone within the puffs of dipteran salivary gland chromosomes (Clever and Ellgaard 1970) is not necessarily at variance with these conclusions. The site of histone acetylation may be a regulator section of the genome distinct from the section actually transcribed.

groups. Thus, in lymphocytes stimulated by phytohemagglutinin, the maximal level of histone acetylation is attained before any marked increase in the rate of gene transcription and may even be observed without concomitant increase in the latter (Ono et al. 1969, Pogo, Allfrey, and Mirsky 1966[b]). An equally dramatic example is afforded by the regenerating liver (Pogo et al. 1968, Pogo, Pogo, and Allfrey 1969). Immediately following partial hepatectomy, there is a large increase in the rate of histone acetylation and a marked decrease in the rate of de-acylation. This results in a large net increase in the level of acetylation of histone IV and smaller increases for histone III and the slightly lysine-rich histones II. The maximal rate of acetylation is attained at 3 to 4 hours after the operation and thereafter declines abruptly to that of a normal liver by 5 hours. These changes precede attainment of the maximal rate of gene transcription and are complete several hours before onset of gene replication. Nevertheless, they are associated with early increases in the template activity of the chromatin (Dahmus and Bonner 1965, Pogo, Allfrey, and Mirsky 1966[a]) and even more dramatic changes in the spectrum of cRNAs transcribed (Church and McCarthy 1967).

Within one hour of the operation there is a remarkable increase in the fraction of the total genome of the cell transcribed. This includes a number of species of highly labile RNAs which are neither transcribed nor present as little as 2 hours later. Indeed, from the first hour the proportion of the total genome transcribed declines steadily to that observed for a normal liver, despite the onset of transcription of hitherto repressed genes after 6 to 12 hours. Further, a number of the species so transiently transcribed are of high molecular weight, and the proportion of such large species of RNA is much greater 1 hour after the operation than at later stages, or in actively developing embryonic liver (Church and McCarthy 1967). *Thus the greatest variety of cRNAs is transcribed before attainment of the maximal rate and level of acetylation of the arginine-rich histones.*

We are, therefore, faced with an apparent paradox, to which the solution will undoubtedly substantially increase our understanding of the mechanisms by which the arginine-rich histones serve as adjunct repressors in conditional repression. On the one hand, enhanced transcription of a new selection of mRNAs is preceded by intense acetylation of arginine-rich histones. On the other hand, maximal rates and levels of this acetylation are preceded by transitory transcription of a wide variety of short-lived species of cRNA. It seems probable that many of the latter are species with a regulatory function and presumably include any new species of chromatin-associated caRNAs which may be required to specify new segments of the genome to be transcribed as mRNAs during the course of regeneration. Perhaps intense acetylation of arginine-rich histone in active regions of the chromosome serves as a general mechanism facilitating access of other macromolecules, including caRNAs, to the DNA of the chromatin. In effect, this would provide a "scanning" mechanism whereby sequences of regulatory RNAs could be matched against conditionally repressed segments of the genome, and unusable species discarded. However, detailed speculation upon this possibility is not warranted at this juncture, and particularly insofar as it does not account for the transient early transcription of a large variety of cRNAs.

PHOSPHORYLATION OF LYSINE-RICH HISTONES AND RELEASE FROM CONDITIONAL REPRESSION

Early studies which directed attention to the phosphoproteins of the chromatin as potential regulators of gene transcription (Allfrey et al. 1966, Kleinsmith, Allfrey, and Mirsky 1966[a]) tended to divert attention from the observation that there was also some phosphorylation of the histone fraction. However, attention has recently been directed to the roles of histone phosphorylation by the finding of a variety of histone and protamine kinase enzymes distinct from the phosphoprotein kinases active on acidic proteins (e.g., Jergil and Dixon 1970, Kuo and Greengard 1970, Langan 1968[b]), and of corresponding specific phosphatases (Dixon et al. 1969, Meisler and Langan 1969).

A number of different types of histone kinase enzyme have now been found in a wide variety of tissues and shown to be active on all classes of histone *in vitro*. Some of these enzymes preferentially phosphorylate the arginine-rich histones (e.g., Kuo and Greengard 1970, Tao, Salas, and Lipmann 1970), whereas others are more active with the lysine-rich and slightly lysine-rich histones as substrates (e.g., Langan 1968[b]). Most of the histone kinases examined to date show a marked stimulation of the rate of histone phosphorylation in response to low concentrations of cyclic 3′·5′ AMP (cAMP) (e.g., Kuo and Greengard 1970, Langan 1969[a]).

At the present time, the only system for which there is unambiguous evidence of the precise functions of such kinase enzymes is that of maturing fish spermatids (Sung and Dixon 1970), considered in Chapter 10. Indeed, analysis of the patterns of histone phosphorylation in different tissues of normal rats over a period of only one hour (Gutierrez and Hnilica 1967) presents a formidable problem.

Nevertheless, there are a few isolated observations in the literature which indicate that phosphorylation of the lysine-rich histones is probably a factor in release from conditional repression. In particular, the level of phosphorylation of the lysine-rich histones of rat liver rises markedly shortly after injection of cAMP (Langan 1969[a]). Similarly, the level of phosphorylation rises markedly within one hour of administration of glucagon or insulin, but is unaffected by hormones for which the physiological action is not mediated by cAMP (Langan 1969[b]). Equally dramatic early increases in the rate of turnover of phosphate groups of the lysine-rich histone fraction have been reported for phytohemagglutinin-stimulated lymphocytes (Cross and Ord 1970, Handmaker and Graef 1970). On the other hand, the rate of phosphorylation of lysine-rich histones in regenerating liver does not rise markedly until just before an increased rate of synthesis of DNA becomes evident (Stevely and Stocken 1968). Thus, no firm general conclusion can yet be made regarding the role of histone phosphorylation in release from conditional repression.

TISSUE SPECIFIC DIFFERENCES IN HISTONE COMPLEMENT

Comparisons of homologous active and inactive chromatins indicate that they differ substantially in the content of total histone relative to that of the DNA (Bonner et al. 1968[b], Frenster 1965[a]), but not in the relative proportions of the various classes of histone (Comings 1967, Frenster 1965[a]). Nevertheless, quite marked differences have been observed in the proportions of the various classes of histones present in chromatins isolated from different tissues of the same organism (e.g., Dastugue et al. 1970, Fambrough, Fujimura, and Bonner 1968, Kinkade 1969, Leclerc et al. 1969). These differences may be extreme in highly specialized cells in which the entire genome appears to be unconditionally repressed and include the presence of entirely novel species of histone or even total replacement of histone by more basic protamines (e.g., Bloch 1969, Dixon et al. 1969, Neelin et al. 1964).

However, for most types of tissue, organ-specific differences in the complement of histones present appear to be limited to the lysine-rich fraction (e.g., Bustin and Cole 1968, Kinkade 1969, Nelson and Yunis 1969, Panyim and Chalkely 1969, Stellwagen and Cole 1969[a]). Moreover, purified histone I fractions isolated from different tissues of a single animal species vary markedly in the sites and extents of phosphorylation by a single preparation of protamine or histone kinase (Jergil, Sung, and Dixon 1970). It seems possible that these phenotypic, tissue-specific differences in the histone I fraction may reflect roles of this class of histone in conditional repression of gene transcription.

On the other hand, as noted earlier, it has been suggested that the histone V specific to avian erythrocytes functions as a super-repressor of gene transcription, and also is the component primarily responsible for the condensation of the erythrocyte chromatin (Seligy and Neelin 1970). Similar roles are envisaged for the protamines which replace histone during maturation of fish spermatids (see Chapter 10) and other highly basic species of histone found specifically in the sperm of other species. It seems equally probable that the lysine-rich histones act in similar, though less dramatic, roles as adjunct repressors of unconditional repression and in the condensation of heterochromatic regions of the chromosomes. Thus, organ-specific differences in the complement of lysine-rich histones probably reflect the differences in proportions of the genome which are derepressed, conditionally repressed, and unconditionally repressed, respectively, although this remains to be established in a nontrivial sense.

MODELS OF MECHANISMS REGULATING GENE TRANSCRIPTION IN METAZOANS

The success with which the concept of the operon (Jacob and Monod 1961[a]) provided a unified interpretation of regulatory phenomena in procaryotes (Chapters 3 to 5) prompted attempts to explain the complex patterns of control in metazoans in terms of analogous operons. Some early success was achieved in devising abstract schemes based on "interlocking operons," in which gene products specified by one operon served as apo-repressors or inducers of one (or more) other operon(s) (Jacob and Monod 1963). However, attempts to apply such schemes to the interpretation of specific regulatory phenomena were considerably less successful. For example, consideration of the "switch" from synthesis of fetal hemoglobin to that of adult hemoglobins which occurs early in human neonatal life led to the postulation of complex multisectioned operons, and even of operons exhibiting properties diametrically opposed to those explicitly defined for operons (e.g., Motulsky 1962, Zuckerkandl 1964, see also Baglioni 1963). Further, as noted by Waddington (1969[b]), such schema provided no obvious basis for interpretation of the important embryological phenomenon of determination.

Nevertheless, the essential elements in the only two detailed hypothetical models which have since been presented are complex gene aggregates formally related to bacterial operons, and the concept of interlocking operons is either explicitly or tacitly retained.

THE BRITTEN-DAVIDSON MODEL

The essential feature of a model (Figure 7.3) formulated by Britten and Davidson (1969) is one of interaction between two distinct types of organized gene complex, each of which shows substantial similarity to the operons of procaryotes. In this model transcription of messenger RNAs (and also ribosomal and transfer RNAs) is determined and regulated by gene complexes of type I. These consist of one structural gene known as a "*producer*" gene and an adjacent segment of regulator genes known as "*receptor*" genes. The latter serve to control transcription from the producer gene in a manner analogous to, but different from, that of the operator-promoter segment of the operon of procaryotes (Chapters 3 and 5). Thus the features proposed for this complex differ only in three respects from those defined for segments of the hypothetical dispersed operon postulated in early attempts to account for the coordinate regulation of synthesis of enzymes of the arginine pathway of *E. coli* (see Chapter 4). First, the complex *may* be envisaged as containing more than one regulatory segment, although this is not an essential requirement. Second, derepression is envisaged as a positive control process in which a regulatory molecule combines with the receptor gene itself to actively displace a repressor. Third, the regulatory molecules are envisaged as RNAs

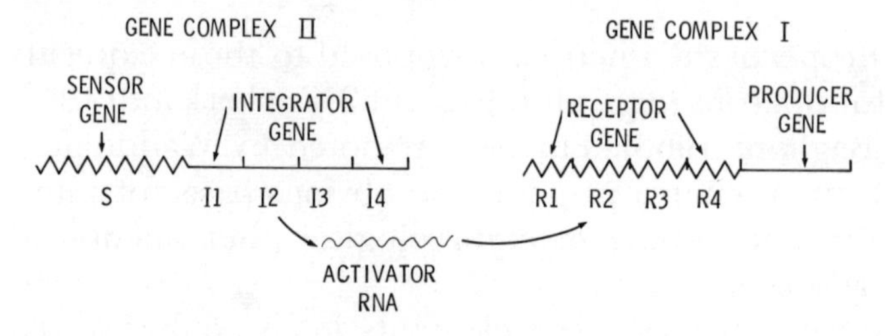

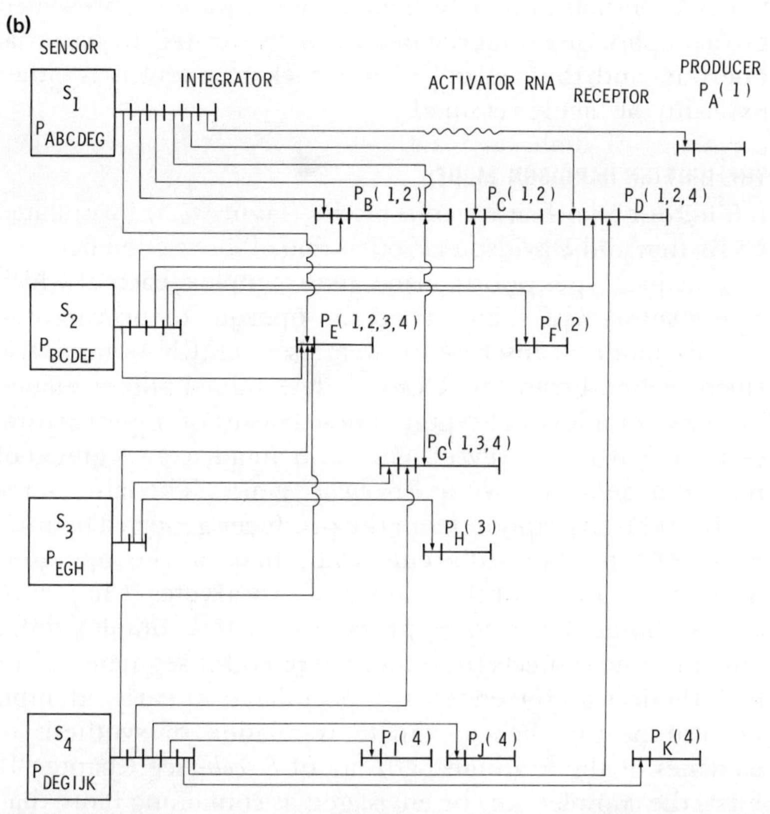

FIGURE 7.3. Regulation of gene transcription. The integrative model of Britten and Davidson (1969)

(a) The basic sensor-integrator and receptor-producer gene complexes. (b) A model integrated system, showing various types of interaction between the two classes of gene complex yielding overlapping gene batteries (see text).

Reproduced with the permission of the authors and the editors of *Science*. Copyright 1969 by the American Association for the Advancement of Science.

transcribed from the structural gene components of the other type of gene complex II.

This latter gene complex II consists of a series of structural genes known as "*integrator*" genes and a single adjacent regulatory segment known as the "*sensor*" gene. The integrator genes of this type of complex consist exclusively of genes specifying the structure of "activator RNAs" of regulatory function. Sensor genes also function in a manner analogous to the operator-promoter segments of bacterial operons. They control transcription of the adjacent integrator genes and are the site of action of inducer molecules such as hormones or embryonic inducers. The features proposed for this type of gene complex II differ from those explicitly defined for bacterial operons only in two respects. First, the RNAs transcribed from the structural integrator genes consist exclusively of species which regulate the activity of gene complexes of type I, by combining specifically with homologous receptor genes.[13] Second, the precise mechanism by which sensor genes exert their control of transcription of the integrator genes of this "operon-like" complex is not specified but is postulated to involve direct combination of the "inducing agent" with the DNA of the sensor gene.

Thus the *mechanism* whereby the spectrum of mRNAs transcribed in the metazoan cell is *controlled* is envisaged as a two-step process. A primary stimulus acts at the level of one (or more) sensor gene(s) to regulate transcription from adjacent integrator genes of an appropriate selection of activator RNAs. In turn, these interact with corresponding receptor genes and thereby control transcription from the producer genes of a highly specific selection of mRNAs.

A second feature of the model provides a mechanism for *regulation* of the *selection* of the highly diverse spectra of

[13] The mechanism postulated here requires an identity of base sequence in the strand of the receptor gene with which the activator RNA complexes and in the integrator strand from which it is transcribed. Such a situation would invalidate determinations of the fraction of the cell genome transcribed by DNA-RNA hybridization experiments.

genes transcribed in different cells of the same organism, in different tissues, at different stages of development, and in response to a variety of physiological stimuli (see Chapter 6). Namely, that the highly reiterated and apparently redundant genes present in the genome (Britten and Kohne 1968; see Chapter 2) afford the possibility of an infinite variety of distinct patterns of interaction between the two types of gene complex. Thus two complexes of type I might contain identical producer genes, but totally different sequences of receptor genes, and be activated independently of each other under the control of different sensor genes. Conversely, two such complexes might contain different producer genes, but common receptor genes, and be controlled by a common sensor gene. The situation actually envisaged is one in which two complexes of type I have some producer genes in common, but differ in others. Similarly, different complexes of type II may have some integrator sequences in common and differ in others.

Thus various degrees of coordination of the transcription of one given pair of specific mRNAs may be envisaged. Entirely coordinate transcription could be obtained from a series of producer genes controlled by a common receptor gene sequence, by coordinate transcription of the corresponding activator RNAs from a series of integrator genes linked to, or controlled by, one common sensor gene. Similarly, various levels of interaction could control concurrent, but noncoordinate, transcription of a series of cRNAs, or regulate their transcription totally independently. Examples of these various complex patterns of interaction are illustrated in Figure 7.3, reproduced from Britten and Davidson (1969).

Stimulation of one given sensor gene and subsequent transcription of the linked integrator genes will lead to activation of a large number of unlinked groups of producer genes. These are collectively defined as the "*battery of genes*" regulated by that sensor. Each distinct sensor gene will regulate the transcription of a distinct battery of genes, defined by the integrator genes linked to the sensor gene and the corresponding receptor genes linked to producer genes. Any one given producer gene may comprise part of a number of distinct batteries. Further, additional copies of any reiterated producer gene may comprise part of additional batteries. Thus any change in the status of one sensor gene may affect the rate or extent of transcription of a large number of producer genes. On the other hand, it is possible to envisage a change in the number of activated sensor genes without any parallel change in the *total number of species* of producer gene available for transcription in the cell. Rather, there would be a change in the *extent* of transcription of certain producer genes in response to various stimuli, and even a qualitative change in the ability to transcribe them under particular restricted sets of environmental conditions. Indeed, the consequences of irreversible inactivation of one of the postulated sensor genes would be formally equivalent to the determination of a new level of differentiation.

This highly speculative proposal contains a number of important and interesting features. In particular, it assigns explicit roles to the many highly reiterated base sequences present in the genomes of metazoan cells, and it provides a basis for interpretations of the phenomenon of determination. At the present time, it would be rashly premature to attempt to make a definitive appraisal of the degree of correspondence between the model and the actual situation in real cells. However, evidence already available poses special problems of interpretation for the model to be retained without modification. The most obvious of these problems is that none of the many species of RNA found in eucaryote cells *appear* to have all the minimum essential properties (Britten and Davidson 1969) of the postulated activator RNAs. Indeed, the only species of nuclear RNAs which may play an obvious role in determining the spectrum of cRNAs synthesized are the disputed chromatin-associated caRNAs.

These may be transcribed from highly reiterated base sequences in the DNA, but appear to select regions of the genome in dehistoned chromatin from which transcription is *repressed* by acidic protein. As suggested earlier in the chapter, this may not be the mechanism of action in intact chromatin; but this remains to be determined. Additional problems are posed by the observation that the template activity and RNA content of the chromatin may change in opposite directions during the course of response of a target organ to exposure to hormone.

Recent studies of the base sequences (Southern 1970), intrachromosomal distribution (Rae 1970, Yasmineh and Yunis 1970, 1971), and species specificities (Walker 1968) of rodent satellite DNAs indicate that these extremely redundant segments of the genome do not function as any of the types of gene postulated in the model. This presents substantially less of a problem but does, however, emphasize the need for caution in assigning specific functions to other reiterated segments of the genome.

THE GEORGIEV MODEL

Georgiev (1969[b]) has proposed a mechanism for the regulation of gene transcription in metazoan cells based upon a model consisting of a "giant" compound operon. This model was formulated primarily to account for properties of highly labile, giant species of cRNAs found only in the cell nucleus, which are considered more fully in Chapter 8. These giant cRNAs contain base sequences which appear to be transcribed from highly reiterated segments of the genomes and are not found in the cell cytoplasm, plus other base sequences either similar to, or identical with, those present in the mRNA fraction isolated from cytoplasmic polysomes.

Georgiev (1969[b]) proposes that the giant species of cRNA are polycistronic transcripts from compound operons. Each such operon (Figure 7.4) is composed of an "*acceptor*" region containing a series of regulator gene loci and an "*informative*" zone of contiguous structural genes. Transcription of this operon commences from a promoter located at the end of the acceptor region and proceeds sequentially toward the informative region of the operon. Interaction of any one of the regulator acceptor gene loci with a repressor of gene transcription will either block or reduce the rate of transcription of the operon beyond that regulator locus. Thus unimpeded transcription of any of the structural genes of the informative zones of the operon would require prior tran-

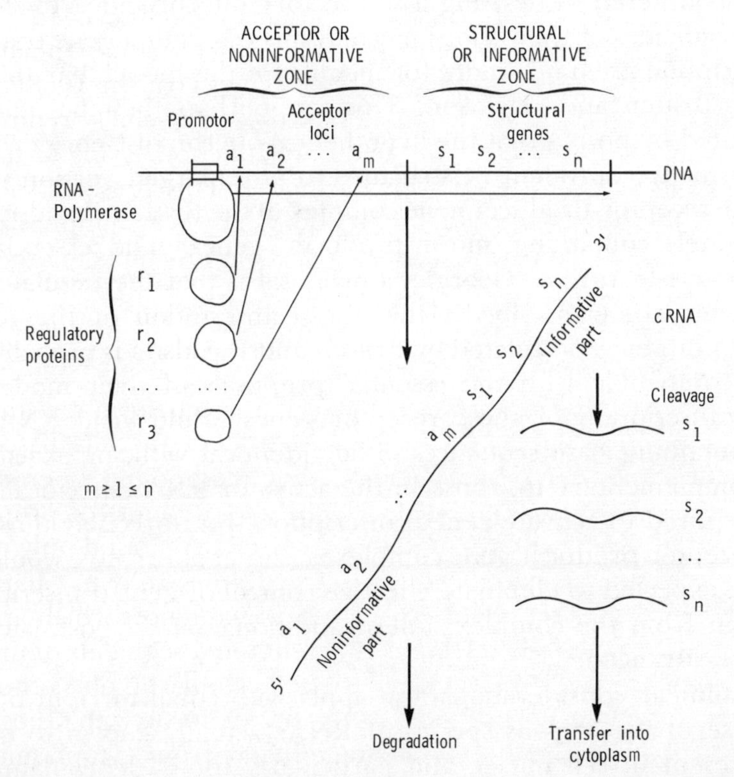

FIGURE 7.4. Regulation of gene transcription. The operon model of Georgiev (1969[b])

Reproduced from the *Journal of Theoretical Biology* with permission of the author and Academic Press, holders of the copyright.

scription of the entire sequence of regulator gene loci. Once the completed giant cRNA is released from the coding strand of the DNA on which it is transcribed, small monocistronic mRNAs are excised and individually transported into the cell cytoplasm, whereas the RNA segment transcribed from the regulator gene segment of the operon is degraded within the nucleus.

Georgiev's proposals do appear to provide one satisfactory interpretation of the data currently available regarding the properties of the giant cRNAs. However, the problems encountered in devising a satisfactory interpretation of the specificities of the various mechanisms regulating gene transcription are much more formidable for this model than that of Britten and Davidson. This can perhaps best be illustrated by noting that the hypothetical operon of Georgiev is formally equivalent in structure to an enlarged version of the receptor-producer gene complex of the Britten-Davidson model, containing more producer genes. The essential central feature of Georgiev's proposal is that the regulator genes are transcribed. However, transcription of the receptor genes postulated by Britten and Davidson is probably incompatible with the essential proposals of their model. Transcription of these receptor genes would yield RNAs containing base sequences either identical with, or exactly complementary to, those in the activator RNAs specifically required to activate gene transcription from precisely those receptor-producer gene complexes. Either alternative would be expected to eliminate effective control of gene transcription from the complex, unless additional *ad hoc* postulates are advanced.

Similar considerations also apply with equal force in the case of the various species of RNAs actually known to be present in cell nuclei, and particularly for the chromatin-associated caRNAs which have been postulated to control the spectrum of genes actually transcribed. On the other hand, a mechanism in which each of the postulated regulatory gene loci is derepressed through displacement of histone by a corresponding locus-specific protein would appear to require the presence in the cell nucleus of an inordinately large number of distinct molecular species of nonhistone protein. These reservations do not, of course, require the conclusion that the Georgiev model is invalid. Rather, they serve to re-emphasize the considerable extent of our present ignorance of the mechanisms which regulate gene transcription in metazoan cells.

A model recently elaborated by Tsanev and Sendov (1971) is essentially a modified version of the Georgiev model. They postulate that the promoter (or "recognition site") of the compound operon is not defined by the base sequence of the DNA, but rather by the structural arrangement of the particular combination of histones which has become associated with it fortuitously during the course of evolution. This histone complex combines with and "blocks" the promoter. The latter is "deblocked" by nonhistone protein which recognizes the specific histone combination and complexes with it, thereby unmasking the site of initiation of gene transcription. It is further proposed that the deblocking protein also binds firmly with the DNA of the promoter, thereby maintaining the operon in a deblocked (or conditionally repressed) state until "cleared" in the process of gene replication. This model compounds problems particular to the specific proposal with those of the more general Georgiev model.

Since, then, one is earlier and another later, does the one make the other, and does the later part exist on account of the part which is next to it, or rather does the one come into being only after the other? —Aristotle, Fourth Century B.C.

For what grows from a thing must have its nature, as an offshoot of its substance. —Claudius Galenus, Second Century B.C.

CHAPTER 8: TRANSCRIPTION AND SELECTIVE EXTRANUCLEAR TRANSMISSION OF GENETIC INFORMATION IN METAZOANS

Analysis of the mechanisms controlling formation of messenger RNAs in metazoan cells is a very intricate task and is currently only in the preliminary stages. It requires discrimination between the contributions of several complex phenomena to any change in the complement of a large variety of species of RNA. Moreover, the analysis is further complicated by three additional factors. One is the presence in the genome of families of very highly reiterated (mean frequency *ca.* 10^6/genome) and highly reiterated (mean frequency 10^3 to 10^4/genome) DNA base sequences (Britten and Kohne 1968, see Chapter 2). The bulk of the cRNA formed *in vivo* in cells of both young and adult animals, or of that synthesized *in vitro* on a template of isolated chromatin or deproteinized DNA, is transcribed from these large families of related base sequences (e.g., Georgiev 1968, Melli and Bishop 1969, 1970, McCarthy and Church 1970). A considerable proportion of the cRNAs synthesized in maturing oocytes (Davidson and Hough 1969) and in fetal tissues (Gelderman, Rake, and Britten 1971) may be transcripts from unique nonreiterated genes. However, the remainder of the cRNAs formed in these cases also are transcripts of the highly reiterated segments of the genome. To date, almost all studies of the processes involved in formation of functional mRNAs in metazoan cells have been limited to examination of the preponderant species of cRNAs transcribed from the highly reiterated base sequences. Therefore, all conclusions which may be made at this time in analysis of the mechanisms controlling mRNA formation in metazoans are qualified by this restriction.

A second complication has arisen from a detailed study of hybrid formation between DNA isolated from rat liver and the cRNAs transcribed from it *in vitro* by the RNA polymerase enzyme of *M. luteus,* formerly known as *M. lysodeikticus* (Melli and Bishop 1970). The data obtained indicate that the very highly reiterated fraction of *deproteinized rat DNA* was transcribed much less efficiently than the highly reiterated fraction. Further study will be required to determine whether all components of the fraction are poor templates for synthesis of RNAs, or whether only a small proportion (about 10 to 20 percent) of the base sequences in the fraction are transcribed. It also remains to be determined whether the very highly reiterated segments of the rat genome are transcribed equally inefficiently from purified rat DNA by homologous rat RNA polymerase enzymes. However, this may well prove to be the case, for the spectra of RNAs transcribed *in vitro* from isolated chromatins by exogenous bacterial RNA polymerase enzymes of different specificities for initiation sites (see Chapter 5) are not readily distinguishable from those transcribed from the same chromatins *in vivo* (Paul and Gilmour 1968, Smith, Church, and McCarthy 1969). Thus it is *possible* that either:

(a) efficient transcription of very highly reiterated base sequences requires accessory factors additional to those required for efficient transcription of the highly reiterated sequences; or

(b) the large majority of the very highly reiterated base sequences are "noncoding" segments of the genome which *normally can neither be transcribed as corresponding RNAs nor play any direct role in regulation of gene expression.*

Present knowledge of the composition, intrachromosomal distribution, and species specificities of the very highly reiterated base sequences present in "satellite DNAs" are in accord with the latter suggestion (e.g., Jones 1970, Rae 1970, Southern 1970, Walker 1968, Yasmineh and Yunis 1970, 1971). As yet, we do not know whether other very highly reiterated base sequences which may be present in metazoan genomes are ever transcribed efficiently *in vivo.* Therefore, the presence of very highly reiterated segments in the genome creates additional problems in the analysis of mechanisms regulating gene transcription in metazoan cells.

Finally, recent studies indicate that some components of both the nuclear cRNA and polysomal mRNA fractions of (at least) mammalian cells contain 3′-terminal segments of polyadenylic acid (Darnell, Wall, and Tushinski 1971, Edmonds, Vaughan, and Nakazato 1971, Kates 1970, Lee, Mendecki, and Brawerman 1971, Lim and Canellakis 1970). At present it is not known whether these segments are transcribed directly from the genome, but it seems more probable that they are attached by specific polymerase enzymes (e.g., see Twu and Bretthauer 1971) after termination of chain transcription. The significance and role of the segments can only be speculated upon at this time.

RNA POLYMERASE ENZYMES OF EUCARYOTES

The nuclei of eucaryote cells contain at least two distinct molecular species of RNA polymerase enzyme. These have been designated polymerases I and II by Roeder and Rutter (1969) and polymerases A and B by Kedinger et al. (1970). For many years analysis of the nuclear complement of enzymes catalyzing gene transcription was hampered by lack of procedures for obtaining all of the enzymic activity in a "solubilized" form. However, techniques developed recently have permitted resolution of two enzymes with different physicochemical properties from rat liver nuclei (Roeder and Rutter 1969, 1970[a]), calf thymus (Kedinger et al. 1970), ovaries of *Xenopus laevis* (Tocchini-Valentini and Crippa 1970), and nuclei of the coconut (Mondal, Mandal, and Biswas 1970). Homologous enzymes plus a third polymerase III have been resolved from preparations isolated from sea urchin embryos (Roeder and Rutter 1969, 1970[b]) and from the aquatic fungus *Blastocladiella emersonii* (Horgen and Griffin 1971).[1] The RNA polymerase I fractions of rat liver and sea urchin embryos have been shown to contain two components, but these are probably different interconvertible configurations of the same species of enzyme molecule (Chesterton and Butterworth 1971, Roeder and Rutter 1970[b]).

RNA polymerases I are nucleolar enzymes and transcribe precursors of the ribosomal RNAs, whereas polymerases II and III are in the nucleoplasm and transcribe cRNAs (Roeder and Rutter 1970[a], Tocchini-Valentini and Crippa 1970). Activity of the nucleolar enzymes appears to be sensitive to inhibition by cycloheximide but insensitive to the α-amanitin toxin from the mushroom *Amanita phalloides.*[2] Conversely, the RNA polymerases II are sensitive to α-amanitin but unaffected by cycloheximide (Horgen and Griffin 1971, Jacob, Sajdel, and Munro 1970, Kedinger et al. 1970, Muramatsu, Shimada, and Higashinakagawa 1970, Novello and Stirpe 1970, Willems, Penman, and Penman 1969). The additional polymerase III of *B. emersonii* is in-

[1] The presence of a third RNA polymerase in rat liver has recently been reported. This highly labile species is insensitive to inhibition by α-amanitin and appears to be a component of the nucleoplasm (Blatti et al. 1970).

[2] The concentrations of cycloheximide at which inhibition is observed are higher than those usually required for the virtually total inhibition of protein synthesis in eucaryote cytoplasm.

sensitive to both cycloheximide and α-amanitin but is inhibited by high concentrations of rifampicin, and it may be derived from organelles of the cell cytoplasm (Horgen and Griffin 1971). Stimulation of the activity of all three classes by manganese is maximal at concentrations of 1 to 4 millimolar in Mn^{2+}, but responses of the enzymes to magnesium and ammonium sulfate differ markedly (Horgen and Griffin 1971, Mondal, Mandal, and Biswas 1970, Roeder and Rutter 1969).

Separation of two, or more, distinct species of enzyme by procedures which provide convincing evidence of *total* separation is a recent accomplishment (Kedinger et al. 1970, Roeder and Rutter 1969). It is, therefore, fortunate that the procedures and criteria which have served in most of the studies to date do in practice very largely define the actual properties of the resolved enzymes.

The first clear indication of the presence of two species of RNA polymerase in rat liver came from an analysis of the mechanism by which high concentrations of ammonium sulfate stimulate the *in vitro* activity of isolated nuclei. Widnell and Tata (1964, 1966) observed that the increased activity was associated with marked changes in the response to divalent metal ions and to pH. In particular, the activity observed in the absence of ammonium sulfate was markedly stimulated by magnesium ions, whereas the additional enzymic activity unmasked by the ammonium sulfate was much more markedly stimulated by manganese than magnesium ions. These observations formed the basis for a differential assay of the two types of activity. In turn, this led to the custom of specifying a "magnesium-dependent" and a "manganese-dependent" (or "manganese-ammonium sulfate-activated") activity. Subsequent attempts to isolate and purify the RNA polymerase enzyme(s) of metazoans were repeatedly found (e.g., Akamatsu et al. 1969, see also Roeder and Rutter 1969) to yield a "readily-solubilized" fraction characterized by high manganese-dependent activity, and an "aggregate" preparation showing substantial magnesium-dependent activity. Analysis of the properties of the fully separated enzymes indicates that differential assay of manganese- and magnesium-dependent activities largely differentiates between the two distinct species of enzyme. Conversely, separation of readily solubilized and aggregate fractions is essentially the separation of the two molecular species of enzyme (Kedinger et al. 1970, Roeder and Rutter 1969). Therefore, at least temporarily and with appropriate reservation, we may equate the criteria which have been used to delineate two enzymic activities with the actual enzymes, as indicated in Table 8.1.

A variety of lines of evidence had led to provisional identification of magnesium-dependent enzyme (I) as a component of the nucleolus and responsible primarily for the transcription of the large ribosomal RNAs. Indeed, Widnell and Tata (1966), and more recently Johnson et al. (1969), postulated this site and function of this species of enzyme from their analysis of the RNAs transcribed by isolated nuclei under various experimental conditions. Conversely, the manganese-dependent enzyme (II) was suggested to be an enzyme of the nucleoplasm, involved primarily in the transcription of cRNAs. Examination of pulse-labeled rat liver nuclei by application of the combined techniques of autoradiography and electron-microscopy also indicated the magnesium-dependent enzyme to be nucleolar (Maul and Hamilton 1967), but damage to the tissue at high ionic strength precluded this approach to location of the other

TABLE 8.1. Comparison of nomenclatures used to define RNA polymerase enzymes of metazoans

Enzyme 1	Enzyme 2	Reference
I	II	Roeder and Rutter (1969)
A	B	Kedinger et al. (1970)
Aggregate	Solubilized	Akamatsu et al. (1969)
Magnesium-dependent	Manganese-dependent	Widnell and Tata (1964)

enzyme (Jacob, Sajdel, and Munro 1968). The marked increases in (nucleolar) synthesis of ribosomal RNA in the response of the uterus to estradiol (Hamilton, Teng, and Means 1968) and of the liver to hydrocortisone (Jacob, Sajdel, and Munro 1969) or triiodothyronine (Tata 1967, Widnell and Tata 1964)[3] are all associated with increases in activity of the magnesium-dependent enzyme. Further, direct analysis of isolated nucleoli has indicated their RNA polymerase to be largely aggregate, magnesium-dependent enzyme (e.g., Akamatsu et al. 1969).

Re-examination of certain of these phenomena by the newer isolation procedures is formally required. However, it seems most probable that major conclusions already made will be, at least qualitatively, substantiated. Indeed, as noted above, the distribution of RNA polymerases I and II is essentially in accord with prior conclusions. Moreover, the change in the program of gene transcription which occurs at the blastula stage of sea urchin development (see Chapter 6) is associated with a change in the relative rates of increase in levels of the three polymerases (Roeder and Rutter 1970[b]).[4]

At present the basis for the different roles of the various species of nuclear RNA polymerase enzymes is unknown. Chesterton and Butterworth (1971) have recently suggested that RNA polymerase I may be converted to the polymerase II by addition of the so-called "*A-factor*" peptide chain which is responsible for the sensitivity to α-amanitin. This interesting proposal has not yet been critically examined. Nevertheless, if such interconversion does occur, the observed distribution of these two species of enzyme would appear to indicate that the presence of *A*-factor in polymerase II

[3] Thyroid hormone does not stimulate gene transcription by a direct action on the chromatin. Rather, interaction with mitochondria leads to synthesis by the latter of a small, labile material which stimulates gene transcription (Sokoloff 1968).

[4] The proportions of RNA polymerases I and II also change as the levels of total activity increase in the regenerating liver and in the response of target organs to exposure to hormones (Blatti et al. 1970).

either reflects or determines the intranuclear location of the enzyme.

The specific activities of the isolated enzymes *in vitro* are at least one order of magnitude lower than necessary to account for the rates of RNA transcription which occur *in vivo* (Roeder and Rutter 1970[a], 1970[b]). Moreover, the base compositions of the RNAs transcribed *in vitro* from native calf thymus DNA by each of the enzymes from sea urchin embryos and by that of *E. coli* are all very similar and do not reflect the marked differences in their functions *in vivo* (Roeder and Rutter 1969). The RNA polymerases isolated from calf thymus also show many similarities to the core enzyme of *E. coli* (see Chapter 5), although they are not stimulated by the specific σ factor for the latter enzyme (Gniadowski et al. 1970). Thus it seems possible that the procedures currently used to isolate RNA polymerases from eucaryote cells remove homologous σ factors which recognize specific sites in the genome from which synthesis of RNA chains is normally initiated.

Indeed, protein factors which stimulate specifically the initiation of new RNA chains from native DNA by preparations of eucaryote RNA polymerase enzymes have been partially purified from calf thymus (Stein and Hausen 1970[a], 1970[b]) and coconut nuclei (Mondal, Mandal, and Biswas 1970). Further, injection of the σ factor of *E. coli* RNA polymerase enzyme into developing oocytes of *X. laevis* markedly stimulates the incorporation of uridine into RNA (Tocchini-Valentini and Crippa 1970). Nevertheless, present evidence indicates that such factors nonselectively stimulate the activities of both polymerases I and II to approximately equal extents.[5] Therefore, at present there is

[5] The "**S** factor" of Stein and Hausen (1970[a], 1970[b]) does not stimulate the activity of isolated RNA polymerase I (Seifart 1970, Stein and Hausen 1970[b]). However, Chesterton and Butterworth (1971) have argued that **S** factor remains associated with the enzyme during isolation.

no indication that the RNA polymerase enzymes of metazoan nuclei selectively transcribe different segments of the genome because of differences in associated factors analogous to the σ factors of procaryotes.[6] It remains possible that future studies will demonstrate a role of such σ factors in regulation of transcription of metazoan genomes. However, it seems equally probable that selection of the genes transcribed by each species of enzyme is determined by its location within the nucleus.

The very marked differences in responses of the enzymes from a single tissue to changes in the composition of the assay medium may reflect a mechanism which coordinates the activities of the different species of polymerase enzymes in transcribing RNAs from discrete selections of genes. Selective or differential stimulation (or inhibition) of the activity of one of the enzymes by a change in the intracellular concentration of a regulatory metabolite could lead to substantial changes in the relative proportions of the RNAs transcribed by the different enzymes. The polyamines putrescine, spermine, and spermidine may play roles in such a discriminatory control mechanism. Variations in the activities of eucaryote RNA polymerases *in vitro* in response to changes in the concentrations of these metabolites are qualitatively similar to (but quantitatively different from) those observed on varying the concentration of Mg^{2+} or Mn^{2+}. Moreover, there is an increasing body of evidence that the RNA content or level of total RNA polymerase activity in metazoan cells is correlated with the levels and rates of formation of the polyamines (e.g., Herbst and Bachrach 1970, Russell 1971). Extensive further study will be required to determine whether variations in either the concentrations of individual polyamines, or in their relative concentrations, have differential effects upon activities of the different RNA

[6] The extensive similarities between mitochondria and procaryote cells warrant speculation that σ and ρ factors may be required for effective transcription of the mitochondrial genome.

polymerases *in vivo.* However, the correlation observed by Russell (1971) between the rate of synthesis of ribosomal RNAs and the cell content of spermidine in developing embryos of *X. laevis* indicates that such differential effects may be anticipated.

STABLE CYTOPLASMIC RNAs AND LABILE NUCLEAR RNAs

The mRNAs present in the polysomes of metazoan cells are of approximately the same range of size as those of procaryotes. In sucrose density gradients the majority sediment in the region of S values 10 to 20. A few species are smaller, and a small proportion sediment with S values as large as 30 (e.g., Darnell et al. 1963, Di Girolamo, Henshaw, and Hiatt 1964, Samarina et al. 1965). However, in comparison with their bacterial counterparts, the mRNAs of metazoan polysomes are relatively stable. Even the most labile species have half-lives of about one hour (e.g., Tschudy, Marver, and Collins 1965), and the majority of the labile species have half-lives of upwards of 3 hours (e.g., Scott and Bell 1965, Tomkins et al. 1969). Many species of mRNA are substantially more stable (e.g., Scott and Bell 1965) and may be transcribed hours, or even days, before corresponding protein is synthesized thereon (e.g., De Bellis, Gluck, and Marks 1964, Doyle and Laufer 1969, Ilan 1968, Kafatos and Reich 1968; see also Chapters 6 and 9).

On the other hand, the bulk of the rapidly labeled cRNA is highly labile. As early as 1959, Harris (1959, 1963, 1968) reported that the majority of the rapidly synthesized RNA was equally rapidly degraded, without ever leaving the nucleus, and suggested that the highly labile RNA might play a role in regulating the activities of the nucleus.[7] Despite con-

[7] In his early papers Harris suggested that *none* of the nuclear cRNAs left the nucleus. This possibility was briefly revived by Bell's (1969) report that the bulk of the cytoplasmic pulse-labeled mRNAs is transcribed from cytoplasmic particles containing DNA, termed "I-somes." However, the latter have been shown to be artifacts (Fromson and Nemer 1970, Müller, Zahn, and Beyer 1970, Williamson 1970).

siderable early scepticism, this important observation was accorded recognition following the discovery (Spirin, Belitsina, and Aitkhozhin 1964) in young loach embryos of very large species of pulse-labeled RNA which were never present in the cell cytoplasm. In the past few years Harris's observation has been repeatedly confirmed for a very wide variety of tissues. Only a small fraction of the nuclear cRNA is metabolically stable and can be shown to be transferred to the cell cytoplasm (e.g., Aronson and Wilt 1969, Samarina et al. 1965, Scherrer and Marcaud 1968). Indeed, it is now recognized that rapid degradation of the bulk of the rapidly labeled RNA within the nucleus is a characteristic feature of gene transcription in metazoan cells. It is aptly expressed in Clever's (1968) aphorism that "differential gene expression is not synonymous with differential gene activity."

Precise relationships of the various species of pulse-labeled nuclear RNAs to each other and to the cytoplasmic polysome mRNAs are still matters of continuing debate (e.g., Daneholt et al. 1969[c], Georgiev 1969[b], Penman, Rosbash, and Penman 1970, Perry and Kelley 1970, Scherrer and Marcaud 1968, Shearer and McCarthy 1970, Soeiro and Darnell 1970, Waddington 1969[a]). Analysis of these relationships is complicated by the presence in the nucleus of a large variety of species of RNAs additional to messenger RNAs, or possible precursors thereof.[8] In particular, they include several species of precursors of the large ribosomal rRNAs, which are of the same range of sizes (S values) as the bulk of the cRNA molecules. Further, at least superficially,

[8] In addition to precursors of the large ribosomal RNAs, these include:

1. 5S ribosomal RNA and possible precursor (e.g., Comb and Katz 1964, Comb and Zehavi-Willner 1967), which will not be further considered.
2. Transfer RNAs and precursors (e.g., Chipchase and Birnstiel 1963, Woods and Zubay 1965), which will be further considered in Chapter 9.
3. Chromatin-associated caRNAs (e.g., Bonner et al. 1968[b]), considered in Chapter 7.
4. A series of small species of RNA of unknown function (e.g., Clason and Burdon 1969, Dingman and Peacock 1968, Enger and Walters 1970, Moriyama et al. 1969, Weinberg and Penman 1969), which will not be further discussed.

there are some similarities between the known processes by which these precursors are "processed" to yield stable rRNAs and phenomena observed to occur with cRNAs.

These similarities form the basis for some interpretations of the relationships between various species of nuclear cRNAs and polysomal mRNAs to be considered below. Therefore, it seems appropriate to digress briefly and consider the processes by which "mature" rRNAs are formed.

TRANSCRIPTION AND MATURATION OF RIBOSOMAL RNAs

Stable 28S and 18S rRNAs of the cytoplasmic ribosomes are formed from larger unstable precursors (e.g., Brown 1967, Darnell 1968, Perry 1967). The latter are transcribed from highly reiterated, or redundant, base sequences which are clustered together in the segments of the genome known as nucleolar organizers (Birnstiel et al. 1968, Huberman and Attardi 1967, Ritossa et al. 1966). This is the fraction of the genome which is amplified during the dramatic increase in number of nucleoli contained in maturing oocytes of *X. laevis* (see Chapter 2). Analyses made with *Drosophila melanogaster* (Quagliarotti and Ritossa 1968) and with *X. laevis* (Birnstiel et al. 1968, Brown and Weber 1968[b]) have shown that the base sequences of both the rRNAs are transcribed within a common precursor RNA. We believe that both types of rRNA are similarly transcribed in a single precursor molecule in all species of metazoans.

The large precursor RNA is processed to yield the stable *mature* rRNAs by sequential cleavages which form a series of distinct intermediates (Figure 8.1). In most types of cell which have been examined all processing of the rRNA precursors appears to occur within the nucleolus (e.g., Darnell 1968, Das et al. 1970). However, Ringborg et al. (1970[a], 1970[b]) have obtained evidence that the final stage of *maturation* of rRNAs in salivary glands of *Chironomus tentans* may be extranucleolar and occur in association with the chromosomes. Details of the processes of transcription and pro-

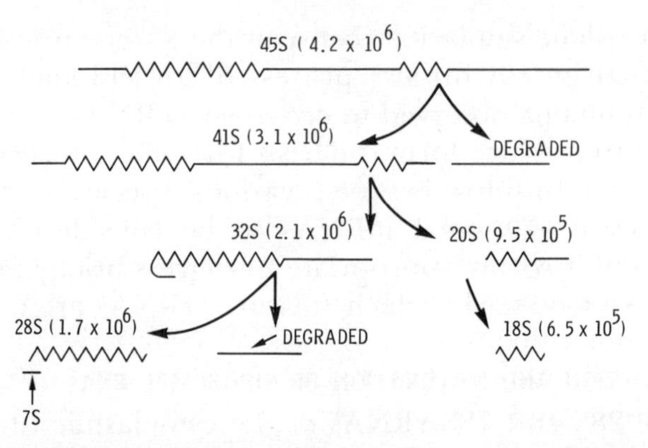

FIGURE 8.1. Maturation of ribosomal RNAs

Based on Egawa, Choi, and Busch (1971) and Weinberg and Penman (1970).

cessing have been studied most extensively for mammalian cells, and I shall therefore use data obtained with such cells for illustrative purposes. Data obtained for other types of metazoan cell, both plant and animal, are similar, except that for most phyla the various types of rRNA precursor molecule are substantially smaller in size (e.g., Applebaum, Ebstein, and Wyatt 1966, Greenberg 1969, Leaver and Key 1970, Loening, Jones, and Birnstiel 1969, Perry et al. 1970, Ringborg et al. 1970[a], Rogers, Loening, and Fraser 1970).[9]

The rRNA cistrons of mammalian cells are transcribed as a 45S RNA, of molecular weight about 4.2 x 10^6. In the processing of this RNA to yield one molecule of each of 28S and 18S RNA (molecular weights approximately 1.7 and 0.65 x 10^6) about half of the base sequences of the precursor molecule are discarded (e.g., Jeanteur and Attardi 1969, McConkey and Hopkins 1969, Weinberg and Penman 1970). The 45S RNA is methylated within the nucleolus, either during or immediately after transcription (Greenberg and

[9] Similar rRNA precursors have also been found in some species of unicellular eucaryotes. However, the process of rRNA formation in other species appears to differ from that in metazoans and resemble that observed in bacteria (see Prescott, Stevens, and Lauth 1971).

Penman 1966, Muramatsu and Fujisawa 1968, Ozban, Tandler, and Sirlin 1964, Retèl, Van den Bos, and Planta 1969). Virtually all of the methyl groups introduced are retained in the mature rRNAs (Wagner, Penman, and Ingram 1967). Further methylation at a late stage of the maturation has been reported for the 18S rRNA of HeLa cells (Zimmerman 1968), but such secondary methylation is not usually observed. In contrast, the discarded portions of the RNA precursor molecule are not methylated (Weinberg et al. 1967), and in addition they contain markedly different proportions of guanine and cytosine than the mature rRNAs (Perry et al. 1970, Willems et al. 1968). These two features presumably serve to define these segments of the precursor molecule as regions serving as substrate for the action of nucleolar endonuclease enzyme(s). Maturation proceeds through a well-defined series of RNA intermediates (Figure 8.1), of which some normally accumulate only in very small quantity, and is not inhibited by suppression of transcription of the precursor (e.g., Penman 1966, Perry 1967, Scherrer, Latham, and Darnell 1963, Warner and Soeiro 1967, Weinberg and Penman 1970, Weinberg et al. 1967). The 7S RNA fragment shown in Figure 8.1 as hydrogen-bonded to the 28S rRNA is not a precursor of the 5S RNA (Brown 1967, Pene, Knight, and Darnell 1968, Perry and Kelley 1968[b], Wimber and Steffensen 1970).

Newly transcribed rRNA precursor complexes with protein to form nucleoprotein particles which are precursors of the ribosome subunits. Maturation of the rRNAs occurs within these particles, and particles of a given S value may contain different rRNA precursors (e.g., Liau and Perry 1969, Vaughan, Warner, and Darnell 1967, Warner and Soeiro 1967, Willems, Penman, and Penman 1969, Yoshikawa-Fukada 1967). Some of the protein incorporated into these various precursor particles is derived from a pool of pre-existing protein (Warner et al. 1966). However, continued protein synthesis is required both for formation of mature ribosome subunits and also the maturation of the rRNA precursors (Willems, Penman, and Penman 1969, Yoshikawa-Fukada 1967). On the other hand, maturation of the rRNAs within the particles appears also to involve discarding some of the protein, with an attendant increase in the relative proportion of RNA (Liau and Perry 1969).

The rate of formation and content of rRNAs in cells of somatic tissues appear to be largely independent of the number of nucleoli present in the nucleus (Brown and Dawid 1968, Kiefer 1968, Ritossa, Atwood, and Spiegelman 1966).[10] However, at present we have little knowledge of the mechanisms which control transcription and processing of the rRNA precursors. Formation of mature rRNAs and accumulation of ribosomal proteins *appear* to be regulated coordinately (Craig 1971, Hallberg and Brown 1969). Nevertheless, the cytoplasmic synthesis of ribosomal proteins occurs independently of the intranucleolar formation of mature rRNAs. Rather, it appears that the ribosomal proteins do not accumulate in large excess, and most of those not incorporated into ribosomes are degraded (Craig and Perry 1971). Conversely, it appears probable that transcription and processing of the rRNA precursors are not directly inhibited by impairment of the synthesis of all protein, whereas the nascent 18S rRNA or its immediate precursor is rapidly degraded (e.g., Ennis 1966, Shields and Korner 1970, Smulson 1970, Smulson and Thomas 1969, Soeiro, Vaughan, and Darnell 1968, Vaughan et al. 1967).[11]

Degradation of preformed precursor appears to be the dominant feature under the more normal physiological conditions of cells in a stationary phase of growth. Thus in

[10] Obvious exceptions to this generalization are the anucleolate and partially nucleolate stock of *X. laevis* (e.g., Knowland and Miller 1970, Miller and Knowland 1970).

[11] There is some ambiguity in data obtained in studies of effects upon rRNA formation of inhibitors of protein synthesis or amino acid deprivation. This may be due in part to indirect effects upon the synthesis of proteins required for rRNA formation and in part to direct inhibition of RNA polymerase activity by cycloheximide.

resting human lymphocytes the synthesis and early stages of processing of rRNAs are indistinguishable from those in cells which have been stimulated to proliferate by treatment with phytohemagglutinin. However, in the later stages of processing in resting cells a large proportion of either the nascent 18S rRNA or its immediate precursor is rapidly degraded. Reduction in the extent of this "wastage" is one of the early features in the response to exposure to phytohemagglutinin (Cooper 1969, 1970, Rubin 1968). Similarly, epithelial cells of the bovine lens which are in the stationary growth phase transcribe and process rRNA precursor, but form no detectable quantities of stable rRNAs (Papaconstantinou and Julku 1968). Regulation of rRNA formation at the stage of maturation of precursors is also indicated by demonstrations that processing of the precursors is arrested before cessation of transcription of the large precursor in such unrelated situations as development of insect larvae (Armelin, Meneghini and Lara 1969) and viral infection of mammalian cells (Ascione and Woude 1969).

However, regulation of the maturation of rRNA precursors is not the factor causing transition from a phase of active growth to a stationary phase. Nor is impairment of the maturation process the only effect upon the protein synthesizing system of eucaryotes which arises on transition to a stationary growth phase. In particular, the proportion of ribosomes present in polysomes declines. There have been very few studies of the mechanisms responsible for this fall in polysome content, and it would be premature to draw general conclusions.

SELECTIVE TRANSMISSION OF NUCLEAR cRNAs TO THE CYTOPLASM

Techniques for forming DNA-RNA hybrids (Gillespie 1968; see Chapter 5) have been used in many comparisons of pulse-labeled nuclear and cytoplasmic cRNAs.[12] This type

[12] Most studies have been limited to comparisons of those cRNAs transcribed frequently from highly reiterated segments of the genome.

of analysis shows that all the base sequences evident in the population of cytoplasmic cRNAs are also present in the nuclear cRNA fraction. However, not all the base sequences contained in the various species of nuclear cRNAs are found in the complement of cytoplasmic cRNAs. Further, the proportion of the nuclear sequences not manifest in the cytoplasmic cRNA fraction varies markedly between different tissues. The proportion of base sequences common to both the cytoplasm and the nucleus is very high in the case of certain tumors (e.g., Chiarugi 1969, Church, Luther, and McCarthy 1969, Drews, Brawerman, and Morris 1968). Yet, for other established tumors only 20 to 30 percent of the base sequences present in the nuclear cRNA fraction are also manifest in the cytoplasm (e.g., Georgiev 1967, Shearer and McCarthy 1967). Similarly, for mouse liver (Smith, Church, and McCarthy 1969), mouse L cells (Shearer and McCarthy 1970), and rat brain (Stévenin, Mandel, and Jacob 1969) approximately 60 percent of the base sequences in the nuclear cRNAs are evident in the cytoplasm, whereas the corresponding proportion for rat liver is variously reported to be between 20 and 40 percent (e.g., Chiarugi 1969, Drews, Brawerman, and Morris 1968). Perhaps the most extreme situation is seen in the case of maturing erythroblasts, where only 10 percent of the base sequences contained in the nuclear cRNA fraction are present in the mRNAs of the polysomes (Scherrer and Marcaud 1968). Thus some process additional to selection of the portion of the genome to be transcribed further restricts the selection of cRNAs transmitted to the cytoplasm (Church, Luther, and McCarthy 1969, Drews, Brawerman, and Morris 1968, Shearer and McCarthy 1967).

A large proportion of the rapidly synthesized nuclear cRNA consists of species which are, or contain, base sequences transcribed from a limited number of highly reiterated segments of the genome (Georgiev 1967, Hoyer, McCarthy, and Bolton 1963, Shearer and McCarthy 1967).

In most tissues the cytoplasmic cRNA fraction is relatively deficient in RNAs containing these base sequences. Thus, despite the large proportion of base sequences common to both the nuclear and cytoplasmic cRNAs of mouse liver, the former is twelve times more effective in competition for binding sites in mouse DNA (Smith, Church, and McCarthy 1969). Similarly, the polysomal mRNA of duck erythroblasts competes very much less efficiently for homologous binding sites in duck DNA than the nuclear cRNA (Scherrer and Marcaud 1968). Thus, the mechanism which determines the selection of cRNAs transmitted to the cytoplasm appears to discriminate against those RNA base sequences which are transcribed from the most highly reiterated segments of the genome.

At the present time the nature of that mechanism is not known. It appears to be susceptible to regulation by hormones, for following *in vivo* administration of 17β-estradiol the cytoplasm of target organs contains species of cRNAs found only in the nucleus of corresponding unstimulated organs (Church and McCarthy 1970). In addition, there are two isolated observations which may be indicative of a requirement for a component of the nucleolus in the transfer of cRNA into the cytoplasm. One has been made with interspecies heterocarya formed by introducing nuclei of avian erythrocytes into mouse L cells or HeLa cells. The hitherto "dormant" erythrocyte nuclei rapidly resume active synthesis of RNA. However, none of this RNA leaves the nuclei, and no protein specified by the avian genome is synthesized, until the erythrocyte nuclei develop nucleoli (Harris 1970, Harris and Cook 1969, Sidebottom and Harris 1969). The second observation is that embryos of the loach (*Misgurnus fossilis*) either lack the mechanism, or do not develop a high degree of selectivity, until after the early gastrula stage (Rachkus et al. 1969[b]), which is attained before the rate of rRNA formation increases (Aitkhozin, Belitsina, and Spirin 1964). Nevertheless, pulse-labeled mRNAs may associate with preexisting ribosomes to form functional polysomes in anucleolate stocks of *X. laevis* (Gurdon and Ford 1967) and in young sea urchin embryos before development of the nucleoli (Kedes et al. 1969[a]). Therefore, possible functions of components of normal nucleoli in the restriction mechanism cannot be speculated upon profitably at this time.

GIANT LABILE NUCLEAR cRNAs

Giant species of cRNAs, or HnRNAs, were first recognized in preparations of pulse-labeled RNAs isolated from loach (*M. fossilis*) embryos (Belitsina et al. 1964, Spirin, Belitsina, and Aitkhozin 1964). Similar large pulse-labeled cRNAs have now been found in a wide variety of metazoan cells. They include chromosomal cRNAs of a dipteran insect (Daneholt 1970, Edström and Daneholt 1967) and cRNAs present in the embryos of a number of invertebrates (e.g., Aronson and Wilt 1969, Collier and Yuyama 1969, Gould 1969). They have been isolated from amphibian embryos (Brown and Gurdon 1966, Landesman and Gross 1968), avian erythropoietic cells (Attardi et al. 1966, Scherrer et al. 1966[a], 1966[b]), and several types of mammalian cells, including cells of human origin (e.g., Arion, Mantieva, and Georgiev 1967, Houssais and Attardi 1966, Schütz, Gallwitz, and Sekeris 1968, Soeiro, Birnboim, and Darnell 1966, Yoshikawa-Fukada 1966). In most cases these giant cRNAs have been formally demonstrated to be confined exclusively to the nucleus. Similar large cRNAs also appear to be present in tissues of higher plants (e.g., Leaver and Key 1970, Rogers, Loening, and Fraser 1970). However, no convincing evidence has yet been obtained of the existence of giant cRNAs in unicellular eucaryotes (see Prescott, Stevens, and Lauth 1971).

As isolated, this class of cRNAs consists of a very heterogeneous mixture of RNAs which sediment in sucrose density gradients with S values ranging from 30 to 80, or even higher, corresponding to molecular weights from 2×10^6 to more than 10^7. They are distinct molecular species, rather than fortuitous aggregates of smaller molecules. Indeed, the giant cRNAs from duck erythroblasts have been observed by electron microscopy as single linear molecules some 4 to 6 μ in length. This is equivalent to some 20,000 to 30,000 nucleotide bases (Granboulan and Scherrer 1969, Scherrer et al. 1966[b]). These huge cRNAs very readily form DNA-RNA hybrids with homologous DNAs (e.g., Arion, Mantieva, and Georgiev 1967, Collier and Yuyama 1969, Yoshikawa-Fukada 1966). This characteristic property reflects the presence in the cRNA molecule of a high proportion of reiterated base sequences, transcribed from a very highly reiterated segment of the genome. Indeed, preparations of the large cRNA fractions isolated from mouse tissues hybridize even more efficiently with mouse satellite DNA than with the total nuclear DNA (Harel et al. 1968).

A second characteristic feature of the heterodisperse fraction of giant nuclear cRNAs is that they are very rapidly degraded. The half-lives of some of these species of RNA have been determined to be of the order of 3 to 10 minutes (Gross and Goldwasser 1969, Houssais and Attardi 1966, Soeiro et al. 1968). Those of the avian erythropoietic cells appear somewhat more stable and have an average half-life of 15 to 30 minutes (Attardi et al. 1966, Scherrer et al. 1966[b]). Nevertheless, even these species are degraded markedly more rapidly than the metabolically stable mRNAs of the cytoplasmic polysomes. Early analyses of the kinetics of labeling of nuclear and cytoplasmic fractions of eucaryote cells by radioactive precursors (e.g., Georgiev et al. 1961, Harris 1959) served to establish that this degradation of rapidly labeled nuclear cRNA occurs within the nucleus itself. More recent studies indicate that the process may commence while the giant cRNA is still complexed with the DNA on which it was transcribed by a process of scission at specific sites within the molecule. A recent modification to a method developed by Georgiev and Mantieva (1962) for selective extraction of different fractions of the nuclear RNAs permits substantial separation of the highly labile giant cRNAs from the metabolically stable smaller species of cRNA. The latter are readily extracted by treatment with phenol at 63°C, whereas temperatures up to 85°C are required for extraction of the larger cRNAs (Arion, Mantieva, and Georgiev 1967). Thus, the giant cRNAs appear to be more intimately associated with the nucleoprotein complex of the genome than other species of cRNAs.

Moreover, pulse-labeled RNA isolated from a single Balbiani ring, or active gene locus, of a salivary gland chromosome from *Chironomus tentans* larvae contains a heterogeneous population of molecules. This consists of a heterodisperse fraction of the same range in size (10 to 90S) as that of the total complement of nuclear cRNAs, together with a variable small proportion of smaller (4 to 5S) species of RNAs (Daneholt 1970, Daneholt et al. 1969[a], 1969[c], Edström and Daneholt 1967). Further, the modal size of the molecules in the heterodisperse fraction appears to be related to the rate of transcription of RNA from the gene locus (Daneholt et al. 1969[c]). Daneholt and colleagues (1969[b], 1969[c]) have suggested that this heterogeneity in size of the cRNA fraction associated with a single gene locus is due to the transcription of molecules of different lengths.[13] However, it seems more probable that the heterogeneity arises from incomplete molecules, and possibly also from the action of a nuclear exonuclease enzyme (e.g., Lazarus and Sporn 1967) on a giant species of cRNA while it is still associated with the chromosome. Indeed, it seems possible that degradation of giant cRNAs may be initiated while they are still complexed with the DNA from which they were transcribed by the action of an enzyme homologous with the RNase H of calf thymus. This enzyme is associated with the RNA polymerase and active on the RNA component of

[13] This suggestion is based on the observation that preparations of pulse-labeled RNAs *isolated from chromosomes* before and after a period of incubation of labeled salivary glands with actinomycin contain molecules of the same range in size and the same average size (Daneholt et al. 1969[b]). However, there was a very marked loss of labeled material from the chromosomes, and the size distribution of this fraction was not determined.

DNA-RNA hybrids (Hausen and Stein 1970, Stein and Hausen 1969).

ROLE(S) OF THE GIANT cRNAs

When pre-labeled cells are further incubated under "chase" conditions the degradation of large cRNAs leads to a change in the spectrum of pulse-labeled RNAs present in the nucleus. The remaining fraction now consists of smaller species of RNAs which are predominantly of a size (8 to 20S) appropriate for a cytoplasmic polysomal mRNA (e.g., Brown and Gurdon 1966, Scherrer and Marcaud 1968). These RNAs are metabolically stable, are more similar to polysomal mRNAs in base sequence, and (relative to the giant cRNAs) are deficient in base sequences transcribed from the very highly reiterated segments of the genome (Arion, Mantieva, and Georgiev 1967, Georgiev 1967). Thus it seems probable that this population of metabolically stable nuclear RNAs consists largely of mRNAs, together with some small chromatin-associated RNAs (see Chapter 7) and other small species of RNA of unknown function. They represent only a small proportion of the total pulse-labeled nuclear cRNAs transcribed, and the exact relationship between them and the giant cRNAs is not yet established.

There are still two major problems to be resolved. One is determining whether:

(a) All stable species of nuclear cRNAs are transcribed within the base sequences of giant cRNAs and then excised during degradation of the latter, as postulated by Scherrer and Marcaud (1968) and Georgiev (1969[b]); or

(b) some or all of the stable nuclear cRNAs are transcribed directly as discrete molecules, but as such a small proportion of the total nuclear cRNA that their presence is masked by the large excess of larger species, as reported by Penman, Rosbach, and Penman (1970) and implied in the model of Britten and Davidson (1969; see Chapter 7).[14]

Evidence currently available indicates that the mRNAs probably are transcribed within the base sequences of large cRNAs and subsequently released in the degradation process. However, as yet there has been no rigorous demonstration that the functional mRNA molecules of the cytoplasmic polyribosomes are excised from the larger cRNA molecules of the nucleus. Therefore, the possibility that some (and in the strictest sense even all) species of mRNAs are individually transcribed as discrete molecules has not yet been eliminated.

The most persuasive evidence that the mRNAs are transcribed within the larger cRNAs is that the two classes of RNA compete for the same sites in the formation of DNA-RNA hybrids. Thus the extent of hybrid formation between the nuclear cRNAs of all sizes and the DNA of mouse L-cells is markedly reduced in the presence of a large excess of the corresponding (total) cytoplasmic RNA. Indeed, the latter fraction appears to contain species which collectively are homologous with approximately two thirds of the base sequences present in the giant (>48S) cRNA molecules (Shearer and McCarthy 1970). Conversely, prior hybridization of HeLa cell DNA with the giant (>45S) nuclear cRNA fraction reduces the extent of hybrid formation with homologous mRNA isolated from the cytoplasmic polyribosomes (Soeiro and Darnell 1970).

Additional evidence in accord with the hypothesis that the mRNAs are transcribed within the base sequences of larger cRNA molecules has been obtained by other approaches. The one which has been used most frequently is based upon the observation that pulse-labeled nuclear cRNA of S values 9 to 18 isolated from rat liver has activity *in vitro* as a template

[14] The recent demonstration (Penman, Rosbach, and Penman 1970) that cordycepin (3′-deoxyadenosine) may suppress incorporation of precursors into polysomal mRNA, yet permit continued incorporation into nuclear cRNA, is amenable to various interpretations.

for the incorporation of amino acid into protein (Hadjivassiliou and Brawerman 1965, 1967). It consists of determining the size distribution of nuclear RNAs with such template activity. Examination of either the total complement of nuclear RNA, or of the resolved chromatin and nucleoplasmic RNAs, shows that the bulk of the template activity is associated with the smaller species (6 to 18S) of RNA. Nevertheless, activity is also found in the RNA fraction of S values in excess of 28 (e.g., Akino and Amano 1970, DiGirolamo et al. 1966, Jacob and Busch 1967). Moreover, the specific activity of these species as templates is as high as, or higher than, that of the smaller RNAs (Akino and Amano 1970).

Unfortunately, however, results obtained to date by this approach are inconclusive. In similar analyses of RNAs isolated from nucleoli a substantial proportion of the template activity is found in the fractions containing precursors of the ribosomal RNAs (e.g., Akino, Amano, and Mitsui 1969, Brentani and Brentani 1969, Jacob and Busch 1967). Therefore, it seems probable that the latter may fortuitously serve as artificial templates for the *in vitro* incorporation of amino acid into polypeptide. The extent of the possible contribution of rRNA precursors to the template activity which has been observed in most studies with nuclear cRNAs cannot be assessed. Moreover, in most cases, the assay of template activity has been made under conditions for which there is no assurance that the polypeptide formed *in vitro* corresponds to any peptide chain(s) normally synthesized *in vivo*.

Two further types of analysis which provide additional presumptive evidence that the mRNA molecules are excised from larger cRNAs have recently been reported.[15] Ryskov

[15] Some additional supporting evidence is afforded by recent studies of the cRNAs isolated from human carcinoma cells infected with herpes virus (Wagner and Roizman 1969). Polysomal mRNAs (10-20S) compete equally well for binding sites in viral DNA with all sizes of nuclear cRNAs (10-80S). In this case, however, the giant cRNAs appear to be largely "tandem repeats" of the polysomal mRNAs, for the latter contain approximately 90 percent of the base sequences present in the nuclear cRNAs.

Similarly, the virus-specific nuclear cRNAs of cells transformed with the simian virus 40 are larger than the homologous mRNAs of the polysomes, and some are larger than the entire sequences of viral genome (Lindberg and Darnell 1970).

and Georgiev (1970) have examined the distribution among the nuclear cRNAs of the 5′-terminal nucleoside triphosphate groups from which RNA chain synthesis is initiated and have found them to be present only in the very large (>35S) species of cRNAs which hybridize with highly reiterated segments of the genome (Ryskov et al. 1971). Perry and Kelley (1970) have presented evidence of an inverse relationship between the concentration of actinomycin required to inhibit (by 63 percent) the formation by L cells of the various classes of pulse-labeled nuclear cRNAs and the size (molecular weight) of the RNAs. On the other hand, the sensitivity to inhibition of formation of the different classes of polysomal mRNAs varies over a wide range, as though these species of RNAs are derived from larger precursors of diverse lengths (Perry and Kelley 1970).

The second major problem which remains is determining the function, if any, of those base sequences of the large cRNAs which are transcribed from the very highly reiterated segments of the genome and not transferred to the cell cytoplasm. Indeed, it has been suggested that they have no function (e.g., Daneholt et al. 1969[c]).

Different roles have been assigned to these portions of the cRNA molecules in various speculative proposals which have been formulated. Thus in the model of Britten and Davidson (1969; see Chapter 7) the large cRNA molecules are the source of activator RNAs which specify the selection of mRNAs transcribed in the smaller species of cRNAs. On the other hand, Georgiev (1969[b]; see Chapter 7) has formulated a model in which the highly reiterated segments of the genome represent regulator gene loci within complex operons. It is postulated that the structural gene segment of such an operon can be transcribed only in the elongation of the RNA chain formed by prior transcription from all of the preceding regulatory loci.[16] In yet another proposal, Scher-

[16] Schema can be devised in which small RNAs transcribed from one region of a polycistronic segment of the genome are transposed to, and activate, a second region of the same segment. Therefore, the two models are not necessarily mutually exclusive.

rer and Marcaud (1968) have developed the analogies between the degradation of giant cRNA molecules and the processing of large rRNA precursors as a scheme of "*cascade regulation*," in which the cRNAs are degraded via a series of discrete smaller derivatives. At each stage in the process some "selection mechanism" serves to discriminate between those species destined for ultimate conversion into mRNA molecules and those which will be destroyed. It is suggested that the latter may play a structural role in defining the specificity of interaction between the large cRNA and informofer protein (see below), nuclease enzymes, and/or molecules regulating the operation of the selection mechanism.

However, there is as yet no evidence that the highly reiterated base sequences of the giant cRNAs do, in fact, play any of the roles envisaged in these various models. Sivolap and Bonner (1970) have recently reported that chromatin-associated caRNAs (see Chapter 7) of the pea plant hybridize preferentially with the very highly reiterated fraction of the homologous DNA. Nevertheless, Shearer and McCarthy (1970) find no evidence of similar caRNAs in the stable fraction of the pulse-labeled RNAs of mouse L cells. It remains possible that caRNAs hybridize only with the untranscribed strands of the very highly reiterated base sequences of the genome, or that only a very small proportion of unstable caRNA precursors transcribed from these sequences is actually converted to stable caRNAs. However, it appears equally possible that the hybridization of caRNAs with these particular base sequences *in vitro* is entirely fortuitous and does not reflect a normal function of these segments of the genome *in vivo*.

INFORMATORS AND INFORMOSOMES

Analysis of the mechanisms controlling formation and extranuclear transfer of mRNAs is further complicated by the fact that the bulk of the cRNA fraction in nuclear extracts appears to be present in the form of ribonucleoprotein complexes (e.g., Köhler and Arends 1968, Mantéva,

Avakyan, and Georgiev 1969, Samarina, Krichevskaya, and Georgiev 1966).

These complexes were first recognized in extracts of nuclei from rat liver or Ehrlich ascites cells as an apparently homogeneous population of small particles (S value 30) of uniform buoyant density, and containing most of the nuclear cRNA (Samarina, Asrijan, and Georgiev 1965, Samarina, Krichevskaya, and Georgiev 1966, Samarina et al. 1967[b]). They consisted of approximately 80 percent of globular protein with 20 percent of small cRNA (10 to 18S) and were sensitive to digestion by RNase enzyme (Samarina et al. 1967[a]). Subsequent analysis of material isolated from rat liver in the presence of inhibitors of RNase activity has shown that the 30S particles are derived from larger complexes, which have been named "*informators*" (Samarina, Lukanidin, and Georgiev 1967, Samarina et al. 1968).

These informator complexes are heterogeneous in size, and they sediment in sucrose density gradients as a series of distinct fractions with S values ranging from 30 to 200. Each fraction appears to consist of complexes containing cRNAs of a narrow range in size (Table 8.2), and all are rapidly converted to a homogeneous population of 30S particles by the action of traces of RNase enzyme. The complexes have been observed in the electron microscope and appear to consist

TABLE 8.2. Correlations between sedimentation constants of particles containing cRNAs and the cRNAs

(a) Nuclear informators of loach embryos

S value of particle	30	45	60	75		120	
S value of contained RNA	8.5	14	17	21		"heterogeneous 32"	

(b) Cytoplasmic informosomes of rat liver

S value of particle	20	30	40	50	55	65	75
S value of contained RNA	4	6	10	15	21	26	35

The S values are those of the peak of the fraction as reported by (a) Samarina et al. (1968) and (b) Ovchinnikov et al. (1969).

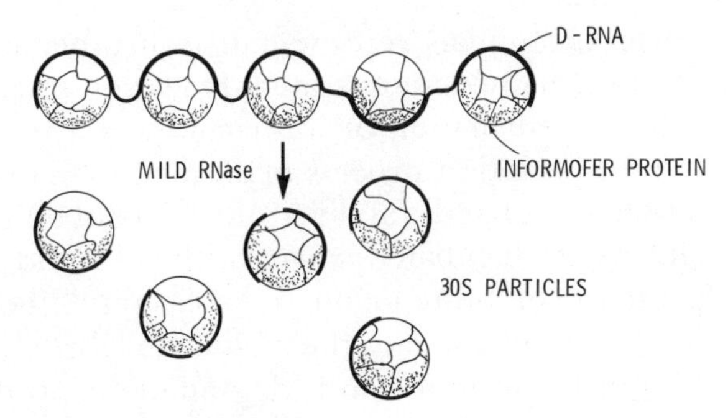

FIGURE 8.2. Structure of the nuclear informator

The model of Samarina et al. (1968).

Reproduced from the *Journal of Molecular Biology*, with permission of the authors and Academic Press, holders of the copyright.

(Figure 8.2) of a strand of cRNA along which are "closely packed" a number (up to 18) of identical particles of globular protein, known as "*informofers*" (Samarina, Lukanidin, and Georgiev 1967, 1968, Samarina et al. 1968, see also Sadowski and Howden 1968).[17]

Both giant cRNAs and small stable cRNAs are found in the nuclear informator particles (Mantéva, Avakijan, and Georgiev 1969). Thus both degradation of the large cRNAs and selection of species of mRNAs to be transferred to the polysomes appear to occur primarily after incorporation into these particles. Indeed, Niessing and Sekeris (1970) have obtained evidence that the cRNA molecule is probably cleaved by the informofer protein at the sites of attachment.

Distinctive particles similar to the informators of rat liver have been isolated recently from rat brain (Stevenin, Mandel, and Jacob 1970) and sheep thyroid (Cartouzou, Poirée, and Lissitzky 1969). The presence of similar particles in the cell

[17] Faiferman, Hamilton, and Pogo (1970) have reported that the smallest, RNase-resistant particle in rat liver nuclei corresponding to the postulated informator subunit has an S value of 60, and they have disputed the existence of "polysome-like" complexes. However, they have recently reported (Faiferman, Hamilton, and Pogo 1971) a value (43) appropriate for an informator subunit (see Moulé and Chauveau 1968).

nuclei of animals representative of other phyla has not yet been formally demonstrated. However, it seems highly probable that formation of informators is a feature common to the transcription process in all metazoan cells.[18]

When isolated rat liver nuclei containing pulse-labeled cRNAs are incubated *in vitro* with ATP and Mg^{2+} ions, in the presence of homologous cytoplasmic "ribonuclease inhibitor," they release into the medium a labeled mixture of labile ribonucleoproteins and degradation products (Ishikawa, Kuroda, and Ogata 1969, 1970, Ishikawa et al. 1970[a]). These ribonucleoproteins are distinct from the informofer-cRNA complexes of the nucleus, and they appear to be related to the cytoplasmic particles known as *informosomes* (see below). The ribonucleoprotein fraction released by the nuclei consists predominantly of particles which sediment in sucrose density gradients with an S value of 45, but it also contains some larger (60S) and smaller (30S) particles. As isolated, the 45S fraction consists of 85 percent protein with 15 percent cRNA and is stable to "mild" treatment with RNase enzyme. However, the particles have essentially the same buoyant density in CsCl density gradients as in the informofer-cRNA complexes (cf. Ishikawa, Kuroda, and Ogata 1969, Samarina et al. 1967[b]), and it seems possible that the fraction contains ribonuclease inhibitor or other additional protein complexed with the particles (cf. Ishikawa, Kuroda, and Ogata 1970, Samarina et al. 1968). The particles contain a heterogeneous population of cRNAs which has activity as a template for the *in vitro* incorporation of amino acid into polypeptide. These cRNAs are of a range in size (4 to 28S) and a modal size (14S) appropriate for polysomal mRNAs (Ishikawa et al. 1970[a]). Moreover, when the postmitochondrial supernatant of a rat liver homogenate is added to the system the cRNA released from the nucleus in 45S particles is found largely in the mRNA-protein frac-

[18] I am not aware of any attempt to seek for the presence of informators in cells of higher plants.

tion recovered from the polysomes (Ishikawa et al. 1970[b]). Thus it seems probable that the particles released from the nuclei *in vitro* are derived from the informators and are either normal intermediates in the transfer of mRNAs to the polysomes or partially degraded derivatives thereof.

Similar particles were first recognized in the cytoplasmic fraction of extracts of young loach (*M. fossilis*) embryos (Belitsina et al. 1964). When centrifuged in sucrose density gradients, the pulse-labeled RNAs sedimented as a series of distinct (20 to 75S) nucleoprotein complexes. As the fractions of lower S values were clearly smaller than ribosome subunits, they were tentatively identified as complexes of mRNAs with protein. Spirin and colleagues (see Spirin 1969[a], Spirin and Nemer 1965) further demonstrated that the buoyant density of the complexes containing pulse-labeled RNAs in cesium chloride density gradients was markedly different from that of ribosome subunits, and that the RNA fraction had activity as a template for *in vitro* incorporation of amino acid into polypeptide. They therefore suggested that these complexes are intermediates in the transfer of mRNAs from the nucleus to the ribosomes, and named them informosomes. Shortly thereafter, Spirin and Nemer (1965) demonstrated the presence of similar complexes in extracts of sea urchin embryos, both as free particles and bound to the ribosomes in the polysome fraction.

The distribution of free informosomes in sucrose density gradients spans the region occupied by the ribosome subunits. This led to some confusion and early tentative identification of the informosomes of other tissues as complexes of mRNAs with the small (40S) ribosome subunits (e.g., Henshaw, Revel, and Hiatt 1965, McConkey and Hopkins 1965). However, informosomes differ markedly from ribosome particles in the content of protein and RNA. This results in marked differences in buoyant density in cesium chloride density gradients and permits ready separation of the two classes of particles (e.g., Spirin 1969[a]). Indeed, the buoyant density of the complexes in this type of density gradient is now accepted as a definitive criterion in their identification.

Free informosomes, and in many cases also ribosome-bound informosomes, have now been isolated from a wide variety of animal cells. They include many types of mammalian cell (e.g., Cartouzou, Attali and Lissitzky 1968, Henshaw and Loebenstein 1970, Kempf et al. 1970, Perry and Kelley 1968[a], Spohr et al. 1970) and insect tissues (Kafatos 1968), in addition to the embryos of loach (Ovchinnikov et al. 1969), and sea urchin embryos (Infante and Nemer 1968, Kedes and Gross 1969[a]).

Properties of the informosomes are very similar to those of the nuclear informators and of the particles released *in vitro* from isolated nuclei. They contain a higher proportion of protein than ribosomal particles and have buoyant densities in the same range as informators (e.g., Burny et al. 1969, Henshaw and Loebenstein 1970, Ovchinnikov et al. 1969, Perry and Kelley 1968[a], Spohr et al. 1970). Both the free and the ribosome-bound populations of informosomes consist of particles of a wide range in size (20 to 80S) and contain cRNAs of sizes (6 to 35S) appropriate for mRNAs (e.g., Burny et al. 1969, Henshaw 1968, Infante and Nemer 1968, Perry and Kelley 1968[a], Spohr et al. 1970). Each of the fractions isolated from sucrose density gradients (Table 8.2) contains cRNAs of a limited range in size (Nemer and Infante 1965, Ovchinnikov et al. 1969, Spirin 1969[a]). The RNAs from these particles and the ribosome-bound particles in polysomes have repeatedly been shown to serve as templates for the incorporation of amino acid into polypeptide. In addition there have been a few demonstrations that the free informosome particles have template activity (Cartouzou, Poirée, and Lissitzky 1969, Chezzi, Grosclaude, and Scherrer 1971, Kempf, Popovic, and Mandel 1970).

However, a recent kinetic analysis by Spohr et al. (1970) indicates that approximately half of the free informosome particles of HeLa cells probably do not contain a functional mRNA molecule. When pulse-labeled cells are further incubated about 50 percent of the labeled cRNA molecules which appear in the free informosome fraction are incorporated into polysomes as mRNAs. The others remain in the cytoplasm as free informosomes and are subsequently degraded at the same rate as the homologous polysomal mRNAs. It has not yet been determined whether the unincorporated cRNAs are defective mRNA molecules or noncoding segments of the large nuclear cRNAs. The observation that the free informosomes of sheep thyroid have low activity as templates for polypeptide formation (Cartouzou, Poirée, and Lissitzky 1969) indicates that this fraction also may contain a high proportion of cRNAs which are not functional mRNA molecules.

RELATIONSHIPS BETWEEN INFORMATORS, INFORMOSOMES, AND POLYSOMES

Evidence currently available is in accord with the hypothesis that (at least in metazoan animals) the nascent cRNAs are incorporated into large informators and then processed to yield smaller informators or "nuclear informosomes." Selected species of the latter are secreted as free informosomes into the cytoplasm, where the functional mRNAs are incorporated into polysomes. Moreover, it seems possible that the mRNAs of the polysomes are present exclusively in the form of informosomes (however, see Blobel 1971, Lee and Brawerman 1971).

However, there are still many problems to be resolved before the relationships between the various types of particle can be fully clarified. One of the most serious is making a definitive determination of whether the informator and informosome particles actually pre-exist in the cell, or are artifacts formed in the preparation of cell homogenates. Formation of similar complexes containing exogenous RNA *in vitro* has been demonstrated in some studies but not found in others (e.g., Baltimore and Huang 1970, Ovchinnikov and Avanesov 1969, Perry and Kelley 1968[a], Spohr et al. 1970). However, it is uncertain what significance should be attributed to either group of observations (see Spohr et al. 1970).

The apparent stability in HeLa cells of the free informosome fraction which does not contain functional mRNAs (Spohr et al. 1970) indicates that these particles probably pre-exist in the intact cells. Demonstrations that exogenous RNAs do not dilute the content of endogenous pulse-labeled cRNAs in the free informosomes (Henshaw and Loebenstein 1970, Ovchinnikov, Avanesov, and Spirin 1969) are also presumptive evidence that these complexes are not experimental artifacts. Further, the release of particles similar to informosomes from isolated nuclei *in vitro* (Ishikawa, Kuroda, and Ogata 1969, 1970, Ishikawa et al. 1970[a]) also supports the hypothesis that both informosomes and nuclear informators are formed *in vivo*. Nevertheless, there is as yet no decisive evidence of the existence of either type of particle, or of ribosome-bound complexes, within intact cells.

A second major problem is determining the role(s) of the protein in the informosome particles. At present this can only be speculated upon. Extracts of metazoan cells contain high levels of RNase activities (e.g., Breillatt and Dickman 1969, Roth 1958, Siler and Fried 1968). Indeed, a number of investigators have presented suggestive evidence that the synthesis of protein in metazoan cells may be regulated in part by fluctuations in the levels of various RNase activities (e.g., Arora and DeLamirande 1967, Imrie and Hutchinson 1965, Kraft and Shortman 1970, Tsukada, Majumdar, and Lieberman 1966). One obvious possible function of the protein component of the informosomes is protection of the cRNAs against degradation by RNase enzymes in the cytoplasm. However, there is no direct evidence in support of this suggestion, and isolated informosomes are very sensitive to RNase digestion (e.g., Samarina, Lukanidin, and Georgiev 1967, 1968). A second obvious possibility is that informosome protein may play a role in control of the selective incorporation of the functional mRNAs into polysomes.

MESSENGER RNAs OF THE CYTOPLASMIC ORGANELLES

To date there have been few studies of the origin of the mRNAs required for synthesis of protein within mitochondria and chloroplasts. The RNAs transcribed in the mitochondria have been shown to include species with activity as templates for *in vitro* incorporation of amino acid into polypeptide (Rakhimbekova and Gaitskhoki 1969), which may be present in particles analogous to informosomes (Vesco and Penman 1969). Moreover, a portion of the RNAs transcribed in the mitochondria may be transported into the membranous fraction of the cytoplasm (Attardi and Attardi 1967, 1969). However, at present we have no knowledge of the mechanisms which regulate either transcription of the mitochondrial DNA or the subsequent transport of any fraction of the transcribed RNA into the cytoplasm.

. . . what evident causes bring about, hidden ones can bring about also.—A. Cornelius Celsus, Second Century.

All this can be recognized where the things are present by means of the signs and good experience.—Theophrastus von Hohenheim (Paracelsus), Sixteenth Century.

CHAPTER 9: **POST-TRANSCRIPTIONAL CONTROL OF PROTEIN SYNTHESIS**

In the preceding chapters I have considered the "*transcriptional level*" mechanisms whereby the selection and concentrations of messenger RNAs present in the polysome complement of the cell are regulated. Control of protein synthesis exerted by these mechanisms may be modified by the operation of additional "*translational level*" or "*post-transcription*" regulatory mechanisms. Indeed, control of protein synthesis at the translational level appears to be a widespread phenomenon, and in eucaryote cells it appears to constitute a mechanism of "fine" control superimposed upon a relatively "coarse" control afforded by regulation at the transcriptional level. Several different actual or potential post-transcription regulatory mechanisms have been formally demonstrated. However, they have not yet been as extensively studied as those regulating the processes of gene transcription, and their relative importance as general mechanisms of regulation of protein synthesis cannot be assessed at this time.

MASKED mRNAs AND PROTEIN SYNTHESIS IN THE EARLY CLEAVAGE STAGES OF EMBRYONIC DEVELOPMENT

Attention was first focused upon post-transcriptional regulatory mechanisms by early attempts to identify the factors responsible for the marked and rapid increase in the rate of protein synthesis which occurs in the early cleavage stages of development of sea urchin embryos (Hultin 1950, 1952). These studies have been extensively considered in a series of recent reviews (e.g., Gross 1967, Spirin 1966, Tyler 1967), and only the major developments will be considered here. The development of a suitable cell-free system permitted Hultin (1961) to demonstrate that the increase in rate of protein synthesis could be largely attributed to a change in the microsome (ribosome-mRNA-membrane complex) fraction of the cleaving embryo. On the other hand, when supplemented with polyuridylic acid as an artificial mRNA, ribosomes isolated from unfertilized eggs were as active as those from cleaving embryos in supporting synthesis of polypeptide (Nemer 1962, Tyler 1963, Wilt and Hultin 1962). Thus the earliest data tended to indicate that the increase in protein synthesis in the early cleavage stages could be attributed to transcription of new mRNAs. Nevertheless, two types of investigation indicated that this was not the case. First, artificially activated enucleate fragments of sea urchin eggs showed as marked increases in rate of amino acid incorporation as either nucleate fragments which had been similarly activated or enucleate fragments obtained from fertilized eggs (Brachet, Ficq, and Tencer 1963, Denny and Tyler 1964, Tyler 1963). Second, eggs fertilized in the presence of sufficient actinomycin D to almost totally suppress RNA synthesis showed enhanced incorporation of amino acid into protein and continued to develop to the stage of apparently normal blastulae (Gross and Cousineau 1963, 1964).

Thus it appeared that the unfertilized egg contained maternal species of mRNA responsible for determining the course of early embryonic development, but in some inactive form not available to the protein-synthesizing system of the unfertilized egg. On the other hand, it was a logical necessity that any such pre-existing maternal mRNA should be protected against degradation by nuclease enzymes. Therefore, it seemed probable that pre-existing mRNAs were present in complexes with some other protective material (probably a coat of protein) as "masked mRNAs." The increased rate of protein synthesis following fertilization or activation of parthenogenetic division could then be attributed to some post-transcription process of activation or "uncoating" of pre-existing maternal masked mRNAs. This interpretation was supported by the subsequent isolation from unfertilized eggs of RNAs with template activity for *in vitro* incorporation of amino acid into protein (Maggio et al. 1964, Slater and Spiegelman 1966[a], 1966[b], 1968). Further support for this interpretation was also provided by the appearance of such endogenous template activity when inactive particulate components of unfertilized eggs were treated briefly with trypsin (Mano and Nagano 1966, Monroy, Maggio, and Rinaldi 1965).

These particles appear to be "heavy bodies" containing ribosomes, in which a "trypsin-like" protease activity develops shortly after fertilization of the egg (Mano 1966, Mano and Nagano 1970). They are presumably the heavy bodies reported by Harris (1967), which contain "ribosome-like" particles and are bounded by annulate lamellae similar in structure to the nuclear membrane. The lamellae of these particles break down and release ribosomes to the cytoplasm simultaneously with breakdown of the nuclear membrane in the first mitotic division of the fertilized egg. However, if both the mRNA and ribosomes are present within these heavy bodies some additional factor must serve to prevent formation of functional polysomes. MacKintosh and Bell (1969) and Stavy and Gross (1969[a], 1969[b]) have suggested that peptide chain initiation (see Chapter 1) is impaired. In accord with this suggestion, Metafora, Felicetti, and Gambino (1971) have reported that ribosomes of the unfertilized egg contain protein which inhibits the formation of (artificial) initiation complexes containing exogenous polyuridylic acid as the template.

Recent studies have also shown that *normal* development of the fertilized sea urchin egg to the stage of the blastula does involve synthesis of protein specified by newly transcribed mRNAs (Kedes and Gross 1969[a], 1969[b], Kedes et al. 1969[a], Kiefer, Entelis, and Infante 1969). Nevertheless, little difference can be seen in the complements of proteins synthesized to the blastula stage by normal and actinomycin-inhibited embryos (Terman 1970). It therefore seems probable that most of the control of protein synthesis between the stages of fertilization and the blastula in sea urchins is exerted at the post-transcriptional level.

To date, technical difficulties have precluded equally extensive analyses of the early cleavage stages of other species of embryo. The available data (e.g., Collier 1966, Gurdon 1968[a], Spirin 1966) are consistent with the hypothesis that, with the possible exception of viviparous species, similar post-transcriptional regulation plays a major role in regulating the early development of most animal species. However, this extensive body of data appears to have been of relatively slight importance in stimulating current interest in post-transcriptional regulatory control mechanisms. This lack of impact can probably be attributed to two features of the data. First, until recently it could legitimately be argued that the process of activation of preformed masked mRNA by removal of protective protein barely qualified as a translational control mechanism. Indeed, Spirin (1966) suggested that the inactive ribosomes of unfertilized eggs were complexes of ribosomes with informosomes (see Chapter 8). The more recent demonstrations of the structure of heavy bodies, and the observation that free ribosomes of the unfertilized egg appear to differ in conformation from those of young embryos (Maggio et al. 1968), serve to establish the reality of the process as one qualitatively distinct from the processes of extranuclear transmission of mRNA observed with more advanced embryonic tissues (see Chapter 8). Second, and probably more important, the process of activation of masked mRNA is not correlated with regulation of the synthesis of one (or few) specific protein(s). Rather, it represents a mechanism for controlling the overall rate of synthesis of all species of protein synthesized by the cleaving embryo, not a mechanism for selective change in the complement of proteins synthesized.

EARLY INDICATIONS OF RATE-LIMITING STEPS IN THE ELONGATION OF HEMOGLOBIN CHAINS

Two types of study served to increase general interest in the possibility of selective regulation of the rate of synthesis of specific proteins by post-transcriptional mechanisms. The first arose from the demonstration that the globin chains of rabbit hemoglobin are assembled from the N-terminal amino acid (Dintzis 1961, Naughton and Dintzis 1962). Subsequent analysis of the rate of incorporation of leucine residues into different positions of the globin chain led to the conclusion that the early portion of the α-chain was assembled rapidly but the C-terminal portion more slowly (Englander and Page 1965). This, in turn, implied the existence of some rate-limiting step in the elongation of the globin chain. Evidence of similar rate-limiting steps in the assembly of the chains of human hemoglobin was reported by Winslow and Ingram (1966). Meanwhile, Weisblum et al. (1965) obtained evidence suggestive of the possibility that different molecular species of tRNA transferred leucine into different positions of the globin chains.[1] It therefore seemed possible that the rate of

[1] Subsequent studies have provided more definitive evidence that different species of "iso-accepting" tRNAs may transfer a given amino acid into different positions of the globin chain (Anderson and Gilbert 1969, Galizzi 1969, Gonano 1967, Sekiya, Takeishi, and Ukita 1969, Weisblum et al. 1967). Moreover, the rate of *in vitro* synthesis of α-chains relative to that of β-chains may be markedly reduced by reduction in the concentration of available tRNA, and probably specifically of one of the alanine-specific tRNAs (Gilbert and Anderson 1970). However, there is no evidence that the rate of globin chain synthesis is limited *in vivo* by the supply of any species of tRNA.

synthesis of complete globin chains might be controlled by the limiting supply of a specific species of tRNA containing the anticodon necessary for complex formation with a so-called *minor* codon of the genetic code (see Jukes 1966), although other types of interpretation were not excluded (e.g., Itano 1966, Winslow and Ingram 1966).

Actually, a number of recent studies indicate that there are no rate-limiting steps beyond which the rates of elongation of the hemoglobin chains are reduced (e.g., Clegg et al. 1968, Hunt, Hunter, and Munro 1968[a], 1969, Luppis, Bargellesi, and Conconi 1970). However, the important point for present purposes is that the apparent existence of rate-limiting steps in the elongation of the globin chains stimulated consideration of their possible nature.

POLARITY AND SUPPRESSOR TRANSFER RNAs

The second type of study which led to consideration of possible post-transcriptional regulatory mechanisms was that of certain polar gene mutations in bacterial operons (see Chapters 3 to 5). These were point mutations within structural genes coding for one of the enzymes controlled by the operon which exhibited three characteristic properties. First, they eliminated synthesis of active enzyme specified by the structural gene containing the site of the mutation. Second, they reduced the level of activity of all enzymes for which the structural genes were located in the operon operator-distal to the site of the mutation. Third, they did not eliminate the coordinate regulation of synthesis of the latter group of proteins. Even today, we have no satisfactory interpretation of all the data obtained in studies of polar mutants and, as noted elsewhere (Chapter 5), the data themselves show some apparent contradictions. However, what has proven to be largely a correct interpretation of the characteristic features of polar mutants was suggested by Ames and Hartman (1963). This interpretation was based upon two postulates which have been amply confirmed in subsequent studies. First, they suggested that a single polycistronic mRNA transcribed from all the structural genes of the operon serves as a common template for the sequential synthesis of all the corresponding proteins. Second, they proposed that a polar mutation created a "modulating triplet" codon in the mRNA at a point corresponding to the mutation. Alignment of amino acid against the modulating triplet required a specific "modulator tRNA" present in the cell in very limiting quantity, and creation of the modulating triplet codon therefore introduced a rate-limiting step in assembly of peptide chains beyond it. This would not, of itself, reduce the steady-state level of these latter peptide chains. However, it was supposed that either ribosomes which failed to acquire modulator tRNA on reaching the modulator triplet dissociated from the mRNA, or that segments of mRNA distal to the modulator triplet would often be degraded before serving as template for the assembly of peptide chains.[2]

We now know that this "modulation hypothesis" is not correct in detail (see Chapter 5). The modulator triplet postulated by Ames and Hartman is, in fact, one of the nonsense codons of the genetic code (see Garen 1968, Jukes 1966). Such a codon normally specifies termination of peptide chain elongation (see Chapter 1). The presence of a nonsense codon at an internal position normally results in premature chain termination and release of a partial peptide chain (e.g., Sarabhai et al. 1964, Webster et al. 1967). Synthesis of any protein corresponding to structural genes of an operon which are operator-distal to a nonsense mutation therefore normally requires re-initiation of protein synthesis distal to the nonsense codon in the mRNA.

The foregoing are the normal responses to the presence of nonsense mutations observed with so-called "nonpermissive" strains of bacteria. However, these responses are

[2] It may also be noted parenthetically that Ames and Hartman (1963) also anticipated the discovery of internal promoters as modulating triplets causing enhanced synthesis of protein from distal regions of the mRNA.

modified in bacterial strains which are designated "permissive." These latter strains contain specific suppressor genes which permit the cell to recognize the nonsense codon as a sense codon specifying a particular amino acid, and to synthesize some apparently normal protein corresponding to the gene containing nonsense codon. A series of different nonsense-suppressor genes are known, which differ in physiological properties. In particular, they differ in the identity of the amino acid inserted into the peptide chain at the position of the nonsense codon (see Garen 1968).

Permissive strains containing the nonsense-suppressor gene *su3*$^+$ appear to compensate for the presence of the nonsense codon in a manner which corresponds exactly with that postulated in the modulator hypothesis of Ames and Hartman (1963). These strains complete the synthesis of some polypeptide chains by inserting a tyrosine residue at the position corresponding to the nonsense codon. The suppressor gene has been recovered in the genome of a transducing strain of bacterial virus derived from the bacteriophage ϕ80 and studied in detail. It is a mutant allele of a structural gene for a tyrosine-specific tRNA in which one base pair of the anticodon region has been changed and codes for a tyrosine-specific tRNA with an anticodon complementary to the amber nonsense codon (Abelson et al. 1969, Andoh and Ozeki 1968, Goodman et al. 1968, Landy et al. 1967). The presence of this species of tRNA in the total complement of tyrosine-specific tRNA of cells containing the suppressor *su3*$^+$ allele is barely detectable (Smith et al. 1966).

Evidence has been obtained that the suppressor gene *su6*$^+$ similarly codes for a minor species of leucine-specific tRNA with an anticodon complementary to the *amber* codon (Gopinathan and Garen 1970). Moreover, it seems probable that the "recessive-lethal" suppressor gene *su7*$^+$ produces a mutant form of a major glutamine-specific tRNA which recognizes only the *amber* codon (Söll and Berg 1969[a],

1969[b]). However, in contrast, the tryptophan-specific tRNA specified by a suppressor gene of the UpGpA nonsense codon contains the normal anticodon, but nevertheless recognizes the nonsense codon efficiently (Hirsh 1970). Present evidence indicates that each of the other nonsense-suppressor genes is also a mutant allele of a structural gene for a transfer RNA (Andoh and Garen 1967, Capecchi and Gussin 1965, Roy and Söll 1968, Wilhelm 1966). Further, some suppressor genes of mis-sense mutants (but certainly not all) control formation of specific tRNAs showing altered patterns of codon recognition (e.g., Carbon, Berg, and Yanofsky 1966, Gupta and Khorana 1966, Hill, Squires, and Carbon 1970).

It is debatable whether determination of the ability to synthesize a specific protein by presence or absence of a species of tRNA specified by a suppressor gene can properly be considered an instance of post-transcriptional regulation of protein synthesis. On the one hand, the suppressor tRNA acts at the translational level. On the other hand, control of the ability to recognize a particular nonsense codon as a "sense" codon is determined in an all-or-none manner solely by the presence or absence of the suppressor gene. However, it is not my intention to debate either case in what is largely a matter of definition. Rather, I wish to emphasize that these studies served to complement those which indicated the possible existence of a rate-limiting step in the synthesis of hemoglobin. The one series raised the question of possible rate-limiting steps in the normal assembly of the major protein synthesized by a normal cell. The other provided conclusive evidence that a rate-limiting step in the limited synthesis of a specific protein in a defective cell was controlled by the presence of a specific minor species of tRNA. Together, they constituted strong circumstantial evidence that modification in the tRNA complement of a cell might serve as a normal mechanism of post-transcriptional regulation of protein synthesis.

CHANGES IN THE PATTERN OF PROTEIN SYNTHESIS CORRELATED WITH VARIATIONS IN THE COMPLEMENT OF TRANSFER RNAs

Recent studies have revealed a large number of situations in which a profound change in the spectrum of proteins synthesized by a given population of cells or tissue appears to be associated with an equally marked change in the cell complement of tRNAs. Nevertheless, it must be recognized that in most cases it has not been shown that the change in pattern of protein synthesis is caused by, or even dependent upon, the change in complement of tRNAs (see Sueoka and Kano-Sueoka 1970). Even in most cases where it seems very (or even extremely) probable that the change in complement of tRNAs does regulate the change in spectrum of proteins synthesized it remains to be determined whether the effect is direct, or the indirect consequence of effects upon synthesis of regulatory species of protein.[3]

Persuasive evidence of direct regulation of protein synthesis by formation of one, or more, new species of tRNA molecule(s) has been obtained for the cuticular protein of mealworm (*Tenebrio molitor*) pupae (Ilan 1968, 1969, Ilan, Ilan, and Patel 1970). Ilan has studied the incorporation of amino acid into this protein by cell-free systems prepared at various stages after onset of pupation, under conditions where synthesis was dependent upon supplements of isolated tRNA. The ribosome fraction isolated at the first day of pupation contained the appropriate mRNA. Homogenates prepared at this stage synthesized cuticular protein when supplemented with both tRNA and aminoacyl-tRNA synthetase enzyme isolated from 7-day pupae. On the other hand, similar homogenates failed to do so when supple-

[3] A most dramatic illustration of the need for caution in interpretation of the significance of changes in complements of tRNAs has recently been reported by Jacobson (1971).

A strain of *D. melanogaster* carrying a suppressor of the vermilion locus affecting eye color *lacks* a species of tyrosine-specific tRNA present in the wild-type strain. This species of tRNA inhibits the activity of the tryptophan pyrrolase enzyme of vermilion flies, the enzyme which catalyzes the first reaction in formation of pigment from tryptophan.

mented with homologous tRNA and enzyme. It therefore seems probable that onset of synthesis of cuticular protein is regulated by the formation of one (or more) new species of tRNA and corresponding "charging" enzyme(s), although all possible alternative interpretations have not been rigorously excluded. At the present time it is not known whether formation of the new species of tRNA(s) requires *de novo*

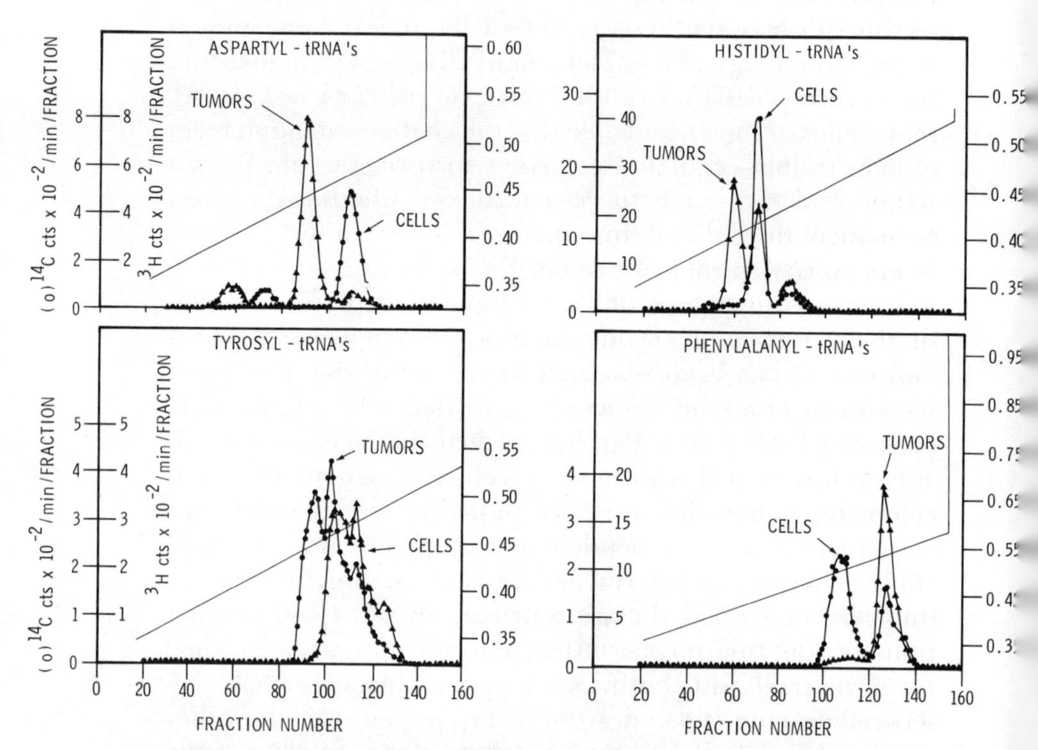

FIGURE 9.1. Demonstration of differences in the complements of tRNAs in LM fibroblasts grown in culture and derived tumors developed from them in mice

Preparations of tRNAs isolated from the two cell populations were differentially labeled with ^{14}C- and ^{3}H-labeled amino acids, mixed, and co-chromatographed on a reversed phase column. In these experiments each preparation of tRNA was acylated by homologous synthetase enzymes. Similar results were obtained when both preparations of tRNA were acylated by the same enzyme preparation.

Data of Yang et al. 1969. Reproduced with permission of the authors and the U.S. National Academy of Science, holders of the copyright.

synthesis, or modification of a pre-existing species of RNA molecule by processes to be considered below.

Onset of the phase of rapid hemoglobin synthesis in the blood islands of young chick blastodiscs also appears to be regulated in part by formation of a minor species of alanine-specific tRNA (Wainwright 1971[a], Wainwright and Thompson 1972, Wainwright, Wainwright, and Tsay 1972; see Chapter 10). At present it is not known whether the effect of this tRNA is directly upon translation of mRNAs for hemoglobin chains, or even whether it is directly upon cells of the erythropoietic series. Formation of this species of tRNA may reflect a "switch" from synthesis of embryonic hemoglobins to that of fetal hemoglobins (Wainwright and Wainwright 1970). A later switch in the ontogeny of hemoglobin formation in the developing chick from synthesis of fetal hemoglobins to that of the adult species is correlated with at least one qualitative difference in the tRNA complements of the corresponding "primitive" and "definitive" lines of erythrocytes (Lee and Ingram 1967). Similarly, differences have been observed in the tRNA complements of erythrocytes of larval and adult bullfrogs (DeWitt 1971).

In most studies the tRNA complements of two types of cell or tissue have been compared by co-chromatography of a mixture of homologous aminoacyl-tRNAs which have been "differentially labeled" by amino acid residues containing different isotopes. Consistent differences between preparations of RNAs from the two sources in the elution profiles from reversed-phase columns (Figure 9.1), or columns of BD-cellulose or MAK, are taken as presumptive evidence of differences in the cell complements of the iso-accepting series of tRNAs for the test amino acid.[4] Such differences were

[4] Reversed-phase columns consist of an organic base adsorbed to an inert support (see Kelmers, Novelli, and Stulberg 1965, Weiss and Kelmers 1967). Procedures for separation of nucleic acids on columns of BD-cellulose, benzoylated diethylaminoethyl-cellulose, were developed by Gillam and colleagues (1967, 1968). Procedures developed by Sueoka and Cheng (1962) have been used extensively in analyses of preparations of aminoacyl-tRNAs on columns of MAK, methylated serum albumin adsorbed to kieselguhr.

demonstrated in the comparisons of preparations of tRNAs isolated from erythrocytes of embryos and adult animals referred to above and have also been observed in a wide variety of analyses.

The complements of various series of iso-accepting tRNAs appear to vary during the early stages of development of sea urchin embryos (Molinaro and Mozzi 1969, Yang and Comb 1968, Zeikus, Taylor, and Buck 1969) and of silkworm (*Bombyx mori* L.) larvae (Garel et al. 1971). Similarly, some differences in tRNA complement have been observed in comparisons of different tissues of a single species of plant or animal (Anderson and Cherry 1969, Holland, Taylor, and Buck 1967, Taylor et al. 1968, Vold and Sypherd 1968, Wevers, Baguley, and Ralph 1966). Yet other differences have been observed in comparisons of normal and regenerating rat liver (Jackson, Irving, and Sells 1970); between tissues of hormone-deficient and hormone-treated animals (e.g., Mäenpää and Bernfield 1969, 1970, Tonoue, Eaton, and Frieden 1969, Yang and Sanadi 1969); or between normal tissue and derived tumors (e.g., Baliga et al. 1969, Gallo and Pestka 1970, Volkers and Taylor 1971, Yang and Novelli 1968).

Even more dramatic changes in the tRNA complement have been observed in cells of *E. coli* infected with the bacteriophages T2 or T4. In one of the earliest studies Kano-Sueoka and Sueoka (1966) reported changes in the complement of iso-accepting tRNAs for leucine as early as 3 minutes after infection, and additional changes at 8 minutes after infection (see also Novelli 1969). Subsequent analysis revealed that the changes reflect both gains and losses of different molecular species of leucine-specific tRNA. An apparently new species of tRNA found within the first 3 minutes following infection is an inactive unstable fragment of pre-existing host tRNA, produced by cleavage during the aminoacylation reaction of a modified molecule of the host species of tRNA (Kano-Sueoka, Nirenberg, and Sueoka 1968, Kano-Sueoka and Sueoka 1968). On the other hand, assays by the technique of DNA-RNA hybrid formation indicate that five new species of tRNAs are transcribed from the viral genome in cells infected with bacteriophage T4 (Daniel, Sarid, and Littauer 1968, 1970, Scherberg and Weiss 1970). These include a leucine-specific tRNA containing thionucleotide (Hsu, Foft, and Weiss 1967, Weiss et al. 1968). Similarly, at least three new species of tRNAs appear to be transcribed from the genome of bacteriophage T2 (Smith and Russell 1970). As many as fourteen new species of tRNAs may be transcribed from the genome of the unrelated bacteriophage T5 (Scherberg and Weiss 1970), and formation of virus-specific tRNAs has also been observed with cells infected by bacteriophage λ (Pearson and Hogness 1971). On the other hand, an early report (Subak-Sharpe and Hay 1965) of the presence in infected mammalian cells of tRNAs transcribed from the genome of herpes simplex virus has not been confirmed (Morris, Wagner, and Roizman 1970). Further, Raška, Frohwirth, and Schlesinger (1970) have found no evidence of the presence of virus-specific tRNAs in cells infected with adenovirus type 2 virus.

Pronounced changes have been observed in the complements of valine-specific and serine-specific tRNAs of *Bacillus subtilis* with onset of sporulation or with marked changes in growth conditions (Arceneaux and Sueoka 1969, Doi, Kaneko, and Goehler 1966, Doi, Kaneko, and Igarashi 1968, Kaneko and Doi 1966). Differences in the complements of isoleucine-specific and phenylalanine-specific tRNAs have also been reported in comparisons of cells of *E. coli* grown aerobically and anaerobically or with limited aeration, respectively (Kwan, Apirion, and Schlessinger 1968, Wettstein and Stent 1968). In contrast, phenomena of induced enzyme synthesis (see Chapter 3) or of enzyme repression (see Chapter 4) have not been correlated with demonstrable changes in tRNA complement (Leisinger and Vogel 1969, Shearn and Horowitz 1969, Smith 1968, Stulberg, Isham, and Stevens 1969).

In none of the foregoing cases is there direct experimental

evidence that the change in tRNA complement serves to regulate the synthesis of protein and does not merely reflect the operation of some other regulatory mechanism. Indeed, the possibility exists that some of the components of the tRNA preparations examined in these studies (e.g., Mäenpää and Bernfield 1970) do not participate in the synthesis of polypeptide chains but serve as intermediates in the formation of aminoacyl-phosphatidyl glycerols (Gould et al. 1968, Nesbitt and Lennarz 1968) or peptidoglycans (Stewart, Roberts, and Strominger 1971). Further, analysis of the profiles of aminoacyl-tRNAs on chromatography columns is subject to complications arising from the possible presence of active dimers of tRNAs (Loehr and Keller 1968, Zachau 1968) or of active precursor molecules lacking substituent groupings of minor nucleotides (e.g., Gefter and Russell 1969, Yang et al. 1969). As mentioned above, the apparently new leucine-specific tRNA found shortly after infection of *E. coli* with bacteriophage T2 is, in fact, an inactive fragment derived from pre-existing tRNA during the aminoacylation reaction (Kano-Sueoka and Sueoka 1968). Tryptophan-specific tRNA of *E. coli* may be isolated in two different readily interconvertible conformations, of which one is biologically inactive (Gartland and Sueoka 1966, Gartland et al. 1969, Ishida and Sueoka 1968). Similar situations have also been reported for other species of tRNA isolated from *E. coli* and from yeast (Fresco et al. 1966, Yegian and Stent 1969[b]). Therefore, it is necessary to consider not only various possible functions of apparently new species of tRNA but also the possibility that they have no distinctive function, and may even be artifacts.

Nevertheless, it is tempting to attribute these various differences in tRNA complements to the presence of species of tRNA required for rate-limiting steps in the assembly of specific proteins, and thereby regulating the synthesis of protein essentially in the manner envisaged by the modulation hypothesis of Ames and Hartman (1963). The changes in tRNA complement would then reflect specific mechanisms either controlling formation and degradation of particular species of tRNA or regulating the interconversion of inactive or partially active species of molecule and fully active species of tRNA. Indeed, there is a substantial body of circumstantial evidence that the reactions which generate the minor nucleotides present in the tRNAs both can and, in fact, may serve as a mechanism for post-transcriptional regulation of protein synthesis.

CHANGES IN THE PATTERN OF PROTEIN SYNTHESIS CORRELATED WITH VARIATIONS OF tRNA-METHYLASE ACTIVITY

Preparations of tRNA isolated from a variety of sources are characterized by a relatively high content of minor bases and nucleotides, especially in the case of preparations isolated from tumors (e.g., Dunn 1959, Hall 1971, Smith and Dunn 1959, Srinivasan and Borek 1964[a]). These minor components are generated by reactions which modify the structure of precursor polynucleotides of larger size, or more open structure, than the corresponding tRNAs (e.g., Bernhardt and Darnell 1969, Burdon, Martin, and Lal 1967, Hurwitz et al. 1963, Kammen and Spengler 1970, Pace, Peterson, and Pace 1970, Srinivasan and Borek 1964[a]). The minor bases include various methylated purines and pyrimidines, thiopyrimidines, and a variety of adenine derivatives with activity (Helgeson 1968) as plant cytokinin hormones. In addition, all tRNAs contain the minor nucleotide pseudouridine and nucleotides in which the 2′ position of the ribose ring is methylated. These various minor bases and nucleotides are considered in depth in a companion volume of this series (Hall 1971), to which the reader is referred for detailed treatment of their effects upon the structure and properties of tRNA molecules. For present purposes, it suffices to note that precursor molecules lacking specific substituent groupings, and tRNA molecules containing supernumerary substituents, may differ from normal tRNA molecules both in ability to accept amino acid and in efficiency of codon recognition (Capra and Peterkofsky 1968, Pillinger, Hay, and Borek 1969, Shugart et al. 1968, Stern et al. 1970).

Methyl groups are transferred to tRNA precursors from S-adenosylmethionine by the action of tRNA-methylase enzymes, first discovered in *E. coli* (Fleissner and Borek 1962, 1963). Extensive analysis has shown that *E. coli* and *Saccharomyces cerevisiae* each contain several distinct species of enzyme differing in specificity for the base methylated, and in nucleotide sequences recognized as sites for methylation (Björk and Svensson 1969, Gold, Hurwitz, and Anders 1963, Hurwitz et al. 1965, Svensson, Björk, and Lundahl 1969). Evidence of a similar multiplicity of enzymes in other types of cell is less direct and is derived from comparisons of the effects of cell extracts upon a series of isolated tRNAs (e.g., Baguley and Staehelin 1969, Srinivasan and Borek 1963, 1964[b]). However, two distinct species of tRNA-methylases which are specific for guanylate residues have been resolved in extracts of rat liver (Kuchino and Nishimura 1970). There is also some preliminary evidence which indicates that there may possibly also be a multiplicity of enzymes for methylation of ribose residues (Gefter 1969).

It seems highly probable that variations in levels of tRNA-methylase activities serve as (part of) a general mechanism for regulation of synthesis of specific proteins, although this remains to be formally established. Many different phenomena which involve modification of the spectrum of proteins synthesized have been correlated with changes in the complement of active tRNA-methylases. These changes are manifested as variations in the level of total methylase activity which are accompanied by apparent modifications in the specificity for sites methylated in the substrate, as evidenced by the extent of methylation of the substrate or qualitative changes in the variety of methylated bases formed (e.g., Borek 1969). Indeed, it seems possible that variation in tRNA methylase activities is as important a factor in regulation of protein synthesis by metazoan cells as are changes in the spectrum of genes transcribed.[5] Especially marked increases in tRNA methylase activities have been reported in a series of comparisons between malignant tumors and the host cells from which they were derived. These comparisons have included tumors of viral origin and those induced chemically, highly distinctive tumors, and minimal deviation hepatomas (e.g., Baguley and Staehelin 1968, Craddock 1970, Hacker and Mandel 1969, Hancock 1967, McFarlane and Shaw 1968, Stewart and Corrance 1969, Tsutsui, Srinivasan, and Borek 1966, Turkington and Riddle 1970[a]). In most cases examined the increase in total methylase activity is associated with a marked change in the pattern of bases methylated.

In general, levels of total methylase activities in various normal tissues of an adult animal are roughly comparable, and no distinct pattern of tissue-specific differences in enzyme content clearly emerges from the data.[6] However, there have been isolated reports of tissue-specific differences in patterns of methylation (e.g., Hacker and Mandel 1969, Kaye and Leboy 1968). In contrast, variations in levels of tRNA-methylase activity have been observed with differentiating systems. A number of workers have reported the activity of vertebrate fetal liver to be markedly higher than that of the corresponding adult tissue (Hancock 1967, Kaye and Leboy 1968, Rennert 1970, Stewart and Corrance 1969). The difference in activities appears to be due primarily to a rapid decline which occurs just prior to birth (Hancock, McFarland, and Fox 1967) caused by formation of an inhibitor of tRNA-

[5] It would not be anticipated that the very rapidly reversible phenomena of induced enzyme synthesis, end-product repression, and catabolite repression (Chapters 3 and 4) would be accompanied by variations in tRNA methylase activity. However, I am not aware of experiments made to examine this prediction.

[6] It should be noted that customary assays of tRNA-methylase activity represent the sum of activities of several enzymes on a mixture of very many potential substrates. A very small apparent net change in activity of one species of tRNA methylase may reflect marked quantitative effects upon one species of substrate.

methylase activity (Kerr 1970). Data of Rennert (1970) indicate that there may also be some change in the pattern of tRNA methylation. Regenerating liver has also been reported to contain a higher level of tRNA-methylase activity (Stewart and Corrance 1969).

Similar decreases in tRNA-methylase activity due (at least partially) to the formation of inhibitor(s) have also been observed during pupation of the mealworm *Tenebrio molitor* (Baliga, Srinivasan, and Borek 1965) and during morphogenesis of the slime mold *Dictyostelium discoideum* (Pillinger and Borek 1969, Sharma and Borek 1970[a]). Variations specific to the uterus in levels of total tRNA-methylase activity following ovariectomy of pigs and in response to subsequent administration of estradiol also appear to be regulated by changes in the tissue content of inhibitor(s) of methylase activity (Sharma and Borek 1970[b]). In contrast, the marked increases of tRNA-methylase activity observed during *in vitro* differentiation of the midterm murine mammary gland in response to hormones appear to be due to (or at least require) the *de novo* synthesis of protein (Turkington 1969[b]).

Selective synthesis of new species of tRNA-methylases has been observed with cells of *E. coli* infected with bacteriophage T2 and upon induction of cells harboring the prophage (see Chapter 4) of the bacterial virus λ. However, this is not a characteristic feature of bacteriophage development (Wainfan 1968, Wainfan, Srinivasan, and Borek 1965). Indeed, infection of *E. coli* with bacteriophage T3 induces formation of S-adenosylmethionine hydrolase enzyme (SAM-ase) (Gold et al. 1964). This leads to impaired methylation of both ribosomal and transfer RNAs and possibly also reduced synthesis of polyamine, and, in turn, to "misreading" of the codons of messenger RNA (Siersma and Lederberg 1970).

INTRODUCTION OF THIOL AND CYTOKININ GROUPS OF TRANSFER RNAs AS POTENTIAL REGULATORY PROCESSES

Many species of tRNA molecules purified from bacterial sources contain thiopyrimidine bases (Carbon, David, and Studier 1968, Lipsett 1965, Lipsett and Peterkofsky 1966). Early studies indicated that these thiopyrimidines are generated in precursor RNA molecules by transfer of thiol groups from β-mercaptopyruvate, formed from cysteine (Hayward and Weiss 1966, Lipsett, Norton, and Peterkofsky 1967). Wong et al. (1970) have purified a sulfurtransferase enzyme of *B. subtilis* which catalyzes the introduction of thiol groups into tRNA from β-mercaptopyruvate. However, Abrell, Kaufman, and Lipsett (1971) have recently demonstrated that *E. coli* contains a different enzyme system which consists of two proteins. This complex transfers thiol groups to tRNA from cysteine. β-Mercaptopyruvate is inactive as either a donor of thiol groups or a cofactor for the system, and the requirement for this compound previously reported (Lipsett, Norton, and Peterkofsky 1967) is now attributed to stabilization of one of the enzyme proteins in crude cell extracts. The precursor molecules, and derivatives of tRNA in which the thiobases have been chemically modified or have formed abnormal intramolecular bonds, may differ from the normal tRNA in ability to accept amino acid or in efficiency of codon recognition (Doi and Goehler 1966, Favre, Yaniv, and Michelson 1969, Harris, Tichener, and Cline 1969, Yaniv, Favre, and Barrell 1969; see also Ohashi et al. 1970).

At the present time there is no evidence to indicate that chemical modification (particularly reversible formation of disulfide bonds) of pre-existing thiopyrimidines plays any role in regulation of protein synthesis. On the other hand, cells of *E. coli* infected with bacteriophage T2 or T4 contain new species of tRNA coded by the viral genome and containing thiol groups (Hsu, Foft, and Weiss 1967, Weiss et al. 1968). Formation of the new species of tRNA is correlated with changes of RNA-sulfurtransferase activity in the cell

(Hsu, Foft, and Weiss 1967), for which the mechanism has not yet been established. To date this is the only phenomenon involving a change in spectrum of proteins synthesized which has been correlated with effects upon the thiolation of tRNAs. Similar changes were not observed in cells of *E. coli* infected with any of the bacteriophages T7, ϕX174, or MS2 (Hsu, Foft, and Weiss 1967), and have not been sought in other experimental systems.

A number of species of tRNAs, from both procaryotes and eucaryotes, contain a cytokinin group as a substituted adenosine base adjacent to the anticodon region (e.g., Burrows et al. 1968, Kline, Fittler, and Hall 1969). This grouping appears to be essential for correct interaction between the anticodon of the tRNA and the corresponding codon of mRNA (Fittler and Hall 1966, Gefter and Russell 1969, Theibe and Zachau 1968). The basic cytokinin structure is generated in a precursor molecule by transfer to adenine of an isopentenyl group from isopentenyl pyrophsophate, derived in turn from mevalonic acid (Bartz, Kline, and Söll 1970, Kline, Fittler, and Hall 1969, Peterkofsky 1968, Vickers and Logan 1970). Additional methyl groups may then be introduced by transmethylation from S-adenosylmethionine, and thiomethyl groups may be derived from cysteine in a reaction not yet characterized (Gefter 1969, Rosenberg and Gefter 1969). At the present time there is no evidence that modifications of the cytokinin content of tRNAs play any role in regulation of protein synthesis.[7] However, as noted by Kline, Fittler, and Hall (1969), restriction of the synthesis of isopentenyl pyrophosphate could serve as a mechanism for control of synthesis of specific protein by regulating the level of a specific tRNA required for a rate-limiting step in assembly of the protein.

[7] The tyrosine-specific tRNA coded by the *su3* suppressor gene of *amber* nonsense mutants contains a cytokinin residue (Goodman et al. 1968). However, in the present context, modification of the spectrum of proteins synthesized due to presence of this tRNA represents a trivial exception to the generalization.

MODIFICATION OF THE AMINOACYL-tRNA SYNTHETASE COMPLEMENT AS A POTENTIAL REGULATORY PROCESS

One potential mechanism for regulation either of the spectrum of proteins synthesized, or of the overall rate of synthesis of all proteins, is that of alteration in the cell complement of aminoacyl-tRNA synthetase enzymes. Onset of synthesis of cuticular protein in mealworm (*Tenebrio molitor*) pupae appears to require either formation of one, or more, new species of aminoacyl-tRNA synthetase, or modification of pre-existing enzyme (Ilan 1968, 1969). Further, apparent differences between two types of cell or tissue in the content of specific aminoacyl-tRNA synthetases have been indicated in comparisons of the maximum levels of acylation of a common preparation of tRNA by enzymes isolated from the two sources and of the elution profiles of the resulting homologous aminoacyl-tRNAs on chromatographic columns (Figure 9.2). Thus there appear to be changes in the complement of lysyl-tRNA and methionyl-tRNA synthetase enzymes during the early stages of development of sea urchin embryos (Ceccarini and Maggio 1969, Ceccarini, Maggio, and Barbata 1967). Similarly, developing frog embryos exhibit a dramatic change in the complement of aminoacyl-tRNA synthetases after development beyond the gastrula stage, which may reflect protein synthesis predominantly in the mitochondria before attaining this stage (Caston 1971). In addition, there is a small number of reports of specific differences in enzyme content between different tissues of the same organism (Anderson and Cherry 1969, Hendley, Hirsch, and Stregler 1969) or between tissues from animals deprived of hormone and the corresponding normal tissues (Germanyuk and Mironenko 1969, Tóth and Mányai 1968).

One particularly important example of a correlation between a regulated change in program of protein synthesis and a modification in the complement of aminoacyl-tRNA synthetase enzymes is afforded by cells of *E. coli* infected with

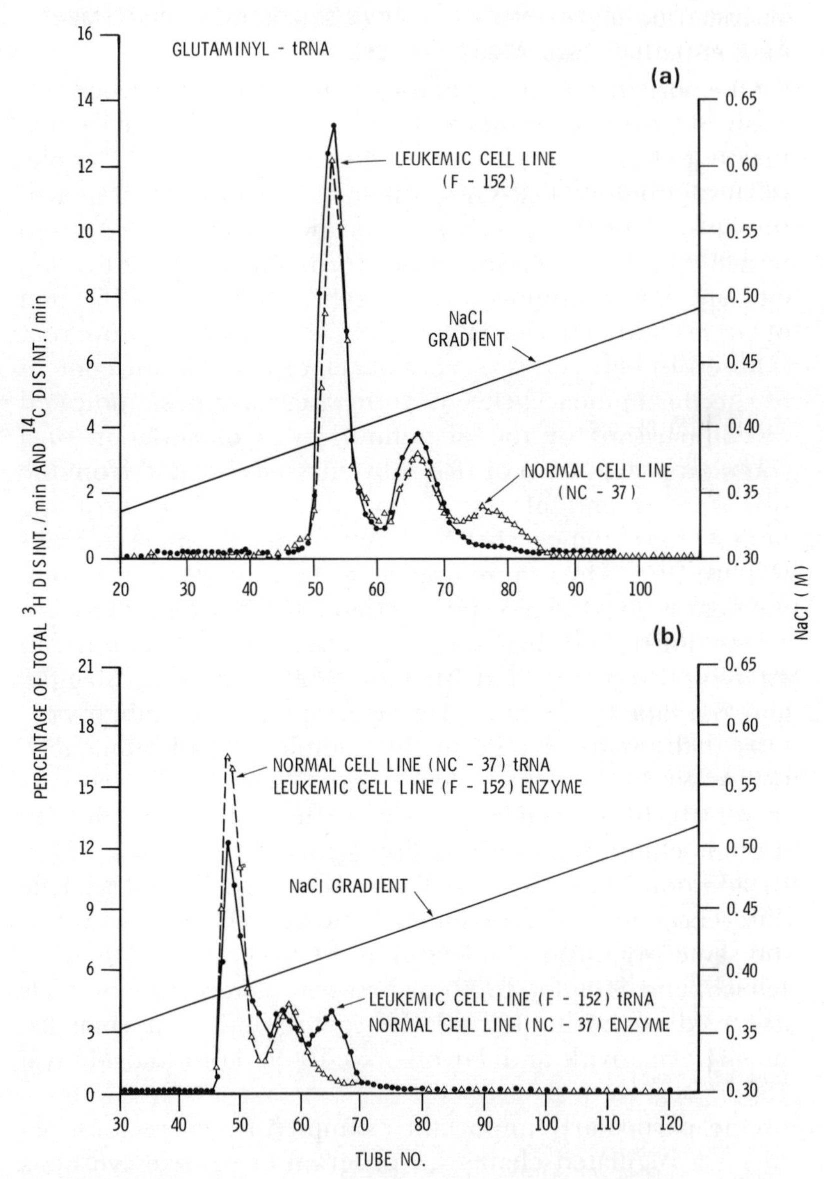

FIGURE 9.2. Demonstration of a probable difference in the complement of glutaminyl-tRNA synthetases of normal and leukemic human lymphoblasts

the bacteriophages T4 and T6. In these cases viral infection is accompanied by a modification of pre-existing host valyl-tRNA synthetase enzyme, which appears to involve addition of further protein specified by the viral genome. The conversion process is specific both to the valyl-tRNA synthetase enzyme and to infection by viruses of the T-even series. One consequence of the modification is a loss of total valyl-tRNA synthetase enzyme activity. Nevertheless, present evidence indicates that the process plays no significant role in regulation of synthesis of the virus proteins (Chrispeels et al. 1968, Neidhardt et al. 1969).

CAUTIO; CAUSE AND COINCIDENCE

The demonstration that modification of pre-existing host valyl-tRNA synthetase enzyme in phage-infected bacterial cells plays no obvious role in regulating the synthesis of viral proteins provides a rather dramatic illustration of the basic problem at the core of all contemporary attempts to identify cell components responsible for specific post-transcriptional regulation of the synthesis of particular defined proteins. There is now a wealth of reports of such specific post-transcriptional regulatory phenomena of various types. These include regulation of formation of specific proteins during the course of cell differentiation (Chapters 6 and 10) and in the response to exposure of target organs to specific hormones (e.g., Cohen 1970, Tomkins et al. 1969, Turkington 1968[a]) or unidentified components of sera (e.g., Gelehrter and Tomkins 1969). In addition, the differential effects of metabolic inhibitors upon the syntheses of histone and of other cell protein indicate the existence of special post-

Figure 9.2. (Facing page) Preparations of tRNAs isolated from the two types of cell were differentially labeled with ^{14}C- and ^{3}H-labeled amino acids, mixed, and co-chromatographed on reversed phase columns.

(a) tRNAs acylated with homologous enzyme.

(b) tRNAs acylated with heterologous enzyme.

Data of Gallo and Pestka (1970). Reproduced with permission of the authors and Academic Press, holders of the copyright.

transcriptional regulatory processes (Malpoix, Zampetti, and Fievez 1969). We also know of a substantial number of ways in which specific components of the protein-synthesizing system are, or may be, modified in ways which could markedly affect the assembly of specific peptide chains. However, there is a paucity of demonstrations that observed post-transcriptional regulation is *caused by* modification of specific factors required for protein synthesis.

The problem is even more difficult in systems in which regulation by transcription of new species of mRNA occurs. This is particularly well illustrated in the case of cells of *E. coli* infected with bacteriophages of the T-even series. Synthesis of the various proteins characteristic of the infected cells is controlled predominantly by the program of transcription of new species of mRNA from the viral genome (see Chapter 5). Concurrent suppression of residual synthesis of protein specified by the host genome appears to be largely, if not entirely, due to a change specified by the phage genome in the complement of initiation factors required for peptide chain initiation (Dube and Rudland 1970, Schedl, Singer, and Conway 1970, Steitz, Dube, and Rudland 1970).[8] Nevertheless, at least seven other components of the protein-synthesizing machinery of the cell and at least three enzymes involved in the insertion of essential substituent groupings of specific tRNAs are modified as a consequence of the viral infection (Table 9.1). It is possible that one, or more, of these modifications imposes some additional regulation upon the synthesis of peptide predetermined by mRNAs transcribed from the phage genome. However, the

[8] Differences in the specificity of mRNA "recognition" have been observed with initiation factors isolated from different tissues of the chick (Heywood 1970[b]) and from larvae of the mealworm at different stages of development (Ilan and Ilan 1971). It seems possible that these factors play a *regulatory* role in *selection* of mRNAs present in cytoplasmic informosomes (see Chapter 8) which are incorporated into polysomes (Spohr et al. 1970). In turn, this raises the possibility that changes in complement of initiation factors are not the consequence of differentiation, but rather play a major role in controlling the course of cell differentiation (see Chapter 2) by regulation of "preferential translation" of particular species of mRNAs.

TABLE 9.1. Changes in complement of transfer RNAs and modifying enzymes of E. coli infected with bacteriophage of the T-even series

Modification	Reference
Loss of host leucine-specific tRNA	Kano-Sueoka and Sueoka (1968).
Transcription of new arginine-specific tRNA Transcription of new glycine-specific tRNA Transcription of new isoleucine-specific tRNA Transcription of new leucine-specific tRNA Transcription of new proline-specific tRNA	Daniel, Sarid, and Littauer (1968). Scherberg and Weiss (1970).
Loss of host tRNA sulfur transferase activity	Hsu, Foft, and Weiss (1967).
Modification of host valyl-tRNA synthetase with net loss of total activity	Neidhardt et al. (1969).
Increase of U-specific tRNA methylase activity Increase of G-specific tRNA methylase activity	Wainfan, Srinivasan, and Borek (1965).

only strong conclusions which can be made at this time are (a) that at least one of these modifications appears to be entirely fortuitous (Neidhardt et al. 1969) and (b) that none are essential for synthesis of certain phage-specified proteins (Salser, Gesteland, and Ricard 1969).

REGULATION OF PROTEIN SYNTHESIS BY CHANGES IN THE CONFORMATION OF NASCENT PEPTIDE CHAINS

A variety of recent studies have led to consideration of possible alternative post-transcriptional mechanisms in which the rate of peptide chain elongation or release from the polysomes is controlled by the conformation of the nascent peptide chain itself. These proposed mechanisms are derived from early proposals by Gruber and Campagne (1965) and Cline and Bock (1966).

Formation of a protein with specific biological properties may involve several processes additional to elongation and subsequent release from polysomes of peptide chains of appropriate amino acid sequence or *primary structure.* One feature common to all individual peptide chains is development of the intramolecular bonds and structure necessary to impose an appropriate configuration upon the completed structure. Early studies of the course of renaturation of denatured proteins, particularly by Anfinsen (1967) and colleagues, tended to indicate that the configuration of the completed peptide chain was determined solely by its primary structure. Denatured preparations of disordered structure spontaneously assumed one specific conformation, which appeared to be that of the native protein,[9] and the process was catalyzed by an enzymic component of liver microsomes.

Thus it appeared that completed nascent peptide chains could readily assume the correct configuration after release from the polysomes and, hence, that the chains were not necessarily in this conformation while bound to the ribosomes. Indeed, some comparisons of the properties of free and ribosome-bound enzymes did furnish some suggestive evidence that the two might differ in structure. Gruber and Campagne postulated that completion of nascent chains, or their release from the polysomes, could be regulated by effects of other cell components upon the conformation of the polysome-bound nascent chain. Specifically, the phenomenon of induced enzyme synthesis (see Chapter 3) was attributed to combination of the inducer with the (actual or putative) catalytic site of the chain, leading to assumption of the correct conformation of the peptide chain to permit chain termination and release. Conversely, phenomena of repression (see Chapter 4) were attributed to induced development of wrong conformations, in which the peptide chain was "glued" to the ribosome and blocked further chain elongation. Concurrent variations in extent of tran-

[9] Recent studies indicate that the situation is more complex (e.g., Deal 1969, Kato and Anfinsen 1969, Mitchell 1968). It now seems more probable that the final configuration evolves progressively during chain elongation by what Mitchell has termed a "sequential residue-by-residue process." This is envisaged as a process in which the newly formed C-terminal segment of a nascent chain assumes a configuration determined primarily by that "frozen" (De Coen 1970) in the previously formed N-terminal portion. Thus, at any given stage in chain elongation, the nascent peptide will be of one specific, predetermined molecular conformation.

scription of the corresponding mRNAs would arise from the obligate coupling of gene transcription with peptide chain synthesis proposed by Stent (1964, 1966). It was also suggested that in the case of multimeric proteins, such as hemoglobin, one peptide chain might similarly serve to regulate the rate of synthesis of the other component chains.

In a similar proposal, Cline and Bock (1966) assumed that growing nascent peptide chains were in compact configurations determined by the primary structure. They noted, however, that individual growing chains progressively increase in size to a point at which they become a potential source of steric hindrance to the various processes involved in chain elongation and release. Additional complexing to the nascent chain of other completed chains of a multimeric protein could drastically alter the extent of the impediment to completion of the chain (Figure 9.3). Thus, the rates of peptide chain elongation and release could be markedly affected by even relatively small changes in conformation of the nascent chain itself.

Subsequent studies of the phenomenon of induced enzyme synthesis, the various types of repression of enzyme

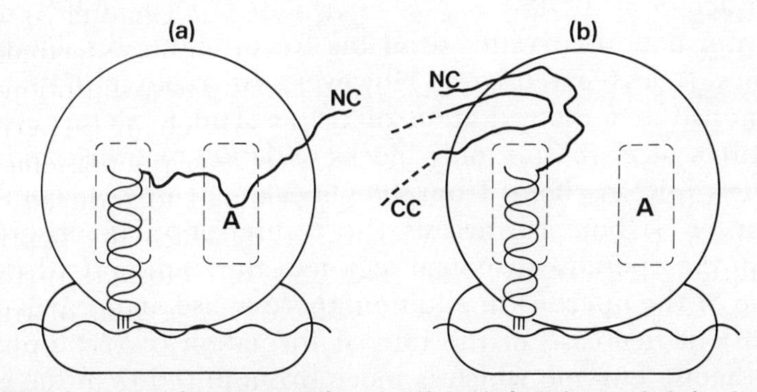

FIGURE 9.3. Schematic representation of self-regulation of rate of chain elongation by a conformation change in nascent peptide chain by complex formation with complementary chain of a multimeric protein.

In (a) nascent chain NC is shown impeding access of incoming aminoacyl-tRNA to the acceptor site A (see Chapter 1). In (b) complex formation with complementary peptide chain (CC) has imposed a change in conformation which removes the impediment at site A.

formation, and regulation of gene transcription in procaryotes (see Chapters 3 to 5) appeared to indicate that both proposals were untenable. However, a number of recent studies form the basis for renewed consideration of the possibility of post-transcriptional repression of protein synthesis by the mechanism envisaged by Gruber and Campagne (1965). In particular, Kovach and co-workers have obtained persuasive evidence that formation in *Salmonella typhimurium* of the enzymes controlled by the *his* operon (see Chapter 4) may be regulated primarily by changes in conformation of nascent chains of the enzyme (phosphoribosyl-ATP-pyrophosphorylase) specified by the "leading" structural gene of the operon. Addition of a pseudo-feedback inhibitor of this enzyme markedly affects the kinetics of repression of the operon by histidine (Kovach et al. 1969[a]). The extent of repression of enzyme formation by the histidine analogue triazole·alanine is drastically altered by mutations which affect the feedback-sensitive site of the enzyme (Kovach et al. 1969[b]). Moreover, histidyl-tRNA, the probable co-repressor for the operon, combines with the enzyme *in vitro* at a site other than the catalytic active site (Kovach et al. 1970).

Information currently available for other repressible enzymes is less extensive.[10] However, Imamoto (1970) has demonstrated that addition of tryptophan to derepressed cultures of *E. coli* not only blocks initiation of transcription of new mRNA chains from the genes of the *trp* operon (see Chapter 4) but also arrests the transcription in progress from the operator-proximal region of the leading structural gene of the operon. In addition, there is also an immediate, transient decrease in the rate of formation of tryptophan synthetase enzyme which is more pronounced than the ex-

[10] It may be recalled that the presumed operator-inactivated mutants of the *lac* operon (Jacob et al. 1960) actually contain point mutations within the *z* structural gene (Beckwith 1964, Brenner and Beckwith 1965). Many similar mutants affected in other operons have been isolated (see Chapters 3 and 4).

tent of the immediate decline in the rate of formation of mRNA chains (Lavallé and DeHauwer 1970). The mechanisms by which tryptophan exerts these additional effects upon expression of the genes of the *trp* operon has not yet been determined, but both are compatible with the mechanism postulated by Gruber and Campagne.

McLellan and Vogel (1970) have obtained compelling evidence that formation of the enzymes specified by the *arg* regulon of *E. coli* (see Chapter 4) is regulated by a post-transcriptional control mechanism. It appears probable that a complex of arginine with a protein gene product of the regulator gene *argR* both prevents translation of mRNAs transcribed from genes of the regulon and also stimulates their degradation. The mechanisms by which the postulated arginine-repressor complex blocks translation of these particular mRNAs has not yet been elucidated. It has been suggested that it may be one of entrapment of the mRNA in "repressive complexes" which contain ribosome-bound nascent (nearly) completed enzyme molecules, the gene product of gene *argR,* arginine, and possibly arginyl-tRNA and cognate synthetase (McLellan and Vogel 1970). Leisinger, Vogel, and Vogel (1969) found that a reversible change of conformation can be induced in preformed acetylornithine transaminase enzyme of intact cells of *E. coli* under conditions causing repression of synthesis of this and other enzymes of the arginine pathway. The change in conformation is manifested by susceptibility to inhibition by high concentrations of magnesium ions. The process requires the presence of the corresponding apo-repressor (see Chapter 4) and appears to be imposed upon the enzyme by interaction with some unknown structure formed in the cell under conditions of repression of enzyme formation. Development and loss of susceptibility to inhibition by magnesium in response to variations in the supply of end-product arginine are extremely slow in comparison with the rapid changes in rate of enzyme formation. Therefore, the relationship between the two phenomena can only be speculated upon at this time. Again the available data are not incompatible with the proposals of Gruber and Campagne. However, they appear equally compatible with the hypothesis that the postulated arginine-repressor complex is one of a different class of translation repressors which may act in an entirely different manner.

REPRESSION OF TRANSLATION

A number of regulatory phenomena have been observed with eucaryote cells in which the translation of one specific mRNA appears to be selectively inhibited by a labile inhibitor. The example which has been studied most extensively is that of formation of tyrosine aminotransferase enzyme in response to glucocorticoid hormone by rat hepatoma cells in tissue culture (see Chapter 6). Present evidence is in accord with the hypothesis that the hormone complexes with and antagonizes a repressor of translation which otherwise combines with the specific mRNA and facilitates its degradation (Martin and Tomkins 1970, Thompson, Granner, and Tomkins 1970, Tomkins et al. 1969). Similar labile repressors may control synthesis of immunoglobulins (Buell and Fahey 1969), retinal glutamine synthetase (Alescio, Moscona, and Moscona 1970, Moscona, Moscona, and Jones 1970), microsomal aryl hydrocarbon hydroxylase of fetal hamster cells (Nebert and Gelboin 1970), and the tyrosinase of *Neurospora crassa* (Horowitz, Feldman, and Pall 1970).

A different category of post-transcriptional regulatory mechanism may control synthesis of proteins coded by the genome of certain small RNA-containing bacteriophages. The "chromosome" of these viruses consists of a molecule of single-stranded RNA. It contains cistrons coding, respectively, for viral coat protein, specific viral RNA-replicase, and a maturation protein (e.g., Gussin 1966, Horiuchi, Lodish, and Zinder 1966). The genome of the virus serves directly as the mRNA for synthesis of the corresponding proteins.

However, unequal numbers of polypeptide chains are synthesized on the three cistronic sections of RNA. Considerably larger numbers of coat protein chains are made than of the other two types, both *in vivo* and, more particularly, *in vitro* (e.g., Lodish 1968, 1969, Nathans et al. 1966, Ohtaka and Spiegelman 1963).

Two different factors may contribute to this imbalance in rates of utilization of the different cistrons as templates for protein synthesis *in vitro*. One is that not all potential sites for peptide chain initiation present in the viral RNA are, in fact, available to participate efficiently in the processes of chain initiation *in vitro* (Bassel 1968, Gesteland and Spahr 1969, Lodish 1969). This appears to reflect structural features of both the ribosomes and the regions of the viral RNA containing the unavailable sites. It has been suggested that polypeptide chain synthesis which is initiated from those sites is due to changes in conformation induced as a consequence of synthesis of protein on the adjacent section of the RNA (Jeppesen et al. 1969, Lodish and Robertson 1969). Indeed, appropriate changes in conformation of the viral RNA can be induced *in vitro* by mild heating (Fukami and Imahori 1971). The other factor is a process of "translation inhibition," in which formation of a complex between viral RNA and completed molecules of coat protein suppresses, or severely impedes, synthesis of other species of protein (e.g., Eggen and Nathans 1969, Skogerson, Roufa, and Leder 1971, Sugiyama and Nakada 1968, Ward, Konings, and Hofschneider 1970). The basic feature in both of these processes is a change in conformation of the template RNA itself, rather than some effect upon the processes of polypeptide synthesis *per se*. It is not clear to what extent they reflect constraints which were necessary to package the genome of the virus within mature particles, or to what extent they may reflect properties of mRNAs in general.

Still less certain is the extent to which these processes actually regulate the synthesis of viral protein *in vivo* within infected cells. In the case of phage **f2**, it has been reported that the bulk of the synthesis of virus-"coded" protein occurs on the nascent strands of progeny RNA present in "replication complexes" and is not directly affected by viral coat protein (Robertson, Webster, and Zinder 1968). On the other hand, the nascent progeny strands appear to be of an unexpectedly small average length. Thus it would appear that regulation of synthesis of virus-coded protein may be largely controlled by the processes of replication of the viral genome, which is the physiological equivalent of gene transcription (Robertson and Zinder 1969).

Selectively enhanced protein synthesis on the mRNA (or cistronic portion) for coat protein of phage T4 has also been inferred from studies of Adesnik and Levinthal (1970). Similarly, disparities in proportions of polypeptide chains synthesized from the operator-proximal and all other segments of a polycistronic mRNA have been reported for two bacterial operons (Brown, Brown, and Zabin 1967, Whitfield et al. 1970). However, this does not appear to be a general property for all bacterial operons (Lee 1971, Wilson and Hogness 1969). Such disparities may reflect the presence in the cell of a series of different specific initiation factors (e.g., Dube and Rudland 1970).

POST-TRANSLATIONAL MODIFICATION OF PEPTIDE CHAIN STRUCTURE AND FACILITATED CHAIN RELEASE

Formation of complete protein molecules involves a variety of secondary processes which may modify the fine structure of polypeptide chains produced by the reactions of protein synthesis. Further, the biological activities of a variety of proteins are controlled by processes which induce changes in their conformation. These various secondary and regulatory processes include:

1. Association of two or more peptide chains to form a larger multimeric protein (e.g., Cook and Koshland 1969, Torriani 1968, Winterhalter 1966).

2. "Induced fit" changes of conformation at the catalytic site of an enzyme through complex formation with substrate, and changes of conformation at other *allosteric* sites through complex formation with *effector* molecules (e.g., see Koshland and Neet 1968, Monod, Wyman, and Changeux 1965).
3. Insertion of nonprotein prosthetic groups, such as heme (e.g., Felicetti, Colombo, and Baglioni 1966, Zucker and Schulman 1967) or the sugar residues of glycoproteins (e.g., Hagopian, Bosmann, and Eylar 1968, Molnar and Sy 1967).
4. Enzymic interconversion of inactive and active species of enzyme protein by insertion and removal of substituent groupings (e.g., see Holzer 1969, Scott and Mitchell 1969).
5. Modification of amino acid residues by insertion of substituent groupings, as, for example, hydroxylation of proline (Udenfriend 1966), or addition of N-terminal amino acid residues (Soffer 1970).
6. Migration into intracellular organelles and subsequent modification (e.g., Gross, McCoy, and Gilmore 1968, Hayashi, Yoda, and Kikuchi 1969, Scholnick, Hammaker, and Marver 1969).
7. Conversion of inactive precursor into functional protein, by cleavage of a larger species of molecule (e.g., Chance, Ellis, and Bromer 1968, Jacobson and Baltimore 1968, Laemmli 1970), or some other mechanism (e.g., Hartwell and Magasanik 1964, Seed and Goldberg 1965).

Each of these various processes may be involved in the formation of several species of protein. However, present evidence indicates that only the first two of these types of process may be involved in regulation of the rate of synthesis of individual peptide chains. It has been repeatedly observed that synthesis of the globin chains of hemoglobin is stimulated by heme, both in intact cells (e.g., Grayzel, Hörchner, and London 1966, Waxman and Rabinovitz 1966) and with cell-free systems (e.g., Adamson, Herbert, and Kemp 1969, Zucker and Schulman 1967). Winslow and Ingram (1966) suggested that this effect of heme might reflect a requirement for insertion of heme in a rate-limiting step in elongation of the peptide chains of human hemoglobin. However, later studies established that the heme group is not introduced until after release of completed peptide chains from the polysomes (Felicetti, Colombo, and Baglioni 1966, Morris and Liang 1968, Zucker and Schulman 1967). On the other hand, the possibility of regulation of chain elongation by insertion of prosthetic groups cannot be summarily dismissed in the case of groups inserted in nascent peptide attached to ribosomes (e.g., Molnar and Sy 1967, Sherr and Uhr 1969). Regulation of polypeptide chain elongation or release by transitory or permanent insertion of substituent groupings (mechanisms 4 and 5) seems highly unlikely but has not been formally eliminated from consideration. The final two types of mechanism can occur only after release of completed peptide chains and thus cannot *directly* regulate prior steps in the synthesis of the individual polypeptide chains.

Probably the most extensively studied group of specific proteins is that of mammalian hemoglobins, including the various normal and mutant human varieties (see, for example, Baglioni 1963, Braunitzer 1966, Ingram 1963, 1967, Itano 1966, Perutz and Lehmann 1968). The vast body of evidence available from studies of these proteins, by a variety of approaches, has led to recognition that synthesis of these respiratory proteins in normal erythroid cells is subject to regulation by a variety of control mechanisms. In particular, normal cells appear to contain a regulatory mechanism which operates to maintain a balance in the rates of synthesis of α-chains and of non-α-chains (i.e., β- and δ-chains in adult humans) such that there is no substantial accumulation of free chains of either type. This regulatory mechanism is disrupted in cells of individuals homozygous for the mutant genes manifested in the clinical conditions of thalassemia. Early studies indicated that in the condition of α-thalassemia synthesis of α-chains is impaired and an excess of non-α-chains accumulates, sometimes as an abnormal species of hemoglobin. Conversely, in the condition of β-thalassemia synthesis of β-chains is impaired but there is no apparent substantial accumulation of α-chains.[11] Thus it seemed possible that synthesis of β-chains was independent of that of α-chains, whereas synthesis of α-chains was in some way regulated by that of the β-chains (e.g., Baglioni 1963).

Colombo and Baglioni (1966) and Weatherall, Clegg, and Naughton (1965) attempted to determine the nature of this regulatory mechanism from an analysis of the globin chains formed by rabbit reticulocytes briefly incubated with radioactive amino acids. The polysome fraction of the cell was found to contain completed α-chains but few or no completed β-chains. On the other hand, the soluble fraction contained a larger proportion of newly formed β-chains than of α-chains. Therefore, in keeping with the belief that free α-chains do not accumulate in erythroid cells, it was originally concluded that α-chains could only be released from the polysomes through combination with a free β-chain. Such a regulatory mechanism by which completed α-chains were "stripped" from the polysomes by free β-chains would be in accord with the basic proposals of Gruber and Campagne (1965) and Cline and Bock (1966). However, re-examination of cells from cases of β-thalassemia revealed that they do, in fact, synthesize an excess of α-chains, which precipitate in inclusion bodies (e.g., Bank 1968, Bank and Marks 1966, Bargellesi et al. 1968, Weatherall 1968). Thus synthesis of α-chains and subsequent release of free chains can occur

[11] Recent data of Schwartz (1970) indicate that the synthesis of globin chains is impaired only in the reticulocytes of thalassemic patients, and not in the nucleated erythroid precursor cells.

without complexing with β-chains. This conclusion, in turn, meant that the data obtained in analyses of the nature of the regulatory process were presumptive evidence of the existence in normal erythroid cells of a pool of excess preformed α-chains. Such pools were, therefore, sought and have been found, or artificially induced (e.g., Honig, Rowan, and Mason 1969, Maleknia, Blum, and Schapira 1969, Shaeffer 1967, Tavill et al. 1968). Consequently, the hypothesis that formation of a dimer of hemoglobin α- and non-α-chains is an *obligate requirement* for release of either type of chain has been abandoned.

Nevertheless, there is an increasing body of evidence that release of completed β-chains, and thereby indirectly synthesis of β-chains, *may normally be facilitated* by complex formation with preformed α-chains of the free pool. Complete α-chains do appear to associate with nascent incomplete β-chains bound to the ribosomes (Baglioni and Campana 1967). Further, addition of exogenous human hemoglobin chains has been reported to stimulate release of the complementary chains from canine reticulocyte polysomes, with formation of hybrid canine-human species of protein (Fischer, Nagel, and Fuhr 1968). Conversely, exogenous human β-chains reduce the extent of release of β-chains from rabbit reticulocyte polysomes (Shaeffer, Trostle, and Evans 1969).[12] Further, there appears to be a step which severely rate-limits either the elongation or release of β-chains in intact rabbit reticulocytes. This reduces the overall rate of formation of individual β-chains to roughly one half that for α-chains and results in an increase in number of ribosomes per polysome engaged in β-chain synthesis (Hunt, Hunter, and Munro 1968[b], 1969) which mimics the effects

[12] Blum, Maleknia, and Schapira (1969, 1970) have also reported that addition of high concentrations of free peptide chain suppresses *in vitro* synthesis of the homologous chain. This observation was not confirmed by Shaeffer, Trostle, and Evans (1969) under their experimental conditions.

of an induced modulation of chain synthesis (Kazazian and Freedman 1968). However, present evidence does not permit identification of the factor(s) regulating this rate-limiting step in the synthesis of the β-chains, and the possibility that release of completed β-chains is facilitated *in vivo* by complex formation with free α-chains remains to be critically assessed.

Later stages in the assembly of the hemoglobin molecule, and the mechanism by which heme stimulates the synthesis of globin chains, have not yet been fully established. Present evidence indicates that the first step in the assembly of the complete molecule is the formation of an $\alpha\beta$-globin dimer, which then combines with the heme prosthetic group and is converted to the final $\alpha_2\beta_2$ tetramer (e.g., Felicetti, Colombo, and Baglioni 1966, London et al. 1967, Tavill et al. 1968, Zucker and Schulman 1967). Experiments with both intact reticulocytes and cell-free systems indicate that heme both promotes the incorporation of ribosomes into polysomes and retards the loss of pre-existing ribosomes (e.g., Adamson, Herbert, and Kemp 1969, Grayzel, Hörchner, and London 1966, Waxman and Rabinovitz 1966, Zucker and Schulman 1967). The heme appears to associate with the ribosomes of the polysome but not with native ribosome subunits, and it is not incorporated into the nascent peptide chains. Therefore, it seems probable that post-transcriptional regulation of globin synthesis by heme is due to some effect upon the later stages (see Chapter 1) of initiation of synthesis of peptide chains (Waxman, Freedman, and Rabinovitz 1967, Zucker and Schulman 1967). Analysis of the effect of heme is, however, complicated by what appears to be an unrelated action of exogenous hemin as an inhibitor of reticulocyte endonuclease activity (Burka 1968, Farkas and Marks 1968, Maxwell and Rabinovitz 1969). Rabinovitz and colleagues (Maxwell and Rabinovitz 1969, Rabinovitz et al. 1969) have suggested that heme prevents the formation of a modified globin derivative acting as an inhibitor of peptide chain initiation,[13] but all alternative possibilities have not been eliminated.

There are many similarities between the processes of formation of immunoglobulins by lymphoid tissue or plasma cell tumors (see Chapter 11) and those of formation of hemoglobin in erythroid cells. The two types of constituent peptide chain are synthesized on different polysomes and subsequently assembled into the complete immunoglobulin molecule. In at least some cases, the cells contain a small pool of free "light" chains. Some of these may complex with nascent incomplete "heavy" chains still attached to ribosomes, and it has been suggested that this process might facilitate completion or release of the heavy chains (Askonas and Williamson 1966, Shapiro et al. 1966[a], 1966[b], Williamson and Askonas 1968). Subsequent study of a variety of different plasma tumor cells by various workers has established that such complex formation is not an obligate requirement for release of completed heavy chains. Nor is formation of a dimer composed of light and heavy chains an obligatory intermediate in the assembly of complete immunoglobulin chains (e.g., Askonas and Williamson 1967, 1968, Cioli and Baglioni 1967, Scharff 1967, Schubert and Cohn 1968). Indeed, there is no evidence that the light chain regulates synthesis of the heavy chain (Cohn 1967). On the other

[13] Fuhr et al. (1969) have presented evidence of the presence of such an inhibitor in reticulocytes of thalassemic patients and its absence from normal reticulocytes. Gilbert et al. (1970) have found no evidence of such an inhibitor. However, Howard, Adamson, and Herbert (1970) have observed the formation of this type of inhibitor during incubation of the ribosome-free supernatant of reticulocyte lysates *in vitro,* and inhibition of the process by hemin. It should also be noted that, unless the inferred inhibitor is, in fact, derived by modification of the globin chain produced in normal amounts, the observations of Fuhr et al. (1969) would imply that there are different inhibitors specific for each of the globin chains. In turn, this would imply: (a) that the defect in thalassemias is the structural gene of either the inhibitor or the enzyme(s) responsible for its formation, and (b) that heme can no longer combine with and inactivate the mutant inhibitor or inhibit the altered enzyme-forming inhibitor.

hand, there is insufficient evidence to warrant total dismissal of the possibility from further consideration.

Thus there is no compelling evidence that synthesis of any specific polypeptide chains is regulated by a process of facilitated release, or stripping, but the possibility remains open. Indeed, Bronskill and Wong (1971) have recently postulated that enzymically active, ribosome-bound partial molecules of the β-galactosidase of *E. coli* are stripped from the ribosome through combination with nascent subunit(s) of the enzyme.

REGULATION OF THE SYNTHESIS OF TOTAL PROTEIN

We know of many potential post-transcriptional mechanisms by which (at least in theory) synthesis of a specific protein could be controlled. We also know of many phenomena in which synthesis of one, or more, specific protein(s) is controlled by a post-transcriptional regulatory mechanism. Yet, in almost no case, including the extensively studied stimulation of globin synthesis by heme, has the actual mechanism been completely established. A similar situation also prevails in the case of post-transcription mechanisms which control rates of total protein synthesis in general.

One of the most extensively studied of these mechanisms is responsible for the marked rise in protein synthesis following fertilization of sea-urchin eggs. This has been considered earlier in the chapter, and the evidence indicating that the major regulatory mechanism is one of activation of masked mRNA in a process associated with proteolysis will not be reconsidered here. Nevertheless, it should be emphasized that additional factors are involved in this regulatory mechanism. The unfertilized egg does synthesize some protein and contains functional polysomes of a size range comparable with that for the polysomes of young embryos (Kedes et al. 1969[a]). Thus some selective process must determine which mRNA molecules formed in the maturing oocyte shall not be masked. Further, the increase in protein synthesis following fertilization is associated with an increased level of transfer factors required for either peptide chain initiation or elongation (Castañeda 1969).[14]

A second control mechanism which has been very extensively studied, but remains almost a total mystery, is that which coordinates the syntheses of all the many components of the protein-synthesizing machinery. Most studies of this type of system have been made with bacterial cells, and particularly by comparisons of strains of *E. coli* with different alleles of the so-called **RC** locus. The early studies have recently been reviewed by Edlin and Broda (1968), and only the essential points will be enumerated.

It has been known for almost a quarter of a century that one feature characteristic of normal cells of *E. coli,* and related species, growing exponentially in (approximately) steady-state conditions is that the rate of protein synthesis is directly proportional to the rate of nucleic acid synthesis (Malmgren and Hedén 1947). With the progressive accumulation of knowledge of the mechanism of protein synthesis and the components required, the nature of this relationship became more precisely defined. The number of ribosomes per cell and (except at low rates of growth) the content of tRNA are approximately proportional to the rate of cell growth, that is, protein synthesis (Gray and Midgley 1970, Maaløe and Kjelgaard 1966). More specifically the number of ribosome subunits which are in the course of assembly is roughly proportional to the mean generation time of the cells (Van Dijk-Salkinoja and Planta 1971). It has also been recently demonstrated (Gordon 1970) that the syntheses of elongation factors **T** and **G** are similarly coordinated. Thus the cell adjusts the rates of formation of three species of

[14] On the basis of studies made with *in vitro* systems, there is some disagreement whether this increased content of transfer factors is, in fact, required to support maximum rates of protein synthesis (e.g., Castañeda 1969, Kedes et al. 1969[b]). However, *in vitro* systems are notoriously inefficient in comparison with the intact cells from which they were derived. Thus greater significance must be accorded to data indicating the requirement.

mature ribosomal RNAs, the various species of ribosomal proteins, many species of transfer RNAs, and all other species of protein in an integrated manner. On the other hand, transcription of mRNAs is regulated totally independently (see Chapter 5).

In normal wild-type cells this coordinating mechanism operates to *stringently* maintain a rough balance in the cell contents of the various components of the protein-synthesizing machinery. Changes in rates of protein synthesis resulting from modifications in either composition of the growth medium or growth conditions are associated with very marked, but transient, changes in rates of synthesis of ribosomes and tRNAs. These result in very rapid readjustment of the balance in cell contents of components required for protein synthesis and of attainment of the new steady-state rate of protein synthesis. Similarly, deprivation of most auxotrophic mutants of sources of amino acid essential for growth arrests the synthesis of protein, and causes a rapid secondary inhibition of synthesis of ribosomal and transfer RNAs. Thus, the increase in rate of protein synthesis on transfer from minimal to rich medium (a "step-up"), or on increasing the limiting supply of required amino acid for a mutant strain, is indirectly regulated by synthesis of RNAs.

An exception to the general rule that starvation of a mutant for required amino acid inhibits synthesis of RNA was found in a methionine-requiring strain studied by Borek and co-workers (Borek, Rockenbach, and Ryan 1956, Borek, Ryan, and Rockenbach 1955). In this strain the coupling of RNA and protein synthesis is *relaxed,* and methionine-deprived cells continue to accumulate RNA. Subsequent analysis revealed that the difference between this strain and others is not a function of the methionine requirement but is due to a mutant allele at a gene locus controlling the coupled regulation of RNA and protein syntheses (Stent and Brenner 1961). This gene locus is named the "**RC**" locus, and the two alleles as $\mathbf{RC}^{\mathrm{stringent}}$ and $\mathbf{RC}^{\mathrm{relaxed}}$, respectively.

Despite a very extensive series of comparisons of strains differing in the **RC** allele present, we still have no real indication of the mechanism or factors responsible for the stringent type of integrated regulation. Some of the early candidates considered as possible regulatory factors include the level of unacylated tRNAs; relative proportions of ribosome subunits, free ribosomes, and polysomes; levels of available nucleoside triphosphates; obligate coupling of gene transcription with peptide chain elongation; and cell content of polyamines (see Edlin and Broda 1968). More recent reports have indicated differences between the two types of strain in the presence of a "protector" substance on one species of leucine-specific tRNA (Yegian and Stent 1969[a]); in systems for protein degradation (Sussman and Gilvarg 1969); and accumulation of lipids and carbohydrates (Sokawa, Nakao-Sato, and Kaziro 1970). There is some disagreement about the role of levels of nucleoside triphosphates (e.g., Edlin and Stent 1969, Gallant and Harada 1969). However, it appears to be generally agreed that none of the foregoing factors is responsible for the stringent type of control mechanism.

The immediate cause of inhibition of the formation of the two larger ribosomal RNAs is accumulation of 5′·diphosphoguanosine-(2′·3′)-diphosphate (ppGpp) in starved cells which contain the stringent **RC** allele (Cashel 1969, Cashel and Gallant 1969, Cashel and Kalbacher 1970). This nucleotide inhibits the action of the psi factor required for transcription of the precursor to these two species of RNAs (Travers, Kamen, and Cashel 1970; see Chapter 5). However, accumulation of the nucleotide may require the prior synthesis of RNA and reflect the operation of the control mechanism rather than be the causal factor (Wong and Nazar 1970). Moreover, transcription of the precursor of the small (5S) ribosomal RNA persists, but processing to the mature species of molecule does not occur (Galibert et al. 1970).

There are a few isolated pieces of recent evidence which, in conjunction with more extensive data for metazoan cells (see Chapter 8), constitute suggestive evidence that the regulatory factor may be the concentration of an unstable immature precursor of the rRNA of small ribosome subunits. In view of the paucity of relevant data they will merely be enumerated as follows:

1. Inhibition of protein synthesis in *E. coli* by deprivation of methionine leads to accumulation of precursor. During recovery there is renewed synthesis of ribosomal RNA, but a pronounced delay in appearance of mature small-subunit RNA (Adesnik and Levinthal 1969). Further, there is a pronounced impairment in the ability of the cells to show induced enzyme synthesis (Javor, Ryan, and Borek 1969). The rRNA precursor has been reported to be unstable (Nakada, Anderson, and Magasanik 1964).
2. Ehrenfeld and Koch (1968) have reported overproduction and degradation of rRNA precursor in spheroplasts of *E. coli.*
3. Virtually all the ribosomal RNA of *B. subtilis* spores has been reported to be in the form of rRNA precursor during very early stages of germination (Bleyman and Woese 1969).

One obvious potential mechanism for nonselective regulation of the rate of protein synthesis is that of variation in the intracellular level of *net* nuclease activity. Indeed, there have been several reports of lower levels of net RNase activity in regenerating liver than in normal liver (e.g., Arora and DeLamirande 1967, Shortman 1962, Tsukada, Majumdar, and Liebermann 1966). Similarly, there have been numerous reports of decreased net RNase activity in specific target organs following the administration of hormones (e.g., Brewer, Foster, and Sells 1969, Imrie and Hutchinson 1965, Sarkar 1969). In addition, an inhibitor of protein synthesis isolated from nonlactating bovine mammary gland cytoplasm was found to contain two inhibitory components, one of which is RNase enzyme (Herrington and Hawtrey 1969). There is, therefore, a substantial body of circumstantial evidence to support the hypothesis that protein synthesis may be regulated by fluctuations in the relative levels of latent and active forms of RNase enzymes and of RNase inhibitors.

On the other hand, there are situations where there is a decline in the cell content of polysomes which is not due to destruction of the mRNA itself (e.g., Ward and Plageman 1969). Detailed studies of such a phenomenon in mycelium of *Neurospora crassa* indicate that the decrease in polysome content is due to a change in activity of the ribosomes (Alberghina et al. 1969). It has been suggested that this change is a conformational change in the ribosomes. However, it seems more probable that the decline in polysome content is due to loss of one, or more, of the factors required for dissociation of the ribosomes into subunits and subsequent reincorporation into new initiation complexes (see Chapter 1). Similar declines in polysome content are observed on adding high concentrations of sodium fluoride to growing cells or to reticulocytes. These effects are due primarily to inhibition of dissociation of the ribosomes into subunits (Colombo, Vesco, and Baglioni 1968, Hogan 1969, Vesco and Colombo 1970). There may also be additional impairment of the association of 60S subunits to initiation complexes containing the 40S subunits (Hoerz and McCarty 1969). Similar progressive decreases in the proportion of ribosomes present in polysomes are also observed during the maturation of reticulocytes (Herzberg, Revel, and Danon 1969, Marks, Rifkind, and Danon 1963, Rifkind, Danon, and Marks 1964). This decrease appears to be due to a reduced ability of the ribosomes to form initiation complexes, probably through loss of initiation factors (Bishop 1966, Herzberg, Revel, and Danon 1969, Rowley and Morris 1967). A similar loss of initiation factors may also be responsible for the decrease in activity of ribosomes of mouse brain in the first few days of neonatal life (Lerner and Johnson 1970), although additional factors also regulate the precipitate decline in protein synthesis (Johnson 1969, Johnson and Luttges 1966).

The studies which have been surveyed to this point do not exhaust the list of potential mechanisms for regulation of the rate of protein synthesis.[15] In theory regulation could be exerted by fluctuations in the level of any component of the protein-synthesizing system, or of any agents which stimulate or inhibit any of the reactions involved in the synthesis of protein. Such fluctuations could be effected not only by synthesis and degradation of the components but also by interconversion between active and inactive configurations or derivatives, and by physical separation in an organized "compartment" within the cell. Most of these possibilities have not been systematically examined or implicated as mechanisms widely used by living cells, and they cannot therefore profitably be discussed. However, there is a growing list of isolated reports of phenomena in which the rate of protein synthesis in general does appear to be affected by a component not considered previously in this chapter as a regulatory factor. Some of these may well prove to be the first reported examples of what is, in fact, a regulatory mechanism of general significance. Among the examples which I find of particular interest are the following:

1. An increase in rate of protein synthesis in the toad fish (*Opsanus tau*) liver on acclimation to cold which is associated with a marked increase in level of transfer factor I (Haschemeyer 1969[a], 1969[b]).

[15] In particular, I have not considered the mechanism of action of interferons, formation of which is elicited by introduction of "foreign" species of nucleic acid into vertebrate cells. Interferons have been reported to have a wide variety of different, not necessarily mutually exclusive, effects upon various components of the protein-synthesizing system. It is beyond the scope of this chapter to review all of these possibilities, and I refer the interested reader to Vilček (1969).

2. Formation of unidentified translation factors which stimulate protein synthesis by cells of target organs as a specific response to exposure to hormone (Sokoloff 1968, Wool, Martin, and Low 1968).
3. Correlations between changes in rate of protein synthesis and cell content of polyamines, and diverse effects of polyamines upon the reactions involved in protein synthesis (e.g., Cohen 1968, Herbst and Bachrach 1970, Pegg, Lockwood, and Williams-Ashman 1970, Takeda 1969[b], Yoshikawa and Maruo 1961).
4. Stimulatory and inhibitory effects of lipid constituents of rat liver upon the reactions of protein synthesis (Hradec and Dušek 1968, Hradec and Štroufová 1960).
5. Reversible inhibition of protein synthesis and dissociation of polysomes, possibly by histone, during the metaphase portion of the cell cycle in some classes of cell (e.g., Salb and Marcus 1965, Scharff and Robbins 1966), but apparently not in all metazoan cells (Fry and Gross 1970).
6. Formation of a translation inhibitor in spores of aquatic fungi (Schmoyer and Lovett 1969).

CONTROL OF SECRETION AND DEGRADATION OF SPECIFIC PROTEINS

The content of the various species of protein present in a cell is not determined solely by the processes which regulate gene transcription and subsequent formation of new protein molecules. Additional mechanisms serve to regulate the processes of active secretion of protein to the exterior of the cell and to control the rates of degradation of specific proteins in the dynamic processes of turnover. We know relatively little about either of these types of mechanism. Synthesis of protein to be exported from the cell appears to occur primarily on membrane-bound polysomes (Ganoza and Williams 1969, Gaye and Denamur 1969) and remains associated with vesicles of membrane until released (Peters 1962[a], 1962[b], Ganoza and Williams 1969). Discontinuous secretion of protein from glandular cells by hormonal stimulation appears to be mediated by cyclic 3′·5′-AMP (Kulka and Sternlight 1968) and is associated with an intense turnover of the phosphate groupings of phosphatides of the membranes of the cell (Hokin and Hokin 1958). However, the relationships between the two processes and also with the processes of secretion from nonglandular cells still remain unclear.

The normal catabolism of protein in undamaged cells is distinct from the autolytic process of proteolysis. Sources of energy and coenzyme A are required (Penn 1960, Simpson 1953, Steinberg, Vaughan, and Anfinsen 1956), and the degradation of a specific protein may be regulated by a variety of factors. These even include specific genes distinct from the structural genes of the protein (Ganschow and Schimke 1969, Recheigl and Heston 1967), possibly related to "architectural genes" which determine intracellular sites of protein accumulation (Ganschow and Paigen 1967). The list of factors affecting rates of degradation of a specific enzyme also includes:

1. Presence of specific substrate (Schimke 1966) and glucose (Jost, Khairallah and Pitot 1968) or, more generally, the nutritional status of the cells (Auricchio, Martin, and Tomkins 1969).
2. Stage of embryonic development (Wagner, Payne, and Briggs 1969).
3. Stage in the cell cycle of individual cells (Martin, Tomkins, and Bresler 1969).
4. Administration of drugs (Kurimiyama et al. 1969). Further, in contrast to the response of cultures of hepatoma cells in rich medium (Auricchio, Martin, and Tomkins 1969), inhibition of (normal) protein synthesis leads to a profound decrease in the rate of degradation of tyrosine-aminotransferase enzyme both in perfused liver and in the liver of an intact fasted animal (Grossman and Mavrides 1967, Kenney 1967, Levitan and Webb 1969, 1970).

STATUS QUO

Thus the efforts of very many people have provided a monumental body of evidence concerning mechanisms potentially able to exert control upon the synthesis and retention of protein by living cells. However, at the present time, we have relatively little real knowledge of the post-transcriptional regulatory mechanisms which *actually control* protein synthesis and the levels of proteins maintained.

The generation of some parts supervenes on others previously existing, from which they become distinct.—William Harvey, 1651.

But that which is remarkable and that which still shows well the close relationships which exist between generation and nutrition, is that there is often need for a special nutrition to cause these evolving transformations.—Claude Bernard, 1872.

CHAPTER 10: **MOLECULAR DIFFERENTIATION IN MODEL SYSTEMS**

Differentiation of a cell of specialized function is marked by sequential changes in the spectrum of proteins synthesized by the cell. These changes both reflect, and appear to be controlled by, differential expression of particular groups of genes selected from the total cell genome.

In the preceding chapters I have considered our current knowledge of actual or potential mechanisms whereby selected portions of the genome of a eucaryote cell may be differentially expressed in the synthesis of corresponding specific proteins. These include:

Primary mechanisms regulating the processes of gene transcription, and also those of gene amplification and elimination.

Secondary mechanisms regulating the processing and extranuclear transport of gene transcripts.

Tertiary mechanisms regulating translation in the cell cytoplasm of polysomal gene transcripts.

Most of these processes have been shown to occur in several different types of differentiating tissue.

However, the course of embryonic development is far more complex than any of the individual phenomena which have proved amenable to analysis at the molecular level. Even the most superficial consideration of the factors involved in the process of organogenesis serves to emphasize the importance of many phenomena of which we have little current understanding. Among these are such major phenomena as cell-cell interactions, cell migration, and development of axial gradients.[1] In one sense these phenomena may themselves be considered as striking examples of the diverse consequences which may follow from the differential expression of appropriate groups of genes during preceding stages of development. It is not my intention to argue otherwise. Rather, I wish to stress that it is a gross oversimplification to consider embryonic development beyond the cleavage stages as merely the summation of processes of cell replication and those required for differentiation of specialized cells from pre-existing populations of precursor cells.

At the present time we do not have sufficient detailed knowledge of the myriad distinct processes involved to make a meaningful assessment of the extent to which the various mechanisms known, or suspected, to control synthesis of specific proteins *actually do regulate* the course of organogenesis or normal embryonic development.[2] It is certainly obvious that prior transcription (processing?) and incorporation of appropriate mRNAs into cytoplasmic polysomes are essential prerequisites for the synthesis of specific proteins characteristic of a particular class of differentiated cell. On the other hand (see Chapters 6, 8, and 9), it is assuredly equally clear that transcription of the mRNA, and in some

[1] It is beyond the scope of this discussion to consider these phenomena or their roles in organogenesis and embryonic development. These topics are presented in many excellent introductory texts. I refer those of my students who have had no formal instruction in embryology to the texts by Balinsky (1970), Ebert (1965), Spratt (1964), and Trinkaus (1969).

[2] The phenomena of gene elimination and gene amplification (see Chapter 2) are obvious exceptions to this generalization.

cases even incorporation into "apparent polysomes," is not necessarily a sufficient condition to ensure initiation of synthesis of the corresponding specific protein. Yet there is a dearth of examples of regulation of protein synthesis at the translational level for which even a tentative identification of the regulatory mechanism has been made.

In view of these uncertainties, it seems appropriate to review the possible roles of the various known and potential regulatory mechanisms in the differentiation of selected model systems. Studies are being made with a variety of such models, each of which possesses certain advantages and disadvantages which make it especially suitable for particular types of study. Some of the results obtained have been considered in appropriate sections of preceding chapters, and I do not propose to recapitulate them here in the form of a catalogue. Rather, I wish to consider a few special systems and summarize the current status of our knowledge of mechanisms which may be involved in controlling their differentiation. Further, my intention is merely to illustrate the complex manner in which differentiation of cells of specialized function may be controlled through the interplay of several different regulatory mechanisms. Many model systems are automatically excluded from further consideration for lack of sufficient information relevant to this particular exercise. The further selection of systems to be considered is a purely arbitrary choice, reflecting my own interests, and does not imply that these are the systems for which most information is available.

REPLACEMENT OF HISTONE BY PROTAMINE IN THE MATURING SPERMATID OF TROUT

Processes characteristic of oocyte maturation in some animal species have been considered in previous chapters (Chapters 2 and 6). Differentiation of spermatogonia to mature spermatozoa also exhibits a characteristic sequence of processes which affect the synthesis of protein by the differentiating cells. The nucleus of the early spermatid is usually a site of intense gene transcription, whereas sperm nuclei are metabolically inert. Conversion of the spermatid into a mature spermatozoan is marked by condensation of the chromatin into the compact structure of the sperm head. In many species of animal this condensation process is associated with replacement of the histones characteristic of their somatic cells by special distinctive species of histone or protamine (e.g., Bloch 1969, Dixon and Smith 1968, Hnilica 1967). These special proteins are in turn displaced from the chromatin by nonhistone protein following fertilization of the oocyte. Indeed, no typical histone is found in the chromatin of developing sea urchin embryos prior to the blastula stage (e.g., Johnson and Hnilica 1970, Ord and Stocken 1968, Orengo and Hnilica 1970), although extranuclear synthesis of "histone-like" protein commences before this stage (e.g., Crane and Villee 1971, Kedes et al. 1969, Orengo and Hnilica 1970). Similarly, the chromatin of developing amphibian embryos lacks typical histone before gastrulation (e.g., Asao 1970, Johnson and Hnilica 1970, Moore 1963). As we believe histones and related proteins to play major roles in regulating gene transcription in eucaryotes (Chapter 7), study of the synthesis of these special proteins provides an opportunity to analyze mechanisms controlling synthesis and mode of action of a key protein component of a control mechanism.

The chromatins of salmonid fish sperm contain protamines in lieu of histone. These are relatively simple, small proteins (or polypeptides), which can be readily separated from histones of the somatic cells on polyacrylamide gels or columns of the resin BioGel-P10 (Ingles and Dixon 1967, Marushige and Dixon 1969). They contain approximately 30 amino acid residues, of which about two-thirds are basic arginine residues and the remainder neutral amino acids. The protamines (clupeines) of Pacific Herring (*Clupea pallasii*) have been resolved into three distinct components and the complete amino acid sequence of each determined (Ando and Sawada 1961, Ando and Suzuki 1966, 1967). From a comparison of these sequences, Black and Dixon (1967) have suggested that the protamine chains contain a fundamental "repeating unit" pentapeptide which may be depicted as:

(neutral amino acid-Arg-Arg-Arg-Arg)

Data for other species are less complete. However, partial sequences determined (Ando and Watanabe 1969) for two of the three protamine fractions isolated from a race of Rainbow Trout (*Salmo irideus*) appear to be in accord with this suggestion (see Dixon et al. 1969). Three discrete protamine fractions have also been isolated recently from fish of another race (species?) of Rainbow Trout (*S. gairdnerii*) by a modified procedure, and designated in order of elution from columns of carboxymethyl-cellulose C_{I}, C_{II}, and C_{III}, respectively. Analyses of the amino acid compositions indicate that each of the fractions is heterogeneous and that they differ from those previously isolated from *S. irideus* (Ling, Jergil, and Dixon 1971). Thus it seems possible that the protamine complement in the sperm of an individual fish consists of a "family" of several distinct proteins of similar primary structures.

The very high arginine content of the protamines permits formation of complexes containing considerably larger proportions of DNA than those of DNA-histone complexes. Further, the arginine content of the protamine present in sperm chromatin is almost, but not completely, equivalent to the content of phosphate groupings in the DNA (Ando and Hashimoto 1958). These two features have led Dixon and Smith (1968) to postulate a basic structure for DNA-protamine complexes (contained in Figure 10.2) which could account for the extremely compact form in the sperm head, estimated (Olins, Olins, and Von Hippel 1968) to contain 2 meters of DNA packed into 10.4 cubic microns.

Dixon and his colleagues (1969) have exploited these

various properties of the protamines and their complexes with DNA to initiate an elegant series of analyses of the factors which regulate onset of protamine synthesis, and subsequent displacement of histone from chromatin, during differentiation of mature sperm in the Rainbow Trout. In most of the studies the differentiation was induced in immature Steelhead Trout, an anadromous variant, by injection of salmon pituitary gonadotrophin but closely paralleled that observed with natural populations during the spawning migration (Ingles et al. 1966).

The immature testis of a 2-year-old fish weighs approximately 0.02 gram, has a DNA content of about 0.07 milligram, and contains only cysts of undifferentiated spermatogonia. About 8 weeks after commencing the course of hormone injections it attains a maximal size of 1.5 to 2 grams (Figure 10.1), has a DNA content of about 85 milligrams, and contains predominantly spermatids. Thereafter, the weight and DNA content of the testis decrease because of the release of sperm (Marushige and Dixon 1969). During the course of the maturation process there is a decline in the content of ribosomes and of total RNA. This becomes very pronounced as the majority of the spermatids mature and the rate of protamine synthesis falls, and it may be accompanied by a small change in overall base composition of the RNA (Ando and Hashimoto 1958, Creelman and Tomlinson 1959, Ling and Dixon 1970). The presence of protamine is first detected toward the end of the growth phase of the testis, when the DNA content has almost reached its maximal level. Thereafter, replacement of chromatin histone by protamine occurs rapidly, and some 8 to 10 weeks after initiating hormone treatment protamine accounts for approximately 30 percent of the acid-soluble protein complement of the entire testis. Of the individual classes, the arginine-rich histones IV are replaced first and the lysine-rich histones I are displaced last (Marushige and Dixon 1969). During the entire period of replacement of the histone approximately half of the protamine synthesized and incorporated into chromatin consists of fraction C_{II}. However, the relative rates of synthesis of the other two fractions vary. In the early phases of the process approximately 20 percent of the protamine synthesized is of fraction C_I and 30 percent of fraction C_{III}. As the testis attains maturity, fraction C_I accounts for only 10 percent of the protamine formed, whereas the rate of synthesis of C_{III} protamine

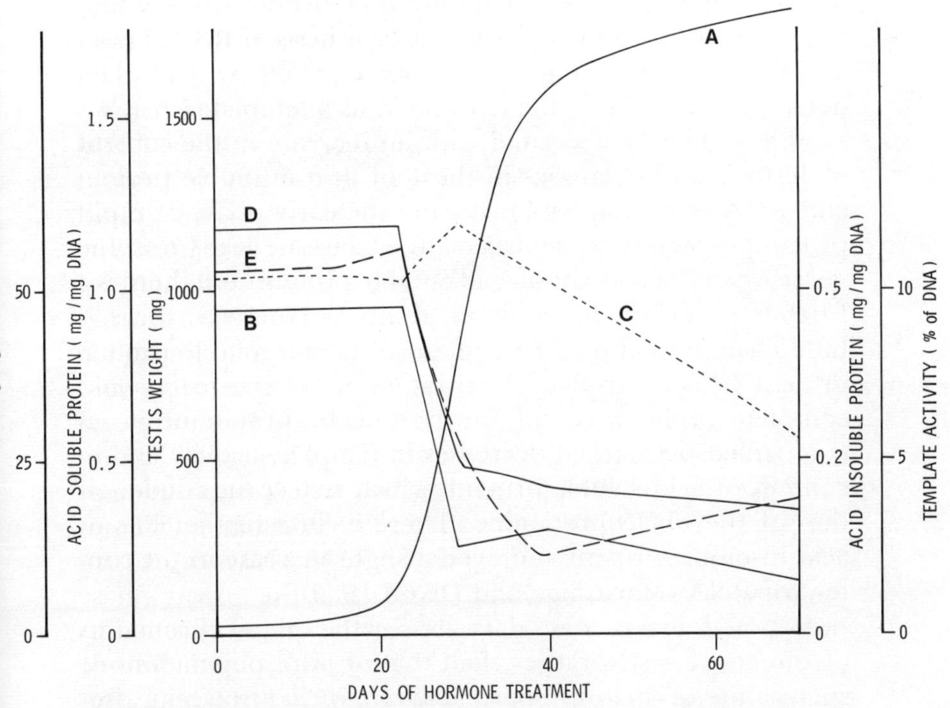

FIGURE 10.1. Changes in properties of trout testis chromatin during development

Curves A. Testis weight.

B. Template activity of the chromatin as a percentage of that of an equivalent weight of purified DNA.

C. Acid-soluble protein.

D. Acid-insoluble protein.

E. RNA.

Based on Marushige and Dixon (1969).

approaches that of the C_I fraction. This marked change in relative rates of synthesis of the two types of protamines is reflected in the protein complement of mature sperm. These contain only traces (2 percent) of fraction C_I and approximately equal amounts of the other two types of protamine (Ling, Jergil, and Dixon 1971).

Changes in the properties of the mixed chromatins isolated from whole testes during this period of development are illustrated in Figure 10.1. No marked changes are seen during the first three weeks in either overall composition or activity as a template for *in vitro* synthesis of RNA. Onset of the rapid growth phase is characterized by a marked decrease in activity of the chromatin as a template for RNA synthesis. This is associated with an increase in the content of histone and decreases in those of acid-insoluble protein and RNA (see Chapter 7). During the early phase of rapid protamine synthesis, replacement of histone leads to some decrease in the content of acid-soluble protein, and there is a further reduction in the level of RNA. However, there is little change in either the content of acid-insoluble protein or activity as a template. Maturation of the spermatids and complete replacement of the histone by protamine is accompanied by marked decreases in template activity and in content of acid-soluble protein, which reflect the condensation of the nucleoprotamine. There is little further loss of acid-insoluble protein and even a slight increase in the content of RNA (Marushige and Dixon 1969).

As noted above, these data are for the mixed chromatins of the entire testis, rather than that of pure populations of spermatids or mature spermatozoa. Nucleoprotamine isolated from the latter has no demonstrable activity as a template for *in vitro* RNA synthesis, and the contents of acid-insoluble protein and RNA are only 20 to 30 percent of those for the mixed testis chromatins in the final stages of differentiation (Marushige and Dixon 1969).

Protamine synthesis occurs in the spermatid cells (Louie,

Lam, and Dixon 1971) and has been studied with intact cell suspensions obtained from testes rich in spermatids and in cell-free homogenates prepared from them (Ingles et al. 1966, Ling, Trevithick, and Dixon 1969, Ling and Dixon 1970). Cell suspensions prepared from testes at the stage of onset of rapid growth (4 weeks) and containing few, if any, spermatids actively synthesize histone but show no detectable synthesis of protamine. On the other hand, cell suspensions rich in spermatids obtained from more mature testes show very active protamine synthesis. These latter suspensions do show some residual synthesis of histone, to an extent which decreases as the spermatids mature. Thus histone synthesis appears to cease as protamine synthesis commences (Ling, Trevithick, and Dixon 1969).

Protamine synthesis takes place in the spermatid cytoplasm on diribosomes (disomes) of a type never found in the fish liver. Further, changes in the rate of protamine synthesis are paralleled by dramatic parallel changes in the proportion of ribosomes present in the disome fraction (Ling, Trevithick, and Dixon 1969, Ling and Dixon 1970). In early experiments, Ingles et al. (1966) found that synthesis of protamine by cell suspensions rich in spermatids was unaffected by concentrations of actinomycin D which inhibited the incorporation of uridine into RNA by approximately 80 percent. This observation indicated that the mRNA of the protamines is metabolically stable and also raised the possibility that onset of protamine synthesis might be regulated at the translational level. In a more recent study Ling and Dixon (1970) have confirmed the metabolic stability of the mRNA and demonstrated that the disomes are remarkably stable, even to digestion by ribonuclease enzyme.[3] However, they also demonstrated that some of the essential mRNA is

[3] Ling and Dixon (1970) have argued persuasively that mRNAs coding for protamines probably have a very high content of guanine and cytosine residues. This could be expected to confer a very high degree of secondary structure upon the molecule in regions not complexed with ribosomes, and this in turn resistance to ribonuclease action.

transcribed during the early period of protamine synthesis. Thus it appears that protamine synthesis commences immediately following incorporation of newly transcribed mRNA into stable disomes. Hence present evidence indicates that onset of the synthesis of protamine in response to the treatment with hormone is regulated primarily by the change in spectrum of cRNAs transcribed which results from the observed changes in composition of the chromatin.

On the other hand, the dissociation of these very stable disomes into monosomes as the rate of protamine synthesis declines (Ling and Dixon 1970) is presumptive evidence of failure to reform initiation complexes (see Chapter 1). It is not yet known whether this is controlled specifically by a special mechanism or is the fortuitous consequence of complex formation between the ribosome particles and protamine formed in excess of the quantity incorporated into chromatin.

All of the seryl residues of the newly synthesized protamine chains are rapidly phosphorylated (Ingles and Dixon 1967). Early studies appeared to indicate that the phosphorylation occurs in the cytoplasm and thereafter the phosphoprotamine enters the cell nucleus (Ling, Trevithick, and Dixon 1969, Marushige, Ling, and Dixon 1969). Indeed, a cAMP-stimulated protamine kinase enzyme has been partially purified from the cytoplasm of testes rich in spermatids (Jergil and Dixon 1970). However, recent studies indicate that most of the phosphorylation occurs in the cell nucleus and that a substantial proportion of the protamine enters the nucleus unphosphorylated (Gilmour and Dixon 1971). Concurrently with the phosphorylation of protamine, histone already present in the chromatin is also both rapidly phosphorylated by a nuclear kinase enzyme (Dixon et al. 1969, Ingles and Dixon 1967, Marushige, Ling, and Dixon 1969) and acetylated (Candido and Dixon 1971, Candido, Sung, and Dixon 1971, Sung and Dixon 1970).

The newly synthesized phosphoprotamine is incorporated

into chromatin before appreciable dephosphorylation occurs (Marushige, Ling, and Dixon 1969). However, the phosphate groupings are subsequently removed as the spermatid matures. Up to three-quarters of the serine residues of protamine isolated from testes shortly after onset of protamine synthesis are present as phosphoserine. As maturation proceeds the proportion of phosphorylated serine residues declines, and that for natural populations of spawning fish is as low as 6 percent (Dixon and Smith 1968, Ingles and Dixon 1967). It remains to be determined whether histone is actually displaced as an acetylated and phosphorylated derivative, whether the substituent groupings are subsequently removed, or whether the modified histone is subject to degradation *in situ*.

The role(s) of protamine phosphorylation is (are) still a matter of speculation. It seemed possible that phosphorylation of protamine might substantially reduce its ability to interact with, and precipitate, nucleic acids. Dixon and coworkers (Dixon et al. 1969, Ingles and Dixon 1967) therefore considered possible consequences of such a reduction in affinity for nucleic acids and proposed three possible roles for protamine phosphorylation. Namely:

1. Modulation of the interaction with DNA of the chromatin to prevent random formation of genetically inert complexes.
2. Facilitation of passage of the newly synthesized protamine from the cell cytoplasm to nucleus.
3. Facilitation of the release of completed protamine chains from the disomes.

These three possibilities are not mutually exclusive and all three could participate in the overall replacement of histone by protamine. Evidence currently available does not support either of the last two proposals but is not incompatible with them.

Some evidence has been obtained in support of the first of

the proposals. The structure of chromatin containing phosphoprotamine appears to be as compact as that of chromatin containing protamine. However, the affinity of the phosphoprotamine in isolated chromatin for the DNA is less than that of unphosphorylated protamine (Marushige, Ling, and Dixon 1969). Affinities for DNA of the modified and unmodified histone in isolated chromatin appear to be very similar. Nevertheless, the possibility remains that the affinities may differ *in situ* in the nucleus of an intact cell. Indeed, Dixon and his colleagues (Marushige, Ling, and Dixon 1969, Sung and Dixon 1970) have formulated an elegantly simple model mechanism whereby the modifications of the histone could facilitate "unzipping" from the complex with DNA.

Present knowledge and concepts of the mechanism by which protamine replaces histone of the chromatin can therefore be summarized by Figure 10.2, which depicts three stages in the replacement of histone IV.[4] These are, respectively:

1. Phosphorylation and acetylation of helices of histone covering the deep groove of the DNA helix.
2. Displacement of the histone by association of phosphoprotamine with the adjoining shallow groove of the DNA helix.
3. Dephosphorylation of the phosphoprotamine and formation of a more compact complex.

As noted above, the fate of the histone is not known, and, in the absence of critical experimental data which permit discrimination between the various possibilities, I have adopted a teleological argument for depicting the release of fully modified histone in the figure.

[4] At present we can only speculate upon the details of the interactions of the various histones and protamines with the DNA of the chromatin. However, selection of the different regions of the DNA helix depicted as "binding sites" in the scheme originally presented by Dixon et al. (1969) is in accord with at least some of the model structures postulated in the literature (see for example Feughelman et al. 1955, Richards and Pardon 1970, Sung and Dixon 1970).

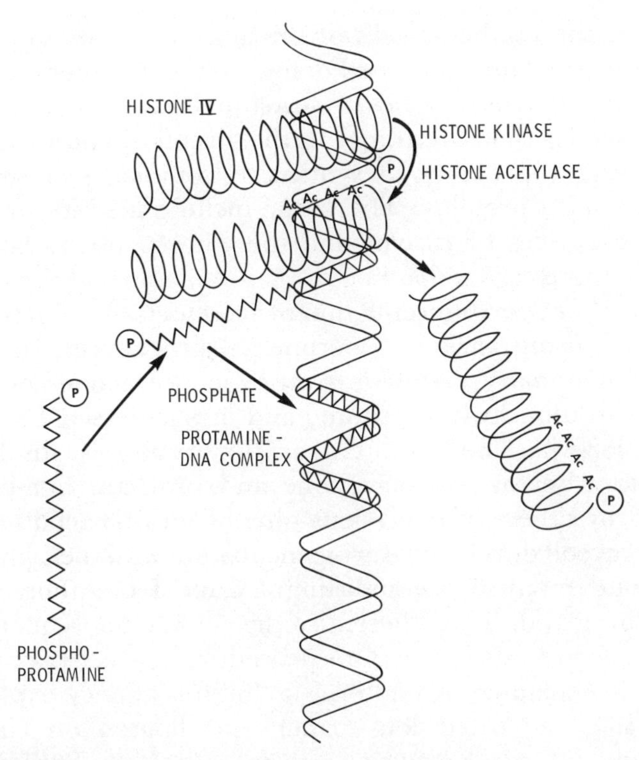

FIGURE 10.2. Model of the mechanism of histone IV replacement by protamine
Based on Dixon et al. (1969), Sung and Dixon (1970).

SYNTHESIS OF MILK PROTEINS BY THE MOUSE MAMMARY GLAND IN ORGAN CULTURE

The mammary gland contains clusters of alveoli formed from epithelial cells and associated epithelial ducts, embedded in fatty tissue. In the nonpregnant mature female mouse the primitive alveoli are small, fairly compact clusters of relatively few cells. During the second half of the gestation period the cells of the alveoli increase both in size and number, and the alveoli assume an ordered structure with a prominent lumen. In the lactating gland, the characteristic proteins and other constituents of milk are synthesized in the epithelial cells of the alveoli and secreted into the lumen. Thus one well-delineated component of the hormone-controlled series of developmental changes during pregnancy is the appearance in the alveoli of epithelial cells possessing the ability to synthesize copious quantities of specific milk proteins. The latter include a series of well-defined species of caseins and the whey proteins α-lactalbumin and β-lactoglobulin.

Similar *in vitro* differentiation of secretory alveoli has been obtained using various experimental procedures, by incubating mammary tissue with medium containing the hormones insulin, hydrocortisone, and prolactin (see Turkington 1968[a]). The requirement for insulin is absolutely specific, whereas hydrocortisone and prolactin can be replaced by others of equivalent physiological function. Cells of the alveoli developed during incubation with the complete hormone mixture are indistinguishable from those of a lactating gland. Thus the entire developmental sequence is amenable to study *in vitro*. One procedure in which mammary tissue is explanted at midterm (a "midpregnancy explant") onto silicone-treated lens paper and floated on culture medium affords particular advantages (see Turkington 1968[a]) and has been extensively exploited.

Explants from virgin mice can synthesize detectable quantities of the caseins and whey proteins, and in those from animals at midterm the rate of synthesis of all these proteins are appreciable (Juergens et al. 1965, Lockwood, Turkington, and Topper 1966). There are further marked increases in the rates of synthesis during both normal development *in situ* of the lactating gland and *in vitro* incubation of midterm explants with medium containing the mixture of hormones (Figure 10.3). Increases in the rates of synthesis of the caseins are essentially coordinate, and the relative rates remain approximately constant throughout the development. Rates of synthesis of β-lactoglobulin, the A protein component of lactose synthetase enzyme and of α-lactal-

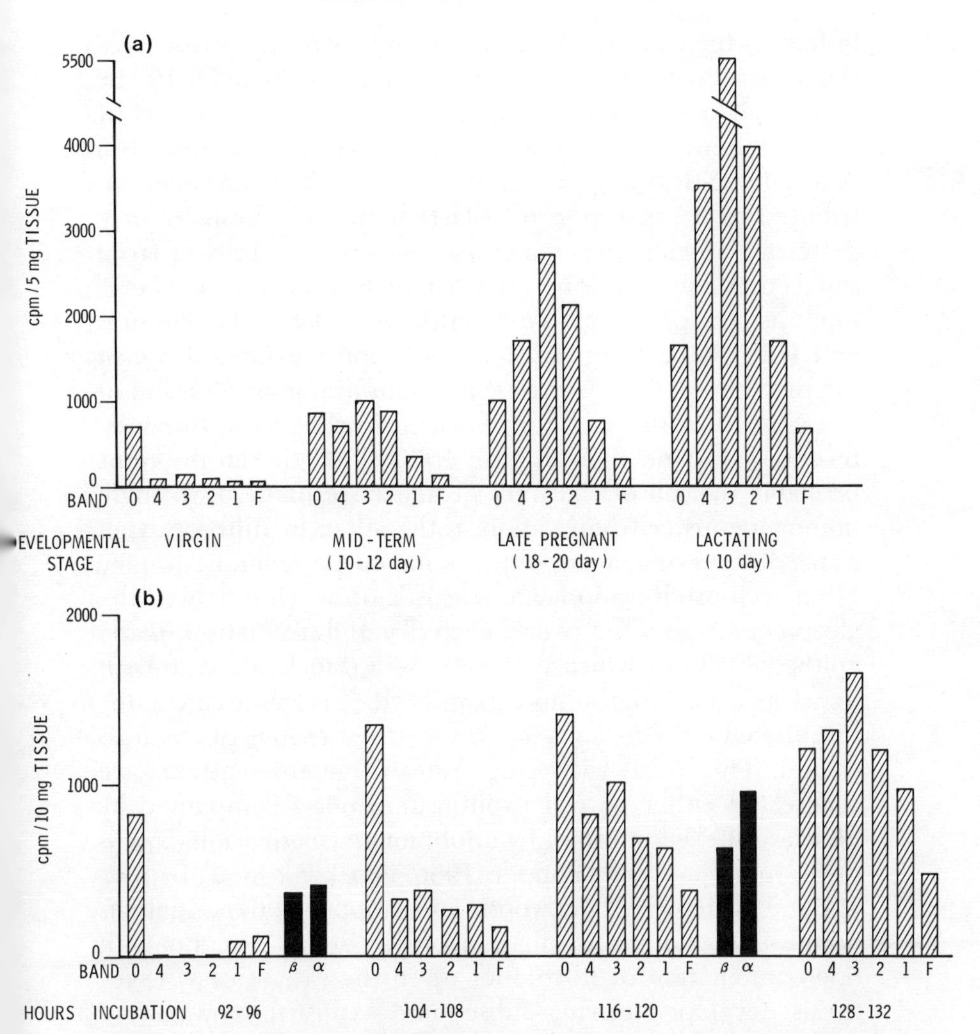

FIGURE 10.3. Increases in rates of synthesis of milk proteins during development of the midterm mouse mammary gland and in response to exposure to hormones

(a) Synthesis of the caseins by explants of glands developing *in situ*. (b) Synthesis of milk proteins by explants cultured in medium with insulin and hydrocortisone for 96 hours and then supplemented with prolactin.

Incorporation of ^{32}P into casein bands 0 (origin), 4, 3, 2, 1 and F (front), and ^{14}C amino acids into α-lactalbumin (α) and β-lactoglobulin (β) during a 4-hour period of incubation was determined.

Based on Turkington 1968(d).

bumin (which is the B protein component of lactose synthetase), increase concurrently but noncoordinately. In the latter half of the gestation period the rate of synthesis of α-lactalbumin remains disproportionately low and then rises markedly with the approach of lactation. This has been attributed to a repression of α-lactalbumin synthesis by progesterone during pregnancy and subsequent release from this repression following parturition (Lockwood, Turkington, and Topper 1966, Turkington et al. 1968, Turkington and Hill 1969). A formal demonstration has been made of the presence of both one of the caseins and α-lactalbumin in virtually all of the secretory epithelial cells of the fully differentiated gland (Turkington 1969[a]). It therefore seems probable that all of the milk proteins are made by a single homogeneous cell population, rather than by different specialized types of cell present in a mixed population.

The cell proliferation characteristic of development of the alveoli is an essential prerequisite for differentiation of the epithelial cells of which it is composed (Stockdale and Topper 1966, Turkington and Topper 1967). Pre-existing differentiated cells do not respond to the presence of the hormones. The initial "base-line" rates of casein synthesis are unaffected either by cell proliferation not accompanied by further differentiation or by inhibition of cell division (Stockdale, Juergens, and Topper 1966, Stockdale and Topper 1966, Turkington, Lockwood, and Topper 1967). Analyses of the effects of preincubation of the organ explants with one combination of hormones upon the rate of casein synthesis developed during subsequent exposure to a second mixture of hormones have permitted identification of the stages of the cell cycle at which each of the hormones is required (Lockwood, Stockdale, and Topper 1967, Turkington 1968[c], Voytovich, Owens, and Topper 1969).

Proliferation of the undifferentiated epithelial cells is induced specifically by insulin, and only explants which have been incubated with this hormone can be further induced to develop mature alveoli or produce casein at an enhanced rate (Lockwood, Stockdale, and Topper 1967, Stockdale and Topper 1966, Turkington 1968[b]). Exposure of the midterm gland to insulin stimulates the rates of synthesis of RNA and protein; induces an increase in the cell content of polyribosomes; stimulates the rates of acetylation of all classes of histone and of phosphorylation of all classes of histone and nuclear nonhistone protein; elicits increases in the levels of RNA- and DNA-polymerase activities; and induces an increase in the overall rate of DNA synthesis in the explant (Lockwood et al. 1967, Marzluff and McCarty 1970, Marzluff, McCarty, and Turkington 1969, Turkington 1968[a], Turkington and Riddle 1969, 1970[b], Turkington and Ward 1969). However, these effects appear primarily to reflect stimulation of the onset of DNA synthesis by cells in the G_1 phase of the cell cycle (Lockwood et al. 1967, Turkington 1968[c]). Although the cells increase in size and number, the alveoli do not show any development of organized structure, and there is no increase above the initial base-line level of casein synthesis (Stockdale, Juergens, and Topper 1966). Further, the progeny cells produced are essentially undifferentiated precursor cells, although they do differ from the parental cells in showing a slight stimulation (less than a doubling) of the rate of casein synthesis in response to prolactin (Juergens et al. 1965).

Explants incubated with insulin plus hydrocortisone (or aldosterone) also show no increase in the base-line level of casein synthesis, and even a small progressive decline in the level on prolonged incubation (Turkington 1968[c], Voytovich, Owens, and Topper 1969). However, the epithelial cells acquire the morphological characters, and the alveoli the structural organization, found in normal mammary glands just before onset of lactation (Stockdale, Juergens, and Topper 1966, Stockdale and Topper 1966, Wellings, Cooper, and Rivera 1966, Wellings, Deome, and Pitelka 1960). Further, the cells have undergone a "covert" differentiation, which is manifested as a response to further exposure to prolactin plus insulin in marked increases in the rates of synthesis of all species of RNA and the various caseins, and by a preferential stimulation of phosphorylation of histones of class IIb (Lockwood, Stockdale, and Topper 1967, Turkington 1968[c], Turkington 1970[a], 1970[b], Turkington and Riddle 1969, 1970[b], Voytovich, Owens, and Topper 1969). The role of hydrocortisone in eliciting this covert differentiation is not clear. Early studies appeared to indicate that it exerts its effect(s) solely during the period of cell proliferation induced by insulin and is unable to induce the differentiation once mitosis has taken place (Lockwood, Stockdale, and Topper 1967, Turkington, Lockwood, and Topper 1967). However, Turkington (1971) has recently reported that the cells do not become responsive to prolactin until they have been exposed to hydrocortisone for some 24 to 36 hours, and that it is not necessary that this prior treatment with the steroid hormone precede mitosis. Hydrocortisone appears to stimulate development of the endoplasmic reticulum of the epithelial cells, but without attendant stimulation of RNA synthesis (Green and Topper 1970, Mills and Topper 1969, 1970). Hydrocortisone also appears to suppress either the formation or the activity of a phosphoprotein phosphatase specifically in cells proliferating in response to insulin (Turkington 1968[d]).

Explants incubated with all three of the required hormones acquire the structure typical of a normal lactating gland and synthesize all of the characteristic milk proteins (Juergens et al. 1965, Stockdale, Juergens, and Topper 1966, Wellings and Deome 1961). Similarly, explants which have been preincubated with insulin and hydrocortisone develop secretory alveoli on subsequent exposure to both prolactin and insulin (Green and Topper 1970, Mills and Topper 1970, Turkington, Lockwood, and Topper 1967). This combination of hormones elicits an increase in RNA polymerase activity and an early marked stimulation of transcription of all classes of RNAs specifically in covertly differentiated cells (Turkington 1968[c], 1969[b], Turkington and Ward 1969). Both hormones exert their effects postmitotically upon the progeny cells in the G_1 phase of the cell cycle (Lockwood, Stockdale, and Topper 1967, Mills and Topper 1970, Voytovich, Owens, and Topper 1969). The mechanism of action of neither hormone is yet known. However, in each case it appears to be one of interaction at the cell surface, for the cells respond equally readily to preparations chemically linked to insoluble particles as they do to the hormones free in solution (Turkington 1970[c]).

Thus the three hormones act in concert to control differentiation of the midterm mammary gland into the secretory lactating gland. Although the mechanisms by which this control is exerted are not yet known in detail, several of the processes known or presumed to regulate the synthesis of specific proteins have already been implicated.

Epithelial cells of the lactating gland contain species of cRNA not detectably present in RNA isolated from the mammary glands of virgin mice, in addition to all of those present in the latter. Many of these species also appear to be already present in the developing gland prior to onset of lactation. However, new species of nuclear cRNAs are formed as the rate of transcription increases following exposure of covertly differentiated cells to insulin and prolactin (Turkington 1970[a], 1970[b]). Further, the cell content of polysomes increases markedly after a lag of 4 to 6 hours, concomitantly with increases in the rates of casein synthesis and phosphorylation (Turkington and Riddle 1970[b]). Thus the primary factors controlling the increase in rate of synthesis of milk proteins appear to be the rates of transcription and incorporation of new species of mRNAs into polysomes. A similar conclusion may be deduced from data obtained in a comparison of nonsecreting glands of pseudo-pregnant rabbits with homologous glands induced to secrete milk by injection of prolactin (Gaye and Denamur

1969). Prolactin induces an increase in the average size of the polysomes and in the proportion of membrane-bound ribosomes, an increase in total activity for amino acid incorporation *in vitro,* and a marked increase in the rate of *in vitro* incorporation of proline relative to that of leucine.

From studies of the chromatins isolated at various stages of development from the mammary glands of rats and rabbits (Stellwagen and Cole 1968, 1969[b]), it seems unlikely that transcription of the new species of cRNA can be correlated with significant differences in the relative rates of synthesis of the various major classes of histone. On the other hand, exposure of midterm mouse mammary glands to all three hormones does lead to some differences in relative rates of synthesis, modification, or displacement from the chromatin, of subclasses of the lysine-rich histones (I) from those observed on incubation with insulin alone (Hohmann and Cole 1969). In addition, exposure to insulin stimulates markedly the rates of phosphorylation of all classes of histone, and prolactin causes a preferential stimulation of phosphorylation of histones of classes IIb1 and IIb2 in cells previously exposed to insulin and hydrocortisone (Turkington and Riddle 1969).

The synthesis of RNA stimulated by exposure of midterm explants to the mixture of all three hormones leads to a disproportionately greater increase in content of transfer RNA relative to ribosomal RNA, which occurs mainly before onset of lactation. This selectively enhanced synthesis of tRNA is paralleled by a threefold increase in the level of tRNA methylase activity, in a process requiring *de novo* protein synthesis. Cell proliferation induced by insulin cannot account for this increase, and insulin must be present throughout the period of exposure to prolactin. On the other hand, the presence of hydrocortisone is not required (Turkington 1969[b]). It remains to be demonstrated that increased activity of these tRNA-methylase enzymes is a prerequisite for synthesis of one, or more, specific milk protein(s).

REGULATION OF SYNTHESIS OF EMBRYONIC HEMOGLOBINS IN THE CHICK BLASTODISC

In contrast to the two preceding systems, the last model I wish to consider is one in which immediate regulation of the synthesis of a specific protein has been demonstrated to be at the translational level (Wilt 1965[a], Wainwright and Wainwright 1966, 1972[a]).

Complete understanding of the mechanisms controlling the formation of hemoglobin in a developing embryonic system will require analysis of at least six different levels at which regulation is imposed. The precise number of known levels is a matter of arbitrary classification, and my listing is given in Table 10.1. For want of a more convenient terminology, at the present time some of these levels can be considered as points of "coarse" control and others as points of "fine" control. I wish to emphasize that use of the terms coarse and fine control in this context is purely a reflection of

TABLE 10.1. Levels of regulation of hemoglobin formation in the developing chick blastodisc

Level	Control
1. Level of available mRNAs	Control of gene transcription, processing of gene transcripts (?), extranuclear transport of mRNA, and degradation of mRNA.
2. Translation of available mRNAs	Control of supply of specific initiation factors and modulating tRNA(s)(?), facilitated release of completed peptide chains from ribosomes, degradation of free polypeptide chains, and breakdown of ribosomes.
3. Differentiation of stem cells into erythrocytes	Control of differential gene activation. This level also includes extrusion of the maturing reticulocyte nucleus in species with anucleate erythrocytes.
4. Ontogeny of types of hemoglobins formed	Control of a switch in program of genes transcribed within one cell, or of type of cell and site in which hemoglobin is synthesized.
5. Cell determination and "commitment"	
6. Cell migration	

the extent of our ignorance and most assuredly does not imply a belief that the processes involved are anything other than exquisitely delicate. Indeed, the most obvious question to be posed is to what extent we may begin to understand the coarse, poorly understood phases or levels of regulation in terms of those fine control mechanisms of which we have some knowledge. It is this question which is my primary concern in the following pages.

Control of hemoglobin synthesis in maturing reticulocytes of adult animals is subject to regulation at two levels of fine control. One set of processes control the levels of appropriate mRNAs available in the cell cytoplasm, and a second set determines the rate or extent of translation of those mRNAs into the corresponding proteins. Collectively, they normally serve to maintain appropriate balances between the rates of synthesis of the α-chains, non-α-chains, and heme of the various hemoglobins formed. With the exception of the human fetus, relatively few studies have been made of these control mechanisms specifically in reticulocytes, or primitive-line red cells, of embryonic tissues. However, we believe that there are no major differences between the types of control mechanism regulating hemoglobin synthesis by individual reticulocytes (or equivalent cells) within embryonic tissues and those found in reticulocytes of mature animals. These various mechanisms have been considered in previous chapters and will not be reviewed in detail here.

At the other extreme of the classification are the very coarse levels of regulation. These determine that certain cells will become committed to develop subsequently into erythropoietic tissue because of the particular position they occupy in the embryo following the early stages of cleavage of the fertilized egg. In addition, they also determine the course of subsequent morphogenetic cell migrations. These phenomena are central to classic, but still contemporary, concepts of embryonic inducers and metabolic gradients. These regulatory processes are, at least to some extent, subject to modification by transplanting the presumptive erythropoietic tissue to another region of the developing embryo, and therefore do not involve irreversible nuclear differentiation. However, there is little more that can profitably be said at this time about these two levels of regulation, and they also will not be considered further here.

Intermediate levels of control regulate differentiation of committed stem cells into maturing reticulocytes, onset of synthesis of hemoglobin(s) in the differentiating cells, switches from synthesis of one species of hemoglobin to that of another species during the course of development, and selective removal of "old" erythrocytes from the circulation. In view of both the complexity and interactions of these mechanisms, analysis of the regulation of synthesis of hemoglobin *in situ* in the developing tissue poses a very challenging problem. Stated in its simplest form, we wish to know to what extent a qualitatively new, or a quantitatively increased, biosynthetic activity represents: (a) an immediate response to a "signal" originating in the responding cell; (b) an immediate response to a signal originating in neighboring cells; and (c) a delayed response as the inevitable outcome of commitment made at earlier stages of development. The developing chick blastodisc is one of the experimental systems particularly suitable for such an analysis and is potentially amenable to approaches using mutant strains of bird. As yet, the analysis is only in the early stages. However, studies by many workers have defined the system and identified some of the regulatory processes probably involved.

Erythrocytes of the so-called primitive line are first observed in clusters of mesodermal cells known as "blood islands." These cells initially migrate into a horseshoe-shaped region of the extraembryonic portion of the developing blastodisc which surrounds the posterior half of the embryo proper. The organized clusters of cells are clearly recognizable by the stage at which the embryo proper has organized the first visible pair of somite blocks, which normally occurs after some 23 to 26 hours of incubation. However, the cells of the cluster do not undergo any recognizable differentiation until the embryo proper has developed two to four pairs of somite blocks, some two to three hours later. The course of the subsequent differentiation of erythropoietic cells of the primitive linc is formally equivalent to that of cells of the definitive line observed with older embryos and adult birds. For more detailed treatments of these stages of chick blastodisc development and of avian hematology, I refer the interested reader to Hamburger and Hamilton (1951), Lucas and Jamroz (1961), Romanoff (1960), and Wilt (1967).

Heme (Granick and Levere 1965) and hemoglobin (see Romanoff 1960, Wilt 1967) can first be detected in the blood islands of the blastodisc *in situ* at the stage of the 6- to 7-somite blastodisc, after incubation for some 27 to 33 hours.[5] This stage also marks the first demonstrable appearance of antigens which cannot be distinguished from the distinctive species of embryonic hemoglobins isolated from the developing blastodisc after 2 to 3 days of incubation (Wilt 1962, 1967). Shortly thereafter the rates of incorporation of amino acids and of iron into material precipitated by a highly specific antiserum against adult chick hemoglobins become appreciable (Wilt 1962, 1965[a]). It therefore seems highly probable that onset of the rapid synthesis of embryonic chick hemoglobins commences at, or just before, the 6- to 7-somite stage of development of the embryo proper.

When young chick blastodiscs are explanted onto a suitable medium essentially normal embryonic development continues for several hours, and hemoglobin is synthesized

[5] The stages of development attained between the first and second day of incubation are customarily defined by the numbers of pairs of somite blocks which are clearly visible. For convenience, I shall give in parentheses the period of incubation normally required to attain this stage of development, from the data of Hamburger and Hamilton (1951). Thus, a 4-somite embryo (26 to 29 hours) is an embryo for which 4 pairs of somites can be seen, a stage of development attained after 26 to 29 hours of incubation *in ovo*.

in the developing blood islands.[6] Further, the presence or absence of visible quantities of hemoglobin, both before and after application of a sensitive stain, provides a convenient semi-quantitative assay of hemoglobin formation for use in exploratory studies. By explanting blastodiscs before the onset of rapid synthesis of embryonic hemoglobin, it is possible to examine the effects of exogenous agents upon the processes which culminate in that rapid synthesis. Thus, for example, inclusion of a sufficient concentration of actinomycin D to prevent synthesis of all RNAs of high molecular weight permits demonstration that onset of rapid synthesis of embryonic hemoglobin is regulated at the translation level (Wilt 1965[a], Wainwright and Wainwright 1966, 1972[a]). Blastodiscs explanted onto this medium at the stage of the primitive streak (18 to 19 hours) fail to synthesize quantities of hemoglobin detectable by the sensitive stain. In contrast, those explanted at the 1-somite stage (23 to 26 hours) form quantities of hemoglobin readily visible to the naked eye if δ-aminolevulinic acid is also added to the medium containing actinomycin (Wainwright and Wainwright 1966, 1972[a]).

Additional useful information was obtained in similar exploratory studies with various metabolic inhibitors and other supplements (Levere and Granick 1965, O'Brien 1959, 1961, Wilt 1965[a], 1966, Wainwright and Wainwright 1967[a], 1967[b], 1967[c]). The results obtained delineated a minimum of four distinct phases controlling onset of rapid synthesis of embryonic hemoglobin and per-

[6] Some of the techniques used for study of explanted young chick blastodiscs (e.g., New 1955) permit completely normal development for periods of at least 36 hours. Most biochemical studies have made use of more convenient techniques in which the blastodisc is explanted onto a solid medium, which retards or prevents lateral expansion of the tissue peripheral to the blood islands. Under these conditions, development is otherwise normal for periods of at least 24 hours on a medium containing egg homogenate or, at a reduced rate, on a simple buffered-glucose-saline medium. However, development and hemoglobin formation are poor on conventional chick tissue-culture media solidified with agar (Wilt 1968, Wainwright and Wainwright, unpublished observations).

mitted provisional identification (Table 10.2) of the dominant (i.e., limiting) regulatory process in each stage (Wainwright and Wainwright 1967[c]). Development of sensitive methods for assay of the small quantities of hemoglobin involved (Levere and Granick 1967, Wainwright 1970) has allowed quantitative confirmation and extension of the major observations made in these exploratory studies. The later analyses have also shown the system to be far more complex than had already been indicated and to be subject to regulation by additional factors. For convenience, these are also listed in Table 10.2 under the regulatory phase in which they are probably most important.

The kinetics of hemoglobin synthesis during the course of development from the stage of the primitive streak (18 to 19 hours) to that of the 10-somite embryo (33 to 38 hours) reflect the complexity of the control mechanisms involved. The blastodisc appears to contain and to synthesize hemoglobin at least as early as the primitive streak stage (Figure 10.4). Thus onset of rapid hemoglobin synthesis at the stage of the 6-somite embryo (27 to 33 hours) represents an increase in the rate of synthesis, rather than initiation of formation of

TABLE 10.2. Phases in regulation of onset of rapid hemoglobin synthesis in chick blastodisc blood islands

Phase 1	Occurs prior to attaining the stage of the definitive primitive streak. Transcription of mRNAs for globin chains of hemoglobin.
Phase 2	Probably commences before development of the definitive primitive streak but is not complete until the 1-somite stage. Transcription of mRNAs for ALA synthetase enzyme and/or other proteins of the mitochondria. Stimulated by 5β-steroids (Levere, Kappas, and Granick 1967).
Phase 3	Extends from the 1- to 6-somite stages. Transcription or modification of tRNA(s) required for formation of ALA synthetase enzyme and/or other proteins of the mitochondrion.
Phase 4	Spans the 6- to 8-somite stages. Onset of rapid hemoglobin synthesis. Stimulated by a lipid component of egg yolk. Stimulated by mitochondrial protein synthesis.

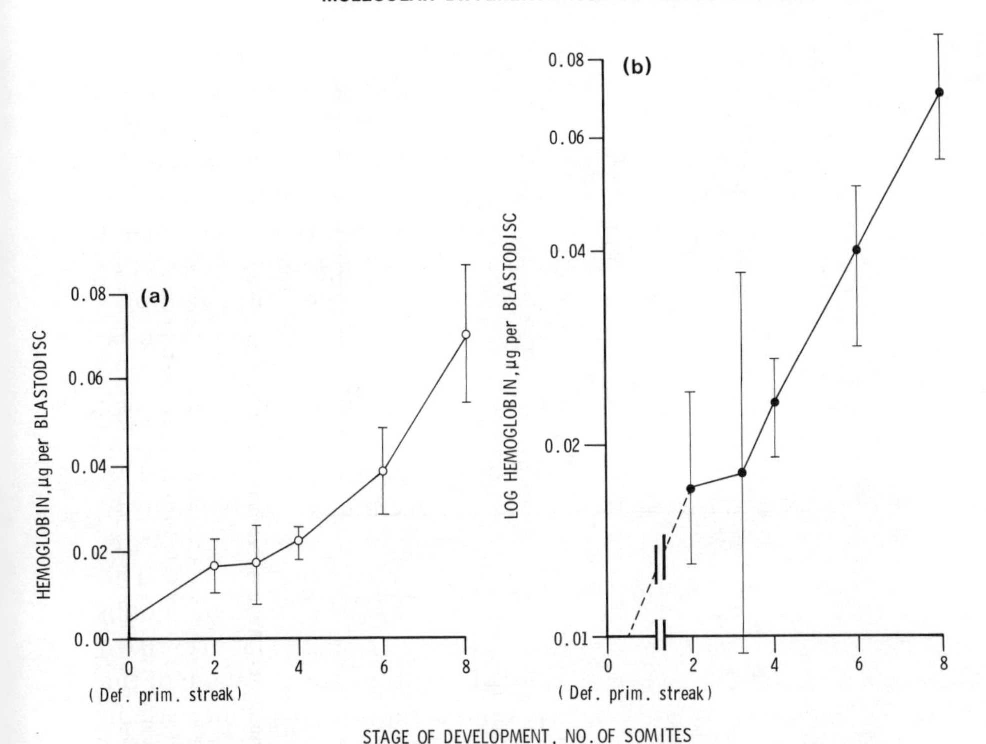

FIGURE 10.4. Kinetics of hemoglobin formation by chick blastodiscs in ovo

Hemoglobin was assayed in pools of blastodiscs collected immediately after dissection. Each value was obtained from a series of 10 pools, each of 20 blastodiscs homogenized to 1.5 or 2 milliliters in saline. (a) Linear plot, (b) semilog plot. Approximately 5 hours are required beyond the stage of the definitive primitive streak for organization of the first pair of somite blocks and a further hour for each additional pair of somites. Plot b is adjusted for the period between 0 and 2 somites to present the data as the time course of hemoglobin formation. The bars indicate one standard deviation.

Reproduced by permission of the National Research Council of Canada from the *Canadian Journal of Biochemistry, 48,* 400 (1970).

the first molecules of hemoglobin. The process may reflect a switch in the species of hemoglobins formed (Wainwright and Wainwright 1970), formally analogous to those extensively studied in other animal species (see Capp, Rigas, and Jones 1970, Craig and Southard 1967, Kleihauer and Stoffler 1968). However, other interpretations of the observed kinetics of hemoglobin formation *in ovo* (Figure 10.4) and

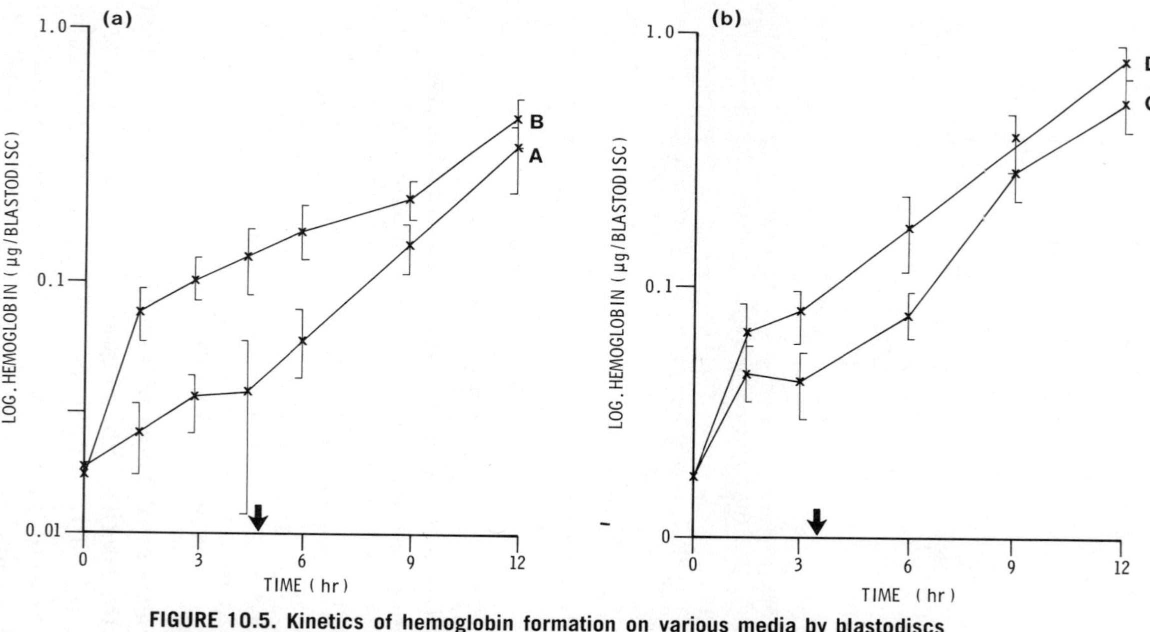

FIGURE 10.5. Kinetics of hemoglobin formation on various media by blastodiscs explanted at the 3-somite stage

(a) Minimal medium without (curve A) and with (curve B) 10^{-3} *M* ALA. (b) Egg homogenate medium without (curve C) and with (curve D) 10^{-3} *M* ALA. The arrows indicate the time at which the blastodisc attains the 6-somite stage.

Reproduced by permission of the National Research Council of Canada from the *Canadian Journal of Biochemistry, 48,* 400 (1970).

with explanted blastodiscs (Figure 10.5) cannot be eliminated at this time.

More important to the present discussion, marked changes in the rates of hemoglobin synthesis are observed at the stages of the 2- to 3-somite embryo (Figure 10.4) and of the 6- to 7-somite embryo (Figure 10.5). These coincide with the onset of two of the regulatory phases defined by the use of metabolic inhibitors (Table 10.2). Further, as also indicated by the exploratory studies, both of these phases are regulated at the translational level. Blastodiscs explanted at the 3-somite stage (25 to 28 hours) onto egg homogenate medium containing sufficient actinomycin to block transcription of mRNAs synthesize as much hemoglobin over a period of 24 hours as controls on normal medium (Wainwright and Wainwright 1972[a]).[7]

The marked increase in rate of hemoglobin synthesis which occurs at the 6- to 7-somite stage is probably regulated primarily, and possibly entirely, by the supply of δ-aminolevulinic acid (ALA). Synthesis of this key intermediate by mitochondrial aminoleuvulinic acid synthetase (ALA synthetase) enzyme is the rate-limiting step in the formation of heme prosthetic groups in many types of tissue, including the erythropoietic cells of young chick blastodiscs (Granick and Levere 1964, 1968, Levere and Granick 1965, 1967). A number of workers have observed stimulation of hemoglobin formation in explanted chick blastodiscs by supplements of ALA (Levere and Granick 1965, Wilt 1966, Wainwright and Wainwright 1967[b]). Analysis of the kinetics of hemoglobin formation (Figure 10.5) shows that this stimulation by ALA occurs primarily before the normal stage of onset of rapid hemoglobin synthesis (Wainwright and Wainwright 1970). There is some slight further stimulation of

[7] This observation does not imply that no further transcription of the appropriate mRNA sequences occurs during the ensuing 24 hours. Rather, this result is due in part to stimulation of hemoglobin formation by the ethanol (final concentration 0.2 percent) used as solvent for actinomycin.

hemoglobin formation beyond the stage of the 6- to 7-somite embryo. However, this is transient and not evident after the blastodiscs have been incubated on rich media for 24 hours (Wilt 1968, Wainwright and Wainwright 1969, 1970).

Thus the onset of rapid hemoglobin formation which occurs normally in the absence of supplements of exogenous ALA probably reflects a rise in the rate of synthesis of endogenous ALA. It is believed that this is due primarily to an increase in the level of ALA synthetase enzyme, but this remains to be established by direct examination.[8] Very marked increases in both the level of ALA synthetase activity and the amount of hemoglobin formed are observed when blastodiscs at the primitive streak to 1-somite stages are treated with natural 5β-H androgens and derived, unconjugated steroid metabolites. These increases are not due to increases in cell number and appear to reflect an induced *de novo* synthesis of ALA synthetase enzyme (Levere, Kappas, and Granick 1967). Extensive studies of a related induced formation of ALA synthetase enzyme in tissue cultures of embryonic chick liver have defined the structural features of steroids active in eliciting this response and provided additional evidence that the steroids stimulate transcription of mRNA for the metabolically labile ALA synthetase enzyme (Granick and Kappas 1967[a], 1967[b], Sassa and Granick 1970).

In contrast, erythropoietin, which has a different mechanism of action from that of the 5β-H steroids (Gordon et al. 1970, Stephenson et al. 1971), appears to have little, or no, effect upon the synthesis of hemoglobin by young chick blastodiscs (Malpoix 1967).

[8] Activity of the ALA synthetase enzyme *in situ* in the cells of the blastodisc, and also as is customarily assayed *in vitro,* is dependent upon formation of the substrate succinyl-coA in the mitochondria. The mitochondria of low ALA synthetase activity in blastodiscs which have not yet attained the 6-somite stage of development may be deficient in cytochromes and cytochrome oxidase activity. This may, in turn, impair the ability of the mitochondria to generate succinyl-coA at rates sufficient to support maximal activity of ALA synthetase enzyme already present.

It is uncertain whether any of the steroids which can elicit increased formation of ALA synthetase enzyme under laboratory conditions plays any role in regulating the normal course of blastodisc development. The blastodisc acquires the ability (at least) to convert testosterone to a very potent inducer of ALA synthetase formation even before attaining the stage of the primitive streak (Parsons 1970). However, no 5β-H androgen or derived metabolite has yet been reported present in the sterol fraction of the egg yolk (see Romanoff 1967).

Egg yolk does contain a lipid fraction which markedly stimulates the formation of hemoglobin by blastodiscs explanted onto a minimal medium up to the stage of the 6-somite embryos, but not by those explanted at later stages (Wainwright and Wainwright 1970, 1972[b]). The active fraction has not yet been identified and may be a mixture of mutually antagonistic components. However, it seems improbable that the stimulation of hemoglobin formation by this fraction can be attributed to stimulation of the transcription of any species of mRNA, for the level of hemoglobin formed during subsequent incubation on a rich egg homogenate medium for 24 hours is not limited by transcription of mRNAs once the 3- to 4-somite stage of development is attained (Wainwright and Wainwright 1972[a]). More probable roles of the lipid fraction include direct stimulation of protoheme ferrolyase enzyme activity (Yoneyama et al. 1969), and particularly as a source of components of intramitochondrial membranes or other proteins.

The formation of hemoglobin by blastodiscs explanted onto minimal medium up to the stage of the 6-somite embryo is markedly inhibited by bacteriostatic concentrations of chloramphenicol. This effect is selective and not accompanied by detectable decreases in the extents of incorporation of leucine into total cell protein or of uridine into polynucleotide. Further, the inhibition is not manifest until several hours after exposure to the antibiotic. Nevertheless, blastodiscs explanted at the 8-somite stage of development form normal levels of hemoglobin in the presence of chloramphenicol. Thus coincident with onset of rapid hemoglobin synthesis the blastodisc shows loss of both responsiveness to yolk lipid fraction and sensitivity to delayed inhibition of hemoglobin formation by chloramphenicol. In contrast, hemoglobin formation by blastodiscs explanted onto the egg homogenate medium remains sensitive to inhibition by chloramphenicol after the 8-somite stage of development is attained (Wainwright and Wainwright 1972[c]).

Extensive studies of the effects of chloramphenicol and erythromycin upon eucaryote cells have shown that they act as selective inhibitors of protein synthesis by mitochondria and chloroplasts (see S. Nass 1969). It therefore seems probable that maintenance specifically of the maximal rate of hemoglobin formation during the phase of rapid synthesis requires the activity of an enzyme synthesized, wholly or in part, within the mitochondria at the stage of onset of rapid hemoglobin formation.[9] This protein may be either the molecular species of ALA synthetase within the mitochondrion itself or an enzyme required for conversion of the species of ALA synthetase formed in the cell cytoplasm (Beattie and Stuchell 1970, Hayaishi, Yoda, and Kikuchi 1970) to that of the mitochondrion.

Wainwright and Wainwright (1967[a]) have postulated that the phase preceding the onset of rapid hemoglobin formation is one involving formation of additional tRNA(s), either by *de novo* synthesis or by modification (see Chapter 9) of pre-existing species. It was clear at the outset that further analysis of this possibility might be hampered by several fac-

[9] Chloramphenicol has, however, also been shown to inhibit the incorporation of amino acid by "nuclear fractions" isolated from a variety of metazoan tissues, including young chick blastodiscs (Shirley and Wainwright 1972, Trevithick and Wainwright 1970). Therefore, caution must be exercised in attributing effects of this antibiotic to actions upon the mitochondria.

tors, including the limited supply of material.[10] Therefore, indirect approaches have been used to provide a rational basis for focusing attention upon one selected series of isoaccepting tRNA. These have indicated that the presumed regulatory species of tRNA is a very minor component of the alanine-specific tRNA complement. This distinct species represents only approximately 1 percent of the total alanine-specific tRNA complement of adult chicken liver (Wainwright 1971[a], 1971[b], Wainwright, Wainwright, and Tsay 1972) and a slightly larger proportion in that of the extraembryonic membranes of the 5- to 6-day embryo (Wainwright, Thompson, Prchal, and Tsay 1972), but is not a component of the mitochondrial complement of tRNAs (Wainwright and Wainwright 1972[d]).

Transfer RNA isolated from the entire blastodisc at the stage of the primitive streak contains a small proportion of this minor alanine-specific species. The proportion of this presumed regulator tRNA in the total alanine-specific complement increases during development to the 8-somite stage, while the proportion of alanine-specific tRNA in the complement of total tRNA appears to remain constant (Wainwright 1971[a], 1971[b], Wainwright and Thompson 1972). Further, the rate of synthesis of tRNA relative to that of other species of RNA appears to increase at the 1- to 2-somite stage of development (Wainwright and Wainwright 1972[e]). It therefore seems possible that this minor species of tRNA does serve to regulate one phase of the developmental sequence which culminates in the onset of rapid hemoglobin formation at the stage of the 6- to 7-somite embryo.

[10] Two factors seemed particularly important as potential sources of difficulty. It is implicit in the concept of "modulating tRNAs" (see Chapter 9) that very marked effects may be elicited in response to formation of very small quantities of the specific tRNA. Further, formation of such modulating tRNA(s) and the response elicited might be restricted to only one (or a few) of the many types of cell comprising the blastodisc.

However, critical examination of this possibility must await development of appropriate chick blastodisc systems for the *de novo* synthesis of hemoglobin *in vitro*. At the present time there is no compelling reason to assume that each of the defined regulatory phases occurs in, or exclusively in, cells of the erythroid line. Formation of organized blood islands within the mesodermal layer of the blastodisc is dependent upon the presence of diffusible, heat-labile material supplied by cells of the endoderm layer (Miura and Wilt 1969, Wilt 1965[b]). It is therefore possible that one, or more, of the regulatory processes is involved in controlling production of this material. The recent development of techniques for obtaining synthesis of hemoglobin in blood islands of cell aggregates formed from cell suspensions of dissociated blastodiscs (Miura and Wilt 1970, Wainwright and Wainright, unpublished) should allow appraisal of this possibility.

Again it is surely ludicrous to suppose that spirits are standing by at the mating and birth of animals—a numberless number of immortals on the lookout for mortal frames, jostling and squabbling to get in first and establish themselves most firmly. Or is there perhaps an established compact that first come shall be first served, without any trial of strength between spirit and spirit?—Lucretius, First Century B.C.

Moreover, when the actions are known, the faculties and their effects are easily made clear; for, indeed, actions proceed from faculties and effects from actions. —H. Fabricius, 1621.

CHAPTER 11: **SOME ASPECTS OF ANTIBODY FORMATION**

Vertebrate animals possess a number of "defense mechanisms" which may be evoked in response to the introduction of "foreign" substances into their tissues. These include mechanisms controlling formation of highly specific antibodies which selectively combine with particular chemical groupings present in the foreign antigens which elicit their formation. One of these mechanisms, which appears also to be present in at least some invertebrates (Cooper 1969, Evans et al. 1970), is responsible for the phenomenon of homograft rejection. This appears to involve the selective proliferation of lymphocytes containing the specific antibody as a component of the cell surface (e.g., Billingham 1969, Wilson 1969). However, this mechanism is little understood at the present time and will not be considered further. The second mechanism is only well developed in vertebrates and even for the cyclostomes is represented only by responses to particulate antigens observed with members of the subclass of lampreys (Evans et al. 1970, Good and Papermaster 1964). This mechanism controls formation of the circulating antibodies of the serum known as *immunoglobulins.*

The process of assembly of the constituent peptide chains into complete immunoglobulin molecules was considered earlier (see Chapter 9) and will not be dealt with further. In the following pages I shall consider briefly only those properties of the immunoglobulins, and features of the processes of antibody formation, which pose special problems to our understanding of the mechanisms regulating synthesis of these particular proteins. For more extensive treatments, which reflect support for a variety of different "preferred" hypotheses, I refer the interested reader to a selection from the following: Frisch (1967), Herzenberg, McDevitt, and Herzenberg (1968), Hood and Talmage (1970), Killander (1967), Lennox and Cohn (1967), and Putnam (1969).

There appears to be a consensus that any one individual person is potentially capable of producing of the order of some 50,000 to 100,000 different species of antibody molecule. This remarkable potential inherent in the antibody-forming system could be readily accounted for by early "*instructive theories*" of antibody formation. The essential feature of such proposals was that the antibody acquired its specificity during the process of peptide chain formation by being "molded to fit" the antigen which elicited its formation (see Haurowitz 1967). A very large number of distinct antibodies, differing from each other in conformation, could thus be generated from the polypeptide chain specified by a single structural gene.

However, it now seems probable that each distinct type of antibody is a unique chemical molecule and (with the exception of isomers differing in the arrangement of disulfide bridges [Frangione, Milstein, and Pink 1969]) differs from all other antibodies in the primary amino acid sequence of one, or both, of the two types of constituent polypeptide chain. It is now believed, although not rigorously established, that the formation of circulating antibody in response to antigenic stimulation results from the selective proliferation of clones of cells which fortuitously contain appropriate structural genes to specify the formation of antibody capable of complexing with the "inducing" antigen (Burnet 1959, Jerne 1955). These *antigen-sensitive* cells (Kennedy et al. 1965, Till et al. 1967) are believed to contain antibody at receptor sites of the cell surface (e.g., Paraskevas, Lee, and Israels 1970, 1971), and to be stimulated to proliferate by complex formation with the antigen (e.g., Dutton 1967, Jones, Marcuson, and Roitt 1970, Krisch 1969, Lennox and Cohn 1967). Some unidentified process introduces an asymmetry of division during the proliferation of these antigen-sensitive cells. This leads to the development of two lines of progeny: one of further stem cells and one of differentiated *antibody-forming,* or *plasma,* cells (e.g., Dutton and Mishell 1967, Nossal 1967, Papermaster 1967).

Thus the combined genomes of the population of antigen-sensitive cells must contain a sufficient number of structural genes to control formation of the (approximately) 10^5 different species of antibody which the individual is potentially able to produce. The problems posed by the requirement for such a large number of discrete structural genes have been markedly magnified by recent determinations of complete or partial amino acid sequences of a number of immunoglobulin polypeptide chains. However, before further considering mechanisms controlling the formation of specific antibodies, I must first digress briefly to outline some of the essential features of the structure of immunoglobulins.

ELEMENTARY ASPECTS OF THE STRUCTURE OF ANTIBODIES

Every immunoglobulin contains four polypeptide chains,[1] of two distinct types, linked together by disulfide bridges

[1] This generalization is only strictly correct in the case of the so-called 7S-antibodies, which include the IgG-immunoglobulins. However, the IgM-immunoglobulin "19S-antibody," elicited in mammals as the first response to primary exposure to a new antigen, and some other types appear to be polymers formed from 7S-monomers by disulfide-bridge formation (e.g., Miller and Metzger 1965, Parkhouse and Askonas 1969).

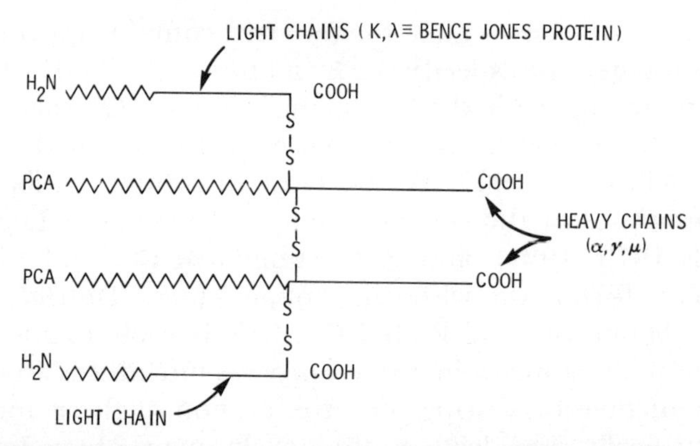

FIGURE 11.1. Diagrammatic representation of immunoglobulin structure

In this diagram constant regions of the chain are depicted as straight lines and variable regions as wavy lines. Intrachain disulfide bridges and interchain disulfide bridges not found in all immunoglobulins have been omitted. The N-terminal groups of the heavy chains show as PCA are cyclized pyrrolidone-carboxylic acid residues. In some immunoglobulins a free amine group is found in this position.

(Figure 11.1). All contain two identical "*heavy*" chains of molecular weight about 52,000 and two identical "*light*" chains of molecular weight approximately 24,000. Each "antigen-combining site" is formed from the N-terminal portions of one chain of each type. The antibodies circulating in the serum of a normal individual comprise a unique heterogeneous mixture, which "records" the history of exposure of this individual to a variety of antigenic stimuli. However, the mixture can be classified into groups of components on the basis of the presence or absence of specific antigenic groups in the immunoglobulins.

Immunoglobulins of any one mammalian species comprise three major *classes* (known as IgG, IgA, and IgM), distinguished on the basis of class-specific antigenic groupings. These groupings are present in the C-terminal portions of the heavy polypeptide chains and serve to define three distinct classes of corresponding heavy chain, known respectively as γ-, α-, and μ-. Additional classes or subclasses may also be recognized. Similarly, the immunoglobulins may be classified into two major *types* on the basis of type-specific antigenic groupings present in the C-terminal portions of the light peptide chains. These define two distinct types of light chain known, respectively, as κ- and λ-. Thus six distinct major categories of immunoglobulin, which differ in biological properties, may be generated from the various possible combinations of light and heavy chains. In general, most antigens elicit formation of some immunoglobulins of all six categories, but with the IgG immunoglobulins predominating.

The immunoglobulins of individual animals may also be classified on the basis of antigenic groups, or "markers," which determine the properties of these proteins as antigens. Certain of these markers are common to all normal individuals of a given species and serve to distinguish between the homologous immunoglobulin chains produced by different species of animal. Thus, for example, some of the antibodies developed in other species against a preparation of rabbit γ-globulin do not distinguish between one rabbit serum and another but do discriminate between rabbit sera and those of other species. These antibodies react with antigenic groups which determine the *isotypic specificities* of the two types of chain in the γ-globulin molecule. In contrast, a second type of antigenic group is not common to all individuals of the same species. These groups, known as *allotypic markers,* define the *allotypic specificities,* or *allotypes,* of the immunoglobulin chains (see Dray et al. 1962). The allotypic markers reflect differences in structure of the immunoglobulin chains which are determined by the alleles present at defined loci of the genome of all cells of the animal, and their inheritance follows standard Mendelian patterns. Thus, for example, all human κ-chains may be classified on the basis of what is termed the *Inv* specificity. This is determined by the identity of the amino acid residue at one specific position of this light chain. One type of chain contains a leucine residue, and the other type contains a valine residue.[2] These allotypic markers (see Figure 11.2) figure prominently in discussion of the mechanism(s) controlling formation of specific antibodies.

The combined effects of differences in antigen specificity and in allotypic specificity are reflected in both the biological and the physicochemical properties of the polypeptide chains. Thus preparations of light immunoglobulins are markedly heterogeneous and are resolved into several distinct protein bands on electrophoresis. Indeed, Cohen and Porter (1964) demonstrated the existence of a minimum of at least 40 distinct forms of human light immunoglobulin chains. This extreme degree of heterogeneity created a severe obstacle to determination of the amino acid composition and sequence of individual immunoglobulin chains of normal antibodies.

However, homogeneous preparations of abnormal immunoglobulins may be isolated from the sera of patients afflicted with multiple myeloma, a tumor of the plasma cells which are the normal site of antibody formation, or cases of macroglobulinemia. Similarly, homogeneous preparations of abnormal light chains, known as "*Bence Jones* proteins," may be isolated from the urine. Proteins homologous with the latter are also secreted by certain types of murine plasma cell tumor. The complete amino acid sequences of a number of these proteins have been determined. Moreover, the pattern which began to emerge from the results obtained in such analyses spurred study of the more complex mixtures of chains obtained from normal immunoglobulins of man, mouse, and rabbit. Complete, or partial, sequences are now also established for a number of these (see, for example, Frisch 1967, Hood and Talmage 1970, Killander 1967, Putnam 1969).

[2] Appella and Perham (1967) report a further type of κ-chain which appears to lack *Inv* specific grouping but also appears to contain a valine residue at this position.

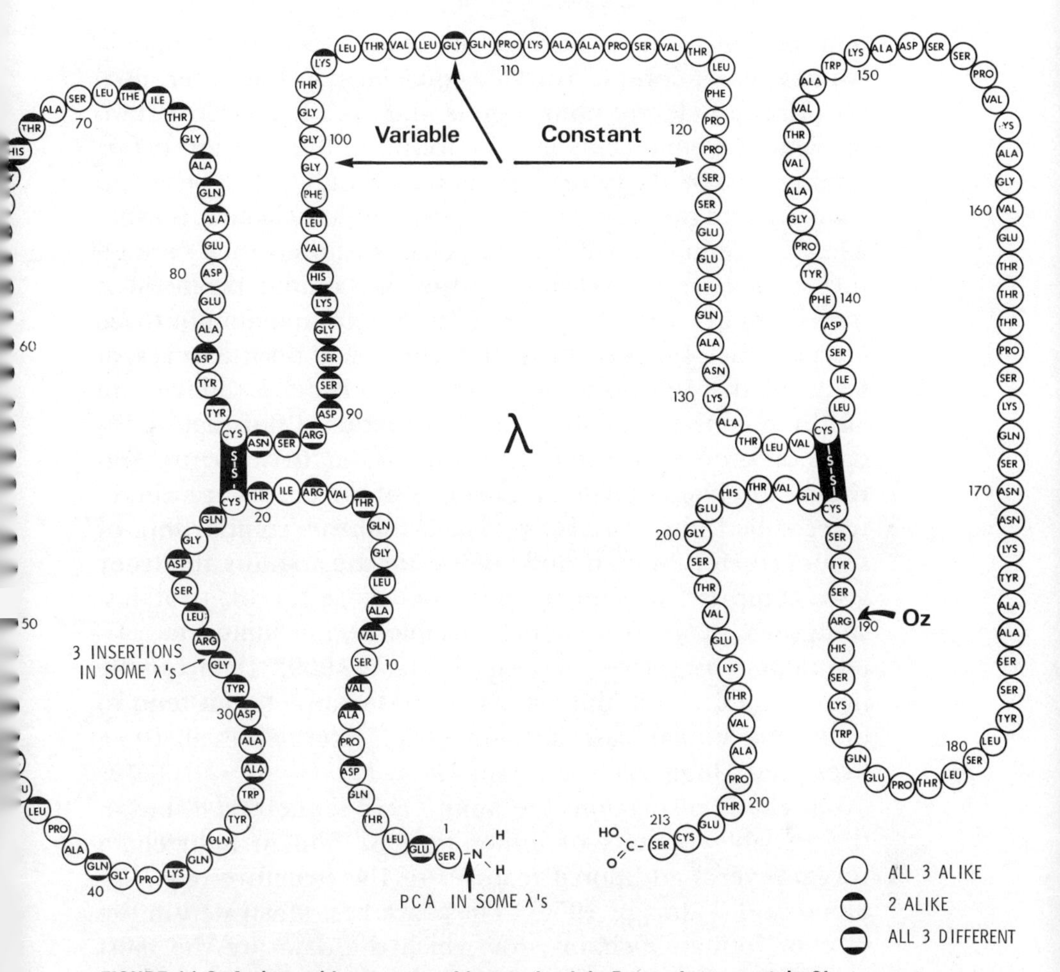

FIGURE 11.2. Amino acid sequence of human lambda Bence Jones protein Sh

Positions indicated by open circles are identical in human lambda Bence Jones Proteins Ha and Bo. Reproduced from Putnam (1967) with permission of the author and Almqvist and Wiksell, Stockholm, holders of the copyright.

CONSTANT AND VARIABLE REGIONS OF IMMUNOGLOBULIN PEPTIDE CHAINS

When the sequences determined for a series of human λ-chains are compared (e.g., see Putnam 1969), it is readily apparent that they constitute a series of homologous proteins of a novel type. Each of the sequences appears to consist of two portions, of roughly equal length. The C-terminal portions are clearly homologous and almost identical. Two alternative amino acids may be found at the position determining *Oz* specificity of the chain, but only three other instances of single amino acid substitutions have been reported. This C-terminal portion of the chains is, therefore, termed the *constant region.* However, the N-terminal portions of the chains are very different. Thus in a comparison of three λ-type Bence Jones proteins the amino acid residues present at 20 of the 108 positions of the sequence are different in all three of the proteins, and at a further 33 of these positions differences are found for one of the three chains (Figure 11.2). The N-terminal portions of the chains are, therefore, called the *variable region.* A similar comparison of sequences for human and mouse κ-chains shows that they also comprise a similar homologous series of proteins, composed of an N-terminal variable region joined to a C-terminal constant region (e.g., Putnam 1969). Data for the heavy immunoglobulin chains are less complete but tend to indicate a similar basic structure (e.g., Edelman et al. 1969, Press and Hogg 1969, Putnam 1969).

On closer inspection, the amino acid sequences found in the variable regions of either type of human light chain reveal several additional features of the structure (e.g., see Hood and Talmage 1970). These are best illustrated in the case of human κ-chains, for which the data are the most extensive. All the sequences can be assigned to one of three, or more, subgroups on the basis of similarities in the overall sequences. The amino acid residue found most frequently at each position of the subgroup may then be said to define a "prototype sequence" for that subgroup. Within any one of these subgroups, the extent of deviation from the prototype sequence is relatively small. At most of the variable positions at which deviation is observed only two alternative amino acid residues have been found, and in the remaining positions the substitutions involve only three acids. Further, with only one exception, the amino acid substitutions observed for each variable position of the sequence can be attributed to a change in only one base of the corresponding codons of the genetic code (see Jukes 1966, Yčas 1969).[3] Thus, at least superficially, the relationship between the various sequences comprising a subgroup and the prototype sequence is the same as that between a series of homologous proteins isolated from different animal species and the hypothetical "ancestral protein" presumed to be the homologue present in an ancestral species, from which the contemporary forms have evolved (see Dayhoff and Eck 1969).

This *apparent* "evolutionary" relationship between N-terminal variable regions also extends to a comparison of the different prototype sequences. The extent of variation is greater than that observed within the subgroups. However, when appropriate gaps are introduced, the prototype sequences show a degree of homology similar to that observed in a comparison of the α- and β-chains of hemoglobin (e.g., Braunitzer 1966) or of other pairs of polypeptide chains coded by different structural genes which are believed to have arisen through duplication of a single "ancestral gene" (e.g., Dayhoff and Eck 1969, Dixon 1966). The *apparent* relationship is reinforced by comparison of the amino acid sequences of homologous chains obtained from different animal species. It is extended in comparisons of the sequences of the light chains with each other and with those of the heavy chains of a single species. Indeed, a persuasive argument can be developed in support of some form of dual hypothesis that postulates:

[3] In the case of human λ-chains there are two series in which more than 3 alternative acids have been found at a single position of the "prototype sequence." However, with one possible exception, the alternatives are all related to the acid of the prototype sequence in the sense that the corresponding codons can all be generated by single base substitutions.

(a) The structural genes coding for all contemporary species of light immunoglobulin chain can be attributed to a combination of (1) divergent evolution of two, or more, copies of each of two types of "primordial gene," with (2) fusion of one of each type of "descendant gene."
(b) The structural genes coding for all contemporary species of heavy immunoglobulin chain can be ascribed to the linear fusion of two of the postulated bipartite light-chain structural genes.

The various versions of this "prototype hypothesis" differ primarily in postulated numbers of discrete copies of the primordial genes generated by gene duplication, and in the stage of evolution at which fusion of descendant genes did, or does, occur. I do not propose either to develop the argument or to debate the relative merits of the various versions in detail, and I refer the interested reader to the reviews noted earlier in the chapter. Rather, I wish to consider briefly the paradoxical problem posed by the observed heterogeneity of the immunoglobulin chains.

THE PROBLEM: ORIGIN OF THE DIVERSITY OF VARIABLE REGIONS

The marked diversity of sequence in the variable regions constitutes strong presumptive evidence that each individual species of immunoglobulin chain is specified by a unique structural gene present in the corresponding committed plasma cell. This, in turn, implies that the enormous potential inherent in the antibody-forming system of one individual animal reflects the presence in the collective, total genome of the population of antigen-sensitive cells of a very large number of different, albeit related, structural genes. On the other hand, the very limited degree of variation seen in the constant regions of immunoglobulins formed by all individuals of a given species is strong presumptive evidence that the structures of the entire complement of antibodies formed within any one individual animal are encoded within a very small number of structural genes. Thus the complex mixture of antibodies present in an individual serum *appears* to be due to fusion of a variety of the (N-terminal) gene products specified by a large number of structural genes (*v*) to a limited number of (C-terminal) gene products specified by a small number of other structural genes (*c*). However, we know of no mechanism whereby such a highly specific process could occur. More important, there is direct experimental evidence that the immunoglobulin chains almost certainly are not formed by a process of end-to-end fusion of two smaller complete peptide chains (Fleischman 1967, Knopf, Parkhouse, and Lennox 1967). Thus the problem posed by the variability observed in the immunoglobulin chains is that of accounting for the generation of a remarkable diversity of sequences in the N-terminal regions of a family of homologous proteins containing an invariant C-terminal region.

A number of quite different and ingenious hypotheses have been advanced to account for this "localized" variability in amino acid sequence. All have three features in common. First, all require the explicit or tacit assumption of at least one *ad hoc* hypothesis for which there is no precedent, to account either for the variability *per se* or for the very marked reduction, or even loss, of it in "committed" antigen-sensitive cells following exposure to specific antigen. Second, each postulated mechanism is extremely uneconomical, in the sense that the vast majority of cells in the potentially antigen-sensitive population will, in fact, produce either immunoglobulin chains devoid of any antibody activity or partial chains. Finally, none of the hypotheses provides both a satisfactory and a complete interpretation of the data. Therefore, it seems highly probable that full understanding of the basis of the heterogeneity in immunoglobulin chains will involve elucidation of novel genetic control mechanisms. Indeed, most of the speculative proposals which have been advanced focus on possible mechanisms whereby the necessary degree of variation can be generated within a restricted section of one structural gene which "codes" for the variable region.

THE SCHUBERT-COHN PROPOSAL: FUSION OF VARIABLE REGION POLYPEPTIDE CHAIN TO NASCENT PARTIAL CONSTANT REGION CHAIN

A number of murine myelomas form a protein which is approximately half the size of the light immunoglobulin chain produced by the cells and serologically related to it (Schubert 1970, Schubert and Cohn 1968, 1970). This protein is not secreted by the cells and was originally believed to be a degradation production of completed chains, similar to materials found (e.g., Berggård and Peterson 1967, Cioli and Baglioni 1967, Solomon and McLaughlin 1969) in the plasma and urine of human patients. However, the kinetics of accumulation of this protein in the intracellular pool of soluble proteins, and of its subsequent disappearance, reveal that it is not derived from completed immunoglobulin chains. Rather, it *could* be an intermediate in the synthesis of the latter (Schubert and Cohn 1970). Indeed, disappearance of pulse-labeled molecules of the protein from the pool of immunoglobulin chains is dependent upon concurrent protein synthesis. Further, both nascent peptide with serological properties of immunoglobulin light chains and the "half-chain protein" are associated with the same fraction of the polysomes. Moreover, the S value of this fraction corresponds to that expected (see Chapter 1) of polysomes containing mRNAs for proteins of the size of half an immunoglobulin light chain. Similarly, the nascent heavy immunoglobulin chains are found in the polysome fraction expected to contain mRNAs specifying proteins only three-quarters of the size of completed heavy chains (Schubert and Cohn 1970).

A similar partial light immunoglobulin chain produced, but not secreted, by cells of a human myeloma has been shown to contain at least a portion of the variable region of the chain (Matsuoka et al. 1969). Nevertheless, immunoglobulin chains are not formed by the union of a variable region peptide chain with a completed constant region peptide chain (Fleischman 1967, Knopf, Parkhouse, and Lennox 1967). Schubert and Cohn (1970) have therefore suggested that completed variable region peptide chains are released into the soluble immunoglobulin pool and subsequently attached to half-completed nascent constant region peptide chains.

All evidence currently available appears compatible with this proposal. It is therefore appropriate to note their additional word of caution. "These studies do not formally establish that the variable and constant regions are linked at the peptide level. Rather they stress that *the generally accepted view that it occurs by translocations at the DNA level should be considered an open question.*" As yet, there is no direct evidence that the half-chain molecules are incorporated into complete immunoglobulin chains or that the C-terminal residues of those molecules in the pool of soluble proteins can be "activated." Moreover, it should be noted that this novel postulated mechanism of chain elongation affords no interpretation of the presence of the very large variety of committed cells believed to be present in the population of potential antibody-forming cells. The diversity of variable portion genes (v) expressed by different cells of this population before exposure to test antigens presumably reflects the operation of some other mechanism which does act "at the DNA level"; that is, a novel genetic mechanism (see below).

POSTULATED MECHANISMS OF DIRECTED MISTRANSLATION

A few interpretations have been advanced in which the heterogeneity in structure of the immunoglobulins arises during the processes of (conventional) polypeptide chain elongation (Campbell 1967, Gyenes 1969, 1970, Mach, Koblet, and Gros 1967, Potter, Appella, and Geisser 1965). These are based on possible mechanisms by which some "error" or "ambiguity" is introduced into the process of selection of the incoming amino acids to be aligned against the codons of the messenger RNA. In most of these proposals it was supposed that the mRNA contains a relatively high proportion of minor codons of the genetic code (see Jukes 1966, Yčas 1969) in the segment coding for the variable region of the chain. It was further supposed that the amino acid specificities of the corresponding tRNAs could be modified by mutational changes in the structure of the tRNA (or in one of the aminoacyl-tRNA synthetase enzymes). This would then lead to mistranslation and insertion of an incorrect amino acid at specific positions of the variable regions. However, even at the outset such proposals failed to account for the fact that the amino acid substitutions observed are restricted to those which can be correlated with single base changes in the corresponding codons. In view of more recent knowledge of the structure of individual tRNAs, it now seems improbable that they could account for the absence of similar heterogeneity in the constant regions of the chains.

As a more recent version of such mistranslation mechanisms, Gyenes (1969, 1970) has formulated an "inverse wobble theory." This is based upon the "wobble hypothesis" developed by Crick (1966[b]) as an explanation of the general nature of the "degeneracy" of the genetic code. In particular, it accounted for the observation (Kellogg et al. 1966, Söll et al. 1966) that alanine-specific tRNAs containing the anticodon CGI can complex with any of the alanine codons GCU, GCC, or GCA. This demonstrated ability of the minor base inosine to form base pairs with three other nucleotide bases is an important factor in the phenomenon of degeneracy. Gyenes proposes that it is also the major factor responsible for the heterogeneity of the variable regions of the various immunoglobulin chains. However, in contrast to the provisions of the wobble hypothesis, it is postulated that the minor base is contained in unusual codons of the mRNAs. These codons could then complex with tRNAs containing any of four different anticodons.[4] The resulting ambiguity could then result in the insertion of a restricted number of different amino acid residues at the corresponding position of the peptide chain. Thus the special feature of heterogeneity in the N-terminal sequence of immunoglobulin chains is attributed to the unusual, and possibly unique, ability of antigen-sensitive cells either to incorporate or to generate inosine (or some other minor base) residues within one section of the mRNA molecules.

Some additional genetic mechanism is required to account for the homogeneity of individual species of immunoglobulin, and Bence Jones proteins, produced by cells of a single clone of plasma cells. Campbell (1967) and Gyenes (1969) have proposed that ambiguities within the mRNAs are effectively eliminated by selective suppression of synthesis of all but one of the iso-accepting series of tRNAs able to complex with each of the ambiguous codons. In support of this proposal, some differences in tRNA complements have been observed between normal and myeloma plasma cells, and between different lines of myeloma cells (Hashizume and Sehon 1970, Mach, Koblet, and Gros 1967, Yang and Novelli 1969). However, for this ingenious proposal to be reconciled with amino acid sequences already determined for human immunoglobulin light chains it is necessary to assume also either (a) that the species of tRNA which complex with the ambiguous codons cannot complex with normal codons, or (b) that the number of distinct subgroups of each type and class of chain is very markedly larger than that already required by the theory. For example, inspection of the N-terminal sequences of κ-chains of subgroup I listed

[4] The inverse wobble theory as postulated by Gyenes (1969) allows possible base pairing between inosine and guanine and the presence of two inosine residues within a single codon. Neither of these two possibilities is permitted in the wobble hypothesis (Crick 1966[b]). Further, it is necessary to assume that the number of distinct subgroups of each type and class is substantially greater than currently believed.

by Hood and Talmage (1970) reveals that synthesis of chains designated *Pap* and *Lay* would require suppression of all species of alanine- and glycine-specific tRNAs capable of complexing with ambiguous codons. Further, in the case of κ-chains of subgroup II, comparison of the sequences at positions 2·3·4 for the protein *Gra* (Met·Val·Met) with those of the prototype protein *Ti* (Ileu·Val·Leu) indicates that the protein *Dob* (Ileu·Ileu·Met) could not be retained within the same subgroup (Table 11.1). It remains possible that present classifications of immunoglobulin chains grossly underestimate the number of distinct subgroups or that some other genetically determined mechanism could circumvent the postulated ambiguity introduced into the mRNAs. However, present evidence indicates this to be unlikely.

GENE FUSION, HYPERMUTATION, AND INTRAGENE COMPLEX RECOMBINATION

All other hypotheses are based upon two common premises. First, it is accepted that the unique amino acid sequence of each distinct species of immunoglobulin chain is unambiguously specified by a corresponding unique structural gene. Second, each of these genes (*v-c* genes) is considered to have arisen by the fusion of two smaller genes coding, respectively, for the variable region (*v* genes) and constant

TABLE 11.1. Standard and inverse wobble codons assigned to amino acid sequences of light immunoglobulin chains of subgroup κ_{II}

Protein	Residue number 2·3·4	Standard codon base sequence *	Common inverse wobble codon base sequence †
Gra	Met·Val·Met	AUG·GUX·AUG	
Ti	Ileu·Val·Leu	AUX·GUX·CUG	AUI·IUX·IUG
Dob	Ileu·Ileu·Met	AUX·AUX·AUG	

* X designates any of the 4 major bases of RNAs.

† I designates inosine, which is postulated to be recognized as any of the 4 major bases of RNAs.

Codon IUX is postulated to be equivalent to both AUI or IUG, but cannot be recognized as either in synthesis of protein Dob. For details see text.

region (*c* genes) of the particular peptide chain. The various hypotheses differ from each other with respect to the number of distinct types of *v*-gene segment present in the germ line of the individual animal, and in the mechanism whereby the diversity of the *v*-gene segments is generated.

At one extreme is the "germ line," or multigene, hypothesis formulated by Dreyer and Bennett (1965). It is supposed that the germ line contains one variable *v* gene for each of the unique antibodies which the animal can synthesize, and one constant *c* gene for each type of common region. During differentiation of the stem cells of the antibody-forming system all, or part, of one *v*-gene base sequence is "spliced" onto the appropriate segment of a *c*-gene base sequence. Thus, according to this hypothesis, all of the *v*-gene base sequences pre-exist in the genome of the animal and are believed to have arisen through the processes of gene duplication and independent mutation. A formal mechanism whereby copies of segments of *v*-gene base sequences are spliced onto copies of *c*-gene sequences in an hypothetical "replisome" has been proposed (Dreyer, Gray, and Hood 1967). In view of the extensive gene redundancy present in eucaryote genomes (Britten and Kohne 1968; see Chapter 2), the *number* of discrete *v* genes required by this type of hypothesis is not excessive (Hood et al. 1970, Hood and Talmage 1970). However, the variable regions of rabbit immunoglobulin heavy chains contain species-specific allotype markers determined by defined gene alleles (Koshland 1967, Wilkinson 1969). Acquisition of the base sequences necessary to specify these markers in each of some thousands of different *v* genes as they have independently evolved seems highly improbable. Conversely, evolution of some thousands of different base sequences from copies of a prototype *v* gene generated by gene amplification after evolution of the distinctive progenitor of contemporary rabbits seems equally highly improbable.

At the other extreme are early "somatic" hypotheses which

supposed that cells of the germ line of the animal contained only one single *v-c* gene for each class or group of immunoglobulin chains. The diverse assortment of various *v-c* genes present in different progeny antibody-forming plasma cells was then generated by "*hypermutation*"; that is, mutation at an unusually high frequency (e.g., Brenner and Milstein 1966, Lederberg 1959). Two interdependent lines of argument are usually advanced for considering mechanisms of this type improbable. The first of these is that no mechanism is known whereby such random mutation could be generated within only one restricted region of a single copy of one gene. Further, the only specific hypothetical mechanism proposed (Brenner and Milstein 1966) seems highly improbable. The second is that any mechanism of this general type would necessarily entail an extremely high degree of "cell waste." Specific antibodies are the products of highly selected groups of plasma cells. Thus the nonrandomness of amino acid sequences observed in the peptide chains does *not necessarily* reflect a similar nonrandomness in the postulated hypermutation process, but rather the dual set of structural restraints required for both activity as an antibody and secretion from the plasma cell. Similarly, the various types of "myeloma" protein are the products of restricted cell populations which have been selectively stimulated to proliferate. The inevitable consequence of a random process of hypermutation would be that a large proportion of the population of potential antigen-sensitive cells would be unable to form any functional immunoglobulin. Thus only a small proportion of "useful" antigen-sensitive cells would be present among a larger population of "waste cells." Estimates of this proportion depend upon the precise assumptions made in developing the argument. They range from an "impossible" low value of 1 in 10^{30}, a number exceeding the total number of cells generated by an average person during the entire life span (Hood and Talmage 1970), to an "acceptable" high value of 1 in 10^3 to 10^4 (Cohn 1967).[5] Either alone, or in conjunction with the absence of an appropriate mechanism of hypermutation, these estimates are generally considered sufficient reason for rejection of this type of extreme somatic hypothesis.

However, it is important that such hypotheses not be totally dismissed from further consideration. As Cohn (1967) has observed, germ-line evolution is itself the outcome of selection of random and very infrequent variants and must also be true for somatic evolution. The primary objection to the extreme type of somatic hypothesis is, therefore, lack of a known mechanism to account for the necessary hypermutation. It would be unfortunate if we lost sight of one of the major objectives in studying this unusual group of proteins, namely, the elucidation of what is generally expected to be a novel genetic regulation mechanism.

In a more central position are hypotheses which are combinations of germ line and somatic theories. These suppose that each distinct class and type of immunoglobulin chain is represented in cells of the germ line by one type of *c*-gene segment and a small number (of the order of 5 to 10) of different *v*-gene segments, which have all arisen from a single common precursor *v* gene by gene duplication and independent evolution. These various *v*-gene segments would show very considerable homology with each other, and extensive variation could be generated by somatic recombination between them. The various hypotheses of this type differ from each other in the nature of the association of the multiple *v*-gene segments.

[5] Modification of the hypothesis to include one additional *v-c* gene for each distinct subgroup of light (Hood and Talmage 1970, Milstein; Milstein, and Feinstein 1969) or heavy (Cunningham et al. 1969, Köhler et al. 1970) chains would require upward revision of all estimated values. However, this would not affect the "unacceptability" of this extreme type of somatic hypothesis.

POSTULATED MECHANISMS OF INTRAGENE COMPLEX RECOMBINATION

In an early proposal of Edelman and Gally (1967) it was supposed that the gene segments corresponding to one distinct type of chain were present in organized gene complexes, each consisting of one *c*-gene segment and a small number of different homologous *v*-gene segments in tandem (see Figure 11.3). A considerable degree of diversity could then be generated during differentiation and maturation of the diploid antigen-sensitive cells by somatic recombination between the *v*-segments of homologous complexes. Subsequently, Edelman (1967) suggested an alternative mechanism in which the somatic cells contained *v*-*c* genes and separate *v*-gene segments. Recombination of a free *v* gene with the fused *v*-gene segment, in a manner analogous to that of prophage insertion into a bacterial chromosome (see Campbell 1969), would yield a complex gene containing two recombinant *v*-gene segments in tandem with the *c*-gene section. Both of these proposals are based upon conventional genetic mechanisms for generating diversity within gene alleles, although at a very much higher frequency than customarily observed.

However, in both cases diversity is generated within a complex gene assembly containing a minimum of two contigous *v*-gene segments. Yet, except for the *v*-gene segment abutting the *c*-gene section, we know of no mechanism which could select only one of those *v*-gene segments and transcribe it continuously with a *c*-gene section from which it is physically separated. It could be supposed that all but one *v*-gene segment is eliminated from the complex by internal recombination between *v*-segments within the complex. Such internal recombination between the first and last *v*-gene sections (Figure 11.3[a]) has been postulated in a related proposal based upon a "cycloid model" of chromosome structure (Whitehouse 1967[a], 1967[b]; see Chapter 2). However, in this case the residual *v*-segment would invariably be a composite of the *c*-gene-proximal portion of the

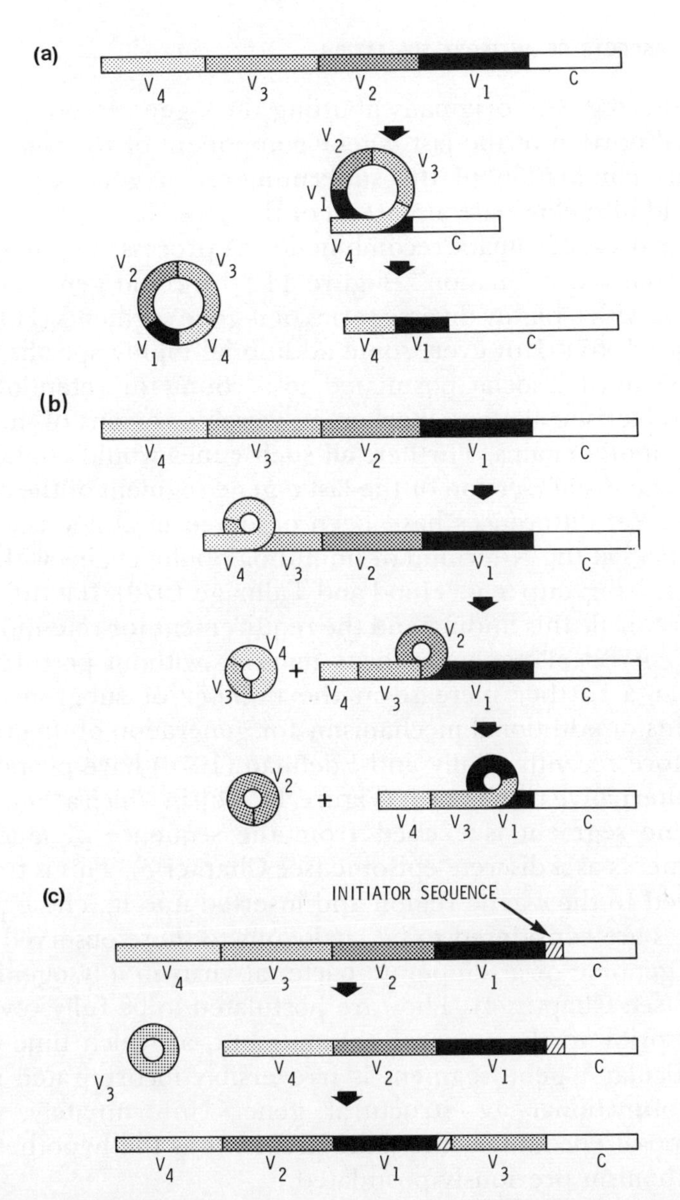

FIGURE 11.3. Regeneration of a single v-c-gene unit from a complex containing multiple tandem v-gene segments

(a) By one internal recombination, based on Whitehouse 1967(a). (b) By multiple internal recombination, based on Hilschmann (1967). (c) By transposition as an episome, based on Gally and Edelman (1970).

v-gene segment originally abutting the *c*-gene section and a distal portion of the last *v*-gene component of the complex. Thus elimination of the supernumerary *v*-gene segments would also eliminate almost all of the diversity generated by the process of somatic recombination. A process of "repetitive internal recombination" (Figure 11.3[b]) could generate *v-c* genes with a highly diverse series of *v*-gene segments (Hilschmann 1967). However; some additional highly specific factor(s) must also be postulated to account for retention of characteristic allotype markers in variable regions of immunoglobulin chains. Further, all such genes would contain a "*c*-gene-distal" section of the last *v*-gene segment of the complex. Yet differences have been observed in amino acid sequences at the N-termini of immunoglobulin chains within a single subgroup (e.g., Hood and Talmage 1970). It is difficult to reconcile this finding and the requirement for retention of the correct allotype specificity markers without postulating either a further increase in the number of subgroups of chains or additional mechanisms for generation of diversity.

More recently, Gally and Edelman (1970) have proposed an alternative mechanism (Figure 11.3[c]) in which a "mixed" *v*-gene segment is excised from the sequence of tandem segments as a discrete episome (see Chapter 3). This is transferred to the *c*-gene region and inserted into it. These processes are considered to be analogous to those observed for the genome of a temperate bacterial virus in a lysogenized cell (see Chapter 4). They are postulated to be fully reversible prior to the stage of commitment, at which time one particular *v*-gene segment is irreversibly incorporated into the functional *v-c* structural gene. Unfortunately, this proposal encounters the same problems as the hypothetical mechanism previously postulated.

An entirely different mechanism for the generation of diversity through somatic recombination between "*antibody gene pairs*" has been suggested by Smithies (1967). As originally formulated, it was supposed that each type of light immunoglobulin chain was represented in the genome by one "*master* (*v-c*) *gene*" and one "*scrambler* (*v*) *gene*" similar to, but not identical with, its homologue in the master gene. In contrast to the mechanism envisaged in Edelman's proposal (1967), recombination between the two homologous *v*-gene segments leads to segregation of "recombinant" base sequences in both the master and scrambler genes (Figure 11.4). Extension of this basic proposal to include one "master-scrambler gene pair" for each subgroup of light and heavy chains would provide a mechanism which would account for generation of almost all the observed amino acid sequences

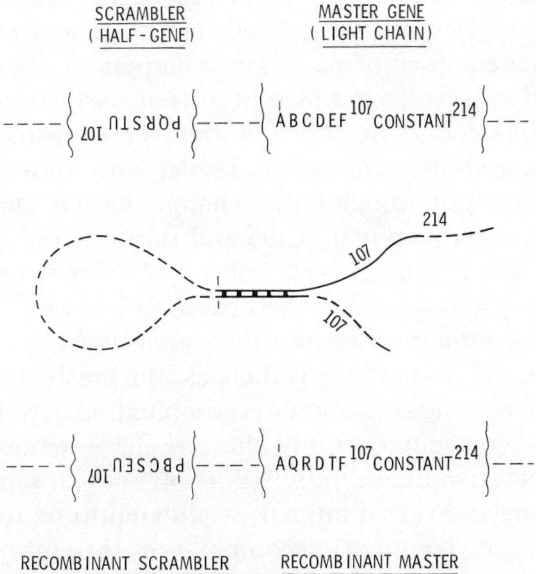

SCRAMBLER (HALF-GENE)
MASTER GENE (LIGHT CHAIN)
ABCDEF 107 CONSTANT 214
AQRDTF 107 CONSTANT 214
RECOMBINANT SCRAMBLER
RECOMBINANT MASTER

FIGURE 11.4. The master-scrambler gene hypothesis

In this illustration both elements of the antibody gene pair are present in one chromosome. The scrambler v gene and homologous v-gene segment of the master gene differ at only six regions; most commonly in only one pair of nucleotide bases, but possibly in more extended runs of bases. These regions are designated by letters.

The diagram illustrates generation of diversity through multiple recombination occurring simultaneously in one event; it could equally arise through single "crossovers" in a series of recombinations.

Reproduced from Smithies (1967) with permission of the author and editors of *Science*. Copyright 1967 by the American Association for the Advancement of Science.

of immunoglobulin chains. The few exceptions could probably be attributed to the additional occurrence of appropriate point mutation.

Nevertheless, despite the attractiveness of the proposal, it has serious shortcomings. Further, additional features must be postulated to account for all aspects of antibody formation. Diversity of base sequence generated through somatic recombination has the same consequences as that generated through other mutational processes, and the proposal is not immune from challenge on the grounds of excessive cell waste (e.g., Hood and Talmage 1970). It seems improbable that heterogeneity at the N-termini of the chains could be generated unless the postulated scrambler genes contain an initiation sequence of bases (see Chapter 1). This could be expected to be reflected by synthesis of partial immunoglobulin chains coded by the scrambler genes with similar kinetics to, and in quantities equimolar with, those of the corresponding immunoglobulin chains. Partial chains which have been observed (e.g., Cioli and Baglioni 1967, Matsuoka et al. 1969, Schubert and Cohn 1970, Solomon and McLaughlin 1969) do not exhibit these properties.

However, the most serious deficiency of the master-scrambler gene pair hypothesis is the lack of a mechanism to regulate the frequency of somatic recombination between homologous *v*-gene segments. On the one hand, generation of a highly heterogeneous population of antigen-sensitive cells within one individual animal would require extensive (and possible very frequent) recombination throughout the entire population of their progenitor cells. On the other hand, equally extensive continued somatic recombination would preclude the persistent replication of an essentially pure clone of plasma cells after removal of inducing antigen. Some mechanism(s) must therefore be postulated to ensure extensive somatic recombination during differentiation (Till et al. 1967) of the committed antigen-sensitive cell, and to then suppress further somatic recombination. One obvious possibility could be the elimination or inactivation of the scrambler genes, but other equally *ad hoc* postulates are possible.

More recently, Smithies (1970) has postulated yet another alternative mechanism in which the essential element is a novel type of complex "branched" gene (Figure 11.5), for which the obvious mechanical analogy is a railroad freight yard. The species of mRNA transcribed from the gene complex is determined by the route followed by the RNA polymerase enzyme as it traverses the "network." In turn, this is determined by protein "switches" which are preset in quasi-stable positions by interaction with small effector molecules. The process of commitment would then consist of the irreversible "setting" of the switches in stable positions and would possibly involve the equivalent of gene elimination (see Chapter 2) in the selective replication of only the two preset branches of the network. It is beyond the scope of this exercise to consider whether the postulated branched gene is compatible with available evidence of the ultrastructure of chromosomes. However, the model would appear to require the existence of as many separate *v*-gene branches as are required by the extreme type of multigene hypothesis. Moreover, the model does not afford ready interpretations of *both* the presence of allotype markers within the variable region of an immunoglobulin chain *and* of differences in amino acid sequences in the N-terminal segments of the chains which precede those markers.

Thus, at present, the special and distinctive mechanism(s) specifically controlling the synthesis of unique species of immunoglobulin peptide chains remains an enigma. None of the various ingenious proposals which have been made provides more than a partial explanation of the observed heterogeneity, and it seems probable that the mode of action of a novel genetic mechanism awaits elucidation.

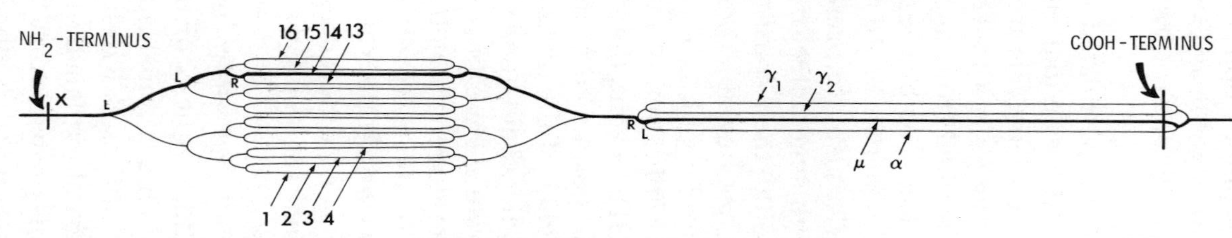

FIGURE 11.5. The branched gene hypothesis

The diagram represents a hypothetical locus for an immunoglobulin heavy chain with 16 variable regions (v_1 to v_{16}) and four constant parts ($c_{\gamma 1}$, $c_{\gamma 2}$, c_α, c_μ). Each line represents a DNA double helix. Commitment to one of the possible proteins coded by the locus is determined by the random setting of its DNA forks in left (L) or right (R) configurations; X represents a position at which allotypic variants could easily be found. The path of an RNA polymerase molecule is indicated by the heavy line; it would transcribe messenger RNA corresponding to v_{14}-c_μ.

Reproduced from Smithies (1970) with permission of the author and editors of *Science*. Copyright 1970 by the American Association for the Advancement of Science.

THE PHENOMENON OF ALLELIC EXCLUSION

Analysis of the mechanism which specifically controls the selective synthesis of a unique species of immunoglobulin chain is further complicated by the effects of at least two additional regulatory mechanisms controlling the formation of specific antibodies. The first of these is manifested in the phenomenon of "*allelic exclusion*" of allotype markers. In animals heterozygous for alleles determining one of the specific allotype markers both of the corresponding antigenic groupings are found in the mixed population of immunoglobulins present in the serum. However, analysis of individual lymphoid cells of such a heterozygote reveals that virtually all of the antibody-forming cells produce immunoglobulin of only one of the two possible allotype specificities (e.g., Pernis et al. 1965, Weiler 1965). Thus some regulatory mechanism determines that only one of the two alleles shall be expressed in each individual antibody-forming cell and the other shall be excluded. Mechanisms of this type have been reviewed previously (Chapter 2) and will not be further considered in detail here. However, the proportions in the serum of immunoglobulins carrying the two types of allotype marker represent a distinctive property of the specific pair of gene alleles and may deviate very markedly from equivalence (e.g., Chou, Cinader, and Dubiski 1967, Dray and Nisonoff 1963). Therefore, either the mechanism of allele exclusion is nonrandom and preferentially excludes one allele of the pair, or some additional regulatory mechanism serves to selectively stimulate proliferation and antibody synthesis in cells in which the "preferred" allele is not excluded.

Similar, or identical, mechanisms also control synthesis of the different types and classes of immunoglobulin. Any particular animal may produce antibodies of each type and class in response to one specific antigen. However, virtually all of the individual lymphoid cells produce only one type of light chain and one class of heavy chain (e.g., Cebra, Colberg, and

Dray 1966, Merchant and Brahmi 1970, Van Furth, Schuit, and Hijmans 1967). Proportions of cells producing the different species of immunoglobulin chain vary for individual animals.

A very small proportion of the cells can be shown to produce both large (19S) immunoglobulins of the IgM class and small (7S) immunoglobulins of the IgG class (e.g., Cosenza and Nordin 1970, Merchant and Brahmi 1970). These are of special interest in two respects, for it has been proposed that they constitute suggestive evidence: (a) that the mechanism of exclusion may be one of transient, rather than permanent, gene inactivation (Merchant and Brahmi 1970) and, more particularly, (b) that the ontogeny of the primary antibody response is regulated by an intracellular molecular switch, rather than a process of intercellular selection (Wang et al. 1970).

THE BIPHASIC PRIMARY ANTIBODY RESPONSE AND ROLE OF MACROPHAGE RNA

A primary antibody response to a new species of antigen characteristically consists of two phases. The first specific antibody molecules elicited are large (19S) IgM immunoglobulins. Synthesis of this class of antibody then declines as a second phase of synthesis of 7S-IgG immunoglobulin commences. The mechanism(s) regulating this sequence are not known. However, four unrelated types of evidence serve to focus attention upon *possible intracellular* switch mechanisms. One is the presence of both an IgM and an IgG immunoglobulin in some individual cells, referred to above. A second is the demonstration that in individual rabbits all the immunoglobulins elicited contain the same antigenic markers in the variable regions (e.g., Prahl et al. 1970, Todd 1963). Further evidence has been furnished by studies of the immunoglobulins of a patient with a multiple myeloma secreting both an IgM and an IgG protein. Both proteins contain the same light chain and appear to contain very similar or identical variable regions in the heavy chains

(Levin et al. 1971, Wang et al. 1970). Finally, when 13-day chick embryos are injected with antiserum prepared against μ-type heavy chains, to eliminate IgM-producing cells from the bursa of Fabricius, subsequent bursectomy at hatching suppresses the synthesis of IgG immunoglobulins. In contrast, when the antiserum is administered after bursectomy of newly hatched chicks the synthesis of IgG immunoglobulins is unaffected. Therefore, it has been concluded that the IgG-producing cells arise from cells of the bursa which have already synthesized IgM immunoglobulin (Kincade et al. 1970). However, each of these various observations is amenable to alternative interpretations. Moreover, it has recently been reported that precursors of plasma cells which are present in murine bone marrow are already committed to develop the ability to synthesize one particular class of immunoglobulin when subsequently exposed to heterologous erythrocytes (Miller and Cudkowicz 1971). Therefore, it is still not established whether the biphasic primary response reflects a change in the class of immunoglobulin produced by individual plasma cells, or a change in the proportions of different types of plasma cell in the population of antibody-forming cells.

However, there is some evidence to indicate that the response *may* be regulated through synthesis of specific types of RNA in macrophage cells. It is beyond the scope of this discussion to consider in detail the possible role(s) of macrophages, and other types of cell, in the synthesis of antibody. However, the salient features can be outlined very briefly. Early studies indicated that very little, if any, of the inducing antigen was associated with the antibody-forming lymphoid cells, and the bulk of it was taken up by macrophages (e.g., Ada et al. 1967, Fishman and Adler 1967, Nossal, Ada, and Austin 1965). This led to demonstrations of the formation of specific antibodies by "virgin" lymphocytes unexposed to inducing antigen in response to preparations of RNA isolated from macrophages which had been incubated with antigen (e.g., Bishop, Pisciotta, and Abramoff 1967, Fishman and Adler 1963). The nature of this "immunogenic RNA" is currently a matter of some dispute. Fishman and Adler (1967) have reported that it contains two distinct components. One consists of a newly synthesized, small (8-10S) species of RNA which elicits formation of a 19S IgM antibody. The other consists of a larger (23-28S) ribonucleoprotein complex containing pre-existing macrophage RNA with fragments of antigen protein and eliciting formation of IgG antibody. On the other hand, Gottlieb (1969[a], 1969[b], Gottlieb and Strauss 1969) reports that the material is a small (molecular weight 12,000) ribonucleoprotein complex of pre-existing macrophage RNA and fragments of antigen containing residual antibody-combining sites. Both groups of workers used the same antigen, and the basis for the discrepancy in findings has not been established. The mechanism of action of the immunogenic RNA can only be speculated upon at this time (e.g., Lurie 1969). A formal similarity with a "superantigen" (Askonas and Rhodes 1965) can be postulated, in which the RNA is analogous to a hapten (Ramming and Pilch 1970). This would permit reconciliation of the experimental observations with the belief that antigen serves to stimulate proliferation of cells already committed to the synthesis of corresponding antibody. An alternative possibility might be the postulation of a formal analogy with inducers of differentiation.

EPITOME: THE ANTICLIMAX

Much of the current interest in mechanisms regulating the synthesis of specific antibodies stemmed from the hope that it might serve as a model system with which to analyze mechanisms regulating the processes of cellular differentiation. Unfortunately, it now appears that regulation of antibody synthesis is anything but a simple model system. Indeed, it exhibits special features which are not as obviously implicated in regulation of some of the differentiating systems for which it was intended to serve as the model.

There was the Door to which I found no Key:
There was the Veil through which I might not see:
—Omar Khayyam, Twelfth Century

CHAPTER 12: **ADDENDUM: RECENT DEVELOPMENTS**

One component of the current information explosion has been the anticipated publication of a large body of new data and concepts since compilation of the material considered in the preceding chapters. The bulk of this recent information is in accord with the overall picture which has been developed. Incorporation of this material into their own evolving concepts will pose no problem for the students to whom this volume is directed. Therefore it would be futile to attempt to summarize all the developments of the past few months within a few paragraphs.

However, some of the recent observations appear to presage the direction of refinement of our concepts in areas that are currently vague, or at least indicate the trend of studies in the immediate future. Moreover, certain of them have led me to modify the emphasis of my own presentation of some topics considered herein. Therefore in these closing pages I propose to note briefly a small group of studies which I consider particularly important to the presentation that has been made. My list follows the same sequence as that in the preceding chapters and the bibliography is given separately at the end of the regular bibliography.

THE MECHANISM OF PROTEIN BIOSYNTHESIS (CHAPTER 1)

Several studies have contributed further knowledge of the processes involved in formation of initiation complexes. Hunter (1971) has reported that formation of initiation complexes containing the mRNA of hemoglobin peptide chains is inhibited specifically by two-stranded RNAs. Moreover, Lebowitz, Weissman, and Radding (1971) have determined the base sequence of a 6S RNA transcribed *in vitro* from the DNA of phage λ and found it to be analogous to the N-terminal initiation sequences of the small RNA-phages discussed previously. Thus it now appears probable that an essential feature of the initiation sequence(s) in an mRNA which is "recognized" by the appropriate specific initiation factor(s) $\mathbf{F}_3$ is the presence of one, or more, segments with particular base-paired hairpin structures.

There is now some evidence that the initiator N-formyl-methionyl-tRNA is first incorporated into the procaryote initiation complex at an entry site with properties distinct from those of either the acceptor or donor sites of a complete 70S ribosome (Tanaka, Lin, and Okuyama 1971, Thach and Thach 1971, Zagorska et al. 1971). However, this entry site could be that portion of a site present in the completed ribosome which is derived from the 30S component subunit. Recent studies by Groner and Revel (1971) indicate that the mechanism by which initiation factor $\mathbf{F}_2$ stimulates incorporation of both initiator N-formyl-methionyl-tRNA and GTP into the initiation complex may involve prior formation of a complex between it and initiation factor $\mathbf{F}_3$ already bound to the 30S subunit. Indeed, they have postulated a formal analogy between an $\mathbf{F}_2$-$\mathbf{F}_3$ complex and the **Tu-Ts** complex of elongation factors, and envisage release of factor $\mathbf{F}_3$ from the initiation complex concurrently with hydrolysis of bound GTP. In this connection, it may also be noted that Weissbach, Redfield, and Brot (1971) have suggested that elongation factor **Ts** is a phosphotransferase enzyme which catalyzes the reaction:

$$\mathbf{Tu}\text{-GDP} + \text{GTP} \rightleftharpoons \mathbf{Tu}\text{-GTP} + \text{GDP}.$$

Raacke (1971) has postulated a two-stage mechanism of formation of the peptide bond in which the small 5S RNA component of the large ribosome subunit plays a pivotal role. Specifically, it is proposed that the nascent peptide chain is transferred to the 3′-terminal of the 5S RNA by the action of the peptidyl transferase enzyme of the 50S subunit. From there it is transferred to the incoming aminoacyl-tRNA by a second transferase enzyme of the cell sap which has different properties from that of the ribosomal enzyme. Thach and Thach (1971) have made an elegant demonstration that the subsequent translocation step is correlated with the expected movement of the mRNA relative to the acceptor site of the ribosome. Evidence has also been presented of roles of an additional factor **Z** (Phillips 1971) and of the dissociation factor $\mathbf{F}_3$ (Herzog, Ghysen, and Bollen 1971) in the processes of peptide chain termination.

The existence of tissue-specific initiation factors required for formation of initiation complexes in eucaryote systems has been further reported by Naora, Kodaira, and Pritchard (1971). Nevertheless, the degree of species and tissue specificity which may have been anticipated from the early reports previously considered has not been observed in various studies. For example, the 9S mRNA isolated from mouse reticulocytes is readily translated in the synthesis of mouse globin chains in cell-free systems derived from rabbit reticulocytes (Lockard and Lingrel 1971) and Krebs II ascites tumor cells (Mathews, Osborn, and Lingrel 1971), and, more particularly, within intact oocytes of *X. laevis* with very high efficiency (Lane, Marbaix, and Gurdon 1971). Nevertheless, the exogenous 9S mRNA does not readily *compete* with the endogenous mRNA of the oocyte for incorporation into functional polysomes (Moar et al. 1971). This may reflect a selective rather than absolute specificity of the initiation factors comparable with that already observed for the homologous factors of *E. coli* (e.g., Berissi, Groner, and Revel 1971). Indeed, it could not be expected that translation of each species of mRNA required the presence of a corresponding unique species of initiation factor.

Demonstration of the formation by yeast of various aminoacyl-puromycins in reactions insensitive to inhibition by cycloheximide (Melcher 1971) markedly reduces the significance of acetyl-N-glycyl-puromycin formation by hen oviduct minces, discussed previously. Conversely, it serves to strengthen our belief that all peptide chains are initiated from N-methionyl-$tRNA_{F*}^{Met}$ in eucaryote cytoplasm.

Incorporation of initiation complexes into membrane-bound polysomes appears to result from association with large 60S subunits already attached to the membrane (Baglioni, Bleiberg, and Zauderer 1971). However, the mechanism which selects the initiation complexes to be bound remains to be determined. One obvious possibility is that the selection is determined by the specificities of initiation factors $\mathbf{F}_3$ or $\mathbf{M}_3$, respectively.

Hradec and co-workers (1971) have further extended their studies and report the reversible inactivation of eucaryote transfer factors by removal of cholesteryl·14·methyl-decanoate.

DETERMINATION, GENE AMPLIFICATION, AND THE EXTRANUCLEAR TRANSPORT OF DNA (CHAPTER 2)

It remains an attractive hypothesis that differentiation may be initiated reversibly by limited amplification of a very restricted portion of the genome, either by replication or by the action of an RNA-dependent DNA polymerase ("reverse transcriptase") of (nearly) absolute specificity for the corresponding cRNA templates. Subsequent extranuclear transport of the additional gene copies and continued replication in the cytoplasm as plasmids could serve to "fix" (i.e., determine and maintain) the differentiated state. Indeed, as noted previously, there is evidence compatible with the hypothesis that immunoglobulin synthesis is correlated with gene amplification in the antibody-forming cells. Moreover, suggestive evidence has now been obtained that formation of the supplementary nucleoli in maturing oocytes of *X. laevis* may require the activity of a reverse transcriptase (Crippa and Tocchini-Valentini 1971, Ficq and Brachet 1971).

A recent study of the DNA polymerase enzymes of rat liver by Baril et al. (1971) therefore assumes considerable importance. These workers report that the membrane fraction of the cell contains a DNA polymerase which, at least after partial purification, differs in properties from the homologous enzymes of the nucleus and mitochondria or associated with the ribosome fraction. The significance of this observation has yet to be established, for the possibility that it is an artifact, and particularly a partially degraded species of molecule attached to a fragment of nuclear membrane, has not been eliminated.

Additional reports have appeared of the presence in eucaryote cytoplasm of membrane-bound, non-mitochondrial DNA, and of the extranuclear transport of DNA replicated within the cell nucleus (e.g., Bell 1971, Koch and Von Pfeil 1971, Lerner, Meinke, and Goldstein 1971). However, as yet there is no persuasive evidence that any of these materials are other than segments of nuclear DNA attached to fragments of nuclear membrane. Moreover, although the "I-somes" reported present in the cytoplasm of embryonic chick muscle (Bell 1969) contain a DNA segment large enough to serve as a message for a peptide chain of some 120–150 amino acid residues, they are far too small to code for the large myosin heavy chains.

INDUCED ENZYME SYNTHESIS AND PHENOMENA OF REPRESSION (CHAPTERS 3 AND 4)

The most marked advances in the analysis of phenomena of induced enzyme synthesis and catabolite repression have come from development of appropriate *in vitro* systems. Consideration of these systems is deferred to the following section. However, it may be noted here that the repressor of the *gal* operon (Parks et al. 1971[b]) and the regulatory protein of the *ara* operon (Wilcox et al. 1971) of *E. coli* have both been partly purified and shown to complex specifically with DNAs containing the corresponding operons.

Formation of the repressor protein required for establishment of the lysogenic state in cells of *E. coli* exposed to bacteriophage λ appears to be controlled initially at a different promoter from that which regulates maintenance of the state of lysogeny (Reichardt and Kaiser 1971). The process also appears to be controlled by the complex of cAMP with cAMP-binding protein (see Hong, Smith, and Ames 1971).

TRANSCRIPTION AND DEGRADATION OF mRNA IN PROCARYOTES (CHAPTER 5)

Fabricant and Kennell (1971) have reported that ghosts of the bacteriophage T4 cause an abrupt and almost total arrest of protein synthesis in cells of *E. coli* for about 30 minutes. Therefore it is not necessary to assume that translation inhibitors and new species of initiation factors found in extracts of infected cells are pre-early proteins.

The existence of RNase V has recently been challenged. Bothwell and Apirion (1971) have found that under all experimental conditions the preparations of *E. coli* ribosomes contain at least as much residual RNase II activity as apparent RNase V activity. On the other hand, Holmes and Singer (1971) failed to find any activity with properties of RNase V in preparations from cells infected with phage T7. (However, see earlier discussion.) Both groups have therefore suggested that RNase V activity represents the combined action of an endonuclease and RNase II. Nevertheless, this suggestion does not eliminate the special features and requirements of RNase V activity.

The possibility that individual mRNA molecules have a fixed life-span has been further examined by analysis of the kinetics of induced synthesis of β-galactosidase by *E. coli* (Coffman, Norris, and Koch 1971, Jacquet and Kepes 1971). It has been concluded that degradation of individual mRNA molecules is initiated at random, either by the action of an endonuclease or by incorporation into the polysome of a particular type of "killer" ribosome. In view of the extreme precision required in analyses of this type, I find the evidence presented inadequate to eliminate either the "ticket" model or the hypothesis of random inactivation. Moreover, the translated operator-proximal portion of a polycistronic mRNA containing a polar nonsense codon is degraded equally readily by extracts of wild-type cells and those of a mutant lacking the endonuclease which normally attacks the untranslated operator-distal portion of that mRNA (Kuwano, Schlessinger, and Morse 1971). Therefore, I consider that the question whether each mRNA molecule has a fixed life-span remains open.

The observation that uptake of proflavine by cells of *E. coli* is inhibited in the presence of cAMP (Conde, Del Campo, and Ramirez 1971) invalidates analyses of the functions of cAMP based upon use of proflavine as an inhibitor of gene transcription. Thus, there is only the indirect evidence of Yudkin (1969) to indicate that cAMP *may* stimulate the translation of mRNA of the *lac* operon of *E. coli* independently of the stimulation of transcription of the mRNA.

Further analyses have partially resolved some of the difficulties posed by early failures to demonstrate regulation of transcription of the *E. coli lac* operon by the cognate repressor *in vitro*. These failures still remain to be explained fully. However, control of *in vitro* transcription of this operon by the interplay of repressor, inducer, cAMP, and cAMP-binding protein is now consistently obtained with preparations of DNA containing the *lac* operon isolated from a different strain of phage λ (de Crombrugghe et al. 1971, Eron and Block 1971). The difference between the two strains of virus is in the orientation of the *lac* operon in the genome, and hence in the DNA strand from which the *lac* mRNA is transcribed.

Regulated *in vitro* transcription and synthesis of the enzymes of the *gal* operon of *E. coli* have been reported (Nissley et al. 1971, Parks et al. 1971[a], Wetekam, Staack, and Ehring 1971). Induced *in vitro* synthesis of enzymes of the *ara* operon of *E. coli* has also been reported (Greenblatt and Schleif 1971, Zubay, Gielow, and Englesberg 1971). However, under the experimental conditions which appear to yield the larger quantity of enzyme there was no consistent requirement for the protein specified by the regulator gene *ara C*.

Additional evidence has been obtained in support of the hypothesis that σ factor may facilitate localized denaturation of promoter regions of an RNA (Ishihama et al. 1971, Kosaganov et al. 1971), and of the course of modification of the *E. coli* RNA polymerase during development of phage T4 (Seifert, Rabussay, and Zillig 1971).

DEVELOPMENT OF X. LAEVIS AND EFFECTS OF HORMONES (CHAPTER 6)

Recent analyses show that transcription of the 5S RNA of the ribosomes of *X. laevis* is concurrent with that of the precursor of the larger rRNAs but not coordinate, and the former is synthesized in large excess (Abe and Yamana 1971, Ford 1971).

Development of a method for the assay of a specific eucaryote mRNA in the cytoplasm of *X. laevis* oocytes (Gurdon et al. 1971, Lane, Marbaix, and Gurdon 1971) has permitted a conclusive demonstration that the rate of protein synthesis in the oocyte is normally limited by the level of available endogenous mRNA (Moar et al. 1971).

In mice (at least) the X-linked condition of testicular feminization is correlated with a depressed level of testosterone-receptor protein specifically in cells of the kidney (Dofuku, Tettenborn, and Ohno 1971, Gehring, Tomkins, and Ohno 1971). From this observation, Ohno (1971) has postulated that hormone-receptor proteins may be allosteric proteins with two important regulatory roles. It is suggested that in the absence of hormone the protein serves as a cytoplasmic repressor of translation, whereas when complexed with hormone it serves as an activator of RNA polymerase action.

The suspicion that the mechanism of repression of enzyme formation in eucaryote cells by glucose is different from that in procaryote cells has been justified in the demonstration that the increase in hepatic level of cAMP elicited in response to glucagon is not depressed by glucose concurrently with the suppression of induced synthesis of hepatic serine dehydratase and tyrosine aminotransferase (Sudilovsky et al. 1971).

CHROMOSOME STRUCTURE AND REGULATION OF DIFFERENTIAL GENE ACTIVITY (CHAPTER 7)

Crick (1971) has recently presented an important speculative model structure for chromosomes of higher organisms. It is proposed that they contain small segments of "fibrous" 2-stranded DNA coding for specific proteins, and larger segments of "globular" unpaired regions which serve regulatory functions. These are manifest in the giant polytene chromosomes of dipteran salivary glands as the interbands and bands, respectively. This model has major implications in the interpretation of observations made with these organelles.

Note may also be made of further conflicting reports that chromatin-associated RNA may serve as a regulator of gene transcription (Kanehisha et al. 1971) or may be entirely artifact (Artman and Roth 1971, Hill, Poccia, and Doty 1971). In addition, Davidson and Britten (1971) have extended their model of gene regulation to suggest that the phenomenon of commitment reflects the activity of a "master integrator gene set."

REGULATION OF mRNA FORMATION IN EUCARYOTES (CHAPTER 8)

Several further studies of the RNA polymerases of eucaryotes have been reported. Two distinct species of nucleoplasmic enzyme have been isolated from rat liver and calf thymus, and the content of large subunits determined (Chesterton and Butterworth 1971[b], Kedinger, Nuret, and Chambon 1971, Mandel and Chambon 1971). These and other studies (Chesterton and Butterworth 1971[a]) serve to indicate that the nucleoplasmic species of enzyme are not derived from the nucleolar enzyme by addition of a sigma-like factor. One of the RNA polymerases previously considered to be of type III appears to be of mitochondrial origin (Horgen and Griffin 1971). At least in the case of *N. crassa,* the mitochondrial species consists of a single peptide chain (Küntzel and Schäfer 1971).

Darnell et al. (1971) have now obtained evidence that the polyA-rich sequences of nuclear cRNAs are attached after completion of transcription of the RNA, and suggest that these sequences define those species of cRNAs which will be processed to yield functional mRNAs. In keeping with this suggestion, the bulk of the nuclear cRNAs probably do not contain segments of polyA, whereas much of the polysomal mRNA does contain such a sequence (Lee, Mendecki, and Brawerman 1971, Lim and Canellakis 1970). Moreover, the polyA sequences appear to be partially removed before or during transport of the mRNAs to the cytoplasmic polysomes (Edmonds, Vaughan, and Nakazato 1971), and a cytoplasmic 17S RNA may possibly be a precursor of the 9S mRNAs of hemoglobin peptide chains (Maroun, Driscoll, and Nardone 1971). Nevertheless, the low specificity of the polyA polymerase enzyme for primer RNAs (Mans and Walter 1971, Twu and Bretthauer 1971) would appear to require that selection of cRNAs to be processed into mRNAs precede attachment of the polyA segment. Rather, this base sequence may serve either to protect the processed mRNA sequences against further nuclease action, or facilitate the extranuclear transport of informosomes (see also Philipson et al. 1971).

Additional evidence has been obtained that mRNAs are transported to the polysomes as informosomes (Faiferman, Cornudella, and Pogo 1971). However, the efficient translation of isolated mRNA in the cytoplasm of *X. laevis* oocytes or fertilized eggs (Gurdon et al. 1971) indicates that the primary function of the informosome protein is probably not one of protecting the mRNA against nuclease action.

REGULATION OF TRANSLATION BY CHANGE IN CONFORMATION OF NASCENT PEPTIDE (CHAPTER 9)

Roodman and Greenberg (1971[a], [b]) have observed an apparent effect of the conformation of the nascent peptide chain upon formation of thymidylate synthetase by a temperature-sensitive mutant strain of *E. coli.* When cells growing at a "permissive" temperature (25°C) are raised to the non-permissive temperature (37°C) the formation of active enzyme is markedly impaired. After the return to the lower temperature there is an initial burst of formation of enzyme which consists primarily of preformed polypeptide. However, the process requires peptide bond formation, and it appears probable that the non-permissive temperature causes almost-completed peptide chains to assume a conformation which precludes chain completion and thereby "jams" the polysomes.

Additional evidence has been obtained that histidyl-tRNA induces a conformational change in the enzyme specified by the leading gene transcribed from the *his* operon of *S. typhimurium* (Blasi et al. 1971).

CELL REPLICATION IN DIFFERENTIATION OF THE MAMMARY GLAND (CHAPTER 10)

Synthesis of caseins and α-lactalbumin by mid-term mammary glands differentiating in organ culture is blocked when the proliferating epithelial cells are exposed to 5·bromodeoxyuridine during one cell cycle. This appears to indicate that only the newly replicated, daughter "sense" strand is expressed in the subsequent formation of milk proteins (Turkington, Majumder, and Riddle 1971). The nature of the difference between daughter and parental chromosomes responsible for this differential gene expression is not known, but it is not due to any detectable change in the pattern of methylation of the DNA bases (Turkington and Spielvogel 1971).

ROLE OF IMMUNOGENIC RNA (CHAPTER 11)

A more attractive hypothesis of the mode of action of immunogenic RNA is raised by recent claims of the presence of "reverse transcriptase" enzyme in normal cells, and reports of its probable role in formation of supplementary nucleoli in maturing oocytes of *X. laevis* (see above). The antibody formed by lymphocytes exposed to the immunogenic RNA exhibits the allotype specificity characteristic of the variable region of immunoglobulin chains synthesized by the donor animal from which it was isolated (Adler et al. 1966, Bell and Dray 1969). Such RNAs may represent mRNA segments "coding" for the variable portions of antibody chains (see also Duke et al. 1971). The possibility exists that a corresponding segment of DNA is transcribed within the recipient cell by the action of a reverse transcriptase and then incorporated into an appropriate site within the genome, in a manner analogous to that envisioned for oncogenic RNA-viruses. Such a process would both account for the phenomena of commitment and antigenic "memory," and concur with evidence that antibody formation is correlated with gene amplification.

GLOSSARY

Introductory courses in biochemistry and biology are prerequisites for the course on which this book is based. This compilation, therefore, is restricted to the less common terms not defined or explained at the first encounter in the text, and also listed in the index.

Allosteric effect (*transition*) A reversible change in conformation of a protein (e.g., enzyme) caused by binding of a small *effector* molecule to an *allosteric site*. It modifies the interaction of the protein with a third type of molecule (e.g., substrate) at a different site (e.g., catalytic site).

Allotype (*specificity*) is defined by antigenic groups present in immunoglobulin chains produced by some animals of a given species, but not in homologous chains formed by other animals of the same species. Presence or absence of a given allotype marker is controlled by a defined gene allele.

Anadromous fish swim up river to spawn.

Aneuploidy A state in which the cell nuclei do not contain an exact multiple of the haploid number of chromosomes.

Ascites tumors grow as populations of dispersed single cells.

Balbiani ring A "puffed" region of a giant (*polytene*) chromosome in the salivary gland of larvae of some species of Diptera (e.g., *Chironomus tentans*).

Blastodisc A disc-shaped layer of cells formed by the cleavage of avian eggs from which the embryo proper arises.

Blastula An early stage of embryonic development consisting of a hollow sphere of cells.

"*Bobbed mutants*" of *Drosophila melanogaster* form abnormally small bristles and have deletions of genes for ribosomal RNAs from the nucleolar organizer region.

Effector A small molecule which elicits a change in conformation of an allosteric protein through complex formation at a specific allosteric site.

Embryonic organizer A portion of a developing embryo which stimulates another portion of the embryo to undergo determination and differentiation.

Endometrium The glandular lining of the mammalian uterus.

Endomitosis Chromosome replication within a cell nucleus which does not divide.

Exclusion The mechanism by which bacterial cells prevent productive infection by potentially lytic bacterial viruses.

Gastrula The stage of embryonic development at which progenitor cells of the internal organs migrate to approximately their definitive position within the cell aggregate.

Halophilic bacteria can reproduce only in an environment of high salt concentration.

Heterocaryon A multinucleate cell containing nuclei of differing genotype.

Heterogenote A bacterial cell containing one complete genome plus a partial genome bearing different gene alleles.

Immunity region A segment of the genome of bacteriophage λ which spans the sites of initiation of transcription (promoters) of the "pre-early" mRNAs and includes the structural gene "coding" for the bacteriophage repressor which prevents initiation of transcription from those sites.

Intercistronic divides Sequences of untranslated nucleotide bases postulated to be interspersed between the "coding" segments of a polycistronic mRNA.

Interspecies syncaryon A single cell formed by the fusion of two or more cells from different species in which the combined genome is enclosed within a single nucleus.

Lysogenic (lysogenized) bacterium A bacterium containing the prophage genome of a temperate virus as a component of the host cell chromosome.

Merogenote A partially diploid bacterial cell containing one complete genome plus a partial genome carried by an episome sex factor.

Myometrium The muscular tissue of the uterine wall.

Neurula The stage of development of a vertebrate embryo at which the neural axis is fully formed.

Nonpermissive bacterium A bacterium lacking suppressor genes which permit formation of "wild-type" protein corresponding to a structural gene containing a "nonsense" codon.

Permissive bacterium A bacterium containing an appropriate "nonsense"-suppressor gene to permit formation of some "wild-type" protein corresponding to a structural gene containing a "nonsense" codon.

Prophage The genome of a temperate bacteriophage as integrated into the chromosome of the host bacterial cell.

Pseudo-allele A term introduced to describe genes which appear to be alleles unable to complement each other's action but can be separated by gene recombination.

Sex-duction The transfer from one bacterial cell to another by an episome sex factor of genes originally derived from a bacterial chromosome.

Spheroplast A microbial cell from which most, or all, of the cell wall has been removed by enzymic digestion. This term is used in lieu of protoplast when it has not been formally established that the digestion removes all cell wall material.

Temperate (lysogenic) bacteriophage A bacterial virus of which the genome may either be incorporated into the chromosome of the host cell or may support a lytic cycle of virus reproduction, according to the physiological conditions pertaining in the host cell.

Thermophilic bacterium A bacterium which can grow at temperatures in the range of 45 to 65°C.

Transducing phage A defective (lysogenic) bacterial virus which contains bacterial genes in lieu of some viral genes because of an error in the process of excision of the prophage genome from the chromosome of the host cell.

Transduction The transfer of genes from the chromosome of one bacterial cell into the genome of another bacterium by means of a (lysogenic) bacterial virus.

Transformed cell (viral) An eucaryote cell in which viral genes have been incorporated into the host cell chromosomes.

BIBLIOGRAPHY OF QUOTATIONS

Quotations used as epigraphs for preface and chapters are taken from the following sources:

Preface

Claudius Galenus. De Anatomicis Administrationibus. Book IV. Translation by Charles Singer, Publication of the Wellcome Historical Medical Museum, Oxford University Press, London, 1956.

Francis Bacon. Essays, 50, Of studies (first edition). From *Bacon's Essays—with Annotations,* by R. Whately, Longmans, Green, Reader and Dyer, London, 1867.

Chapter 1

Aristotle. De Generatione Animalium. Book II-I. Translation by A. Platt, Oxford University Press, London, First edition, 1912.

Theophrastus von Hohenheim (Paracelsus). Book of Nymphs, Sylphs, Pygmies, Salamanders and Kindred Beings. Translation by Manly P. Hall in *The Mystical and Medical Philosophy of Paracelsus,* the Philosophical Research Society, Inc., Los Angeles, 1964.

Chapter 2

Louis Pasteur. Leçons de chimie professées en 1860. Alembic Club Reprint No. 14. Livingstone, Edinburgh, 1948.

Leonardo da Vinci. In the *Codice Atlantico.* From *Leonardo da Vinci, His Notebooks.* Translation by E. McCurdy, Empire State Book Company, New York, 1923.

Chapter 3

Hippocrates. Aphorisms II-L. Translation by F. Adams, The Sydenham Society, Adlard, London, 1849.

Antony van Leeuwenhoek. Letter 144 in *Antony van Leeuwenhoek and His 'Little Animals.'* Clifford Dobell, Constable and Company, London, 1932, as re-issued by Dover Publications, Inc., New York, 1960.

Chapter 4

Robert Hooke. Micrographia. Observation LI. Facsimile reproductions, Brussels, 1965.

H. Fabricius. De Formato Foetu in *The Embryological Treatises of Fabricius.* Facsimile edition. Translation by H. B. Adelman, Cornell University Press, Ithaca, 1942.

Chapter 5

Luke, "the beloved physician." Luke 7:27.

A. Weismann. Essays upon Heredity and Kindred Biological Problems. Vol. I. From *Great Experiments in Biology,* Editors M. L. Gabriel and S. Fogel, Prentice-Hall, Englewood Cliffs, 1955.

Chapter 6

L. Spallanzani. Expériences pour servir à l'histoire des animaux et des plantes. From *Great Experiments in Biology,* Editors M. L. Gabriel and S. Fogel, Prentice-Hall, Englewood Cliffs, 1955.

Albrecht von Haller. First Lines of Physiology. Vol. 2. Translation by H. A. Weisberg, Johnson Reprint Corporation, New York, 1966.

Chapter 7

H. Fabricius. De Formatione Ovi et Pulli in *The Embryological Treatises of Fabricius.* Facsimile edition. Translation by H. B. Adelman, Cornell University Press, Ithaca, 1942.

X. Bichat. Recherches physiologiques sur la vie et la mort. Third edition. Husson, Paris, 1805.

Chapter 8

Aristotle. De Generatione Animalium. Book II. Translation by A. Platt, Oxford University Press, London, First edition, 1912.

Claudius Galenus. De Anatomicis Administrationibus. Translation of the later books by W. L. H. Duckworth, Editors M. C. Lyons and B. Towers, Cambridge University Press, Cambridge, 1962.

Chapter 9

A. Cornelius Celsus. De medecina. Book III. Translation by W. G. Spencer, Loeb Classical Library, W. Heinemann, London, 1935.

Theophrastus von Hohenheim (Paracelsus). On the Miner's Sickness and Other Miner's Diseases. From *Four Treatises of Theophrastus von Hohenheim called*

Paracelsus. Translation by G. Rosen, Editor H. E. Sigerist, Johns Hopkins Press, Baltimore, 1941.

Chapter 10

William Harvey. Excercitationes de Generatione Animalium. Exercise the Forty-Fifth. From *The Works of William Harvey.* Translation by R. Willis, Johnson Reprint Corporation, New York, 1965.

Claude Bernard. De la physiologie générale. Culture and Civilization Reprint, Brussels, 1965.

Chapter 11

Lucretius. The Nature of the Universe. Book III. Translation by R. E. Latham, Penguin Books, Harmondsworth, 1951.

H. Fabricius. De Formatione Ovi et Pulli. From *The Embryological Treatises of Fabricius.* Facsimile edition. Translation by H. B. Adelman, Cornell University Press, Ithaca, 1942.

Chapter 12

Omar Khayyám. The Rubáiyát. XXXII. Translation by E. Fitzgerald, Outing Publishing Company, New York.

BIBLIOGRAPHY

Aaij, C., Saccone, C., Borst, P. and Gadaleta, M. N. 1970. Biochim. Biophys. Acta *199*, 373.

Abelson, J., Barnett, L., Brenner, S., Gefter, M., Landy, A., Russell, R. and Smith, J. D. 1969. FEBS Letters *3*, 1.

Abelson, J. and Thomas, C. A. 1966. J. Mol. Biol. *18*, 262.

Aboud, M. and Burger, M. 1970. Biochem. Biophys. Res. Commun. *38*, 1023.

Aboud, M. and Burger, M. 1971. Biochem. Biophys. Res. Commun. *43*, 174.

Abrell, J. W., Kaufman, E. E. and Lipsett, M. N. 1971. J. Biol. Chem. *246*, 294.

Acs, G., Reich, E. and Valanja, S. 1963. Biochim. Biophys. Acta *76*, 68.

Ada, G. L., Parish, C. R., Nossal, G. J. V. and Abbot, A. 1967. Cold Spring Harbor Symp. Quant. Biol. *32*, 381.

Adams, G. H. M., Vidali, G. and Neelin, J. M. 1970. Can. J. Biochem. *48*, 33.

Adams, J. M. 1968. J. Mol. Biol. *33*, 571.

Adams, J. M. and Cory, S. 1970. Nature *227*, 570.

Adams, J. M., Jeppesen, P. G. N., Sanger, F. and Barrell, B. G. 1969. Nature *223*, 1009.

Adamson, S. D., Herbert, E. and Godchaux, W. 1968. Arch. Biochem. Biophys. *125*, 671.

Adamson, S. D., Herbert, E. and Kemp, S. F. 1969. J. Mol. Biol. *42*, 247.

Adelman, R. C., Lo, C. H. and Weinhouse, S. 1968. J. Biol. Chem. *243*, 2538.

Adesnik, M. and Levinthal, C. 1969. J. Mol. Biol. *46*, 281.

Adesnik, M. and Levinthal, C. 1970. J. Mol. Biol. *48*, 187.

Adhya, S. and Echols, H. 1966. J. Bact. *92*, 601.

Adhya, S. L. and Shapiro, J. A. 1969. Genetics *62*, 231.

Ahmed, A. 1968. Molec. Gen. Genetics *103*, 185.

Ahmed, A., Case, M. E. and Giles, N. H. 1964. Brookhaven Symp. Biol. *17*, 53.

Aitkhozin, M. A., Belitsina, N. V. and Spirin, A. S. 1964. Biokhimiya (*Trans.*) *29*, 145.

Akamatsu, N., Maeda, H., Kamiyama, T. and Muira, Y. 1969. J. Biochem. *66*, 101.
Akino, T. and Amano, M. 1970. J. Biochem. *67*, 533.
Akino, T., Amano, M. and Mitsui, H. 1969. J. Biochem. *66*, 867.
Alberghina, F. A. M., Sturani, E., Ursino, D. J. and Marrè, E. 1969. Biochim. Biophys. Acta *195*, 576.
Albrecht, J., Stap, F., Voorma, H. O., van Knippenberg, P. H. and Bosch, L. 1970. FEBS Letters *6*, 297.
Alescio, T., Moscona, M. and Moscona, A. A. 1970. Exp. Cell Res. *61*, 342.
Algranati, I. D. 1970. FEBS Letters *10*, 153.
Algranati, I. D., Gonzalez, N. S. and Bade, E. G. 1969. Proc. Natl. Acad. Sci. *62*, 574.
Allende, J. E., Seeds, N. W., Conway, T. W. and Weissbach, H. 1967. Proc. Natl. Acad. Sci. *58*, 1566.
Allfrey, V. G. 1968. *Regulatory Mechanisms for Protein Synthesis in Mammalian Cells*. Editors A. S. Pietro, M. R. Lamborg and F. T. Kenney, Academic Press, New York, p. 65.
Allfrey, V. G., Daly, M. M. and Mirsky, A. E. 1953. J. Gen. Phys. *37*, 157.
Allfrey, V. G., Faulkner, R. and Mirsky, A. E. 1964. Proc. Natl. Acad. Sci. *51*, 786.
Allfrey, V. G. Littau, V. C. and Mirsky, A. E. 1963. Proc. Natl. Acad. Sci. *49*, 414.
Allfrey, V. G. and Mirsky, A. E. 1957. Proc. Natl. Acad. Sci. *43*, 821.
Allfrey, V. G. and Mirsky, A. E. 1962. Proc. Natl. Acad. Sci. *48*, 1590.
Allfrey, V. G., Pogo, B. G. T., Pogo, A. O., Kleinsmith, L. J. and Mirsky, A. E. 1966. *Histones*. Ciba Foundation, Symposium 24. Editors A. V. S. de Reuck and J. Knight, Churchill, London, p. 42.
Aloni, Y. and Attardi, G. 1971(a). J. Mol. Biol. *55*, 251.
Aloni, Y. and Attardi, G. 1971(b). J. Mol. Biol. *55*, 271.
Aloni, Y., Winocour, E. and Sachs, L. 1968. J. Mol. Biol. *31*, 415.
Alpers, D. H. and Tomkins, G. M. 1965. Proc. Natl. Acad. Sci. *53*, 797.
Alpers, D. H. and Tomkins, G. M. 1966. J. Biol. Chem. *241*, 4434.
Ames. B. N. and Garry, B. 1959. Proc. Natl. Acad. Sci. *45*, 1453.
Ames, B. N., Goldberger, R. F., Hartman, P. E., Martin, R. G. and Roth, J. R. 1967. *Regulation of Nucleic Acid and Protein Biosynthesis*. Editors V. V. Koningsberger and L. Bosch, Elsevier, Amsterdam, p. 272.
Ames, B. N. and Hartman, P. E. 1963. Cold Spring Harbor Symp. Quant. Biol. *28*, 349.
Ames, B. N., Hartman, P. E. and Jacob, F. 1963. J. Mol. Biol. *7*, 23.
Ames, B. N., Martin, R. G. and Garry, B. J. 1961. J. Biol. Chem. *236*, 2019.
Ames, B. N. and Whitfield, H. J. 1966. Cold Spring Harbor Symp. Quant. Biol. *31*, 221.
Anderson, J. S., Dahlberg, J. E., Bretscher, M. S., Revel, M. and Clark, B. F. C. 1967. Nature *216*, 1072.
Anderson, M. B. and Cherry, J. H. 1969. Proc. Natl. Acad. Sci. *62*, 202.
Anderson, W. F. and Gilbert, J. M. 1969. Cold Spring Harbor Symp. Quant. Biol. *34*, 585.
Ando, T. and Hashimoto, C. 1958. J. Biochem. *45*, 529.
Ando, T. and Sawada, F. 1961. J. Biochem. *49*, 252.
Ando, T. and Suzuki, K. 1966. Biochim. Biophys. Acta *121*, 427.
Ando, T. and Suzuki, K. 1967. Biochim. Biophys. Acta *140*, 375.
Ando, T. and Watanabe, S. 1969. Int. J. Protein Res. *1*, 221, cited in Ling, Jergil and Dixon.
Andoh, T. and Garen, A. 1967. J. Mol. Biol. *24*, 129.
Andoh, T., Natori, S. and Mizuno, D. 1963. J. Biochem. *54*, 339.

Andoh, T. and Ozeki, H. 1968. Proc. Natl. Acad. Sci. *59*, 792.
Anfinsen, C. B. 1959. *The Molecular Basis of Evolution.* John Wiley and Sons, Inc., New York.
Anfinsen, C. B. 1967. Harvey Lectures *61*, 95.
Antón, D. N. 1968. J. Mol. Biol. *33*, 533.
Appella, E. and Perham, R. N. 1967. Cold Spring Harbor Symp. Quant. Biol. *32*, 37.
Applebaum, S. W., Ebstein, R. P. and Wyatt, G. R. 1966. J. Mol. Biol. *21*, 29.
Arber, W. 1965(a). J. Mol. Biol. *11*, 247.
Arber, W. 1965(b). Ann. Rev. Microbiol. *19*, 365.
Arceneaux, J. L. and Sueoka, N. 1969. J. Biol. Chem. *244*, 5959.
Arditti, R. R., Scaife, J. G. and Beckwith, J. R. 1968. J. Mol. Biol. *38*, 421.
Arion, V. J., Mantieva, V. L. and Georgiev, G. P. 1967. Biochim. Biophys. Acta *138*, 436.
Arlinghaus, R., Shaeffer, J. and Schweet, R. 1964. Proc. Natl. Acad. Sci. *51*, 1291.
Armelin, H. A., Meneghini, R. and Lara, F. J. S. 1969. Genetics *61*, Supp. *1*, 351.
Armentrout, S. A., Schmickel, R. D. and Simmons, L. R. 1965. Arch. Biochem. Biophys. *112*, 304.
Arnaud, M., Beziat, Y., Guilleux, J. C., Hough, A., Hough, D. and Mousseron-Canet, M. 1971. Biochim. Biophys. Acta *232*, 117.
Arnheim, N. and Taylor, C. E. 1969. Nature *223*, 900.
Aronson, A. I. and Wilt, F. H. 1969. Proc. Natl. Acad. Sci. *62*, 186.
Aronson, M. 1952. Arch. Biochem. Biophys. *39*, 370.
Arora, D. J. S. and De Lamirande, G. 1967. Can. J. Biochem. *45*, 1021.
Asao, T. 1970. Exp. Cell Res. *61*, 255.
Ascione, R. and Woude, G. F. V. 1969. J. Virol. *4*, 727.
Ashworth, J. M. and Sackin, M. J. 1969. Nature *224*, 817.
Askonas, B. A. and Rhodes, J. M. 1965. Nature *205*, 470.
Askonas, B. A. and Williamson, A. R. 1966. Nature *211*, 369.
Askonas, B. A. and Williamson, A. R. 1967. *Gamma Globulins. Structure and Control of Biosynthesis.* Nobel Symposium *3*. Editor J. Killander, Almqvist and Wiksell, Stockholm, p. 369.
Askonas, B. A. and Williamson, A. R. 1968. Biochem. J. *109*, 637.
Attardi, B. and Attardi, G. 1967. Proc. Natl. Acad. Sci. *58*, 1051.
Attardi, B. and Attardi, G. 1969. Nature *224*, 1079.
Attardi, G. 1963. *Biological Organization at the Cellular and Supercellular Level.* Editor R. J. C. Harris, Academic Press, London, p. 43.
Attardi, G., Naono, S., Rouvière, J., Jacob, F. and Gros, F. 1963. Cold Spring Harbor Symp. Quant. Biol. *28*, 363.
Attardi, G., Parnas, H., Hwang, M. I. H. and Attardi, B. 1966. J. Mol. Biol. *20*, 145.
Atwood, K. C. Genetics *61*, Supp. *1*, 319.
Aurbach, G., Perlman, R. and Pastan, I. 1971. Cited in Pastan and Perlman 1970.
Auricchio, F., Martin, D. and Tomkins, G. 1969. Nature *224*, 806.
Bachvarova, R., Davidson, E. H., Allfrey, V. G. and Mirsky, A. E. 1966. Proc. Natl. Acad. Sci. *55*, 358.
Bacon, D. F. and Vogel, H. J. 1963. Cold Spring Harbor Symp. Quant. Biol. *28*, 437.
Baglioni, C. 1963. *Molecular Genetics.* Editor J. H. Taylor, Academic Press, New York, Vol. *1*, p. 405.
Baglioni, C. 1970. Biochem. Biophys. Res. Commun. *38*, 212.
Baglioni, C. and Campana, T. 1967. Europ. J. Biochem. *2*, 480.

Baguley, B. C. and Staehelin, M. 1968. Europ. J. Biochem. *6,* 1.
Baguley, B. C. and Staehelin, M. 1969. Biochemistry *8,* 257.
Baker, R. F. and Yanofsky, C. 1968(a). Proc. Natl. Acad. Sci. *60,* 313.
Baker, R. F. and Yanofsky, C. 1968(b). Nature *219,* 26.
Baker, R. and Yanofsky, C. 1970. Cold Spring Harbor Symp. Quant. Biol. *35,* 467.
Baldwin, A. N. and Berg, P. 1966. J. Biol. Chem. *241,* 839.
Baliga, B. S., Borek, E., Weinstein, I. B. and Srinivasan, P. R. 1969. Proc. Natl. Acad. Sci. *62,* 899.
Baliga, B. S., Srinivasan, P. R. and Borek, E. 1965. Nature *208,* 555.
Balinsky, B. I. 1970. *An Introduction to Embryology,* 3rd edition. W. B. Saunders Co., Philadelphia.
Ballard, P. L. and Tomkins, G. M. 1969. Nature *224,* 344.
Ballard, P. L. and Tomkins, G. M. 1970. J. Cell Biol. *47,* 222.
Baltimore, D. and Huang, A. S. 1970. J. Mol. Biol. *47,* 263.
Bank, A. 1968. J. Clin. Invest. *47,* 860.
Bank, A. and Marks, P. A. 1966. Nature *212,* 1198.
Bargellesi, A., Pontremoli, S., Menini, C. and Conconi, F. 1969. Europ. J. Biochem. *7,* 73.
Barker, K. L. 1971. Biochemistry *10,* 284.
Barker, K. L. and Warren, J. C. 1966. Proc. Natl. Acad. Sci. *56,* 1298.
Barnea, A. and Gorski, J. 1970. Biochemistry *9,* 1899.
Barnett, W. E., Brown, D. H. and Epler, J. L. 1967. Proc. Natl. Acad. Sci. *57,* 1775.
Barondes, S. H., Dingman, C. W. and Sporn, M. B. 1962. Nature *196,* 145.
Barr, G. C. and Butler, J. A. V. 1963. Nature *199,* 1170.
Bartz, J. K., Kline, L. K. and Söll, D. 1970. Biochem. Biophys. Res. Commun. *40,* 1481.
Bassel, B. A. 1968. Proc. Natl. Acad. Sci. *60,* 321.
Bauerle, R. H. and Margolin, P. 1966. Proc. Natl. Acad. Sci. *56,* 111.
Bauerle, R. H. and Margolin, P. 1967. J. Mol. Biol. *26,* 423.
Bautz, E. K. F. 1962. Biochem. Biophys. Res. Commun. *9,* 192.
Bautz, E. K. F. 1963(a). Proc. Natl. Acad. Sci. *49,* 68.
Bautz, E. K. F. 1963(b). Cold Spring Harbor Symp. Quant. Biol. *28,* 205.
Bautz, E. K. F. 1966. J. Mol. Biol. *17,* 298.
Bautz, E. K. F. and Bautz, F. A. 1970. Nature *226,* 1219.
Bautz, E. K. F., Bautz, F. A. and Dunn, J. J. 1969. Nature *223,* 1022.
Bautz, E. K. F. and Hall, B. D. 1962. Proc. Natl. Acad. Sci. *48,* 400.
Bautz, E. K. F. and Heding, L. 1964. Biochemistry *3,* 1010.
Bautz, E. K. F., Kasai, T., Reilly, E. and Bautz, F. A. 1966. Proc. Natl. Acad. Sci. *55,* 1081.
Bautz, E. K. F. and Reilly, E. 1966. Science *151,* 328.
Bayley, S. T. and Kushner, D. J. 1964. J. Mol. Biol. *9,* 654.
Bayliss, W. M. and Starling, E. H. 1904(a). Proc. Roy. Soc. B 73, 310.
Bayliss, W. M. and Starling, E. H. 1904(b). Nature *70,* 65.
Beadle, G. W. and Tatum, E. L. 1941. Proc. Natl. Acad. Sci. *27,* 499.
Bear, P. D. and Skalka, A. 1969. Proc. Natl. Acad. Sci. *62,* 385.
Beattie, D. S., Basford, R. E. and Koritz, S. B. 1967. Biochemistry *6,* 3099.
Beattie, D. S. and Stuchell, R. N. 1970. Arch. Biochem. Biophys. *139,* 291.
Becker, Y. and Joklik, W. K. 1964. Proc. Natl. Acad. Sci. *51,* 577.
Beckwith, J. R. 1964. J. Mol. Biol. *8,* 427.
Beckwith, J. R. 1967. Science *156,* 597.
Beckwith, J. R., Signer, E. R. and Epstein, W. 1966. Cold Spring Harbor Symp. Quant. Biol. *31,* 393.

Beckwith, J. R. and Zipser, D. 1970. Editors, *The Lactose Operon*. Cold Spring Harbor Laboratory, New York.
Beermann, W. 1963. Amer. Zool. *3*, 23.
Bekhor, I., Bonner, J. and Dahmus, G. K. 1969. Proc. Natl. Acad. Sci. *62*, 271.
Bekhor, I., Kung, G. M. and Bonner, J. 1969. J. Mol. Biol. *39*, 351.
Belitsina, N. V., Aitkhozhin, M. A., Gavrilova, L. P. and Spirin, A. S. 1964. Biokhimiya (*Trans.*) *29*, 315.
Beljanski, M. 1960. Biochim. Biophys. Acta *41*, 104.
Beljanski, M. 1965. Biochem. Zeit. *342*, 392.
Beljanski, M., Beljanski, M. and Lovigny, T. 1962. Biochim. Biophys. Acta *56*, 559.
Beljanski, M., Fischer-Ferraro, C. and Bourgarel, P. 1968. Europ. J. Biochem. *4*, 184.
Bell, E. 1969. Nature *224*, 326.
Benjamin, W. B. and Goodman, R. M. 1969. Science *166*, 629.
Benjamin, W., Levander, O. A., Gellhorn, A. and De Bellis, R. H. 1966. Proc. Natl. Acad. Sci. *55*, 858.
Benzer, S. 1953. Biochim. Biophys. Acta *11*, 383.
Benzer, S. 1957. *The Chemical Basis of Heredity*. Editors W. D. McElroy and B. Glass, Johns Hopkins Press, Baltimore, p. 70.
Benzinger, R. and Hartman, P. E. 1962. Virology *18*, 614.
Berberich, M. A., Venetianer, P. and Goldberger, R. F. 1966. J. Biol. Chem. *241*, 4426.
Berg, D. and Chamberlin, M. 1970. Biochemistry *9*, 5055.
Berggård, I. and Peterson, P. 1967. *Gamma Globulins, Structure and Control of Biosynthesis*. Nobel Symposium *3*. Editor J. Killander, Almqvist and Wiksell, Stockholm, p. 71.
Bernhardt, D. and Darnell, J. E. 1969. J. Mol. Biol. *42*, 43.
Beutler, E., Yeh, M. and Fairbanks, V. F. 1962. Proc. Natl. Acad. Sci. *48*, 9.
Billeter, M. A., Dahlberg, J. E., Goodman, H. M., Hindley, J. and Weissmann, C. 1969. Nature *224*, 1083.
Billing, R. J., Barbiroli, B. and Smellie, R. M. S. 1969. Biochim. Biophys. Acta *190*, 52.
Billingham, R. E. 1969. *Medicine in the University and Community of the Future*. Proceedings of the Scientific Sessions, Dalhousie University Faculty of Medicine Centennial. Whynot and Associates, Ltd., Halifax, p. 206.
Birnstiel, M., Speirs, J., Purdom, I., Jones, K. and Loening, U. E. 1968. Nature *219*, 454.
Bishop, D. C., Pisciotta, A. V. and Abramoff, P. 1967. J. Immunol. *99*, 751.
Bishop, J. O. 1966. J. Mol. Biol. *17*, 285.
Bishop, J., Leahy, J. and Schweet, R. 1960. Proc. Natl. Acad. Sci. *46*, 1030.
Björk, G. R. and Svensson, I. 1969. Europ. J. Biochem. *9*, 207.
Black, J. A. and Dixon, G. H. 1967. Nature *216*, 152.
Blatti, S. P., Ingles, C. J., Lindell, T. J., Morris, P. W., Weaver, R. F., Weinberg, F. and Rutter, W. J. 1970. Cold Spring Harbor Symp. Quant. Biol. *35*, 649.
Bleyman, M., Kondo, M., Hecht, N. and Woese, C. 1969. J. Bact. *99*, 535.
Bleyman, M. and Woese, C. 1969. J. Bact. *97*, 27.
Blobel, G. 1971. Proc. Natl. Acad. Sci. *68*, 832.
Blobel, G. and Sabatini, D. D. 1970. J. Cell Biol. *45*, 130.
Bloch, D. P. 1969. Genetics *61*, Supp. *1*, 93.
Blum, N., Maleknia, M. and Schapira, G. 1969. Biochim. Biophys. Acta *179*, 448.
Blum, N., Maleknia, M. and Schapira, G. 1970. Biochim. Biophys. Acta *199*, 236.
Boezi, J. A. and Cowie, D. B. 1961. Biophys. J. *1*, 639.

Bolle, A., Epstein, R. H., Salser, W. and Geiduscheck, E. P. 1968. J. Mol. Biol. *33,* 339.
Bolton, E. T. and McCarthy, B. J. 1962. Proc. Natl. Acad. Sci. *48,* 1390.
Bolton, E. T. and McCarthy, B. J. 1964. J. Mol. Biol. *8,* 201.
Bolund, L., Darźynkiewicz, Z. and Ringertz, N. R. 1969. Exp. Cell Res. *56,* 406.
Bolund, L., Ringertz, N. R. and Harris, H. 1969. J. Cell Sci. *4,* 71.
Bondy, S. C., Roberts, S. and Morelos, B. S. 1970. Biochem. J. *119,* 665.
Bonner, D. M. 1964. J. Exp. Zool. *157,* 9.
Bonner, J. 1965. *The Molecular Biology of Development.* Clarendon Press, Oxford.
Bonner, J. 1967. *Regulation of Nucleic Acid and Protein Biosynthesis.* Editors V. V. Koningsberger and L. Bosch, Elsevier, Amsterdam, p. 211.
Bonner, J., Chalkley, G. R., Dahmus, M., Fambrough, D., Fujimura, F., Huang, R. C., Huberman, J., Jensen, R., Marushige, K., Ohlenbusch, H., Olivera, B. and Widholm, J. 1968(a). *Methods in Enzymology,* Vol. XII B. Editors L. Grossman and K. Moldave, Academic Press, New York, p. 3.
Bonner, J., Dahmus, M. E., Fambrough, D., Huang, R. C., Marushige, K. and Tuan, D. Y. H. 1968(b). Science *159,* 47.
Bonner, J. and Huang, R. C. 1963. J. Mol. Biol. *6,* 169.
Bonner, J., Huang, R. C. and Gilden, R. V. 1963. Proc. Natl. Acad. Sci. *50,* 893.
Bonner, J., Huang, R. C. and Maheshwari, N. 1961. Proc. Natl. Acad. Sci. *47,* 1548.
Bonner, J. and Ts'o, P. O. P. (editors). 1964. *The Nucleohistones.* Holden-Day, San Francisco.
Bonner, J. and Widholm, J. 1967. Proc. Natl. Acad. Sci. *57,* 1379.
Bonner, J. T. 1970. Proc. Natl. Acad. Sci. *65,* 110.
Boos, W. 1969. Europ. J. Biochem. *10,* 66.
Borek, E. 1969. *Exploitable Molecular Mechanisms and Neoplasia.* Williams and Wilkins Co., Baltimore, p. 163.
Borek, E., Rockenbach, J. and Ryan, A. 1956. J. Bact. *71,* 318.
Borek, E., Ryan, A. and Rockenbach, J. 1955. J. Bact. *69,* 460.
Borst, P. and Aaij, C. 1969. Biochem. Biophys. Res. Commun. *34,* 358.
Borst, P. and Grivell, L. A. 1971. FEBS Letters *13,* 73.
Borun, T. W., Scharff, M. D. and Robbins, E. 1967. Proc. Natl. Acad. Sci. *58,* 1977.
Botsford, J. L. and DeMoss, R. D. 1971. J. Bact. *105,* 303.
Bourgeois, S., Cohn, M. and Orgel, L. E. 1965. J. Mol. Biol. *14,* 300.
Bøvre, K. and Szybalski, W. 1969. Virol. *38,* 614.
Brachet, J. 1942. Arch. Biol. (Liège) *53,* 207.
Brachet, J. and Denis, H. 1963. Nature *198,* 205.
Brachet, J., Ficq, A. and Tencer, R. 1963. Exp. Cell Res. *32,* 168.
Brachet, P., Eisen, H. and Rambach, A. 1970. Molec. Gen. Genetics *108,* 266.
Bradshaw, W. S. and Papaconstantinou, J. 1970. Biochem. Biophys. Res. Commun. *41,* 306.
Brasch, K., Seligy, V. L. and Setterfield, G. 1971. Exp. Cell Res. *65,* 61.
Brasch, K., Setterfield, G. and Neelin, J. M. 1971. In preparation.
Braun, R. and Behrens, K. 1969. Biochim. Biophys. Acta *195,* 87.
Braunitzer, G. 1966. J. Cell. Phys. *67,* Supp. *1,* 1.
Brawerman, G. 1969. Cold Spring Harbor Symp. Quant. Biol. *34,* 307.
Brawerman, G., Gold, L. and Eisenstadt, J. 1963. Proc. Natl. Acad. Sci. *50,* 630.
Brawerman, G., Revel, M., Salser, W. and Gros, F. 1969. Nature *223,* 957.
Breillatt, J. and Dickman, S. R. 1969. Biochim. Biophys. Acta *195,* 531.
Bremer, H. and Konrad, M. W. 1964. Proc. Natl. Acad. Sci. *51,* 801.
Bremer, H., Konrad, M. W., Gaines, K. and Stent, G. S. 1965. J. Mol. Biol. *13,* 540.
Bremer, H. and Yuan, D. 1968(a). J. Mol. Biol. *34,* 527.

Bremer, H. and Yuan, D. 1968(b). J. Mol. Biol. *38,* 163.
Brenner, S. and Beckwith, J. R. 1965. J. Mol. Biol. *13,* 629.
Brenner, S., Jacob, F. and Meselson, M. 1961. Nature *190,* 576.
Brenner, S. and Milstein, C. 1966. Nature *211,* 242.
Brentani, R. and Brentani, M. 1969. Genetics *61,* Supp. *1,* 391.
Bretscher, M. S. 1966. Cold Spring Harbor Symp. Quant. Biol. *31,* 289.
Bretscher, M. S. 1968(a). Nature *218,* 675.
Bretscher, M. S. 1968(b). J. Mol. Biol. *34,* 131.
Brewer, E. N., Foster, L. B. and Sells, B. H. 1969. J. Biol. Chem. *244,* 1389.
Bridges, C. B. 1936. Science *83,* 210.
Briggs, R., Signoret, J. and Humphrey, R. R. 1964. Dev. Biol. *10,* 233.
Britten, R. J. and Davidson, E. H. 1969. Science *165,* 349.
Britten, R. J. and Kohne, D. E. 1968. Science *161,* 529.
Brody, E. N. and Geiduschek, E. P. 1970. Biochemistry *9,* 1300.
Brody, S. and Yanofsky, C. 1965. J. Bact. *90,* 687.
Bronskill, P. M. and Wong, J. T. F. 1971. J. Bact. *105,* 498.
Brossard, M. and Nicole, L. 1969. Can. J. Biochem. *47,* 226.
Brot, N., Yamasaki, E., Redfield, B. and Weissbach, H. 1970. Biochem. Biophys. Res. Commun. *40,* 698.
Brown, D. D. 1967. *Current Topics in Developmental Biology.* Editors, A. A. Moscona and A. Monroy, Academic Press, New York, Vol. *2,* p. 47.
Brown, D. D. and Dawid, I. B. 1968. Science *160,* 272.
Brown, D. D. and Gurdon, J. B. 1964. Proc. Natl. Acad. Sci. *51,* 139.
Brown, D. D. and Gurdon, J. B. 1966. J. Mol. Biol. *19,* 399.
Brown, D. D. and Littna, E. 1964(a). J. Mol. Biol. *8,* 669.
Brown, D. D. and Littna, E. 1964(b). J. Mol. Biol. *8,* 688.
Brown, D. D. and Littna, E. 1966(a). J. Mol. Biol. *20,* 81.
Brown, D. D. and Littna, E. 1966(b). J. Mol. Biol. *20,* 95.
Brown, D. D. and Monod, J. 1961. Fed. Proc. *20,* 222.
Brown, D. D. and Weber, C. S. 1968(a). J. Mol. Biol. *34,* 661.
Brown, D. D. and Weber, C. S. 1968(b). J. Mol. Biol. *34,* 681.
Brown, D. G. and Abrams, A. 1970. Biochim. Biophys. Acta *200,* 522.
Brown, J. C. and Smith, A. E. 1970. Nature *226,* 610.
Brown, J. L., Brown, D. M. and Zabin, I. 1967. J. Biol. Chem. *242,* 4254.
Brown, S. W. 1969. Genetics *61,* Supp. *1,* 191.
Brownlee, G. G., Sanger, F. and Barrell, B. G. 1967. Nature *215,* 735.
Brownlee, G. G., Sanger, F. and Barrell, B. G. 1968. J. Mol. Biol. *34,* 379.
Bruner, R. and Cape, R. E. 1970. J. Mol. Biol. *53,* 69.
Buck, C. A. and Nass, M. M. K. 1969. J. Mol. Biol. *41,* 67.
Buell, D. N. and Fahey, J. L. 1969. Science *164,* 1524.
Bujard, H. 1969. Proc. Natl. Acad. Sci. *62,* 1167.
Burdick, C. J. and Himes, M. 1969. Nature *221,* 1150.
Burdon, R. H., Martin, B. T. and Lal, B. M. 1967. J. Mol. Biol. *28,* 357.
Burgess, R. R. 1969. J. Biol. Chem. *244,* 6160.
Burgess, R. R., Travers, A. A., Dunn, J. J. and Bautz, E. K. F. 1969. Nature *221,* 43.
Burka, E. R. 1968. Science *162,* 1287.
Burnet, F. M. 1959. *Clonal Selection Theory of Acquired Immunity.* Cambridge University Press, Cambridge.
Burny, A., Huez, G., Marbaix, G. and Chantrenne, H. 1969. Biochim. Biophys. Acta *190,* 228.
Burrows, W. J., Armstrong, D. J., Skoog, F., Hecht, S. M., Boyle, J. T. A., Leonard, N. J. and Occolowitz, J. 1968. Science *161,* 691.

Burstein, C., Cohn, M., Kepes, A. and Monod, J. 1965. Biochim. Biophys. Acta *95,* 634.
Busch, H. 1968. *Methods in Enzymology,* Vol. XII B. Editors L. Grossman and K. Moldave, Academic Press, New York, p. 65.
Bustin, M. and Cole, R. D. 1968. J. Biol. Chem. *243,* 4500.
Butler, J. A. V. 1966. *Histones,* Ciba Foundation Symposium 24. Editors A. V. S. de Reuck and J. Knight, Churchill, London, p. 4.
Buttin, G. 1963(a). J. Mol. Biol. *7,* 164.
Buttin, G. 1963(b). J. Mol. Biol. *7,* 183.
Buttin, G. 1963(c). J. Mol. Biol. *7,* 610.
Byrne, R., Levin, J. G., Bladen, H. A. and Nirenberg, M. W. 1964. Proc. Natl. Acad. Sci. *52,* 140.
Caffier, H., Raskas, H. J., Parsons, J. T. and Green, M. 1971. Nature New Biology *229,* 239.
Callan, H. G. 1966. J. Cell Sci. *1,* 85.
Callan, H. G. 1967. J. Cell Sci. *2,* 1.
Cameron, I. L. and Prescott, D. M. 1963. Exp. Cell Res. *30,* 609.
Cammack, K. A. and Wade, H. E. 1965. Biochem. J. *96,* 671.
Campbell, A. M. 1969. *Episomes.* Harper and Row, New York.
Campbell, J. 1967. J. Theoret. Biol. *16,* 321.
Candido, E. P. M. and Dixon, G. H. 1971. J. Biol. Chem. *246,* 3182.
Candido, E. P. M., Sung, M. T. and Dixon, G. H. 1971. Cited in Sung and Dixon 1970.
Canfield, R. E. and Anfinsen, C. B. 1963. Biochemistry *2,* 1073.
Cannon, M. 1967. Biochem. J. *104,* 934.
Cannon, M., Klug, R. and Gilbert, W. 1963. J. Mol. Biol. *7,* 360.
Cánovas, J. L., Ornston, L. N. and Stanier, R. Y. 1967. Science *156,* 1695.
Capecchi, M. R. 1966. J. Mol. Biol. *21,* 173.
Capecchi, M. R. 1967. Proc. Natl. Acad. Sci. *58,* 1144.
Capecchi, M. R. and Gussin, G. N. 1965. Science *149,* 417.
Capecchi, M. R. and Klein, H. A. 1969. Cold Spring Harbor Symp. Quant. Biol. *34,* 469.
Capecchi, M. R. and Klein, H. A. 1970. Nature *226,* 1029.
Capp, G. L., Rigas, D. A. and Jones, R. T. 1967. Science *157,* 65.
Capra, J. D. and Peterkofsky, A. 1968. J. Mol. Biol. *33,* 591.
Carbon, J., Berg, P. and Yanofsky, C. 1966. Proc. Natl. Acad. Sci. *56,* 764.
Carbon, J., David, H. and Studier, M. H. 1968. Science *161,* 1146.
Carter, T. and Newton, A. 1969. Nature *223,* 707.
Cartouzou, G., Attali, J. C. and Lissitzky, S. 1968. Europ. J. Biochem. *4,* 41.
Cartouzou, G., Poirée, J. C. and Lissitzky, S. 1969. Europ. J. Biochem. *8,* 357.
Cashel, M. 1969. J. Biol. Chem. *244,* 3133.
Cashel, M. and Gallant, J. 1969. Nature *221,* 838.
Cashel, M. and Kalbacher, B. 1970. J. Biol. Chem. *245,* 2309.
Caskey, C. T., Tompkins, R., Scolnick, E., Caryk, T. and Nirenberg, M. 1968. Science *162,* 135.
Caspersson, T. 1941. Naturwissenschaften *28,* 33.
Castañeda, M. 1969. Biochim. Biophys. Acta *179,* 381.
Castles, J. J. and Singer, M. F. 1969. J. Mol. Biol. *40,* 1.
Caston, J. D. 1971. Dev. Biol. *24,* 19.
Catcheside, D. G. 1951. *The Genetics of Micro-organisms,* Pitman and Sons, London.
Cebra, J. J., Colberg, J. E. and Dray, S. 1966. J. Exp. Med. *123,* 547.
Ceccarini, C. and Maggio, R. 1969. Biochim. Biophys. Acta *190,* 556.

Ceccarini, C., Maggio, R. and Barbata, G. 1967. Proc. Natl. Acad. Sci. *58,* 2235.
Chae, Y. B., Mazumder, R. and Ochoa, S. 1969. Proc. Natl. Acad. Sci. *63,* 828.
Chakrabarty, A. M., Gunsalus, C. F. and Gunsalus, I. C. 1968. Proc. Natl. Acad. Sci. *60,* 168.
Chamberlain, J. P. 1971. Cited in Biochim. Biophys. Acta *213,* 183.
Chamberlin, M. and Berg, P. 1962. Proc. Natl. Acad. Sci. *48,* 81.
Chamberlin, M., McGrath, J. and Waskell, L. 1970. Nature *228,* 227.
Chambers, D. A. and Zubay, G. 1969. Proc. Natl. Acad. Sci. *63,* 118.
Chance, R. E., Ellis, R. M. and Bromer, W. W. 1968. Science *161,* 165.
Chandler, B., Hayashi, M., Hayashi, M. N. and Spiegelman, S. 1963. Science *143,* 47.
Chapman, L. F. and Nester, E. W. 1968. J. Bact. *96,* 1658.
Chassy, B. M., Love, L. L. and Krichevsky, M. I. 1969. Proc. Natl. Acad. Sci. *64,* 296.
Cheevers, W. P. and Sheinin, R. 1970. Can. J. Biochem. *48,* 1104.
Chen, D. and Osborne, D. J. 1970. Nature *225,* 336.
Chesterton, C. J. and Butterworth, P. H. W. 1971. FEBS Letters *12,* 301.
Chezzi, C., Grosclaude, J. and Scherrer, K. 1971. Cited in Spohr et al. (1970).
Chiarugi, V. P. 1969. Biochim. Biophys. Acta *179,* 129.
Chipchase, M. I. H. and Birnstiel, M. L. 1963. Proc. Natl. Acad. Sci. *49,* 692.
Chipchase, M. I. H. and Birnstiel, M. L. 1964. Proc. Natl. Acad. Sci. *50,* 1101.
Chou, C. T., Cinader, B. and Dubiski, S. 1967. Cold Spring Harbor Symp. Quant. Biol. *32,* 317.
Chovnick, A., Finnerty, V., Schalet, A. and Duck, P. 1969. Genetics *62,* 145.
Chovnick, A., Lefkowitz, R. J. and Fox, A. S. 1956. Genetics *41,* 589.
Chrispeels, M. J., Boyd, R. F., Williams, L. S. and Neidhardt, F. C. 1968. J. Mol. Biol. *31,* 463.
Church, R. B., Luther, S. W. and McCarthy, B. J. 1969. Biochim. Biophys. Acta *190,* 30.
Church, R. B. and McCarthy, B. J. 1967. J. Mol. Biol. *23,* 459.
Church, R. B. and McCarthy, B. J. 1970. Biochim. Biophys. Acta *199,* 103.
Cioli, D. and Baglioni, C. 1967. *Gamma Globulins. Structure and Control of Biosynthesis.* Nobel Symposium *3.* Editor J. Killander, Almqvist and Wiksell, Stockholm, p. 401.
Citri, N. 1958. Biochim. Biophys. Acta *27,* 277.
Citri, N. and Garber, N. 1958. Biochim. Biophys. Acta *30,* 664.
Citri, N., Garber, N. and Sela, M. 1960. J. Biol. Chem. *235,* 3454.
Clark, R. J. and Felsenfeld, G. 1971. Nature New Biology *229,* 101.
Clark-Walker, G. D. and Linnane, A. W. 1966. Biochem. Biophys. Res. Commun. *25,* 8.
Clark-Walker, G. D. and Linnane, A. W. 1967. J. Cell Biol. *34,* 1.
Clason, A. E. and Burdon, R. H. 1969. Nature *223,* 1063.
Clegg, J. B., Weatherall, D. J., Na-Nakorn, S. and Wasi, P. 1968. Nature *220,* 664.
Clement, A. C. 1962. J. Exp. Zool. *149,* 193.
Clement, A. C. 1968. Dev. Biol. *17,* 165.
Clever, U. 1964. Science *146,* 794.
Clever, U. 1968. Ann. Rev. Genet. *2,* 11.
Clever, U. and Ellgaard, E. G. 1970. Science *169,* 373.
Clever, U. and Rombull, G. 1966. Proc. Natl. Acad. Sci. *56,* 1470.
Cline, A. L. and Bock, R. M. 1966. Cold Spring Harbor Symp. Quant. Biol. *31,* 321.
Cohen, G. N. 1967. *Biosynthesis of Small Molecules.* Harper and Row, New York.
Cohen, G. and Jacob, F. 1959. Comptes Rend. Acad. Sci. (Paris) *248,* 3490.

Cohen, G. N. and Monod, J. 1957. Bact. Rev. *21,* 169.
Cohen, G. N. and Patte, J. C. 1963. Cold Spring Harbor Symp. Quant. Biol. *28,* 513.
Cohen, P. P. 1970. Science *168,* 533.
Cohen, S. N. and Hurwitz, J. 1967. Proc. Natl. Acad. Sci. *57,* 1759.
Cohen, S. N. and Hurwitz, J. 1968. J. Mol. Biol. *37,* 387.
Cohen, S. and Porter, R. R. 1964. Biochem. J. *90,* 278.
Cohen, S. S. 1968. *Virus-Induced Enzymes.* Columbia University Press, New York.
Cohen, S. S. and Raina, A. 1967. *Organizational Biosynthesis.* Editors H. J. Vogel, J. O. Lampen and V. Bryson, Academic Press, New York, p. 157.
Cohen-Bazire, G. and Monod, J. 1951. Comptes Rend. Acad. Sci. (Paris) *232,* 1515.
Cohn, M. 1956. *Enzymes: Units of Biological Structure and Function.* Editor H. O. Gaebler, Academic Press, New York, p. 41.
Cohn, M. 1967. *Gamma Globulins. Structure and Control of Biosynthesis.* Nobel Symposium *3.* Editor J. Killander, Almqvist and Wiksell, Stockholm, p. 615.
Cohn, M., Cohen, G. N. and Monod, J. 1953. Comptes Rend. Acad. Sci. (Paris) *236,* 746.
Cohn, M. and Horibata, K. 1959. J. Bact. *78,* 624.
Cohn, M., Lennox, E. and Spiegelman, S. 1960. Biochim. Biophys. Acta *39,* 255.
Cohn, M. and Monod, J. 1951. Biochim. Biophys. Acta *7,* 153.
Cohn, M. and Monod, J. 1953. Symp. Soc. Gen. Microbiol. *3,* 132.
Cohn, M. and Torriani, A. M. 1951. Comptes Rend. Acad. Sci. (Paris) *232,* 115.
Cohn, M. and Torriani, A. M. 1952. J. Immunol. *69,* 471.
Cohn, M. and Torriani, A. M. 1953. Biochim. Biophys. Acta *10,* 280.
Collier, J. R. 1966. *Current Topics in Developmental Biology.* Editors A. A. Moscona and A. Monroy, Academic Press, New York, Vol. *1,* p. 39.
Collier, J. R. and Yuyama, S. 1969. Exp. Cell Res. *56,* 281.
Collins, J. F. 1964. J. Gen. Microbiol. *34,* 363.
Collins, J. F., Mandelstam, J., Pollock, M. R., Richmond, M. H. and Sneath, P. H. A. 1965. Nature *208,* 841.
Colombo, B. and Baglioni, C. 1966. J. Mol. Biol. *16,* 51.
Colombo, B., Vesco, C. and Baglioni, C. 1968. Proc. Natl. Acad. Sci. *61,* 651.
Comb, D. G. 1965. J. Mol. Biol. *11,* 851.
Comb, D. G. and Katz, S. 1964. J. Mol. Biol. *8,* 790.
Comb, D. G. and Zehavi-Willner, T. 1967. J. Mol. Biol. *23,* 441.
Comings, D. E. 1967. J. Cell Biol. *35,* 699.
Comings, D. E. and Kakefuda, T. 1968. J. Mol. Biol. *33,* 225.
Commerford, S. L. and Delihas, N. 1966. Proc. Natl. Acad. Sci. *56,* 1759.
Contesse, G., Crépin, M. and Gros, F. 1969. Comptes Rend. Acad. Sci. (Paris) *268,* 2301.
Contesse, G., Crépin, M. and Gros, F. 1970. *The Lactose Operon.* Editors J. R. Beckwith and D. Zipser, Cold Spring Harbor Laboratory, New York, p. 111.
Contesse, G., Crépin, M., Gros, F., Ullmann, A. and Monod, J. 1970. *The Lactose Operon.* Editors J. R. Beckwith and D. Zipser, Cold Spring Harbor Laboratory, New York, p. 401.
Contesse, G., Naono, S. and Gros, F. 1966. Comptes Rend. Acad. Sci. (Paris) *263,* 1007.
Cook, P. R. 1970. J. Cell Sci. *7,* 1.
Cook, R. A. and Koshland, D. E. 1969. Proc. Natl. Acad. Sci. *64,* 247.
Cooper, D. and Gordon, J. 1969. Biochemistry *8,* 4289.
Cooper, E. L. 1969. Science *166,* 1414.
Cooper, H. L. 1969. J. Biol. Chem. *244,* 5590.

Cooper, H. L. 1970. Nature *227,* 1105.
Cooper, P. D. and Rowley, D. 1949. Nature *163,* 480.
Cooper, P. D., Rowley, D. and Dawson, I. M. 1949. Nature *164,* 842.
Cosenza, H. and Nordin, A. A. 1970. J. Immunol. *104,* 976.
Coston, M. B. and Loomis, W. F. 1969. J. Bact. *100,* 1208.
Coughlan, M. P., Rajagopalan, K. V. and Handler, P. 1969. Arch. Biochem. Biophys. *131,* 177.
Cove, D. J. and Pateman, J. A. 1969. J. Bact. *97,* 1374.
Cox, R. F. and Mathias, A. P. 1969. Biochem. J. *115,* 777.
Craddock, V. M. 1970. Nature *228,* 1264.
Craig, M. L. and Southard, J. L. 1967. Dev. Biol. *16,* 331.
Craig, N. C. 1971. J. Mol. Biol. *55,* 129.
Craig, N. and Perry, R. P. 1971. Nature New Biology *229,* 75.
Craig, S. P. 1970. J. Mol. Biol. *47,* 615.
Crane, C. M. and Villee, C. A. 1971. J. Biol. Chem. *246,* 719.
Craven, G. R., Voynow, P., Hardy, S. J. S. and Kurland, C. G. 1969. Biochemistry *8,* 2906.
Crawford, I. P. and Gunsalus, I. C. 1966. Proc. Natl. Acad. Sci. *56,* 717.
Creelman, V. M. and Tomlinson, N. 1959. J. Fisheries Res. Board Can. *16,* 421.
Crick, F. H. C. 1966(a). Cold Spring Harbor Symp. Quant. Biol. *31,* 3.
Crick, F. H. C. 1966(b). J. Mol. Biol. *19,* 548.
Crippa, M. 1970. Nature *227,* 1138.
Crippa, M., Davidson, E. H. and Mirsky, A. E. 1967. Proc. Natl. Acad. Sci. *57,* 885.
Crippa, M. and Gross, P. R. 1969. Proc. Natl. Acad. Sci. *62,* 120.
Crompton, B., Jago, M., Crawford, K., Newton, G. G. F. and Abraham, E. P. 1962. Biochem. J. *83,* 52.
Cross, M. E. and Ord, M. G. 1970. Biochem. J. *118,* 191.
Csányi, V., Jacobi, G. and Straub, B. F. 1957. Biochim. Biophys. Acta *145,* 470.
Culp, W. J., McKeehan, W. L. and Hardesty, B. 1969(a). Proc. Natl. Acad. Sci. *63,* 1431.
Culp, W., McKeehan, W. and Hardesty, B. 1969(b). Proc. Natl. Acad. Sci. *64,* 388.
Culp, W. J., Mosteller, R. D. and Hardesty, B. 1968. Arch. Biochem. Biophys. *125,* 658.
Cunin, R., Elseviers, D., Sand, G., Freundlich, G. and Glansdorff, N. 1969. Molec. Gen. Genetics *106,* 32.
Cunningham, B. A., Pflumm, M. N., Rutishauser, U. and Edelman, G. M. 1969. Proc. Natl. Acad. Sci. *64,* 997.
Cuzin, F., Kretchmer, N., Greenberg, R. E., Hurwitz, R. and Chapeville, F. 1967. Proc. Natl. Acad. Sci. 58, 2079.
Dahmus, M. E. and Bonner, J. 1965. Proc. Natl. Acad. Sci. *53,* 1370.
Dahmus, M. E. and McConnell, D. J. 1969. Biochemistry *8,* 1524.
Daneholt, B. 1970. J. Mol. Biol. *49,* 381.
Daneholt, B., Edström, J. E., Egyházi, E., Lambert, B. and Ringborg, U. 1969(a). Chromosoma *28,* 379.
Daneholt, B., Edström, J. E., Egyházi, E., Lambert, B. and Ringborg, U. 1969(b). Chromosoma *28,* 399.
Daneholt, B., Edström, J. E., Egyházi, E., Lambert, B. and Ringborg, U. 1969(c). Chromosoma *28,* 418.
Daniel, V., Sarid, S. and Littauer, U. Z. 1968. Proc. Natl. Acad. Sci. *60,* 1403.
Daniel, V., Sarid, S. and Littauer, U. Z. 1970. Science *167,* 1682.
Darlix, J. L., Sentenac, A., Ruet, A. and Fromageot, P. 1969. Europ. J. Biochem. *11,* 43.

Darnell, J. E. 1968. Bact. Rev. *32,* 262.
Darnell, J. E., Penman, S., Scherrer, K. and Becker, Y. 1963. Cold Spring Harbor Symp. Quant. Biol. *28,* 211.
Darnell, J. E., Wall, R. and Tushinski, R. J. 1971. Proc. Natl. Acad. Sci. *68,* 1321.
Darźynkiewicz, Z., Bolund, L. and Ringertz, N. R. 1969. Exp. Cell Res. *56,* 418.
Das, H. K., Goldstein, A. and Lowney, L. I. 1967. J. Mol. Biol. *24,* 231.
Das, N. K., Micou-Eastwood, J., Ramamurthy, G. and Alfert, M. 1970. Proc. Natl. Acad. Sci. *67,* 968.
Dastugue, B., Tichonicky, L., Penit-Soria, J. and Kruh, J. 1970. Bull. Soc. Chim. Biol. *52,* 391.
Datta, P. 1969. Science *165,* 556.
Davey, P. J., Yu, R. and Linnane, A. W. 1969. Biochem. Biophys. Res. Commun. *36,* 30.
Davidson, E. H. 1968. *Gene Activity in Early Development.* Academic Press, New York.
Davidson, E. H., Allfrey, V. G. and Mirsky, A. E. 1964. Proc. Natl. Acad. Sci. *52,* 501.
Davidson, E. H., Crippa, M., Kramer, F. R. and Mirsky, A. E. 1966. Proc. Natl. Acad. Sci. *56,* 856.
Davidson, E. H., Crippa, M. and Mirsky, A. E. 1968. Proc. Natl. Acad. Sci. *60,* 152.
Davidson, E. H., Haslett, G. W., Finney, R. J., Allfrey, V. G. and Mirsky, A. E. 1965. Proc. Natl. Acad. Sci. *54,* 696.
Davidson, E. H. and Hough, B. R. 1969. Proc. Natl. Acad. Sci. *63,* 342.
Davidson, E. H. and Hough, B. R. 1971. J. Mol. Biol. *56,* 491.
Davidson, R., Ephrussi, B. and Yamamoto, K. 1968. J. Cell Phys. *72,* 115.
Davidson, R. G., Nitowsky, H. M. and Childs, B. 1963. Proc. Natl. Acad. Sci. *50,* 481.
Davies, J. 1966. Cold Spring Harbor Symp. Quant. Biol. *31,* 665.
Davies, J. and Jacob, F. 1968. J. Mol. Biol. *36,* 413.
Davis, R. W. and Hyman, R. W. 1970. Cold Spring Harbor Symp. Quant. Biol. *35,* 269.
Davison, J., Pilarski, L. M. and Echols, H. 1968. Proc. Natl. Acad. Sci. *63,* 168.
Dawid, I. B. 1965. J. Mol. Biol. *12,* 581.
Dawid, I. B. 1966. Proc. Natl. Acad. Sci. *56,* 269.
Dawid, I. B. and Brown, D. D. 1970. Dev. Biol. *22,* 1.
Dawid, I. B., Brown, D. D. and Reeder, R. H. 1970. J. Mol. Biol. *51,* 341.
Dayhoff, M. O. and Eck, R. V. 1969. *Atlas of Protein Sequence and Structure 1968–1969.* National Biomedical Research Foundation, Silver Spring, Md.
Deal, W. C. 1969. Biochemistry *8,* 2795.
DeAngelo, A. B. and Gorski, J. 1970. Proc. Natl. Acad. Sci. *66,* 693.
DeBellis, R. H., Gluck, N. and Marks, P. A. 1964. J. Clin. Invest. *43,* 1329.
DeCoen, J. L. 1970. J. Mol. Biol. *49,* 405.
de Crombrugghe, B., Chen, B., Anderson, W., Nissley, P., Gottesman, M., Pastan, I. and Perlman, R. 1971(a). Nature New Biology *231,* 139.
de Crombrugghe, B., Chen, B., Gottesman, M., Pastan, I., Varmus, H. E., Emmer, M. and Perlman, R. L. 1971(b). Nature New Biology *230,* 37.
de Crombrugghe, B., Periman, R. L., Varmus, H. E. and Pastan, I. 1969. J. Biol. Chem. *244,* 5828.
de Crombrugghe, B., Varmus, H. E., Perlman, R. L. and Pastan, I. H. 1970. Biochem. Biophys. Res. Commun. *38,* 894.
DeFilippes, F. M. 1970. Biochim. Biophys. Acta *199,* 562.
DeLange, R. J., Fambrough, D. M., Smith, E. L. and Bonner, J. 1969. J. Biol. Chem. *244,* 5669.
DeLorenzo, F. and Ames, B. N. 1970. J. Biol. Chem. *245,* 1710.

DelValle, M. R. and Aronson, A. I. 1962. Biochem. Biophys. Res. Commun. *9,* 421.
DeMoss, J. A. 1965. Biochem. Biophys. Res. Commun. *18,* 850.
DeMoss, J. A. and Wegman, J. 1965. Proc. Natl. Acad. Sci. *54,* 241.
Dénes, G. 1961. Biochim. Biophys. Acta *50,* 408.
Denis, H. 1966(a). J. Mol. Biol. *22,* 269.
Denis, H. 1966(b). J. Mol. Biol. *22,* 285.
Denny, P. C. and Reback, P. 1970. J. Exp. Zool. *175,* 133.
Denny, P. C. and Tyler, A. 1964. Biochem. Biophys. Res. Commun. *14,* 245.
DeSombre, E. R., Puca, G. A. and Jensen, V. E. 1969. Proc. Natl. Acad. Sci. *64,* 148.
DeTerra, N. 1969. Intern. Rev. Cytol. *25,* 1.
DeWitt, W. 1971. Biochem. Biophys. Res. Commun. *42,* 267.
Dienert, F. 1900. Ann. Inst. Pasteur *14,* 139.
DiGirolamo, A., DiGirolamo, M., Gaetani, S. and Spadoni, M. A. 1966. Biochim. Biophys. Acta *114,* 195.
DeGirolamo, A., Henshaw, E. C. and Hiatt, H. H. 1964. J. Mol. Biol. *8,* 479.
DiMauro, E., Snyder, L., Marino, P., Lamberti, A., Coppo, A. and Tocchini-Valentini, G. P. 1969. Nature *222,* 533.
Dingman, C. W. and Peacock, A. C. 1968. Biochemistry *7,* 659.
Dingman, C. W. and Sporn, M. B. 1964. J. Biol. Chem. *239,* 3483.
Dintzis, H. M. 1961. Proc. Natl. Acad. Sci. *47,* 247.
Dixon, G. H. 1966. *Essays in Biochemistry.* Editors P. N. Campbell and G. D. Greville, Academic Press, London, Vol. *2,* p. 147.
Dixon, G. H., Ingles, C. J., Jergil, B., Ling, V. and Marushige, K. 1969. Can. Cancer Conf. *8,* 76.
Dixon, G. H. and Smith, M. 1968. *Progress in Nucleic Acid Research and Molecular Biology.* Editors J. N. Davidson and W. E. Cohn, Academic Press, New York, Vol. *8,* p. 9.
Doi, R. H. and Goehler, B. 1966. Biochem. Biophys. Res. Commun. *24,* 44.
Doi, R. H., Kaneko, I. and Goehler, B. 1966. Proc. Natl. Acad. Sci. *56,* 1548.
Doi, R. H., Kaneko, I. and Igarashi, R. T. 1968. J. Biol. Chem. *243,* 945.
Doi, R. H. and Spiegelman, S. 1963. Proc. Natl. Acad. Sci. *49,* 353.
Doolittle, W. F. and Yanofsky, C. 1968. J. Bact. *95,* 1283.
Dove, W. 1968. Ann. Rev. Genet. *2,* 305.
Doyle, D. and Laufer, H. 1969. Exp. Cell Res. *57,* 205.
Dravid, A. R. and Burdman, J. A. 1968. J. Neurochem. *15,* 25.
Dray, S., Dubiski, S., Kelus, A., Lennox, E. S. and Oudin, J. 1962. Nature *195,* 785.
Dray, S. and Nisonoff, A. 1963. Proc. Soc. Expl. Biol. Med. *113,* 20.
Drews, J., Brawerman, G. and Morris, H. P. 1968. Europ. J. Biochem. *3,* 284.
Dreyer, W. J. and Bennett, J. C. 1965. Proc. Natl. Acad. Sci. *54,* 864.
Dreyer, W. J., Gray, W. R. and Hood, L. 1967. Cold Spring Harbor Symp. Quant. Biol. *32,* 353.
Dube, S. K. and Rudland, P. S. 1970. Nature *226,* 820.
Dubnau, D. A. and Pollock, M. R. 1965. J. Gen. Microbiol. *41,* 7.
Dubnoff, J. S. and Maitra, U. 1969. Cold Spring Harbor Symp. Quant. Biol. *34,* 301.
Dubnoff, J. S. and Maitra, U. 1971. Proc. Natl. Acad. Sci. *68,* 318.
Dulbecco, R. 1969. Science *166,* 962.
Dunn, D. B. 1959. Biochim. Biophys. Acta *34,* 286.
Dunn, J. J. and Bautz, E. K. F. 1969. Biochem. Biophys. Res. Commun. *36,* 925.
DuPraw, E. J. 1968. *Cell and Molecular Biology.* Academic Press, New York.
Dutton, R. W. 1967. Advances in Immunology *6,* 253.
Dutton, R. W. and Mishell, R. I. 1967. Cold Spring Harbor Symp. Quant. Biol. *32,* 407.
Dwyer, S. B. and Umbarger, H. E. 1968. J. Bact. *95,* 1680.

Ebert, J. D. 1965. *Interacting Systems in Development.* Holt, Rinehart and Winston, New York.
Ebert, J. D. 1968. *Current Topics in Developmental Biology.* Editors A. A. Moscona and A. Monroy. Academic Press, New York, Vol. *3,* p. xv.
Echols, H., Garen, A., Garen, S. and Torriani, A. 1961. J. Mol. Biol. *3,* 425.
Echols, H., Pilarski, L. and Cheng, P. Y. 1968. Proc. Natl. Acad. Sci. *59,* 1016.
Edelman, G. M. 1967. *Gamma Globulins. Structure and Control of Biosynthesis.* Nobel Symposium *3.* Editor J. Killander, Almqvist and Wiksell, Stockholm, p. 89.
Edelman, G. M., Cunningham, B. A., Gall, W. E., Gottlieb, P. D., Rutishauser, U. and Waxdal, M. J. 1969. Proc. Natl. Acad. Sci. *63,* 78.
Edelman, G. M. and Gally, G. A. 1967. Proc. Natl. Acad. Sci. *57,* 353.
Edlin, G. and Broda, P. 1968. Bact. Rev. *32,* 206.
Edlin, G. and Stent, G. S. 1969. Proc. Natl. Acad. Sci. *62,* 475.
Edlin, G., Stent, G. S., Baker, R. F. and Yanofsky, C. 1968. J. Mol. Biol. *37,* 257.
Edmonds, M., Vaughan, M. H. and Nakazato, H. 1971. Proc. Natl. Acad. Sci. *68,* 1336.
Edström, J. E. and Daneholt, B. 1967. J. Mol. Biol. *28,* 331.
Edwards, J. G., James, D. R., Mathias, A. P. and Evans, M. J. 1970. Biochim. Biophys. Acta *213,* 469.
Edwards, L. J. and Hnilica, L. S. 1968. Experientia *24,* 228.
Egawa, K., Choi, Y. C. and Busch, H. 1971. J. Mol. Biol. *56,* 565.
Eggen, K. and Nathans, D. 1969. J. Mol. Biol. *39,* 293.
Ehrenfeld, E. R. and Koch, A. L. 1968. Biochim. Biophys. Acta *169,* 44.
Eidlic, L. and Neidhardt, F. C. 1965. Proc. Natl. Acad. Sci. *53,* 539.
Eisen, H., Brachet, P., Da Silva, L. P. and Jacob, F. 1970. Proc. Natl. Acad. Sci. *66,* 855.
Eisen, H. A., Fuerst, C. R., Siminovitch, L., Thomas, R., Lambert, L., Pereira Da Silva, L. H. and Jacob, F. 1966. Virology *30,* 224.
Eisenstadt, J. and Lengyel, P. 1966. Science *154,* 524.
Eisenstein, R. B. and Yanofsky, C. 1962. J. Bact. *83,* 193.
Eiserling, F., Levin, J. G., Byrne, R., Karlsson, U., Nirenberg, M. W. and Sjöstrand, F. S. 1964. J. Mol. Biol. *10,* 536.
Elgin, S. C. R. and Bonner, J. 1970. Biochemistry *9,* 4440.
Eliceiri, G. L. and Green, H. 1969. J. Mol. Biol. *41,* 253.
Emerson, C. P. and Humphreys, T. 1971. Science *171,* 898.
Emmer, M., DeCrombrugghe, B., Pastan, I. and Perlman, R. 1970. Proc. Natl. Acad. Sci. *66,* 480.
Enger, M. D. and Walters, R. A. 1970. Biochemistry *9,* 3551.
Englander, S. W. and Page, L. A. 1965. Biochem. Biophys. Res. Commun. *19,* 565.
Englesberg, E., Irr, J., Power, J. and Lee, N. 1965. J. Bact. *90,* 946.
Englesberg, E., Sheppard, D., Squires, C. and Meronk, F. 1969. J. Mol. Biol. *43,* 281.
Englesberg, E., Squires, C. and Meronk, F. 1969. Proc. Natl. Acad. Sci. *62,* 1100.
Ennis, H. L. 1966. Mol. Pharmacol. *2,* 543.
Ennis, H. L. and Gorini, L. 1961. J. Mol. Biol. *3,* 439.
Ephrussi, B. and Weiss, M. C. 1967. *Control Mechanisms in Developmental Processes.* Editor M. Locke, Academic Press, New York, p. 136.
Epler, J. L., Shugart, L. R. and Barnett, W. E. 1970. Biochemistry *9,* 3575.
Epstein, W. and Beckwith, J. R. 1968. Ann. Rev. Biochem. *37,* 411.
Erbe, R. W., Nau, M. M. and Leder, P. 1969. J. Mol. Biol. *39,* 441.
Eron, L., Arditti, R., Zubay, G., Connaway, S. and Beckwith, J. R. 1971. Proc. Natl. Acad. Sci. *68,* 215.

Ertel, R., Brot, N., Redfield, B., Allende, J. E. and Weissbach, H. 1968(a). Proc. Natl. Acad. Sci. *59,* 861.

Ertel, R., Redfield, B., Brot, N. and Weissbach, H. 1968(b). Arch. Biochem. Biophys. *128,* 331.

Evans, E. E., Acton, R. T., Bennett, J. C. and Weinheimer, P. F. 1970. *Protides of the Biological Fluids,* Proc. 17th Colloq., Bruges. Pergamon, Oxford, p. 29.

Faiferman, I., Hamilton, M. G. and Pogo, A. O. 1970. Biochim. Biophys. Acta *204,* 550.

Faiferman, I., Hamilton, M. G. and Pogo, A. O. 1971. Biochim. Biophys. Acta *232,* 685.

Falvey, A. K. and Staehelin, T. 1970(a). J. Mol. Biol. *53,* 1.

Falvey, A. K. and Staehelin, T. 1970(b). J. Mol. Biol. *53,* 21.

Fambrough, D. M. and Bonner, J. 1968. J. Biol. Chem. *243,* 4434.

Fambrough, D. M., Fujimura, F. and Bonner, J. 1968. Biochemistry *7,* 575.

Fan, D. P., Higa, A. and Levinthal, C. 1964. J. Mol. Biol. *8,* 210.

Fangman, W. L., Gross, C. and Novick, A. 1967. J. Mol. Biol. *29,* 317.

Fargie, B. and Holloway, B. W. 1965. Genet. Res. *6,* 284.

Farkas, W. and Marks, P. A. 1968. J. Biol. Chem. *243,* 6464.

Fauman, M., Rabinowitz, M. and Getz, G. S. 1969. Biochim. Biophys. Acta *182,* 355.

Favre, A., Yaniv, M. and Michelson, A. M. 1969. Biochem. Biophys. Res. Commun. *37,* 266.

Feherty, P., Robertson, D. M., Waynforth, H. B. and Kellie, A. E. 1970. Biochem. J. *120,* 837.

Feigelson, P. and Greengard, O. 1961. Biochim. Biophys. Acta *52,* 509.

Felicetti, L., Colombo, B. and Baglioni, C. 1966. Biochim. Biophys. Acta *129,* 380.

Felicetti, L., Tocchini-Valentini, G. P. and DiMatteo, G. F. 1969. Biochemistry *8,* 3428.

Fernandes, J. F., Castellani, O. and Kimura, E. 1969. Genetics *61,* Supp. *1,* 213.

Ferrari, T. E. and Varner, J. E. 1970. Proc. Natl. Acad. Sci. *65,* 729.

Feughelman, M., Langridge, R., Seeds, W. E., Stokes, A. R., Wilson, H. R., Hooper, C. W., Wilkins, M. H. F., Barcley, R. K. and Hamilton, L. D. 1955. Nature *175,* 834.

Fiethen, L. and Starlinger, P. 1970. Molec. Gen. Genetics *108,* 322.

Finch, B. W. and Ephrussi, B. 1967. Proc. Natl. Acad. Sci. *57,* 615.

Fink, G. R. 1966. Genetics *53,* 445.

Fink, G. R. and Roth, J. R. 1968. J. Mol. Biol. *33,* 547.

Fischberg, M. and Blackler, A. W. 1963. *Biological Organization at the Cellular and Supercellular Level.* Editor R. J. C. Harris, Academic Press, London, p. 111.

Fischer, S., Nagel, R. L. and Fuhr, J. 1968. Biochim. Biophys. Acta *169,* 566.

Fishman, M. and Adler, F. L. 1963. J. Exp. Med. *117,* 595.

Fishman, M. and Adler, F. L. 1967. Cold Spring Harbor Symp. Quant. Biol. *32,* 343.

Fittler, F. and Hall, R. H. 1966. Biochem. Biophys. Res. Commun. *25,* 441.

Fleischman, J. B. 1967. Biochemistry *6,* 1311.

Fleissner, E. and Borek, E. 1962. Proc. Natl. Acad. Sci. *48,* 1199.

Fleissner, E. and Borek, E. 1963. Biochemistry *2,* 1093.

Flickinger, R. A. 1969. Exp. Cell Res. *55,* 422.

Flickinger, R. A., Freedman, M. L. and Stambrook, P. J. 1967. Dev. Biol. *16,* 457.

Forsyth, G. W., Carnevale, H. and Jones, E. E. 1969. J. Biol. Chem. *244,* 5226.

Fox, C. F. 1969. Proc. Natl. Acad. Sci. *63,* 850.

Fox, C. F., Carter, J. R. and Kennedy, E. P. 1967. Proc. Natl. Acad. Sci. *57,* 698.

Frangione, B., Milstein, C. and Pink, J. R. L. 1969. Nature *221,* 145.
Franklin, R. M. 1963. Biochim. Biophys. Acta *72,* 555.
Frenster, J. H. 1965(a). Nature *206,* 680.
Frenster, J. H. 1965(b). Nature *206,* 1269.
Frenster, J. H. 1965(c). Nature *208,* 894.
Frenster, J. H. 1965(d). Nature *208,* 1093.
Frenster, J. H., Allfrey, V. G. and Mirsky, A. E. 1963. Proc. Natl. Acad. Sci. *50,* 1026.
Fresco, J. R., Adams, A., Ascione, R., Henley, D. and Lindahl, T. 1966. Cold Spring Harbor Symp. Quant. Biol. *31,* 527.
Freundlich, M., Burns, O. R. and Umbarger, H. E. 1962. Proc. Natl. Acad. Sci. *48,* 1804.
Freundlich, M., Burns, O. R. and Umbarger, H. E. 1963. *Informational Macromolecules.* Editors H. J. Vogel, V. Bryson and J. O. Lampen, Academic Press, New York, p. 287.
Fridlender, B. R. and Wettstein, F. O. 1970. Biochem. Biophys. Res. Commun. *39,* 247.
Frisch, L. 1967. Editor, Cold Spring Harbor Symp. Quant. Biol. Vol. *32.*
Fromson, D. and Nemer, M. 1970. Science *168,* 266.
Fry, B. J. and Gross, P. R. 1970. Dev. Biol. *21,* 105.
Fry, K. T. and Lamborg, M. R. 1967. J. Mol. Biol. *28,* 423.
Fuhr, J., Natta, C., Marks, P. A. and Bank, A. 1969. Nature *224,* 1305.
Fujinaga, K. and Green, M. 1970. Proc. Natl. Acad. Sci. *65,* 375.
Fujinaga, K., Pina, M. and Green, M. 1969. Proc. Natl. Acad. Sci. *64,* 255.
Fujita, S. 1962. Exp. Cell Res. *28,* 52.
Fujita, S. and Takamoto, K. 1963. Nature *200,* 494.
Fukami, H. and Imahori, K. 1971. Proc. Natl. Acad. Sci. *68,* 570.
Fukuhara, H. 1966. J. Mol. Biol. *17,* 334.
Fukuhara, H. and Sels, A. 1966. J. Mol. Biol. *17,* 319.
Gale, E. F. 1943. Bact. Rev. 7, 139.
Galibert, F., Elardari, M. E., Larsen, C. J. and Boiron, M. 1970. Europ. J. Biochem. *13,* 273.
Galizzi, A. 1967. J. Mol. Biol. *27,* 619.
Gallant, J. and Harada, B. 1969. J. Biol. Chem. *244,* 3125.
Gallo, R. C. and Pestka, S. 1970. J. Mol. Biol. *52,* 195.
Gallwitz, D. 1970. Zeit. Physiol. Chem. *351,* 1050.
Gallwitz, D. and Mueller, G. C. 1969. Europ. J. Biochem. *9,* 431.
Gallwitz, D. and Sekeris, C. E. 1969. FEBS Letters *3,* 99.
Gally, J. A. and Edelman, G. M. 1970. Nature *227,* 341.
Gamow, E. I. and Prescott, D. M. 1970. Exp. Cell Res. *59,* 117.
Ganoza, M. C. and Williams, C. A. 1969. Proc. Natl. Acad. Sci. *63,* 1370.
Ganschow, R. and Paigen, K. 1967. Proc. Natl. Acad. Sci. *58,* 938.
Ganschow, R. E. and Schimke, R. T. 1969. J. Biol. Chem. *244,* 4649.
Gardner, R. S. and Tomkins, G. M. 1969. J. Biol. Chem. *244,* 4761.
Garel, J. P., Mandel, P., Chavancy, G. and Daillie, J. 1971. FEBS Letters *12,* 249.
Garen, A. 1960. Symp. Soc. Gen. Microbiol. *10,* 239.
Garen, A. 1968. Science *160,* 149.
Garen, A. and Echols, H. 1962. Proc. Natl. Acad. Sci. *48,* 1398.
Garen, A. and Garen, S. 1963. J. Mol. Biol. *6,* 433.
Garen, A. and Levinthal, C. 1960. Biochim. Biophys. Acta *38,* 470.
Garen, A. and Otsuji, N. 1964. J. Mol. Biol. *8,* 841.
Gariglio, P., Roodman, S. T. and Green, M. H. 1969. Biochem. Biophys. Res. Commun. *37,* 945.

Garren, L. D., Howell, R. R. and Tomkins, G. M. 1964. J. Mol. Biol. *9,* 100.
Garrick-Silversmith, L. and Hartman, P. E. 1970. Genetics *66,* 231.
Gartland, W. J., Ishida, T., Sueoka, N. and Nirenberg, M. W. 1969. J. Mol. Biol. *44,* 403.
Gartland, W. J. and Sueoka, N. 1966. Proc. Natl. Acad. Sci. *55,* 948.
Gaye, P. and Denamur, R. 1969. Biochim. Biophys. Acta *186,* 99.
Gaye, P. and Denamur, R. 1970. Biochem. Biophys. Res. Commun. *41,* 266.
Gefter, M. L. 1969. Biochem. Biophys. Res. Commun. *36,* 435.
Gefter, M. L. and Russell, R. L. 1969. J. Mol. Biol. *39,* 145.
Geiduschek, E. P. and Grau, O. 1970. Lepetit Colloq. Biol. Med. *1,* 190.
Geiduschek, E. P. and Haselkorn, R. 1969. Ann. Rev. Biochem. *38,* 647.
Geiduschek, E. P. and Sklar, J. 1969. Nature *221,* 833.
Geiduschek, E. P., Tocchini-Valentini, G. P. and Sarnat, M. T. 1964. Proc. Natl. Acad. Sci. *52,* 486.
Gelderman, A. H., Rake, A. V. and Britten, R. J. 1971. Proc. Natl. Acad. Sci. *68,* 172.
Gelehrter, T. D. and Tomkins, G. M. 1967. J. Mol. Biol. *29,* 59.
Gelehrter, T. D. and Tomkins, G. M. 1969. Proc. Natl. Acad. Sci. *64,* 723.
Gelehrter, T. D. and Tomkins, G. M. 1970. Proc. Natl. Acad. Sci. *66,* 390.
Gelfand, D. H. and Hayashi, M. 1970. Nature *228,* 1162.
Georgiev, G. P. 1967. *Progress in Nucleic Acid Research and Molecular Biology.* Editors J. N. Davidson and W. Cohn, Academic Press, New York, Vol. *6,* p. 259.
Georgiev, G. P. 1968. *Regulatory Mechanisms for Protein Synthesis in Mammalian Cells.* Editors A. S. Pietro, M. R. Lamborg and F. T. Kenney, Academic Press, New York, p. 25.
Georgiev, G. P. 1969(a). Ann. Rev. Genetics *3,* 155.
Georgiev, G. P. 1969(b). J. Theoret. Biol. *25,* 473.
Georgiev, G. P., Ananieva, L. N. and Kozlov, J. V. 1966. J. Mol. Biol. *22,* 365.
Georgiev, G. P. and Mantieva, V. L. 1962. Biochim. Biophys. Acta *61,* 153.
Georgiev, G. P., Samarina, O. P., Lerman, M. I., Smirnov, M. N. and Severtzov, A. N. 1963. Nature *200,* 1291.
Georgiev, G. P., Samarina, O. P., Mantieva, V. L. and Zbarsky, I. B. 1961. Biochim. Biophys. Acta *46,* 399.
Germaine, G. R. and Rogers, P. 1970. J. Mol. Biol. *47,* 121.
Germanyuk, Y. L. and Mironenko, V. I. 1969. Nature *222,* 486.
Gershey, E. L., Haslett, G. W., Vidali, G. and Allfrey, V. G. 1969. J. Biol. Chem. *244,* 4871.
Gershey, E. L., Vidali, G. and Allfrey, V. G. 1968. J. Biol. Chem. *243,* 5018.
Gesteland, R. F. 1966. J. Mol. Biol. *16,* 67.
Gesteland, R. F. and Salser, W. 1969. Genetics *61,* Supp. *1,* 429.
Gesteland, R. F., Salser, W. and Bolle, A. 1967. Proc. Natl. Acad. Sci. *58,* 2036.
Gesteland, R. F. and Spahr, P. F. 1969. Cold Spring Harbor Symp. Quant. Biol. *34,* 707.
Ghosh, H. P., Söll, D. and Khorana, H. G. 1967. J. Mol. Biol. *25,* 275.
Giacomoni, D. and Spiegelman, S. 1962. Science *138,* 1328.
Gibson, F. and Pittard, J. 1968. Bact. Rev. *32,* 465.
Gierer, A. 1963. J. Mol. Biol. *6,* 148.
Gilbert, J. M. and Anderson, W. F. 1970. J. Biol. Chem. *245,* 2342.
Gilbert, J. M., Thornton, A. G., Nienhuis, A. W. and Anderson, W. F. 1970. Proc. Natl. Acad. Sci. *67,* 1854.
Gilbert, W. 1963. J. Mol. Biol. *6,* 389.
Gilbert, W. and Müller-Hill, B. 1966. Proc. Natl. Acad. Sci. *56,* 1891.
Gilbert, W. and Müller-Hill, B. 1967. Proc. Natl. Acad. Sci. *58,* 2415.

Gill, G. N. and Garren, L. D. 1970. Biochem. Biophys. Res. Commun. *39,* 335.
Gillam, I., Blew, D., Warrington, R. C., Von Tigerstrom, M. and Tener, G. M. 1968. Biochemistry *7,* 3459.
Gillam, I., Millward, S., Blew, D., Von Tigerstrom, M., Wimmer, E. and Tener, G. M. 1967. Biochemistry *6,* 3043.
Gillespie, D. 1968. *Methods in Enzymology,* Vol. XII B. Editors L. Grossman and K. Moldave, Academic Press, New York, p. 641.
Gillespie, D. and Spiegelman, S. 1965. J. Mol. Biol. *12,* 829.
Gilmour, R. S. and Dixon, G. H. 1971. Personal communication.
Gilmour, R. S. and Paul, J. 1969. J. Mol. Biol. *40,* 137.
Gilmour, R. S. and Paul, J. 1970. FEBS Letters *9,* 242.
Gniadowski, M., Mandel, J. L., Gissinger, F., Kedinger, C. and Chambon, P. 1970. Biochem. Biophys. Res. Commun. *38,* 1033.
Goff, C. G. and Minkley, E. G. 1970. Lepetit Colloq. Biol. Med. *1,* 124.
Goff, C. G. and Weber, K. 1970. Cold Spring Harbor Symp. Quant. Biol. *35,* 101.
Goff, C. G. and Weber, K. 1971. Cited in Travers 1971.
Goffeau, A. and Brachet, J. 1965. Biochim. Biophys. Acta *95,* 302.
Gold, L. M. and Schweiger, M. 1969(a). Proc. Natl. Acad. Sci. *62,* 892.
Gold, L. M. and Schweiger, M. 1969(b). J. Biol. Chem. *244,* 5100.
Gold, M., Hausmann, R., Maitra, U. and Hurwitz, J. 1964. Proc. Natl. Acad. Sci. *52,* 292.
Gold, M., Hurwitz, J. and Anders, M. 1963. Biochem. Biophys. Res. Commun. *11,* 107.
Goldberg, I. H. 1961. Biochim. Biophys. Acta *51,* 201.
Goldberg, I. H., Rabinowitz, M. and Reich, E. 1962. Proc. Natl. Acad. Sci. *48,* 2094.
Goldberger, R. F. and Berberich, M. A. 1965. Proc. Natl. Acad. Sci. *54,* 279.
Goldberger, R. F. and Berberich, M. A. 1967. *Organizational Biosynthesis.* Editors H. J. Vogel, J. O. Lampen and V. Bryson, Academic Press, New York, p. 199.
Goldenbaum, P. E. and Dobrogosz, W. J. 1968. Biochem. Biophys. Res. Commun. *33,* 828.
Goldschmidt, E. P., Cater, M. S., Matney, T. S., Butler, M. A. and Greene, A. 1970. Genetics *66,* 219.
Goldstein, A., Kirschbaum, J. B. and Roman, A. 1965. Proc. Natl. Acad. Sci. *54,* 1669.
Goldstein, J. L., Beaudet, A. L. and Caskey, C. T. 1970. Proc. Natl. Acad. Sci. *67,* 99.
Goldstein, J. L. and Caskey, C. T. 1970. Proc. Natl. Acad. Sci. *67,* 537.
Goldstein, J., Milman, G., Scolnick, E. and Caskey, T. 1970. Proc. Natl. Acad. Sci. *65,* 430.
Goldstein, L. and Micou, J. 1959. J. Biophys. Biochem. Cytol. *6,* 1.
Goldstein, L. and Plaut, W. 1955. Proc. Natl. Acad. Sci. *41,* 874.
Gonano, F. 1967. Biochemistry *6,* 977.
Gonano, F. and Baglioni, C. 1969. Europ. J. Biochem. *11,* 7.
Gonzalez, N. S., Bade, E. G. and Algranati, I. D. 1969. FEBS Letters *4,* 331.
Good, R. A. and Papermaster, B. W. 1964. Advances in Immunol. *4,* 1.
Goodman, H. M., Abelson, J., Landy, A., Brenner, S. and Smith, J. D. 1968. Nature *217,* 1019.
Goodman, H. M., Billeter, M. A., Hindley, J. and Weissman, C. 1970. Proc. Natl. Acad. Sci. *67,* 921.
Goodman, H. M. and Rich, A. 1962. Proc. Natl. Acad. Sci. *48,* 2101.
Goodwin, B. C. 1966. Nature *209,* 479.
Goodwin, B. C. 1969. Symp. Soc. Gen. Microbiol. *19,* 223.

Gopinathan, K. P. and Garen, A. 1970. J. Mol. Biol. *47,* 393.
Gordon, A. S., Zanjani, E. D., Levere, R. D. and Kappas, A. 1970. Proc. Natl. Acad. Sci. *65,* 919.
Gordon, J. 1967. Proc. Natl. Acad. Sci. *58,* 1574.
Gordon, J. 1968. Proc. Natl. Acad. Sci. *59,* 179.
Gordon, J. 1969. J. Biol. Chem. *244,* 5680.
Gordon, J. 1970. Biochemistry *9,* 912.
Gorini, L. and Gundersen, W. 1961. Proc. Natl. Acad. Sci. *47,* 961.
Gorini, L., Gundersen, W. and Burger, M. 1961. Cold Spring Harbor Symp. Quant. Biol. *26,* 173.
Gorini, L., Jacoby, G. A. and Breckenridge, L. 1966. Cold Spring Harbor Symp. Quant. Biol. *31,* 657.
Gorini, L. and Maas, W. F. 1957. Biochim. Biophys. Acta *25,* 208.
Gorovsky, M. A. and Woodard, J. 1969. J. Cell Biol. *42,* 673.
Gorski, J. 1964. J. Biol. Chem. *239,* 889.
Gorski, J. and Nelson, N. J. 1965. Arch. Biochem. Biophys. *110,* 284.
Gorski, J., Noteboom, W. D. and Nicolette, J. A. 1965. J. Cell. Comp. Phys. *66,* Supp. *1,* 91.
Gorski, J., Shyamala, G. and Toft, D. 1969. *Current Topics in Developmental Biology.* Editors A. A. Moscona and A. Monroy, Academic Press, New York, Vol. 4, p. 149.
Gottlieb, A. A. 1969(a). Biochemistry *8,* 2111.
Gottlieb, A. A. 1969(b). Science *165,* 592.
Gottlieb, A. A. and Straus, D. S. 1969. J. Biol. Chem. *244,* 3324.
Gould, H. J. 1970. Nature *227,* 1145.
Gould, M. C. 1969. Dev. Biol. *19,* 482.
Gould, R. M., Thornton, M. P., Liepkalns, V. and Lennarz, W. J. 1968. J. Biol. Chem. *243,* 3096.
Graham, C. F. and Morgan, R. W. 1966. Dev. Biol. *14,* 439.
Granboulan, N. and Scherrer, K. 1969. Europ. J. Biochem. *9,* 1.
Granick, S. and Kappas, A. 1967(a). Proc. Natl. Acad. Sci. *57,* 1463.
Granick, S. and Kappas, A. 1967(b). J. Biol. Chem. *242,* 4587.
Granick, S. and Levere, R. D. 1964. Prog. in Hematol. *4,* 1.
Granick, S. and Levere, R. D. 1965. J. Cell Biol. *26,* 167.
Granick, S. and Levere, R. D. 1968. Proc. Internat. Soc. Hematol. XIIth Congress. p. 274.
Granner, D. K., Hayashi, S. I., Thompson, E. B. and Tomkins, G. M. 1968. J. Mol. Biol. *35,* 291.
Granner, D. K., Thompson, E. B. and Tomkins, G. M. 1970. J. Biol. Chem. *245,* 1472.
Grasso, R. J. and Buchanan, J. M. 1969. Nature *224,* 882.
Grau, O., Guha, A., Geiduschek, E. P. and Szybalski, W. 1969. Nature *224,* 1105.
Gray, C. T., Wimpenny, J. W. T. and Mossman, M. R. 1966. Biochim. Biophys. Acta *117,* 33.
Gray, W. J. H. and Midgley, J. E. M. 1970. Biochem. J. *120,* 279.
Grayzel, A. I., Hörchner, P. and London, I. M. 1966. Proc. Natl. Acad. Sci. *55,* 650.
Green, M. H. 1964. Proc. Natl. Acad. Sci. *52,* 1388.
Green, M. R. and Topper, Y. J. 1970. Biochim. Biophys. Acta *204,* 441.
Greenberg, H. and Penman, S. 1966. J. Mol. Biol. *21,* 527.
Greenberg, J. R. 1969. J. Mol. Biol. *46,* 85.
Greengard, O. 1969(a). Biochem. J. *115,* 19.
Greengard, O. 1969(b). Science *163,* 891.

Greengard, O. and Feigelson, P. 1961. J. Biol. Chem. *236,* 158.
Greenshpan, H. and Revel, M. 1969. Nature *224,* 331.
Grogan, D., Desjardins, R. and Busch, H. 1966. Cancer Res. *26,* 775.
Gros, F., Gallant, J., Weisberg, R. and Cashel, M. 1967. J. Mol. Biol. *25,* 555.
Gros, F., Hiatt, H., Gilbert, W., Kurland, C. G., Risebrough, R. W. and Watson, J. D. 1961. Nature *190,* 581.
Gros, F., Kourilsky, P., Luzzati, D. and Naono, S. 1970. Proc. Roy. Soc. B, *176,* 251.
Gros, F., Naono, S., Woese, C., Willson, C. and Attardi, G. 1963. *Informational Macromolecules.* Editors H. J. Vogel, V. Bryson and J. O. Lampen, Academic Press, New York, p. 387.
Gross, M. and Goldwasser, E. 1969. Biochemistry *8,* 1795.
Gross, P. R. 1967. *Current Topics in Developmental Biology.* Editors A. A. Moscona and A. Monroy, Academic Press, New York, Vol. *2,* p. 1.
Gross, P. R. 1968. Ann. Rev. Biochem. *37,* 631.
Gross, P. R. and Cousineau, G. H. 1963. Biochem. Biophys. Res. Commun. *10,* 321.
Gross, P. R. and Cousineau, G. H. 1964. Exp. Cell Res. *33,* 368.
Gross, P. R., Kraemer, K. and Malkin, L. I. 1965. Biochem. Biophys. Res. Commun. *18,* 569.
Gross, P. R., Malkin, L. I. and Hubbard, M. 1965. J. Mol. Biol. *13,* 463.
Gross, P. R., Malkin, L. I. and Moyer, W. A. 1964. Proc. Natl. Acad. Sci. *51,* 407.
Gross, S. R., McCoy, M. T. and Gilmore, E. B. 1968. Proc. Natl. Acad. Sci. *61,* 253.
Grossman, A. and Mavrides, C. 1967. J. Biol. Chem. *242,* 1398.
Gruber, M. and Campagne, R. N. 1965. Koninkl. Nederl. Akad. Van. Wet *C, 68,* 270.
Grunberg-Manago, M., Clark, B. F. C., Revel, M., Rudland, P. S. and Dondon, J. 1969. J. Mol. Biol. *40,* 33.
Guest, J. R. and Yanofsky, C. 1966. Nature *210,* 799.
Guha, A. and Szybalski, W. 1968. Virology *34,* 608.
Guha, A., Tabaczyński, M. and Szybalski, W. 1968. J. Mol. Biol. *35,* 207.
Gunsalus, I. C., Gunsalus, C. F., Chakrabarty, A. M., Sikes, S. and Crawford, I. P. 1968. Genetics *60,* 419.
Gupta, N. K. and Khorana, H. G. 1966. Proc. Natl. Acad. Sci. *56,* 772.
Gupta, S. L., Chen, J., Schaefer, L., Lengyel, P. and Weissman, S. M. 1970. Biochem. Biophys. Res. Commun. *39,* 883.
Gurdon, J. B. 1968(a). *Essays in Biochemistry.* Editors P. N. Campbell and G. D. Greville, Academic Press, London, Vol. *4,* p. 25.
Gurdon, J. B. 1968(b). J. Emb. Exp. Morph. *20,* 401.
Gurdon, J. B. 1969. Dev. Biol. Supp. *3,* 59.
Gurdon, J. B. and Ford, P. J. 1967. Nature *216,* 666.
Gurdon, J. B. and Laskey, R. A. 1970. J. Emb. Exp. Morph. *24,* 227.
Gurdon, J. B. and Uehlinger, V. 1966. Nature *210,* 1240.
Gurdon, J. B. and Woodland, H. R. 1969. Proc. Roy. Soc., London, Series B, *173,* 99.
Gussin, G. N. 1966. J. Mol. Biol. *21,* 435.
Guthrie, C. and Nomura, M. 1968. Nature *219,* 232.
Gutierrez, R. M. and Hnilica, L. S. 1967. Science *157,* 1324.
Guttman, B. S. and Novick, A. 1963. Cold Spring Harbor Symp. Quant. Biol. *28,* 373.
Gyenes, L. 1969. Rev. Can. Biol. *28,* 179.
Gyenes, L. 1970. *Protides of the Biological Fluids.* Proc. 17th Colloq. Bruges. Pergamon, Oxford, p. 63.
Hacker, B. and Mandel, L. R. 1969. Biochim. Biophys. Acta *190,* 38.

Hadjivassiliou, A. and Brawerman, G. 1965. Biochim. Biophys. Acta *103,* 211.
Hadjivassiliou, A. and Brawerman, G. 1967. Biochemistry *6,* 1934.
Hadorn, E. 1965. Brookhaven Symp. Biol. *18,* 148.
Hadorn, E. 1966. *Major Problems in Developmental Biology.* Editor M. Locke, Academic Press, New York, p. 85.
Haenni, A. L. and Lucas-Lenard, J. 1968. Proc. Natl. Acad. Sci. *61,* 1363.
Hagopian, A., Bosmann, H. B. and Eylar, E. H. 1968. Arch. Biochem. Biophys. *128,* 387.
Hall, B. D., Green, M., Nygaard, A. P. and Boezi, J. 1963. Cold Spring Harbor Symp. Quant. Biol. *28,* 201.
Hall, B. D. and Spiegelman, S. 1961. Proc. Natl. Acad. Sci. *47,* 137.
Hall, R. H. 1971. *The Modified Nucleosides in Nucleic Acids.* Columbia University Press, New York.
Hallberg, R. L. and Brown, D. D. 1969. J. Mol. Biol. *46,* 393.
Hallick, L., Boyce, R. P. and Echols, H. 1969. Nature *223,* 1239.
Hamada, K., Yang, P., Heintz, R. and Schweet, R. 1968. Arch. Biochem. Biophys. *125,* 598.
Hamburger, V. and Hamilton, H. L. 1951. J. Morphol. *88,* 49.
Hamerton, J. L., Giannelli, F., Collins, F., Hallett, J., Fryer, A., McGuire, V. M. and Short, R. V. 1969. Nature *222,* 1277.
Hamilton, T. H. 1963. Proc. Natl. Acad. Sci. *49,* 373.
Hamilton, T. H. 1964. Proc. Natl. Acad. Sci. *51,* 83.
Hamilton, T. H. 1968. Science *161,* 649.
Hamilton, T. H. and Teng, C. S. 1969. Genetics *61,* Supp. *1,* 381.
Hamilton, T. H., Teng, C. S. and Means, A. R. 1968. Proc. Natl. Acad. Sci. *59,* 1265.
Hamilton, T. H., Widnell, C. C. and Tata, J. R. 1965. Biochim. Biophys. Acta *108,* 168.
Hamilton, T. H., Widnell, C. C. and Tata, J. R. 1968. J. Biol. Chem. *243,* 408.
Hancock, R. L. 1967. Can. J. Biochem. *45,* 1513.
Hancock, R. L., McFarland, P. and Fox, R. R. 1967. Experientia *23,* 806.
Handmaker, S. D. and Graef, J. W. 1970. Biochim. Biophys. Acta *199,* 95.
Hanocq-Quertier, J., Baltus, E., Ficq, A. and Brachet, J. 1968. J. Emb. Exp. Morph. *19,* 273.
Hardin, J. W., O'Brien, T. J. and Cherry, J. H. 1970. Biochim. Biophys. Acta *224,* 667.
Harel, J., Hanania, N., Tapiero, H. and Harel, L. 1968. Biochem. Biophys. Res. Commun. *33,* 696.
Harris, C. L., Titchener, E. B. and Cline, A. L. 1969. J. Bact. *100,* 1322.
Harris, H. 1959. Biochem. J. *73,* 362.
Harris, H. 1963. *Progress in Nucleic Acid Research and Molecular Biology.* Editors J. N. Davidson and W. Cohn, Academic Press, New York, Vol. *2,* p. 19.
Harris, H. 1967. J. Cell Sci. *2,* 23.
Harris, H. 1968. *Nucleus and Cytoplasm.* Clarendon Press, Oxford.
Harris, H. 1970. *Cell Fusion.* Clarendon Press, Oxford.
Harris, H. and Cook, P. R. 1969. J. Cell Sci. *5,* 121.
Harris, H., Watkins, J. F., Ford, C. E. and Schoefl, G. I. 1966. J. Cell Sci. *1,* 1.
Harris, P. 1967. Exp. Cell. Res. *48,* 569.
Hartman, P. E., Loper, J. C. and Šerman, D. 1960. J. Gen. Microbiol. *22,* 323.
Hartman, P. E., Rusgis, C. and Stahl, R. C. 1965. Proc. Natl. Acad. Sci. *53,* 1332.
Hartmann, J. F. and Comb, D. G. 1969. J. Mol. Biol. *41,* 155.
Hartwell, L. H. and Magasanik, B. 1963. J. Mol. Biol. *7,* 401.

Hartwell, L. H. and Magasanik, B. 1964. J. Mol. Biol. *10,* 105.
Haschemeyer, A. E. V. 1969(a). Proc. Natl. Acad. Sci. *62,* 128.
Haschemeyer, A. E. V. 1969(b). Comp. Biochem. Physiol. *28,* 535.
Haselkorn, R., Vogel, M. and Brown, R. D. 1969. Nature *221,* 836.
Hashizume, H. and Sehon, A. H. 1970. Proc. Can. Fed. Biol. Soc. *13,* 36.
Hatfield, D., Hofnung, M. and Schwartz, M. 1969(a). J. Bact. *98,* 559.
Hatfield, D., Hofnung, M. and Schwartz, M. 1969(b). J. Bact. *100,* 1311.
Hatfield, G. W. and Burns, R. O. 1970. Proc. Natl. Acad. Sci. *66,* 1027.
Hattman, S. and Hofschneider, P. H. 1968. J. Mol. Biol. *35,* 513.
Haurowitz, F. 1967. Cold Spring Harbor Symp. Quant. Biol. *32,* 559.
Hausen, P. and Stein, H. 1970. Europ. J. Biochem. *14,* 278.
Hausmann, R. and Gold, M. 1966. J. Biol. Chem. *241,* 1985.
Hawtrey, A. O. 1969. Biochem. J. *113,* 643.
Hayashi, M., Hayashi, M. N. and Spiegelman, S. 1963. Proc. Natl. Acad. Sci. *50,* 664.
Hayashi, M. and Spiegelman, S. 1961. Proc. Natl. Acad. Sci. *47,* 1564.
Hayashi, M., Spiegelman, S., Franklin, N. C. and Luria, S. E. 1963. Proc. Natl. Acad. Sci. *49,* 729.
Hayashi, N., Yoda, B. and Kikuchi, G. 1969. Arch. Biochem. Biophys. *131,* 83.
Hayashi, N., Yoda, B. and Kikuchi, G. 1970. J. Biochem. *67,* 859.
Hayes, W. H. 1964. *The Genetics of Bacteria and Their Viruses.* John Wiley and Sons Inc., New York.
Hayward, R. S. and Weiss, S. B. 1966. Proc. Natl. Acad. Sci. *55,* 1161.
Hayward, W. S. and Green, M. H. 1969. Proc. Natl. Acad. Sci. *64,* 962.
Hayward, W. S. and Green, M. H. 1970. Biochim. Biophys. Acta *204,* 58.
Hegeman, G. D. 1966(a). J. Bact. *91,* 1140.
Hegeman, G. D. 1966(b). J. Bact. *91,* 1161.
Heinemann, S. F. and Spiegelman, W. G. 1970. Proc. Natl. Acad. Sci. *67,* 1122.
Helgeson, J. P. 1968. Science *161,* 974.
Henderson, A. and Ritossa, F. 1970. Genetics *66,* 463.
Hendler, R. W. 1963. Science *142,* 402.
Hendler, R. W. 1965. Nature *207,* 1053.
Hendley, D. D., Hirsch, G. P. and Stregler, B. L. 1969. Biochim. Biophys. Acta *190,* 312.
Hennig, W. 1968. J. Mol. Biol. *38,* 227.
Henning, U., Dennert, G., Hertel, R. and Shipp, W. S. 1966. Cold Spring Harbor Symp. Quant. Biol. *31,* 227.
Henning, U. and Herz, G. 1964. Zeit. Vererbungsl. *95,* 260.
Henning, U., Herz, C. and Szolyvay, K. 1964. Zeit. Vererbungsl. *95,* 236.
Henning, U., Szolyvay, K. and Herz, C. 1964. Zeit. Vererbungsl. *95,* 276.
Henshaw, E. C. 1968. J. Mol. Biol. *36,* 401.
Henshaw, E. C. and Loebenstein, J. 1970. Biochim. Biophys. Acta *199,* 405.
Henshaw, E. C., Revel, M. and Hiatt, H. H. 1965. J. Mol. Biol. *14,* 241.
Herbst, E. J. and Bachrach, U. 1970. Editors. *Metabolism and Biological Functions of Polyamines.* Ann. N.Y. Acad. Sci. *171,* 691.
Herman, R. K. and Pardee, A. B. 1965. Biochim. Biophys. Acta *108,* 513.
Herrington, M. D. and Hawtrey, A. O. 1969. Biochem. J. *115,* 671.
Hershey, J. W. B., Dewey, K. F. and Thach, R. E. 1969. Nature *222,* 944.
Hershey, J. W. B. and Thach, R. E. 1967. Proc. Natl. Acad. Sci. *57,* 759.
Herskowitz, I. and Signer, E. R. 1970. J. Mol. Biol. *47,* 545.
Herzberg, M., Lelong, J. C. and Revel, M. 1969. J. Mol. Biol. *44,* 297.
Herzberg, M., Revel, M. and Danon, D. 1969. Europ. J. Biochem. *11,* 148.
Herzenberg, L. A., McDevitt, H. O. and Herzenberg, L. A. 1968. Ann. Rev. Genetics *2,* 209.

Hestrin, S. and Lindegren, C. C. 1951. Nature *168,* 913.
Heyden, H. W. V. and Zachau, H. G. 1971. Biochim. Biophys. Acta *232,* 651.
Heywood, S. M. 1969. Cold Spring Harbor Symp. Quant. Biol. *34,* 799.
Heywood, S. M. 1970(a). Proc. Natl. Acad. Sci. *67,* 1782.
Heywood, S. M. 1970(b). Nature *225,* 696.
Heywood, S. M. and Nwagwu, M. 1968. Proc. Natl. Acad. Sci. *60,* 229.
Hill, C. W., Squires, C. and Carbon, J. 1970. J. Mol. Biol. *52,* 557.
Hill, R. N. and Yunis, J. J. 1967. Science *155,* 1120.
Hill, T. L. 1969. Proc. Natl. Acad. Sci. *64,* 267.
Hill, W. E., Rossetti, G. P. and Van Holde, K. E. 1969. J. Mol. Biol. *44,* 263.
Hilschmann, N. 1967. *Gamma Globulins. Structure and Control of Biosynthesis.* Nobel Symposium *3.* Editor J. Killander, Almqvist and Wiksell, Stockholm, p. 33.
Hindley, J. and Staples, D. H. 1969. Nature *224,* 964.
Hinkle, D. C. and Chamberlin, M. 1970. Cold Spring Harbor Symp. Quant. Biol. *35,* 65.
Hiraga, S. 1969. J. Mol. Biol. *39,* 159.
Hiraga, S., Ito, K., Hamada, K. and Yura, T. 1967. Biochem. Biophys. Res. Commun. *26,* 522.
Hirschfield, I. N., De Deken, R., Horn, P. C., Hopwood, D. A. and Maas, W. K. 1968. J. Mol. Biol. *35,* 83.
Hirsch-Kauffmann, M. and Sauerbier, W. 1968. Molec. Gen. Genetics *102,* 89.
Hirsh, D. 1970. Nature *228,* 57.
Hirvonen, A. P. and Vogel, H. J. 1970. Biochem. Biophys. Res. Commun. *41,* 1611.
Hishizawa, T., Lessard, J. L. and Pestka, S. 1970. Proc. Natl. Acad. Sci. *66,* 523.
Hitzeroth, H., Klose, J., Ohno, S. and Wolf, U. 1968. Biochem. Genet. *1,* 287.
Hnilica, L. S. 1967. *Progress In Nucleic Acid Research and Molecular Biology.* Editors, J. N. Davidson and W. E. Cohn, Academic Press, New York, Vol. 7, p. 25.
Hnilica, L., Johns, E. W. and Butler, J. A. V. 1962. Biochem. J. *82,* 123.
Hnilica, L. S. and Johnson, A. W. 1971. Exp. Cell Res. *63,* 261.
Hoerz, W. and McCarty, K. S. 1969. Proc. Natl. Acad. Sci. *63,* 1206.
Hoerz, W. and McCarty, K. S. 1971. Biochim. Biophys. Acta. *228,* 526.
Hogan, B. L. M. 1969. Biochim. Biophys. Acta *182,* 264.
Hogan, B. L. M. and Korner, A. 1968. Biochim. Biophys. Acta *169,* 139.
Hogg, R. W. and Englesberg, E. 1969. J. Bact. *100,* 423.
Hogness, D. S., Cohn, M. and Monod, J. 1955. Biochim. Biophys. Acta *16,* 99.
Hogness, D. S. and Simmons, J. R. 1964. J. Mol. Biol. *9,* 411.
Hohmann, P. and Cole, R. D. 1969. Nature, *223,* 1064.
Hokin, L. E. and Hokin, M. R. 1958. J. Biol. Chem. *233,* 805.
Holder, J. W. and Lingrel, J. B. 1970. Biochim. Biophys. Acta *204,* 210.
Holland, J. J., Taylor, M. W. and Buck, C. A. 1967. Proc. Natl. Acad. Sci. *58,* 2437.
Holmes, R. S. and Markert, C. L. 1969. Proc. Natl. Acad. Sci. *64,* 205.
Holmes, R., Sheinin, R. and Crocker, B. F. 1961. Canad. J. Biochem. Physiol. *39,* 45.
Holt, P. G. and Oliver, I. T. 1969. Biochemistry *8,* 1429.
Holzer, H. 1969. Adv. Enzymol. *32,* 297.
Honig, G. R. and Rabinovitz, M. 1966. J. Biol. Chem. *241,* 1681.
Honig, G. R., Rowan, B. Q. and Mason, R. G. 1969. J. Biol. Chem. *244,* 2027.
Hood, L., Eichmann, K., Lackland, H., Krause, R. M. and Ohms, J. J. 1970. Nature *228,* 1040.
Hood, L. and Talmage, D. W. 1970. Science *168,* 325.
Horgen, P. A. and Griffin, D. H. 1971. Proc. Natl. Acad. Sci. *68,* 338.
Horiuchi, K., Lodish, H. F. and Zinder, N. D. 1966. Virology *28,* 438.
Horiuchi, K. and Matsuhashi, S. 1970. Virology *42,* 49.
Horowitz, N. H. 1948. Genetics *33,* 612.

Horowitz, N. H. 1965. *Evolving Genes and Proteins.* Editors V. Bryson and H. J. Vogel, Academic Press, New York, p. 15.
Horowitz, N. H., Feldman, H. M. and Pall, M. L. 1970. J. Biol. Chem. *245,* 2784.
Horowitz, N. H. and Metzenberg, R. L. 1965. Ann. Rev. Biochem. *34,* 527.
Hosoda, J. and Mathews, E. 1968. Proc. Natl. Acad. Sci. *61,* 997.
Housman, D., Jacobs-Lorena, M., Rajbhandary, U. L. and Lodish, H. F. 1970. Nature *227,* 913.
Houssais, J. F. and Attardi, G. 1966. Proc. Natl. Acad. Sci. *56,* 616.
Howard, G. A., Adamson, S. D. and Herbert, E. 1970. Biochim. Biophys. Acta *213,* 237.
Howell, R. R., Loeb, J. N. and Tomkins, G. M. 1964. Proc. Natl. Acad. Sci. *52,* 1241.
Hoyer, B. H., McCarthy, B. J. and Bolton, E. T. 1963. Science *140,* 1408.
Hradec, J. and Dušek, Z. 1968. Biochem. J. *110,* 1.
Hradec, J. and Štroufova, A. 1960. Biochim. Biophys. Acta *40,* 32.
Hradecna, Z. and Szybalski, W. 1967. Virology *32,* 633.
Hradecna, Z. and Szybalski, W. 1969. Virology *38,* 473.
Hsie, A. W., Rickenberg, H. V., Schulz, D. W. and Kirsch, W. M. 1969. J. Bact. *98,* 407.
Hsu, T. C., Schmid, W. and Stubblefield, E. 1964. *The Role of Chromosomes in Development.* Editor M. Locke, Academic Press, New York, p. 83.
Hsu, W. T., Foft, J. W. and Weiss, S. B. 1967. Proc. Natl. Acad. Sci. *58,* 2028.
Hsu, W. T. and Weiss, S. B. 1969. Proc. Natl. Acad. Sci. *64,* 345.
Hu, A. S. L. 1969. Biochim. Biophys. Acta *174,* 387.
Huang, R. C. and Bonner, J. 1965. Proc. Natl. Acad. Sci. *54,* 960.
Huang, R. C. and Huang, P. C. 1969. J. Mol. Biol. *39,* 365.
Huberman, J. A. and Attardi, G. 1967. J. Mol. Biol. *29,* 487.
Hudson, J., Goldstein, D. and Weil, R. 1970. Proc. Natl. Acad. Sci. *65,* 226.
Hultin, T. 1950. Exp. Cell Res. *1,* 599.
Hultin, T. 1952. Exp. Cell Res. *3,* 494.
Hultin, T. 1961. Exp. Cell Res. *25,* 405.
Hunt, T., Hunter, T. and Munro, A. 1968(a). J. Mol. Biol. *36,* 31.
Hunt, T. R., Hunter, A. R. and Munro, A. J. 1968(b). Nature *220,* 481.
Hunt, T., Hunter, T. and Munro, A. 1969. J. Mol. Biol. *43,* 123.
Hunter, G. D. and Goodsall, R. A. 1961. Biochem. J. *78,* 564.
Hunter, G. D. and James, A. T. 1963. Nature *198,* 789.
Hurwitz, J., Anders, M., Gold, M. and Smith, I. 1965. J. Biol. Chem. *240,* 1256.
Hurwitz, J., Evans, A., Babinet, C. and Skalka, A. 1963. Cold Spring Harbor Symp. Quant. Biol. *28,* 59.
Hurwitz, J., Furth, J. J., Anders, M., Ortiz, P. J. and August, J. T. 1961. Cold Spring Harbor Symp. Quant. Biol. *26,* 91.
Hütter, R. and DeMoss, J. A. 1967. J. Bact. *94,* 1896.
Huxley, H. E. and Zubay, G. 1960. J. Mol. Biol. *2,* 10.
Iaccarino, M. and Berg, P. 1971. J. Bact. *105,* 527.
Iida, Y., Kameyama, T., Ohshima, Y. and Horiuchi, T. 1970. Molec. Gen. Genetics *106,* 296.
Ilan, J. 1968. J. Biol. Chem. *243,* 5859.
Ilan, J. 1969. Biochemistry *8,* 4825.
Ilan, J. and Ilan, J. 1971. Dev. Biol. *25,* 280.
Ilan, J., Ilan, J. and Patel, N. 1970. J. Biol. Chem. *245,* 1275.
Imamoto, F. 1968(a). Nature *220,* 31.
Imamoto, F. 1968(b). Proc. Natl. Acad. Sci. *60,* 305.
Imamoto, F. 1970. Molec. Gen. Genetics *106,* 123.

Imamoto, F. and Ito, J. 1968. Nature *220,* 27.
Imamoto, F., Ito, J. and Yanofsky, C. 1966. Cold Spring Harbor Symp. Quant. Biol. *31,* 235.
Imamoto, F., Morikawa, N. and Sato, K. 1965. J. Mol. Biol. *13,* 169.
Imamoto, F. and Yanofsky, C. 1967(a). J. Mol. Biol. *28,* 1.
Imamoto, F. and Yanofsky, C. 1967(b). J. Mol. Biol. *28,* 25.
Imrie, R. C. and Hutchison, W. C. 1965. Biochim. Biophys. Acta *108,* 106.
Imsande, J. 1970. J. Bact. *101,* 173.
Infante, A. A. and Nemer, M. 1968. J. Mol. Biol. *32,* 543.
Ingles, C. J. and Dixon, G. H. 1967. Proc. Nat. Acad. Sci. *58,* 1011.
Ingles, C. J., Trevithick, J. R., Smith, M. and Dixon, G. H. 1966. Biochem. Biophys. Res. Commun. *22,* 627.
Ingram, V. M. 1963. *The Hemoglobins in Genetics and Evolution.* Columbia University Press, New York.
Ingram, V. M. 1967. Harvey Lectures *61,* 43.
Ippen, K., Miller, J. H., Scaife, J. and Beckwith, J. 1968. Nature *217,* 825.
Ishida, T. and Sueoka, N. 1968. J. Biol. Chem. *243,* 5329.
Ishikawa, K., Kuroda, C. and Ogata, K. 1969. Biochim. Biophys. Acta *179,* 316.
Ishikawa, K., Kuroda, C. and Ogata, K. 1970. Biochim. Biophys. Acta *213,* 505.
Ishikawa, K., Kuroda, C., Ueki, M. and Ogata, K. 1970(a). Biochim. Biophys. Acta *213,* 495.
Ishikawa, K., Ueki, M., Nagai, K. and Ogata, K. 1970(b). Biochim. Biophys. Acta *213,* 542.
Ishitsuka, H. and Kaji, A. 1970. Proc. Natl. Acad. Sci. *66,* 168.
Itano, H. A. 1966. J. Cell. Phys. *67,* Supp. *1,* 65.
Ito, J. and Crawford, C. 1965. Genetics *52,* 1303.
Ito, J. and Imamoto, F. 1968. Nature *220,* 441.
Ito, K., Hiraga, S. and Yura, T. 1969. Genetics *61,* 521.
Itzhaki, R. F. 1971. Cited in Biochem. Biophys. Res. Commun. *41,* 25.
Iwasaki, K., Sabol, S., Wahba, A. J. and Ochoa, S. 1968. Arch. Biochem. Biophys. *125,* 542.
Izawa, M., Allfrey, V. G. and Mirsky, A. E. 1963. Proc. Natl. Acad. Sci. *49,* 544.
Jackson, C. D., Irving, C. C. and Sells, B. H. 1970. Biochim. Biophys. Acta *217,* 64.
Jackson, R. and Hunter, T. 1970. Nature *227,* 672.
Jacob, F. and Adelberg, E. A. 1959. Comptes Rend. Acad. Sci. (Paris) *249,* 189.
Jacob, F., Fuerst, C. R. and Wollman, E. 1957. Ann. Inst. Pasteur *93,* 724.
Jacob, F. and Monod. J. 1961(a). J. Mol. Biol. *3,* 318.
Jacob, F. and Monod, J. 1961(b). Cold Spring Harbor Symp. Quant. Biol. *26,* 193.
Jacob, F. and Monod, J. 1963. *Cytodifferentiation and Macromolecular Synthesis.* Editor M. Locke, Academic Press, New York, p. 30.
Jacob, F., Perrin, D., Sanchez, C. and Monod, J. 1960. Comptes Rend. Acad. Sci. (Paris) *250,* 1727.
Jacob, F., Sussman, R. and Monod, J. 1962. Comptes Rend. Acad. Sci. (Paris) *254,* 4214.
Jacob, F., Ullman, A. and Monod, J. 1964. Comptes Rend. Acad. Sci. (Paris) *258,* 3125.
Jacob, F., Ullmann, A. and Monod, J. 1965. J. Mol. Biol. *13,* 704.
Jacob, F. and Wollman, E. 1961. *Sexuality and the Genetics of Bacteria.* Academic Press, New York.
Jacob, S. T. and Busch, H. 1967. Biochim. Biophys. Acta *138,* 249.
Jacob, S. T., Sajdel, E. M. and Munro, H. N. 1968. Biochim. Biophys. Acta *157,* 421.
Jacob, S. T., Sajdel, E. M. and Munro, H. N. 1969. Europ. J. Biochem. *7,* 449.

Jacob, S. T., Sajdel, E. M. and Munro, H. N. 1970. Biochem. Biophys. Res. Commun. *38,* 765.
Jacobson, K. B. 1971. Nature New Biology *231,* 17.
Jacobson, M. F. and Baltimore, D. 1968. Proc. Natl. Acad. Sci. *61,* 77.
Jacobson, R. A. and Bonner, J. 1968. Biochem. Biophys. Res. Commun. *33,* 716.
Jacoby, G. A. and Gorini, L. 1967. J. Mol. Biol. *24,* 41.
Jacoby, G. A. and Gorini, L. 1969. J. Mol. Biol. *39,* 73.
Jacquemin-Sablon, A. and Richardson, C. C. 1970. J. Mol. Biol. *47,* 477.
Jacquet, M. and Kepes, A. 1969. Biochem. Biophys. Res. Commun. *36,* 84.
Jaenisch, R., Jacob, E. and Hofschneider, P. H. 1970. Nature *227,* 59.
Jakoby, W. B. and Bonner, D. M. 1953. J. Biol. Chem. *205,* 699.
James, D. W., Rabin, B. R. and Williams, D. J. 1969. Nature *224,* 371.
Javor, G. T., Ryan, A. and Borek, E. 1969. Biochim. Biophys. Acta *190,* 442.
Jayaraman, K., Müller-Hill, B. and Rickenberg, H. V. 1966. J. Mol. Biol. *18,* 339.
Jayaraman, R. and Goldberg, E. B. 1969. Proc. Natl. Acad. Sci. *64,* 198.
Jayaraman, R. and Goldberg, E. B. 1970. Cold Spring Harbor Symp. Quant. Biol. *35,* 197.
Jeanteur, P. and Attardi, G. 1969. J. Mol. Biol. *45,* 305.
Jenkins, R. G. and Drabble, W. T. 1969. Nature *223,* 296.
Jensen, E. V. 1966. Can. Cancer Conf. *6,* 143.
Jensen, E. V., De Sombre, E. R. and Jungblut, P. W. 1967. Proc. Second Int. Cong. Hormonal Steroids, Excerpta Medica Foundation, Holland, p. 492.
Jensen, E. V., Suzuki, T., Kawashima, T., Stumpf, W. E., Jungblut, P. W. and De Sombre, E. R. 1968. Proc. Natl. Acad. Sci. *59,* 632.
Jeppesen, P. G. N., Steitz, J. A., Gesteland, R. F. and Spahr, P. F. 1969. Nature *226,* 230.
Jerez, C., Sandoval, A., Allende, J., Henes, C. and Ofengand, J. 1969. Biochemistry *8,* 3006.
Jergil, B. and Dixon, G. H. 1970. J. Biol. Chem. *245,* 425.
Jergil, B., Sung, M. and Dixon, G. H. 1970. J. Biol. Chem. *245,* 5867.
Jerne, N. K. 1955. Proc. Natl. Acad. Sci. *41,* 849.
Joel, P. B. and Hagerman, D. D. 1969. Biochim. Biophys. Acta *195,* 328.
John, H. A., Birnstiel, M. L. and Jones, K. W. 1969. Nature *223,* 582.
Johns, E. W. 1964. Biochem. J. *92,* 55.
Johns, E. W. and Forrester, S. 1969. Europ. J. Biochem. *8,* 547.
Johns, E. W., Phillips, D. M. P., Simson, P. and Butler, J. A. V. 1960. Biochem. J. *77,* 631.
Johns, E. W., Phillips, D. M. P., Simson, P. and Butler, J. A. V. 1961. Biochem. J. *80,* 189.
Johnson, A. W. and Hnilica, L. S. 1970. Biochim. Biophys. Acta *224,* 518.
Johnson, F. H. 1938. J. Cell. Comp. Physiol. *12,* 281.
Johnson, J. D., Jant, B. A., Sokoloff, L. and Kaufman, S. 1969. Biochim. Biophys. Acta *179,* 526.
Johnson, R. T. and Harris, H. 1969. J. Cell Sci. *5,* 625.
Johnson, T. C. 1969. J. Neurochem. *16,* 1125.
Johnson, T. C. and Luttges, M. W. 1966. J. Neurochem. *13,* 545.
Jones, G., Marcuson, E. C. and Roitt, I. M. 1970. Nature *227,* 1051.
Jones, K. W. 1970. Nature *225,* 912.
Jones, O. W., Dieckmann, M. and Berg, P. 1968. J. Mol. Biol. *31,* 177.
Jones, T. C. 1969. Molec. Gen. Genetics *105,* 91.
Jones-Mortimer, M. C. 1968(a). Biochem. J. *110,* 589.
Jones-Mortimer, M. C. 1968(b). Biochem. J. *110,* 597.

Jordan, E. and Saedler, H. 1967. Molec. Gen. Genetics *100,* 283.
Jordan, E., Saedler, H. and Starlinger, P. 1968. Molec. Gen. Genetics *102,* 353.
Jorgensen, S., Buch, L. B. and Nierlich, D. P. 1969. Science *164,* 1067.
Jost, J. P., Hsie, A., Hughes, S. D. and Ryan, L. 1970. J. Biol. Chem. *245,* 351.
Jost, J. P., Hsie, A. W. and Rickenberg, H. V. 1969. Biochem. Biophys. Res. Commun. *34,* 748.
Jost, J. P., Khairallah, E. A. and Pitot, H. C. 1968. J. Biol. Chem. *243,* 3057.
Juergens, W. G., Stockdale, F. E., Topper, Y. J. and Elias, J. J. 1965. Proc. Natl. Acad. Sci. *54,* 629.
Jukes, T. H. 1966. *Molecules and Evolution.* Columbia University Press, New York.
Kabach, H. R. 1970. Ann. Rev. Biochem. *39,* 561.
Kabat, D. 1968. J. Biol. Chem. *243,* 2597.
Kabat, D. 1970. Biochemistry *9,* 4160.
Kabat, D. 1971. Biochemistry *10,* 197.
Kabat, D. and Rich, A. 1969. Biochemistry *8,* 3742.
Kadenbach, B. 1967(a). Biochim. Biophys. Acta *134,* 430.
Kadenbach, B. 1967(b). Biochim. Biophys. Acta *138,* 651.
Kadenbach, B. 1969. Europ. J. Biochem. *10,* 312.
Kaempfer, R. 1969. Nature *222,* 950.
Kaempfer, R. 1970. Nature *228,* 534.
Kaempfer, R. O. R., Meselson, M. and Raskas, H. J. 1968. J. Mol. Biol. *31,* 277.
Kafatos, F. C. 1968. Proc. Natl. Acad. Sci. *59,* 1251.
Kafatos, F. C. and Reich, J. 1968. Proc. Natl. Acad. Sci. *60,* 1458.
Kaiser, A. D. 1957. Virology, *3,* 42.
Kaiser, A. D. 1962. J. Mol. Biol. *4,* 275.
Kaji, H., Suzuka, I. and Kaji, A. 1965. J. Biol. Chem. *241,* 1251.
Kalckar, H. M., Kurahashi, K. and Jordan, E. 1959. Proc. Natl. Acad. Sci. *45,* 1776.
Kalman, S. M. and Opsahl, P. M. 1961. Biochim. Biophys. Acta *49,* 614.
Kaltschmidt, E. and Whittmann, H. G. 1970. Proc. Natl. Acad. Sci. *67,* 1276.
Kaminskas, E. and Magasanik, B. 1970. J. Biol. Chem. *245,* 3549.
Kamiyama, M. and Wang, T. Y. 1971. Biochim. Biophys. Acta *228,* 563.
Kammen, H. O. and Spengler, S. J. 1970. Biochim. Biophys. Acta *213,* 352.
Kan, Y. W., Golini, F. and Thach, R. E. 1970. Proc. Natl. Acad. Sci. *67,* 1137.
Kane, J. F. and Jensen, R. A. 1970. J. Biol. Chem. *245,* 2384.
Kaneko, I. and Doi, R. H. 1966. Proc. Natl. Acad. Sci. *55,* 564.
Kano, Y., Matsushiro, A. and Shimura, Y. 1968. Molec. Gen. Genetics *102,* 15.
Kano-Sueoka, T., Nirenberg, M. and Sueoka, N. 1968. J. Mol. Biol. *35,* 1.
Kano-Sueoka, T. and Spiegelman, S. 1962. Proc. Natl. Acad. Sci. *48,* 1942.
Kano-Sueoka, T. and Sueoka, N. 1966. J. Mol. Biol. *20,* 183.
Kano-Sueoka, T. and Sueoka, N. 1968. J. Mol. Biol. *37,* 475.
Karam, J. D. and Speyer, J. F. 1970. Virology *42,* 196.
Karlsson, J. L. and Barker, H. A. 1948. J. Biol. Chem. *175,* 913.
Karlström, O. and Gorini, L. 1969. J. Mol. Biol. *39,* 89.
Karström, H. 1938. Ergebn. Enzymforsch. *7,* 350.
Kates, J. 1970. Cold Spring Harbor Symp. Quant. Biol. *35,* 743.
Kates, J. R. and McAuslan, B. R. 1967. Proc. Natl. Acad. Sci. *57,* 314.
Kates, J. R. and McAuslan, B. R. 1968. Proc. Natl. Acad. Sci. *58,* 134.
Kato, I. and Anfinsen, C. B. 1969. J. Biol. Chem. *244,* 1004.
Kaye, A. M. and Leboy, P. S. 1968. Biochim. Biophys. Acta *157,* 289.
Kaye, E. M. and Sheratzky, D. 1969. Biochim. Biophys. Acta *190,* 527.
Kazazian, H. H. and Freedman, M. L. 1968. J. Biol. Chem. *243,* 6446.
Kaziro, Y. and Inoue, N. 1968. J. Biochem. *64,* 423.

Kedes, L. H. and Gross, P. R. 1969(a). J. Mol. Biol. *42,* 559.
Kedes, L. H. and Gross, P. R. 1969(b). Nature *223,* 1335.
Kedes, L. H., Gross, P. R., Cognetti, G. and Hunter, A. L. 1969(a). J. Mol. Biol. *45,* 337.
Kedes, L. H., Hogan, B., Cognetti, G., Selvig, S., Yanover, P. and Gross, P. R. 1969(b). Cold Spring Harbor Symp. Quant. Biol. *34,* 717.
Kedinger, C., Gniadowski, M., Mandel, J. L., Gissinger, F. and Chambon, P. 1970. Biochem. Biophys. Res. Commun. *38,* 165.
Keep, E. 1962. Can. J. Genet. Cytol. *4,* 206.
Keller, E. B., Zamecnik, P. C. and Loftfield, R. B. 1954. J. Histochem. Cytochem. *2,* 378.
Kelley, W. S. and Schaechter, M. 1970. J. Mol. Biol. *50,* 171.
Kellogg, D. A., Doctor, B. P., Loebel, J. E. and Nirenberg, M. W. 1966. Proc. Natl. Acad. Sci. *55,* 912.
Kelly, M. and Clarke, P. H. 1962. J. Gen. Microbiol. *27,* 305.
Kelmers, A. D., Novelli, G. D. and Stulberg, M. P. 1965. J. Biol. Chem. *240,* 3979.
Kemp, M. B. and Hegeman, G. D. 1968. J. Bact. *96,* 1488.
Kempf, J., Popovic, D. and Mandel, P. 1970. FEBS Letters *9,* 141.
Kempf, J., Zahnd, J. P., Mandel, P. and Pfleger, N. 1970. Europ. J. Biochem. *17,* 124.
Kennedy, J. C., Till, J. E., Siminovitch, L. and McCulloch, E. A. 1965. J. Immunol. *94,* 715.
Kennell, D. 1970. J. Virol. *6,* 208.
Kenney, F. T. 1962. J. Biol. Chem. *237,* 3495.
Kenney, F. T. 1967. Science *156,* 525.
Kenney, F. T. and Kull, F. J. 1963. Proc. Natl. Acad. Sci. *50,* 493.
Kenney, F. T., Reel, J. R., Hager, C. B. and Wittliff, J. L. 1968. *Regulatory Mechanisms for Protein Synthesis in Mammalian Cells.* Editors A. S. Pietro, M. R. Lamborg and F. T. Kenney, Academic Press, New York, p. 119.
Kenney, F. T., Wicks, W. D. and Greenman, D. L. 1965. J. Cell Comp. Phys. *66,* Supp. *1,* 125.
Kepes, A. 1960. Biochim. Biophys. Acta *40,* 70.
Kepes, A. 1963. Biochim. Biophys. Acta *76,* 293.
Kerjan, P. and Szulmajster, J. 1969. FEBS Letters *5,* 288.
Kerr, S. J. 1970. Biochemistry *9,* 690.
Kessler, E. and Oesterheld, H. 1970. Nature *228,* 287.
Khawaja, J. A. and Raina, A. 1970. Biochem. Biophys. Res. Commun. *41,* 512.
Khesin, R. B. 1970. Lepetit Colloq. Biol. Med. *1,* 167.
Khesin, R. B. and Shemyakin, M. F. 1963. Biokhimiya *27,* (*Trans.*) 647.
Kidson, C. and Kirby, K. S. 1964. J. Mol. Biol. *10,* 187.
Kiefer, B. I. 1968. Proc. Natl. Acad. Sci. *61,* 85.
Kiefer, B. I., Entelis, C. F. and Infante, A. A. 1969. Proc. Natl. Acad. Sci. *64,* 857.
Kiho, Y. and Rich, A. 1965. Proc. Natl. Acad. Sci. *54,* 1751.
Killander, J. 1967. Editor, *Gamma Globulins. Structure and Control of Biosynthesis.* Nobel Symposium *3.* Almqvist and Wiksell, Stockholm.
Kincade, P. W., Lawton, A. R., Bockman, D. E. and Cooper, M. D. 1970. Proc. Natl. Acad. Sci. *67,* 1918.
King, J. L. and Jukes, T. H. 1969. Science *164,* 788.
King, T. J. and Briggs, R. 1956. Cold Spring Harbor Symp. Quant. Biol. *21,* 271.
Kinkade, J. M. 1969. J. Biol. Chem. *244,* 3375.
Kinsey, J. D. 1967. Genetics *55,* 337.
Kleihauer, E. and Stöffler, G. 1968. Molec. Gen. Genetics *101,* 59.

Kleinsmith, L. J. and Allfrey, V. G. 1969(a). Biochim. Biophys. Acta *175,* 123.
Kleinsmith, L. J. and Allfrey, V. G. 1969(b). Biochim. Biophys. Acta *175,* 136.
Kleinsmith, L. J., Allfrey, V. G. and Mirsky, A. E. 1966(a). Proc. Natl. Acad. Sci. *55,* 1182.
Kleinsmith, L. J., Allfrey, V. G. and Mirsky, A. E. 1966(b). Science *154,* 780.
Kleinsmith, L. J., Heidema, J. and Carroll, A. 1970. Nature *226,* 1025.
Klem, E. B., Hsu, W. T. and Weiss, S. B. 1970. Proc. Natl. Acad. Sci. *67,* 696.
Klevecz, R. R. 1969. J. Cell Biol. *43,* 207.
Kline, L. K., Fittler, F. and Hall, R. H. 1969. Biochemistry *8,* 4361.
Kloos, W. E. and Pattee, P. A. 1965. J. Gen. Microbiol. *39,* 195.
Kloppstech, K., Schumann, G. and Klink, F. 1969. Zeit. Physiol. Chem. *350,* 1027.
Knopf, P. M., Parkhouse, R. M. E. and Lennox, E. S. 1967. Proc. Natl. Acad. Sci. *58,* 2288.
Knowland, J. and Miller, L. 1970. J. Mol. Biol. *53,* 321.
Knox, R. and Pollock, M. R. 1944. Biochem. J. *38,* 299.
Knox, W. E. 1966. Advances in Enzyme Regulation *4,* 287.
Knox, W. E. and Greengard, O. 1965. Advances in Enzyme Regulation *3,* 247.
Knox, W. E. and Mehler, A. H. 1950. J. Biol. Chem. *187,* 419.
Knox, W. E. and Mehler, A. H. 1951. Science *113,* 237.
Kogut, M., Pollock, M. R. and Tridgell, E. J. 1956. Biochem. J. *62,* 391.
Köhler, H., Shimizu, A., Paul, C. and Putnam, F. W. 1970. Science *169,* 56.
Köhler, K. and Arends, S. 1968. Europ. J. Biochem. *5,* 500.
Kohler, R. E., Ron, E. Z. and Davis, B. D. 1968. J. Mol. Biol. *36,* 71.
Kolakofsky, D., Dewey, K. and Thach, D. E. 1969. Nature *223,* 694.
Kolodny, G. M. and Gross, P. R. 1969. Exp. Cell Res. *56,* 117.
Konijn, T. M., Van de Meene, J. G. C., Bonner, J. T. and Barkley, D. S. 1967. Proc. Natl. Acad. Sci. *58,* 1152.
Konings, R. N. H., Ward, R., Francke, B. and Hofschneider, P. H. 1970. Nature *226,* 604.
Konrad, M. 1970. J. Mol. Biol. *53,* 389.
Korenman, S. G. and Rao, B. R. 1968. Proc. Natl. Acad. Sci. *61,* 1028.
Kornberg, H. L. 1966. Biochem. J. *99,* 1.
Koshland, D. E. and Neet, K. E. 1968. Ann. Rev. Biochem. *37,* 359.
Koshland, M. E. 1967. Cold Spring Harbor Symp. Quant. Biol. *32,* 119.
Koslov, Y. V. and Georgiev, G. P. 1970. Nature *228,* 245.
Kössel, H. 1970. Biochim. Biophys. Acta *204,* 191.
Kössel, H. and Rajbhandary, U. L. 1968. J. Mol. Biol. *35,* 539.
Kourilsky, P., Bourguignon, M. F., Bouquet, M. and Gros, F. 1970. Cold Spring Harbor Symp. Quant. Biol. *35,* 305.
Kourilsky, P., Marcaud, L., Portier, M. M., Zamansky, M. H. and Gros, F. 1969. Bull. Soc. Chim. Biol. *51,* 1429.
Kourilsky, P., Marcaud, L., Sheldrick, P., Luzzati, D. and Gros, F. 1968. Proc. Natl. Acad. Sci. *61,* 1013.
Kovach, J. S., Berberich, M. A., Venetianer, P. and Goldberger, R. F. 1969(a). J. Bact. *97,* 1283.
Kovach, J. S., Phang, J. M., Blasi, F., Barton, R. W., Ballesteros-Olmo, A. and Goldberger, R. F. 1970. J. Bact. *104,* 787.
Kovach, J. S., Phang, J. M., Ference, M. and Goldberger, R. F. 1969(b). Proc. Natl. Acad. Sci. *63,* 481.
Kraft, N. and Shortman, K. 1970. Biochim. Biophys. Acta *217,* 164.
Krakow, J. S. and Canellakis, E. S. 1961. Fed. Proc. *20,* 361.
Krakow, J. S., Daley, K. and Karstadt, M. 1969. Proc. Natl. Acad. Sci. *62,* 432.

Kreil, G. and Kreil-Kiss, G. 1967. Biochem. Biophys. Res. Commun. *27,* 275.
Krisch, R. E. 1969. Nature *222,* 1295.
Kroeger, H. 1963. J. Cell. Comp. Phys. *62,* Supp. *1,* 45.
Krueger, R. G. and McCarthy, B. J. 1970. Biochem. Biophys. Res. Commun. *41,* 944.
Kruh, J., Rosa, J., Dreyfus, F. C. and Schapira, G. 1961. Biochim. Biophys. Acta *49,* 509.
Kruh, J., Tichonicky, L. and Wajcman, H. 1969. Biochim. Biophys. Acta *195,* 549.
Kubinski, H., Opara-Kubinska, Z. and Szybalski, W. 1966. J. Mol. Biol. *20,* 313.
Kuchino, Y. and Nishimura, S. 1970. Biochem. Biophys. Res. Commun. *40,* 306.
Kulka, R. G. and Sternlicht, E. 1968. Proc. Natl. Acad. Sci. *61,* 1123.
Kumar, S., Bøvre, K., Guha, A., Hradecna, Z., Maher, V. M. and Szybalski, W. 1969. Nature *221,* 823.
Kumar, S. and Szybalski, W. 1969. J. Mol. Biol. *40,* 145.
Kumar, S. and Szybalski, W. 1970. Virology *41,* 665.
Küntzel, H. 1969. Nature *222,* 142.
Kuo, J. F. and Greengard, P. 1969. Proc. Natl. Acad. Sci. *64,* 1349.
Kuo, J. F. and Greengard, P. 1970. Biochim. Biophys. Acta *212,* 434.
Kuriki, Y. and Kaji, A. 1968. Proc. Natl. Acad. Sci. *61,* 1399.
Kurimiyama, Y., Omura, T., Siekevitz, P. and Palade, G. E. 1969. J. Biol. Chem. *244,* 2017.
Kurland, C. G. 1970. Science *169,* 1171.
Kuwabara, S. 1970. Biochem. J. *118,* 457.
Kuwabara, S. and Abraham, E. P. 1969. Biochem. J. *115,* 859.
Kuwabara, S., Adams, E. P. and Abraham, E. P. 1970. Biochem. J. *118,* 475.
Kuwano, M., Apirion, D. and Schlessinger, D. 1970(a). J. Mol. Biol. *51,* 453.
Kuwano, M., Apirion, D. and Schlessinger, D. 1970(b). Science *168,* 1225.
Kuwano, M., Ishizawa, M. and Endo, H. 1968. J. Mol. Biol. *33,* 513.
Kuwano, M., Kwan, C. N., Apirion, D. and Schlessinger, D. 1969. Proc. Natl. Acad. Sci. *64,* 693.
Kuwano, M. and Schlessinger, D. 1970. Proc. Natl. Acad. Sci. *66,* 146.
Kuwano, M., Schlessinger, D. and Apirion, D. 1970(a). Nature *226,* 514.
Kuwano, M., Schlessinger, D. and Apirion, D. 1970(b). J. Mol. Biol. *51,* 75.
Kwan, C. N., Apirion, D. and Schlessinger, D. 1968. Biochemistry *7,* 427.
Labrie, F. and Korner, A. 1968. J. Biol. Chem. *243,* 1116.
Lacroute, F. and Stent, G. S. 1968. J. Mol. Biol. *35,* 165.
Laemmli, U. K. 1970. Nature *227,* 680.
Lamb, A. J., Clark-Walker, G. D. and Linnane, A. W. 1968. Biochim. Biophys. Acta *161,* 415.
Lamfrom, H. 1961. J. Mol. Biol. *3,* 241.
Landesman, R. and Gross, P. R. 1968. Dev. Biol. *18,* 571.
Landy, A., Abelson, J., Goodman, H. M. and Smith, J. D. 1967. J. Mol. Biol. *29,* 457.
Landy, A. and Spiegelman, S. 1968. Biochemistry *7,* 585.
Lane, B. G. and Tamaoki, T. 1969. Biochim. Biophys. Acta *179,* 332.
Langan, T. A. 1967. *Regulation of Nucleic Acid and Protein Biosynthesis.* Editors V. V. Koningsberger and L. Bosch, Elsevier, Amsterdam, p. 233.
Langan, T. A. 1968(a). Science *162,* 579.
Langan, T. A. 1968(b). *Regulatory Mechanisms for Protein Synthesis in Mammalian Cells.* Editors A. S. Pietro, M. R. Lamborg and F. T. Kenney, Academic Press, New York, p. 101.
Langan, T. A. 1969(a). J. Biol. Chem. *244,* 5763.

Langan, T. A. 1969(b). Proc. Natl. Acad. Sci. *64,* 1276.
Lanni, Y. T. 1969. J. Mol. Biol. *44,* 173.
Lanni, Y. T., McCorquodale, D. J. and Wilson, C. M. 1964. J. Mol. Biol. *10,* 19.
Laskey, R. A. and Gurdon, J. B. 1970. Nature *228,* 1332.
Lavallé, R. and DeHauwer, G. 1968. J. Mol. Biol. *37,* 269.
Lavallé, R. and DeHauwer, G. 1970. J. Mol. Biol. *51,* 435.
Lawford, G. R. 1969. Biochem. Biophys. Res. Commun. *37,* 143.
Laycock, D. G. and Hunt, J. A. 1969. Nature *221,* 1118.
Lazarus, H. M. and Sporn, M. B. 1967. Proc. Natl. Acad. Sci. *57,* 1386.
Leader, D. P., Wool, I. G. and Castles, J. J. 1970. Proc. Natl. Acad. Sci. *67,* 523.
Leaver, C. J. and Key, J. L. 1970. J. Mol. Biol. *49,* 671.
Leavitt, R. I. and Umbarger, H. E. 1961. J. Biol. Chem. *236,* 2486.
Leavitt, W. W., Friend, J. P. and Robinson, J. A. 1969. Science *165,* 496.
LeClerc, J., Martinage, A., Moschetto, Y. and Biserte, G. 1969. Europ. J. Biochem. *11,* 261.
Leder, P., Bernardi, A., Livingston, D., Loyd, B., Roufa, D. and Skogerson, L. 1969. Cold Spring Harbor Symp. Quant. Biol. *34,* 411.
Leder, P. and Bursztyn, H. 1966. Cold Spring Harbor Symp. Quant. Biol. *31,* 297.
Lederberg, E. M. 1952. Genetics *37,* 469.
Lederberg, J. 1947. Genetics *32,* 505.
Lederberg, J. 1950. J. Bact. *60,* 381.
Lederberg, J. 1959. Science *129,* 1649.
Lederberg, J. and Tatum, E. L. 1946. Nature *158,* 558.
Lee, J. C. and Ingram, V. M. 1967. Science *158,* 1330.
Lee, K. L., Reel, J. R. and Kenney, F. T. 1970. J. Biol. Chem. *245,* 5806.
Lee, N. 1971. From Whitfield et al. (1970).
Lee, S. Y. and Brawerman, G. 1971. Biochemistry *10,* 510.
Lee, S. Y., Mendecki, J. and Brawerman, G. 1971. Proc. Natl. Acad. Sci. *68,* 1331.
Leisinger, T. and Vogel, H. J. 1969. Biochim. Biophys. Acta. *182,* 572.
Leisinger, T., Vogel, R. H. and Vogel, H. J. 1969. Proc. Natl. Acad. Sci. *64,* 686.
Leitner, F., Sweeney, H. M., Martin, T. F. and Cohen, S. 1963. J. Bact. *86,* 717.
Leive, L. and Kollin, V. 1967. J. Mol. Biol. *24,* 247.
Lelong, J. C., Grunberg-Manago, M., Dondon, J., Gros, D. and Gros, F. 1970. Nature *226,* 505.
Lembach, K. J. and Buchanan, J. M. 1970. J. Biol. Chem. *245,* 1575.
Lembach, K. J., Kuninaka, A. and Buchanan, J. M. 1969. Proc. Natl. Acad. Sci. *62,* 446.
Lengyel, P. and Söll, D. 1969. Bact. Rev. *33,* 264.
Lennox, E. S. and Cohn, M. 1967. Ann. Rev. Biochem. *36,* 365.
Lerner, M. P. and Johnson, T. C. 1970. J. Biol. Chem. *245,* 1388.
Lester, G. and Bonner, D. M. 1952. J. Bact. *63,* 759.
Lester, G. and Yanofsky, C. 1961. J. Bact. *81,* 81.
Levere, R. D. and Granick, S. 1965. Proc. Natl. Acad. Sci. *54,* 134.
Levere, R. D. and Granick, S. 1967. J. Biol. Chem. *242,* 1903.
Levere, R. D., Kappas, A. and Granick, S. 1967. Proc. Natl. Acad. Sci. *58,* 985.
Levin, A. P. and Hartman, P. E. 1963. J. Bact. *86,* 820.
Levin, A. S., Fudenber, H. H., Hopper, J. E., Wilson, S. K. and Nisonoff, A. 1971. Proc. Natl. Acad. Sci. *68,* 169.
Levinthal, C., Keynan, A. and Higa, A. 1962. Proc. Natl. Acad. Sci. *48,* 1631.
Levinthal, C., Signer, E. R. and Fetherolf, K. 1962. Proc. Natl. Acad. Sci. *48,* 1230.
Levitan, I. B. and Webb, T. E. 1969. J. Biol. Chem. *244,* 4684.
Levitan, I. B. and Webb, T. E. 1970. J. Mol. Biol. *48,* 339.

Lewis, E. B. 1951. Cold Spring Harbor Symp. Quant. Biol. *16,* 159.
Lewis, E. B. 1963. Amer. Zool. *3,* 33.
Liau, M. C. and Perry, R. P. 1969. J. Cell. Biol. *42,* 272.
Liew, C. C., Haslett, G. W. and Allfrey, V. G. 1970. Nature *226,* 414.
Lim, L. and Canellakis, E. S. 1970. Nature *227,* 710.
Lima-De-Faria, A., Birnstiel, M. and Jaworska, H. 1969. Genetics *61,* Supp. *1,* 145.
Lin, E. C. C. and Knox, W. E. 1957. Biochim. Biophys. Acta *26,* 85.
Lin, S. Y. and Riggs, A. D. 1970. Nature *228,* 1184.
Lindberg, U. and Darnell, J. E. 1970. Proc. Natl. Acad. Sci. *65,* 1089.
Lindegren, C. C. 1945. Ann. Mo. Bot. Gard. *32,* 107.
Lindegren, C. C., Spiegelman, S. and Lindegren, G. 1944. Proc. Natl. Acad. Sci. *30,* 346.
Ling, V. 1971. Biochem. Biophys. Res. Commun. *42,* 82.
Ling, V. and Dixon, G. H. 1970. J. Biol. Chem. *245,* 3035.
Ling, V., Jergil, B. and Dixon, G. H. 1971. J. Biol. Chem. *246,* 1168.
Ling, V., Trevithick, J. R. and Dixon, G. H. 1969. Can. J. Biochem. *47,* 51.
Lingrel, J. B. and Borsook, H. 1963. Biochemistry *2,* 309.
Lipmann, F. 1969. Science *164,* 1024.
Lipmann, F., Nishizuka, Y., Gordon, J., Lucas-Lenard, J. and Gottesman, M. 1967. *Organizational Biosynthesis.* Editors H. J. Vogel, J. O. Lampen and V. Bryson, Academic Press, New York, p. 131.
Lipsett, M. N. 1965. J. Biol. Chem. *240,* 3975.
Lipsett, M. N., Norton, J. S. and Peterkofsky, A. 1967. Biochemistry *6,* 855.
Lipsett, M. N. and Peterkofsky, A. 1966. Proc. Natl. Acad. Sci. *55,* 1169.
Littau, V. C., Allfrey, V. G., Frenster, J. H. and Mirsky, A. E. 1964. Proc. Natl. Acad. Sci. *52,* 93.
Littau, V. C., Burdick, C. J., Allfrey, V. G. and Mirsky, A. E. 1965. Proc. Natl. Acad. Sci. *54,* 1204.
Little, J. R. and Donahue, H. A. 1970. Proc. Natl. Acad. Sci. *67,* 1299.
Livingston, D. M. and Leder, P. 1969. Biochemistry *8,* 435.
Llanes, B. and McFall, E. 1969(a). J. Bact. *97,* 217.
Llanes, B. and McFall, E. 1969(b). J. Bact. *97,* 223.
Lloyd, P. H. and Peacocke, A. R. 1970. Biochem. J. *118,* 467.
Löbbecke, E. A., Schultze, B. and Maurer, W. 1969. Exp. Cell Res. *55,* 176.
Lockard, R. E. and Lingrel, J. B. 1969. Biochem. Biophys. Res. Commun. *37,* 204.
Lockwood, A. H., Hattman, S. and Maitra, U. 1969. Cold Spring Harbor Symp. Quant. Biol. *34,* 433.
Lockwood, D. H., Stockdale, F. E. and Topper, Y. J. 1967. Science, *156,* 945.
Lockwood, D. H., Turkington, R. W. and Topper, Y. J. 1966. Biochim. Biophys. Acta, *130,* 493.
Lockwood, D. H., Voytovich, A. E., Stockdale, F. E. and Topper, Y. J. 1967. Proc. Natl. Acad. Sci. *58,* 658.
Lodish, H. F. 1968. Nature *220,* 345.
Lodish, H. F. 1969. Nature *224,* 867.
Lodish, H. F. 1970(a). J. Mol. Biol. *50,* 689.
Lodish, H. F. 1970(b). Nature *226,* 705.
Lodish, H. F. and Robertson, H. D. 1969. Cold Spring Harbor Symp. Quant. Biol. *34,* 655.
Loeb, J. N. and Hubby, B. G. 1968. Biochim. Biophys. Acta *166,* 745.
Loehr, J. G. and Keller, E. B. 1968. Proc. Natl. Acad. Sci. *61,* 1115.
Loening, U. E., Jones, K. W. and Birnstiel, M. L. 1969. J. Mol. Biol. *45,* 353.
Loftfield, R. B. 1963. Biochem. J. *89,* 82.

Loftfield, R. B. and Eigner, E. A. 1965. J. Biol. Chem. *240 PC* 1482.
Loftfield, R. B., Hecht, L. I. and Eigner, E. A. 1963. Biochim. Biophys. Acta *72,* 383.
Logan, R. 1957. Biochim. Biophys. Acta *26,* 227.
London, I. M., Tavill, A. S., Vanderhoff, G. A., Hunt, T. and Grayzel, A. I. 1967. *Control Mechanisms in Developmental Processes.* Editor M. Locke, Academic Press, New York, p. 227.
Loomis, W. F. 1969. J. Bact. *100,* 417.
Loomis, W. F. 1970. Exp. Cell Res. *60,* 285.
Loomis, W. F. and Magasanik, B. 1964. J. Mol. Biol. *8,* 417.
Loomis, W. F. and Magasanik, B. 1967. J. Mol. Biol. *23,* 487.
Losick, R., Shorenstein, R. G. and Sonenshein, A. L. 1970. Nature *227,* 910.
Louie, A., Lam, D. and Dixon, G. H. 1971. Cited in Ling and Dixon (1970).
Lozeron, H. A. and Szybalski, W. 1969. Virology *39,* 373.
Lucas, A. M. and Jamroz, C. 1961. U.S. Dept. Agriculture Monograph *25.*
Lucas-Lenard, J. and Haenni, A. L. 1968. Proc. Natl. Acad. Sci. *59,* 554.
Lucas-Lenard, J. and Haenni, A. L. 1969. Proc. Natl. Acad. Sci. *63,* 93.
Lucas-Lenard, J. and Lipmann, F. 1966. Proc. Natl. Acad. Sci. *55,* 1562.
Luck, D. N. and Barry, J. M. 1964. J. Mol. Biol. *9,* 186.
Lukács, I. and Sekeris, C. E. 1967. Biochim. Biophys. Acta *134,* 85.
Luppis, B., Bargellesi, A. and Conconi, F. 1970. Biochemistry *9,* 4175.
Luria, S. E. 1970. Science *168,* 1166.
Lurie, M. 1969. J. Theoret. Biol. *23,* 380.
Lwoff, A. 1953. Bact. Rev. *17,* 269.
Lwoff, A. 1966. Science *152,* 1216.
Lyon, M. F. 1961. Nature *190,* 372.
Lyon, M. F. 1962. Amer. J. Human Genet. *14,* 135.
Lyon, M. F. 1968. Ann. Rev. Genet. *2,* 31.
Lysenko, T. 1948. *The Science of Biology Today.* International Publishers Co., Inc., New York.
Maaløe, O. and Kjeldgaard, N. O. 1966. *Control of Macromolecular Synthesis.* W. A. Benjamin, Inc., New York.
Maas, W. K. 1961. Cold Spring Harbor Symp. Quant. Biol. *26,* 183.
Maas, W. K. and Clark, A. J. 1964. J. Mol. Biol. *8,* 365.
Maas, W. K., Maas, R., Wiame, J. M. and Glansdorff, N. 1964. J. Mol. Biol. *8,* 359.
McCarthy, B. J. and Bolton, E. T. 1964. J. Mol. Biol. *8,* 184.
McCarthy, B. J. and Church, R. B. 1970. Ann. Rev. Biochem. *39,* 131.
McCarthy, B. J. and Hoyer, B. H. 1964. Proc. Natl. Acad. Sci. *52,* 915.
McConaughy, R. L., Laird, C. D. and McCarthy, B. J. 1969. Biochemistry *8,* 3289.
McConkey, E. H. and Hopkins, J. W. 1964. Proc. Natl. Acad. Sci. *51,* 1197.
McConkey, E. H. and Hopkins, J. W. 1965. J. Mol. Biol. *14,* 257.
McConkey, E H. and Hopkins, J. W. 1969. J. Mol. Biol. *39,* 545.
McCorquodale, D. J. and Buchanan, J. M. 1968. J. Biol. Chem. *243,* 2550.
McCorquodale, D. J. and Lanni, Y. T. 1964. J. Mol. Biol. *10,* 10.
McCorquodale, D. J. and Lanni, Y. T. 1970. J. Mol. Biol. *48,* 133.
McCorquodale, D. J. and Mueller, G. C. 1958. J. Biol. Chem. *232,* 31.
McFall, E. 1964(a). J. Mol. Biol. *9,* 746.
McFall, E. 1964(b). J. Mol. Biol. *9,* 754.
McFall, E. and Bloom, F. R. 1971. J. Bact. *105,* 241.
McFall, E. and Mandelstam, J. 1963(a). Nature *197,* 880.
McFall, E. and Mandelstam, J. 1963(b). Biochem. J. *89,* 391.
McFarlane, E. S. and Shaw, G. J. 1968. Can. J. Microbiol. *14,* 499.
MacFarlane, M. G. 1962. Nature *196,* 136.

Mach, B., Koblet, H. and Gros, D. 1967. Cold Spring Harbor Symp. Quant. Biol. *32,* 269.
McKeehan, W. L. and Hardesty, B. 1969. J. Biol. Chem. *244,* 4330.
MacKintosh, F. R. and Bell, E. 1969. J. Mol. Biol. *41,* 365.
McLellan, W. I. and Vogel, H. J. 1970. Proc. Natl. Acad. Sci. *67,* 1703.
McMacken, R., Mantei, N., Butler, B., Joyner, A. and Echols, H. 1970. J. Mol. Biol. *49,* 639.
McMaster-Kaye, R. and Taylor, J. H. 1958. J. Biophys. Biochem. Cytol. *4,* 5.
Mäenpää, P. H. and Bernfield, M. R. 1969. Biochemistry *8,* 4926.
Mäenpää, P. H. and Bernfield, M. R. 1970. Proc. Natl. Acad. Sci. *67,* 688.
Magasanik, B. 1957. Ann. Rev. Microbiol. *11,* 221.
Magasanik, B. 1961. Cold Spring Harbor Symp. Quant. Biol. *26,* 249.
Magasanik, B. 1963. *Informational Macromolecules.* Editors H. J. Vogel, V. Bryson and J. O. Lampen, Academic Press, New York, p. 271.
Maggio, R., Vittorelli, M. L., Caffarelli-Mormino, I. and Monroy, A. 1968. J. Mol. Biol. *31,* 621.
Maggio, R., Vittorelli, M. L. Rinaldi, A. M. and Monroy, A. 1964. Biochem. Biophys. Res. Commun. *15,* 436.
Maio, J. J. and Schildkraut, C. L. 1969. J. Mol. Biol. *40,* 203.
Maitra, U. 1970. Biochem. Biophys. Res. Commun. *41,* 1255.
Maitra, U. and Hurwitz, J. 1965. Proc. Natl. Acad. Sci. *54,* 815.
Maitra, U. and Hurwitz, J. 1967. J. Biol. Chem. *242,* 4897.
Makman, R. S. and Sutherland, E. W. 1965. J. Biol. Chem. *240,* 1309.
Malamy, M. H. 1966. Cold Spring Harbor Symp. Quant. Biol. *31,* 189.
Malamy, M. H. and Horecker, B. L. 1964. Biochemistry *3,* 1889.
Maleknia, N., Blum, N. and Schapira, G. 1969. Comptes Rend. Acad. Sci. (Paris) *268,* 2631.
Malmgren, B. and Hedén, C. G. 1947. Acta Path. et Microbiol. Scandinav. *24,* 448.
Malpoix, P. 1967. Biochim. Biophys. Acta *145,* 181.
Malpoix, P., Zampetti, F. and Fievez, M. 1969. Biochim. Biophys. Acta *182,* 214.
Malva, C., Razzino, G. and Calef, E. 1969. Virology *38,* 358.
Mandelstam, J. 1957. Nature *179,* 1179.
Mandelstam, J. 1958(a). Biochem. J. *69,* 103.
Mandelstam, J. 1958(b). Biochem. J. *69,* 110.
Mandelstam, J. 1960. Bact. Rev. *24,* 289.
Mandelstam, J. 1961. Biochem. J. *79,* 489.
Mandelstam, J. 1962. Biochem. J. *82,* 489.
Mandelstam, J. and Halvorson, H. 1960. Biochim. Biophys. Acta *40,* 43.
Mandelstam, J. and Jacoby, G. A. 1965. Biochem. J. *94,* 569.
Mangiarotti, G. and Schlessinger, D. 1966. J. Mol. Biol. *20,* 123.
Mano, Y. 1970. Dev. Biol. *22,* 433.
Mano, Y. and Nagano, H. 1966. Biochem. Biophys. Res. Commun. *25,* 210.
Mano, Y. and Nagano, H. 1970. J. Biochem. *67,* 611.
Manor, H., Goodman, D. and Stent, G. S. 1969. J. Mol. Biol. *39,* 1.
Manson, E. E. D., Pollock, M. R. and Tridgell, E. J. 1954. J. Gen. Microbiol. *11,* 493.
Mansour, A. M. and Nass, S. 1970. Nature *228,* 665.
Mantéva, V. L., Avakiyan, E. R. and Georgiev, G. P. 1969. Molek. Biol. *3* (Trans.), 428.
Marcaud, L., Portier, M. M., Kourilsky, P., Barrell, B. G. and Gros, F. 1971. J. Mol. Biol. *57,* 247.
Marcker, K. A., Clark, B. F. C. and Anderson, J. S. 1966. Cold Spring Harbor Symp. Quant. Biol. *31,* 279.

Marcker, K. and Sanger, F. 1964. J. Mol. Biol. *8,* 835.
Margulies, L., Remeza, V. and Rudner, R. 1970. J. Bact. *103,* 560.
Mark, K. K. 1970. Virology *42,* 20.
Markert, C. L. 1964. *The Role of Chromosomes in Development.* Editor M. Locke, Academic Press, New York, p. 1.
Marks, P. A., Rifkind, R. A. and Danon, D. 1963. Proc. Natl. Acad. Sci. *50,* 336.
Marks, P. A., Willson, C., Kruh, J. and Gros, F. 1962. Biochem. Biophys. Res. Commun. *8,* 9.
Marmur, J. and Greenspan, C. M. 1963. Science *142,* 387.
Martelo, O. J., Woo, S. L. C., Reimann, E. M. and Davie, E. W. 1970. Biochemistry *9,* 4807.
Martin, D. W. and Tomkins, G. M. 1970. Proc. Natl. Acad. Sci. *65,* 1064.
Martin, D. W., Tomkins, G. M. and Bresler, M. A. 1969. Proc. Natl. Acad. Sci. *63,* 842.
Martin, D., Tomkins, G. M. and Granner, D. 1969. Proc. Natl. Acad. Sci. *62,* 248.
Martin, R. G. 1963(a). J. Biol. Chem. *238,* 257.
Martin, R. G. 1963(b). Cold Spring Harbor Symp. Quant. Biol. *28,* 357.
Marushige, K., Brutlag, D. and Bonner, J. 1968. Biochemistry *7,* 3149.
Marushige, K. and Dixon, G. H. 1969. Dev. Biol. *19,* 397.
Marushige, K., Ling, V. and Dixon, G. H. 1969. J. Biol. Chem. *244,* 5953.
Marver, D., Berberich, M. A. and Goldberger, R. F. 1966. Science *153,* 1655.
Marzluff, W. F. and McCarty, K. S. 1970. J. Biol. Chem. *245,* 5635.
Marzluff, W. F., McCarty, K. S. and Turkington, R. W. 1969. Biochim. Biophys. Acta, *190,* 517.
Masters, M. and Donachie, W. D. 1966. Nature *209,* 476.
Mathews, M. B. 1970. Nature *228,* 661.
Matsukage, A., Murakami, S. and Kameyama, T. 1969. Biochim. Biophys. Acta *179,* 145.
Matsuoka, Y., Yagi, Y., Moore, G. E. and Pressman, D. 1969. J. Immunol. *103,* 962.
Matsushiro, A., Kida, S., Ito, J., Sato, K. and Imamoto, F. 1962. Biochem. Biophys. Res. Commun. *9,* 204.
Matsushiro, A., Sato, K., Ito, J., Kida, S. and Imamoto, F. 1965. J. Mol. Biol. *11,* 54.
Matthysse, A. G. 1970. Biochim. Biophys. Acta *199,* 519.
Matthysse, A. G. and Abrams, M. 1970. Biochim. Biophys. Acta *199,* 511.
Maul, G. G. and Hamilton, T. H. 1967. Proc. Natl. Acad. Sci. *57,* 1371.
Maurer, H. R. and Chalkley, G. R. 1967. J. Mol. Biol. *27,* 431.
Maxwell, C. R. and Rabinovitz, M. 1969. Biochem. Biophys. Res. Commun. *35,* 79.
Mayol, R. F. and Thayer, S. A. 1970. Biochemistry *9,* 2484.
Mazen, A. and Champagne, M. 1969. FEBS Letters *2,* 248.
Mazumder, R., Chae, Y. B. and Ochoa, S. 1969. Proc. Natl. Acad. Sci. *63,* 98.
Means, A. R. and Hamilton, T. H. 1966. Proc. Natl. Acad. Sci. *56,* 686.
Mee, B. J. and Lee, B. T. O. 1967. Genetics *55,* 709.
Meisler, M. H. and Langan, T. A. 1969. J. Biol. Chem. *244,* 4961.
Melli, M. and Bishop, J. O. 1969. J. Mol. Biol. *40,* 117.
Melli, M. and Bishop, J. O. 1970. Biochem. J. *120,* 225.
Menninger, J. R., Mulholland, M. C. and Stirewalt, W. S. 1970. Biochim. Biophys. Acta *217,* 496.
Merchant, B. and Brahmi, Z. 1970. Science *167,* 69.
Merriam, R. W. 1969. J. Cell Sci. *5,* 333.
Meselson, M. and Yuan, R. 1968. Nature *217,* 1110.
Metofara, S., Felicetti, L. and Gambino, R. 1971. Proc. Natl. Acad. Sci. *68,* 605.
Miall, S. H., Kato, T. and Tamaoki, T. 1970. Nature *226,* 1050.

Michaelis, G. and Starlinger, P. 1967. Molec. Gen. Genetics *100,* 210.
Michels, C. A. and Reznikoff, W. S. 1971. J. Mol. Biol. *55,* 119.
Midgley, J. E. M. 1962. Biochim. Biophys. Acta *61,* 513.
Midgley, J. E. M. and McCarthy, B. J. 1962. Biochim. Biophys. Acta *61,* 696.
Milanesi, G., Brody, E. N., Grau, O. and Geiduschek, E. P. 1970. Proc. Natl. Acad. Sci. *66,* 181.
Miller, D. L. and Weissbach, H. 1969. Arch. Biochem. Biophys. *132,* 146.
Miller, F. and Metzger, H. 1965. J. Biol. Chem. *240,* 3325.
Miller, H. C. and Cudkowicz, G. 1971. Science *171,* 913.
Miller, J. H., Ippen, K., Scaife, J. G. and Beckwith, J. R. 1968. J. Mol. Biol. *38,* 413.
Miller, L. and Knowland, J. 1970. J. Mol. Biol. *53,* 329.
Miller, M. J. and Wahba, A. J. 1971. Personal communication.
Miller, M. J., Zasloff, M. and Ochoa, S. 1969. FEBS Letters *3,* 50.
Miller, O. L. 1965. Natl. Cancer Inst. Monograph. *18,* 79.
Miller, O. L. 1966. Natl. Cancer Inst. Monograph. *23,* 53.
Miller, O. L. and Beatty, B. R. 1969(a). Science *164,* 955.
Miller, O. L. and Beatty, B. R. 1969(b). J. Cell. Phys. *74,* Supp. *1,* 225.
Miller, O. L., Hamkalo, B. A. and Thomas, C. A. 1970. Science *169,* 392.
Mills, E. S. and Topper, Y. J. 1969. Science *165,* 1127.
Mills, E. S. and Topper, Y. J. 1970. J. Cell Biol. *44,* 310.
Milman, G., Goldstein, J., Scolnick, E. and Caskey, T. 1969. Proc. Natl. Acad. Sci. *63,* 183.
Milstein, C., Milstein, C. P. and Feinstein, A. 1969. Nature *221,* 151.
Minson, A. C. and Creaser, E. H. 1969. Biochem. J. *114,* 49.
Mintz, B. 1964. J. Exp. Zool. *157,* 85.
Mintz, B. 1967. Proc. Natl. Acad. Sci. *58,* 344.
Mintz, B. and Palm, J. 1969. J. Exp. Med. *129,* 1013.
Mirsky, A. E., Burdick, C. J., Davidson, E. H. and Littau, V. C. 1968. Proc. Natl. Acad. Sci. *61,* 592.
Mitchell, W. M. 1968. Proc. Natl. Acad. Sci. *61,* 742.
Mitchison, J. M. 1969. Science *165,* 657.
Mittwoch, U. 1967. *Sex Chromosomes.* Academic Press, London.
Mittwoch, U. 1969. Nature *221,* 446.
Miura, Y. and Wilt, F. H. 1969. Dev. Biol. *19,* 201.
Miura, Y. and Wilt, F. H. 1970. Exp. Cell Res. *59,* 217.
Miyamoto, E., Kuo, J. F. and Greengard, P. 1969. J. Biol. Chem. *244,* 6395.
Mizuno, N. S., Stoops, C. E. and Sinha, A. A. 1971. Nature New Biology *229,* 24.
Moav, B. and Harris, T. N. 1967. Biochem. Biophys. Res. Commun. *29,* 773.
Molinaro, M. and Mozzi, R. 1969. Exp. Cell. Res. *56,* 163.
Molnar, J. and Sy, D. 1967. Biochemistry *6,* 1941.
Monard, D., Janeček, J. and Rickenberg, H. V. 1969. Biochem. Biophys. Res. Commun. *35,* 584.
Monard, D., Janĕcek, J. and Rickenberg, H. V. 1970. *The Lactose Operon.* Editors J. R. Beckwith and D. Zipser, Cold Spring Harbor Laboratory, New York, p. 393.
Mondal, H., Mandal, R. K. and Biswas, B. B. 1970. Biochem. Biophys. Res. Commun. *40,* 1194.
Monier, R., Naono, S., Hayes, D., Hayes, F. and Gros, F. 1962. J. Mol. Biol. *5,* 311.
Monjardino, J. P. P. V. and MacGillivray, A. J. 1970. Exp. Cell Res. *60,* 1.
Monod, J. 1942. *Recherches sur la croissance des cultures bactériennes.* Hermann, Paris.
Monod, J. 1943. Ann. Inst. Pasteur *69,* 179.
Monod, J. 1944(a). Ann. Inst. Pasteur *70,* 57.
Monod, J. 1944(b). Ann. Inst. Pasteur *70,* 60.

Monod, J. 1944(c). Ann. Inst. Pasteur *70,* 381.
Monod, J. 1945. Ann. Inst. Pasteur *71,* 37.
Monod, J. 1947. Growth *11,* 223.
Monod, J. 1949. *Unités biologiques douées de continuité génétique.* C.N.R.S., Paris, p. 181.
Monod, J. 1950. Ann. Inst. Pasteur *79,* 390.
Monod, J. 1956. *Enzymes: Units of Biological Structure and Function.* Editor O. H. Gaebler, Academic Press, New York, p. 7.
Monod, J. and Andureau, A. 1946. Ann. Inst. Pasteur *72,* 868.
Monod, J., Changeux, J. P. and Jacob, F. 1963. J. Mol. Biol. *6,* 306.
Monod, J. and Cohen-Bazire, G. 1953(a). Comptes Rend. Acad. Sci. (Paris) *236,* 417.
Monod, J. and Cohen-Bazire, G. 1953(b). Comptes Rend. Acad. Sci. (Paris) *236,* 530.
Monod, J., Cohen-Bazire, G. and Cohn, M. 1951. Biochim. Biophys. Acta *7,* 585.
Monod, J. and Cohn, M. 1952. Advances in Enzymology *13,* 67.
Monod, J. and Jacob, F. 1961. Cold Spring Harbor Symp. Quant. Biol. *26,* 389.
Monod, J., Pappenheimer, A. M. and Cohen-Bazire, G. 1952. Biochim. Biophys. Acta *9,* 648.
Monod, J., Wyman, J. and Changeux, J. P. 1965. J. Mol. Biol. *12,* 88.
Monro, R. E., Cerná, J. and Marcker, K. A. 1968. Proc. Natl. Acad. Sci. *61,* 1042.
Monroy, A., Maggio, R. and Rinaldi, A. M. 1965. Proc. Natl. Acad. Sci. *54,* 107.
Moon, H. M., Collins, J. F. and Maxwell, E. S. 1970. Biochem. Biophys. Res. Commun. *41,* 170.
Moore, B. C. 1963. Proc. Natl. Acad. Sci. *50,* 1018.
Moore, L. D. and Umbreit, W. W. 1965. Biochim. Biophys. Acta *103,* 466.
Moore, R. J. and Hamilton, T. H. 1964. Proc. Natl. Acad. Sci. *52,* 439.
Morey, K. S. and Litwack, G. 1969. Biochemistry *8,* 4813.
Morikawa, N. and Imamoto, F. 1969. Nature *223,* 37.
Moriyama, Y., Hodnett, J. L., Prestayko, A. W. and Busch, H. 1969. J. Mol. Biol. *39,* 335.
Morris, A. J. 1966. Biochem. Biophys. Res. Commun. *22,* 498.
Morris, A. J. and Liang, K. 1968. Arch. Biochem. Biophys. *125,* 468.
Morris, D. W. and Kjeldgaard, N. O. 1968. J. Mol. Biol. *31,* 145.
Morris, T. 1968. Genet. Res. *12,* 125.
Morris, V. L., Wagner, E. K. and Roizman, B. 1970. J. Mol. Biol. *52,* 247.
Morse, D. E. 1971. J. Mol. Biol. *55,* 113.
Morse, D. E., Baker, R. F. and Yanofsky, C. 1968. Proc. Natl. Acad. Sci. *60,* 1428.
Morse, D. E., Mosteller, R., Baker, R. F. and Yanofsky, C. 1969. Nature *223,* 40.
Morse, D. E. and Yanofsky, C. 1968. J. Mol. Biol. *38,* 447.
Morse, D. E. and Yanofsky, C. 1969(a). Nature *224,* 329.
Morse, D. E. and Yanofsky, C. 1969(b). J. Mol. Biol. *41,* 317.
Morse, D. E. and Yanofsky, C. 1969(c). J. Mol. Biol. *44,* 185.
Moscona, A. A., Moscona, M. and Jones, R. E. 1970. Biochem. Biophys. Res. Commun. *39,* 943.
Moses, V. and Calvin, M. 1965. J. Bact. *90,* 1205.
Moses, V. and Prevost, C. 1966. Biochem. J. *100,* 336.
Moses, V. and Sharp, P. B. 1970. Biochem. J. *118,* 481.
Moses, V. and Yudkin, M. D. 1968. Biochem. J. *110,* 135.
Mosteller, R. D., Culp, W. J. and Hardesty, B. 1968(a). Biochem. Biophys. Res. Commun. *30,* 631.
Mosteller, R. D., Culp, W. J. and Hardesty, B. 1968(b). J. Biol. Chem. *243,* 6343.
Mosteller, R. D. and Yanofsky, C. 1971. J. Bact. *105,* 268.

Motulsky, A. G. 1962. Nature *194,* 607.
Moulé, Y. and Chauveau, J. 1968. J. Mol. Biol. *33,* 465.
Moyed, H. S. 1960. J. Biol. Chem. *235,* 1098.
Moyed, H. S. 1961. J. Biol. Chem. *236,* 2261.
Moyer, R. W. and Buchanan, J. M. 1969. Proc. Natl. Acad. Sci. *64,* 1249.
Mukherjee, B. B., Wright, W. C., Ghosal, S. K., Burkholder, G. D. and Mann, K. E. 1968. Nature *220,* 714.
Mukundan, M. A., Hershey, J. W. B., Dewey, K. F. and Thach, R. E. 1968. Nature *217,* 1013.
Müller, W. E. G., Zahn, R. K. and Beyer, R. 1970. Nature *227,* 1211.
Müller-Hill, B. 1966. J. Mol. Biol. *15,* 374.
Müller-Hill, B., Crapo, L. and Gilbert, W. 1968. Proc. Natl. Acad. Sci. *59,* 1259.
Müller-Hill, B., Rickenberg, H. V. and Wallenfels, K. 1964. J. Mol. Biol. *10,* 303.
Munro, A. J., Jackson, R. J. and Korner, A. 1964. Biochem. J. *92,* 289.
Muramatsu, M. and Fujisawa, T. 1968. Biochim. Biophys. Acta *157,* 476.
Muramatsu, M., Shimada, N. and Higashinakagawa, T. 1970. J. Mol. Biol. *53,* 91.
Murray, K. 1966. J. Mol. Biol. *15,* 409.
Murray, K. 1969. J. Mol. Biol. *39,* 125.
Murray, K., Bradbury, E. M., Crane-Robinson, C., Stephens, R. M., Haydon, A. J. and Peacocke, A. R. 1970. Biochem. J. *120,* 859.
Murty, C. N. and Hallinan, T. 1968. Biochim. Biophys. Acta *157,* 414.
Mushynski, W. E. and Spencer, J. H. 1970(a). J. Mol. Biol. *52,* 91.
Mushynski, W. E. and Spencer, J. H. 1970(b). J. Mol. Biol. *52,* 107.
Nakada, D. 1965. J. Mol. Biol. *12,* 695.
Nakada, D., Anderson, I. A. C. and Magasanik, B. 1964. J. Mol. Biol. *9,* 472.
Nakada, D. and Magasanik, B. 1964. J. Mol. Biol. *8,* 105.
Nakamoto, T. and Kolakofsky, D. 1966. Proc. Natl. Acad. Sci. *55,* 606.
Nanney, D. L. 1964. *The Role of Chromosomes in Development.* Editor M. Locke, Academic Press, New York, p. 253.
Naono, S. and Tokuyama, K. 1970. Cold Spring Harbor Symp. Quant. Biol. *35,* 375.
Naora, H. 1968. J. Theoret. Biol. *19,* 183.
Naora, H. and Kodaira, K. 1969. Biochim. Biophys. Acta *182,* 469.
Narita, K., Tsuchida, I. and Ogata, K. 1968. Biochem. Biophys. Res. Commun. *33,* 430.
Narita, K., Tsuchida, I., Tsunazawa, S. and Ogata, K. 1969. Biochem. Biophys. Res. Commun. *37,* 327.
Nass, G., Poralla, K. and Zähner, H. 1969. Biochem. Biophys. Res. Commun. *34,* 84.
Nass, M. M. K. 1969. Science *165,* 25.
Nass, M. M. K. and Buck, C. A. 1969. Proc. Natl. Acad. Sci. *62,* 506.
Nass, M. M. K. and Buck, C. A. 1970. J. Mol. Biol. *54,* 187.
Nass, S. 1969. Intern. Rev. Cytol. *25,* 55.
Nasser, D., Henderson, G. and Nester, E. W. 1969. J. Bact. *98,* 44.
Nathans, D. 1965. J. Mol. Biol. *13,* 521.
Nathans, D. and Lipmann, F. 1961. Proc. Natl. Acad. Sci. *47,* 497.
Nathans, D., Notani, G., Schwartz, J. H. and Zinder, N. D. 1962. Proc. Natl. Acad. Sci. *48,* 1424.
Nathans, D., Oeschger, M. P., Eggen, K. and Shimura, Y. 1966. Proc. Natl. Acad. Sci. *56,* 1844.
Nathans, D., Oeschger, M. P., Polmar, S. K. and Eggen, K. 1969. J. Mol. Biol. *39,* 279.
Natori, S. and Garen, A. 1970. J. Mol. Biol. *49,* 577.

Naughton, M. A. and Dintzis, H. M. 1962. Proc. Natl. Acad. Sci. *48,* 1822.
Nazario, M. 1967. Biochim. Biophys. Acta *145,* 138.
Nebert, D. W. and Gelboin, H. V. 1968. J. Biol. Chem. *243,* 6250.
Nebert, D. W. and Gelboin, H. V. 1970. J. Biol. Chem. *245,* 160.
Neelin, J. M. 1968. Can. J. Biochem. *46,* 241.
Neelin, J. M., Callahan, P. X., Lamb, D. C. and Murray, K. 1964. Can. J. Biochem. Physiol. *42,* 1743.
Neidhardt, F. C. 1960. J. Bact. *80,* 536.
Neidhardt, F. C. 1966. Bact. Rev. *30,* 701.
Neidhardt, F. C. and Fraenkel, D. G. 1961. Cold Spring Harbor Symp. Quant. Biol. *26,* 63.
Neidhardt, F. C. and Magasanik, B. 1956. Nature *178,* 801.
Neidhardt, F. C. and Magasanik, B. 1957. J. Bact. *73,* 253.
Neidhardt, F. C., Marchin, G. L., McClain, W. H., Boyd, R. F. and Earhart, C. F. 1969. J. Cell. Phys. *74,* Supp. *1,* 87.
Nelson, R. D. and Yunis, J. J. 1969. Exp. Cell Res. *57,* 311.
Nemer, M. 1962. Biochem. Biophys. Res. Commun. *8,* 511.
Nemer, M. and Infante, A. A. 1965. Science *150,* 217.
Nemer, M. and Lindsay, D. T. 1969. Biochem. Biophys. Res. Commun. *35,* 156.
Nesbitt, J. A. and Lennarz, W. J. 1968. J. Biol. Chem. *243,* 3088.
Nester, E. W. 1968. J. Bact. *96,* 1649.
Nester, E. W., Jensen, R. A. and Nasser, D. S. 1969. J. Bact. *97,* 83.
Neubauer, Z. and Calef, E. 1970. J. Mol. Biol. *51,* 1.
Neupert, W., Sebald, W., Schwab, A. J., Massinger, P. and Bücher, Th. 1969. Europ. J. Biochem. *10,* 589.
New, D. A. T. 1955. J. Emb. Exp. Morph. *3,* 326.
Newell, P. C. and Sussman, M. 1970. J. Mol. Biol. *49,* 627.
Newton, A. 1969. J. Mol. Biol. *41,* 329.
Newton, W. A., Beckwith, J. R., Zipser, D. and Brenner, S. 1965. J. Mol. Biol. *14,* 290.
Nichols, J. L. 1970. Nature *225,* 147.
Nichols, J. L. and Lane, B. G. 1966. Can. J. Biochem. *44,* 1633.
Nichols, J. L. and Lane, B. G. 1967. Can. J. Biochem. *45,* 937.
Niessing, J. and Sekeris, C. E. 1970. Biochim. Biophys. Acta *209,* 484.
Nijkamp, H. J. J., Bøvre, K. and Szybalski, W. 1970. J. Mol. Biol. *54,* 599.
Ning, C. and Stevens, A. 1962. J. Mol. Biol. *5,* 650.
Nirenberg, M. W. and Matthaei, J. H. 1961. Proc. Natl. Acad. Sci. *47,* 1588.
Nishizuka, Y. and Lipmann, F. 1966. Arch. Biochem. Biophys. *116,* 344.
Noall, M. W. and Allen, W. M. 1961. J. Biol. Chem. *236,* 2987.
Nomura, M. 1970. Bact. Rev. *34,* 228.
Nomura, M., Hall, B. D. and Spiegelman, S. 1960. J. Mol. Biol. *2,* 306.
Nomura, M., Witten, C., Mantei, N. and Echols, H. 1966. J. Mol. Biol. *17,* 279.
Nossal, G. J. V. 1967. *Gamma Globulins. Structure and Control of Synthesis.* Nobel Symposium *3.* Editor J. Killander, Almqvist and Wiksell, Stockholm, p. 429.
Nossal, G. J. V., Ada, G. L. and Austin, C. M. 1965. J. Exp. Med. *121,* 945.
Nossal, G. J. V., Shortman, K. D., Miller, J. A. F. P., Mitchell, G. F. and Haskill, J. S. 1967. Cold Spring Harbor Symp. Quant. Biol. *32,* 369.
Noteboom, W. D. and Gorski, J. 1963. Proc. Natl. Acad. Sci. *50,* 250.
Noteboom, W. D. and Gorski, J. 1965. Arch. Biochem. Biophys. *111,* 559.
Notides, A. and Gorski, J. 1966. Proc. Natl. Acad. Sci. *56,* 230.
Novelli, G. D. 1969. J. Cell. Phys. *74,* Supp. *1,* 121.
Novello, F. and Stirpe, F. 1969. Biochem. J. *112,* 721.

Novick, A. and Szilard, L. 1950. Proc. Natl. Acad. Sci. *36,* 708.
Novick, A. and Weiner, M. 1957. Proc. Natl. Acad. Sci. *43,* 553.
Novick, R. P. and Richmond, M. H. 1965. J. Bact. *90,* 467.
Nygaard, A. P. and Hall, B. D. 1963. Biochem. Biophys. Res. Commun. *12,* 98.
O'Brien, B. R. A. 1959. Nature *184,* 376.
O'Brien, B. R. A. 1961. J. Emb. Exp. Morph. *9,* 202.
O'Connor, P. J. 1969. Biochem. Biophys. Res. Commun. *35,* 805.
Oda, K. I. and Joklik, W. K. 1967. J. Mol. Biol. *27,* 395.
Oda, K., Sakakibara, Y. and Tomizawa, J. 1969. Virology *39,* 901.
Ogawa, T. and Tomizawa, J. I. 1968. J. Mol. Biol. *38,* 217.
Ohashi, Z., Saneyoshi, M., Harada, F., Hara, H. and Nishimura, S. 1970. Biochem. Biophys. Res. Commun. *40,* 866.
Ohba, Y. 1966. Biochim. Biophys. Acta *123,* 76.
Ohlenbusch, H. H., Olivera, B. M., Tuan, D. and Davidson, N. 1967. J. Mol. Biol. *25,* 299.
Ohno, S., Christian, L., Stenius, C., Castro-Sierra, E. and Muramoto, J. 1969. Biochem. Genet. *2,* 361.
Ohno, S., Stenius, C., Christian, L. C. and Harris, C. 1968. Biochem. Genet. *2,* 197.
Ohno, S., Wolf, U. and Atkin, N. B. 1968. Hereditas *59,* 169.
Ohshima, Y., Horiuchi, T., Iida, Y. and Kameyama, T. 1970. Molec. Gen. Genetics *106,* 307.
Ohta, T. and Thach, R. E. 1968. Nature *219,* 238.
Ohtaka, Y. and Spiegelman, S. 1963. Science *142,* 493.
Okinaka, R. T. and Dobrogosz, W. J. 1967. J. Bact. *93,* 1644.
Olins, D. E. 1969. J. Mol. Biol. *43,* 439.
Olins, D. E., Olins, A. D. and von Hippel, P. H. 1968. J. Mol. Biol. *33,* 265.
O'Malley, B. W., Sherman, M. R. and Toft, D. O. 1970. Proc. Natl. Acad. Sci. *67,* 501.
Ono, T., Terayama, H., Takaku, F. and Nakao, K. 1969. Biochim. Biophys. Acta *179,* 214.
Ono, Y., Skoultchi, A., Klein, A. and Lengyel, P. 1968. Nature *220,* 1304.
Ono, Y., Skoultchi, A., Waterson, J. and Lengyel, P. 1969(a). Nature *222,* 645.
Ono, Y., Skoultchi, A., Waterson, J. and Lengyel, P. 1969(b). Nature *223,* 697.
Oppenheim, A. B., Neubauer, Z. and Calef, E. 1970. Nature *226,* 31.
Ord, M. J. 1969. Nature *221,* 964.
Ord, M. J. and Stocken, L. A. 1968. Biochem. J. *116,* 415.
Orengo, A. and Hnilica, L. S. 1970. Exp. Cell Res. *62,* 331.
Orenstein, J. M. and Marsh, W. H. 1968. Biochem. J. *109,* 697.
Ornston, L. N. 1966. J. Biol. Chem. *241,* 3800.
Ovchinnikov, L. P., Aitkhozhin, M. A., Bystrova, T. E. and Spirin, A. S. 1969. Molek. Biol. *3* (*Trans.*), 449.
Ovchinnikov, L. P. and Avanesov, A. C. 1969. Molek. Biol. *3* (*Trans.*), 711.
Ovchinnikov, L. P., Avanesov, A. C. and Spirin, A. S. 1969. Molek. Biol. *3* (*Trans.*), 465
Ozato, K. 1969. Embryologia *10,* 297.
Ozban, N., Tandler, C. J. and Sirlin, J. L. 1964. J. Emb. Exp. Morph. *12,* 373.
Ozer, J. H., Lowery, D. L. and Saz, A. K. 1970. J. Bact. *102,* 52.
Pace, B., Peterson, R. L. and Pace, N. R. 1970. Proc. Natl. Acad. Sci. *65,* 1097.
Paigen, K., Williams, B. and McGinnis, J. 1967. J. Bact. *94,* 493.
Painter, T. S. 1940. Proc. Natl. Acad. Sci. *26,* 95.
Painter, T. S. and Reindorp, E. 1939. Chromosoma *1,* 276.
Palleroni, N. J. and Stanier, R. Y. 1964. J. Gen. Microbiol. *35,* 319.

Palmer, J. and Moses, V. 1967. Biochem. J. *103,* 358.
Panyim, S. and Chalkley, R. 1969. Biochem. Biophys. Res. Commun. *37,* 1042.
Paoletti, R. A. and Huang, R. C. C. 1969. Biochemistry *8,* 1615.
Papaconstantinou, J. and Julku, E. M. 1968. J. Cell. Phys. *72,* Supp. *1,* 161.
Papermaster, B. W. 1967. Cold Spring Harbor Symp. Quant. Biol. *32,* 447.
Paraskevas, F., Lee, S. T. and Israels, L. G. 1970. Nature *227,* 395.
Paraskevas, F., Lee, S. T. and Israels, L. G. 1971. J. Immunol. *106,* 160.
Pardee, A. B. and Beckwith, J. R. 1962. Biochim. Biophys. Acta *60,* 452.
Pardee, A. B., Jacob, F. and Monod, J. 1959. J. Mol. Biol. *1,* 165.
Pardee, A. B. and Prestidge, L. S. 1961. Biochim. Biophys. Acta *49,* 77.
Pardue, M. L. and Gall, J. G. 1969. Proc. Natl. Acad. Sci. *64,* 600.
Parenti-Rosina, R., Eisenstadt, A. and Eisenstadt, J. M. 1969. Nature *221,* 363.
Parkhouse, R. M. E. and Askonas, B. A. 1969. Biochem. J. *115,* 163.
Parmeggiani, A. 1968. Biochem. Biophys. Res. Commun. *30,* 613.
Parsons, I. C. 1970. Steroids, *16,* 59.
Pastan, I. and Perlman, R. L. 1968. Proc. Natl. Acad. Sci. *61,* 1336.
Pastan, I. and Perlman, R. L. 1969(a). J. Biol. Chem. *244,* 2226.
Pastan, I. and Perlman, R. L. 1969(b). J. Biol. Chem. *244,* 5836.
Pastan, I. and Perlman, R. 1970. Science *169,* 339.
Patte, J. C., Le Bras, G. and Cohen, G. N. 1967. Biochim. Biophys. Acta *136,* 245.
Patterson, B. C. and Davies, D. D. 1969. Biochem. Biophys. Res. Commun. *34,* 791.
Paul, J. and Gilmour, R. S. 1966. Nature *210,* 992.
Paul, J. and Gilmour, R. S. 1968. J. Mol. Biol. *34,* 305.
Pavan, C. 1965. Brookhaven Symp. Biol. *18,* 222.
Pavan, C. and DaCunha, A. B. 1969. Genetics *61,* Supp. *1,* 289.
Pawlowski, P. J. and Berlowitz, L. 1969. Exp. Cell Res. *56,* 154.
Pazur, J. H. 1953. Science *117,* 355.
Peacock, W. L. 1965. Natl. Cancer Inst. Monograph *18,* 101.
Pearlman, R. and Bloch, K. 1963. Proc. Natl. Acad. Sci. *50,* 533.
Pearson, and Hogness. 1970. Cited in Daniel et al. 1970.
Pegg, A. E., Lockwood, D. H. and Williams-Ashman, H. G. 1970. Biochem. J. *117,* 17.
Pene, J. J., Knight, E. and Darnell, J. E. 1968. J. Mol. Biol. *33,* 609.
Penman, S. 1966. J. Mol. Biol. *17,* 117.
Penman, S., Rosbash, M., and Penman, S. 1970. Proc. Natl. Acad. Sci. *67,* 1878.
Penman, S., Scherrer, K., Becker, Y. and Darnell, J. E. 1963. Proc. Natl. Acad. Sci. *49,* 654.
Penn, N. W. 1960. Biochim. Biophys. Acta *37,* 55.
Perkowska, E., MacGregor, H. C. and Birnstiel, M. L. 1968. Nature *217,* 649.
Perlman, R. L., DeCrombrugghe, B. and Pastan, I. 1969. Nature *223,* 810.
Perlman, R. L. and Pastan, I. 1968. J. Biol. Chem. *243,* 5420.
Perlman, R. L. and Pastan, I. 1969. Biochem. Biophys. Res. Commun. *37,* 151.
Perlman, R. and Pastan, I. 1971. Cited in Pastan and Perlman (1970).
Pernis, B., Chiappino, G., Kelus, A. S. and Gell, P. G. H. 1965. J. Exp. Med. *122,* 853.
Pero, J. 1970. Virology *40,* 65.
Perrin, D., Bussard, A. and Monod, J. 1959. Comptes Rend. Acad. Sci. (Paris) *249,* 778.
Perry, R. P. 1967. *Progress in Nucleic Acid Research and Molecular Biology.* Editors J. N. Davidson and W. Cohn, Academic Press, New York, Vol. *6,* p. 219.
Perry, R. P., Chen, T. Y., Freed, J. J., Greenberg, J. R., Kelley, D. E. and Tartof, K. D. 1970. Proc. Natl. Acad. Sci. *65,* 609.

Perry, R. P. and Kelley, D. E. 1968(a). J. Mol. Biol. *35,* 37.
Perry, R. P. and Kelley, D. E. 1968(b). J. Cell. Phys. *72,* 235.
Perry, R. P. and Kelley, D. E. 1970. J. Cell. Phys. *76,* 127.
Perry, R. P., Srinivasan, P. R. and Kelley, D. E. 1964. Science *145,* 504.
Perutz, M. F. and Lehmann, H. 1968. Nature *219,* 902.
Pestka, S. 1968. Proc. Natl. Acad. Sci. *61,* 726.
Pestka, S. 1969(a). Biochem. Biophys. Res. Commun. *36,* 589.
Pestka, S. 1969(b). J. Biol. Chem. *244,* 1533.
Peterkofsky, A. 1968. Biochemistry *7,* 472.
Peterkofsky, B. and Tomkins, G. M. 1968. Proc. Natl. Acad. Sci. *60,* 222.
Peters, T. 1962(a). J. Biol. Chem. *237,* 1181.
Peters, T. 1962(b). J. Biol. Chem. *237,* 1186.
Pettijohn, D. E., Stenington, O. G. and Kossman, C. R. 1970. Nature *228,* 235.
Phillips, L. A., Hotham-Iglewski, B. and Franklin, R. M. 1969. J. Mol. Biol. *45,* 23.
Piatigorsky, J. 1968. Biochim. Biophys. Acta *166,* 142.
Pillinger, D. and Borek, E. 1969. Proc. Natl. Acad. Sci. *62,* 1145.
Pillinger, D. J., Hay, J. and Borek, E. 1969. Biochem. J. *114,* 429.
Pine, M. J. 1969. Biochim. Biophys. Acta *174,* 359.
Pine, M. J., Gordon, B. and Sarimo, S. S. 1969. Biochim. Biophys. Acta *179,* 439.
Pinheiro, P., Leblond, C. P. and Droz, B. 1963. Exp. Cell Res. *31,* 517.
Pironio, M. and Ghysen, A. 1970. Molec. Gen. Genetics *108,* 374.
Pirrotta, V. and Ptashne, M. 1969. Nature *222,* 541.
Pitot, H. C. and Cho, Y. S. 1961. Biochim. Biophys. Acta *50,* 197.
Pitot, H. C. and Jost, J. P. 1968. *Regulatory Mechanisms for Protein Synthesis in Mammalian Cells.* Editors A. S. Pietro, M. R. Lamborg and F. T. Kenney, Academic Press, New York, p. 283.
Pogo, A. O., Allfrey, V. G. and Mirsky, A. E. 1966(a). Proc. Natl. Acad. Sci. *56,* 550.
Pogo, B. G. T., Allfrey, V. G. and Mirsky, A. E. 1966(b). Proc. Natl. Acad. Sci. *55,* 805.
Pogo, B. G. T., Allfrey, V. G. and Mirsky, A. E. 1967. J. Cell Biol. *35,* 477.
Pogo, B. G. T., Pogo, A. O., Allfrey, V. G. and Mirsky, A. E. 1968. Proc. Natl. Acad. Sci. *59,* 1337.
Pogo, B. G. T., Pogo, A. O. and Allfrey, V. G. 1969. Genetics *61,* Supp. *1,* 373.
Pollack, Y., Groner, Y., Aviv, H. and Revel, M. 1970. FEBS Letters *9,* 218.
Pollock, M. R. 1946. Brit. J. Exp. Path. *27,* 419.
Pollock, M. R. 1950. Brit. J. Exp. Path. *31,* 739.
Pollock, M. R. 1952. Brit. J. Exp. Path. *33,* 587.
Pollock, M. R. 1953. Symp. Soc. Gen. Microbiol. *3,* 150.
Pollock, M. R. 1956(a). J. Gen. Microbiol. *14,* 90.
Pollock, M. R. 1956(b). J. Gen. Microbiol. *15,* 154.
Pollock, M. R. 1957(a). Biochem. J. *66,* 419.
Pollock, M. R. 1957(b). Ciba Foundation Symposium on Drug Resistance in Microorganisms, p. 78.
Pollock, M. R. 1958. Proc. Roy. Soc. Series B, *148,* 340.
Pollock, M. R. 1959. *The Enzymes,* 2nd Edition. Editors P. D. Boyer, H. Lardy and K. Myrbäck, Academic Press, New York, Vol. *1,* p. 619.
Pollock, M. R. 1964. *Antimicrobial Agents and Chemotherapy.* Amer. Soc. Microbiol., Ann Arbor, Mich., p. 292.
Pollock, M. R. 1967. Bull. Soc. Chim. Biol. *49,* 633.
Pollock, M. R. and Kramer, M. 1958. Biochem. J. *70,* 665.
Pollock, M. R. and Perret, C. J. 1951. Brit. J. Exp. Path. *32,* 387.
Polz, G. and Kreil, G. 1970. Biochem. Biophys. Res. Commun. *39,* 516.

Potter, M., Appella, E. and Geisser, S. 1965. J. Mol. Biol. *14*, 361.
Power, J. 1967. Genetics *55*, 557.
Pragnell, I. B. and Arnstein, H. R. V. 1970. FEBS Letters *9*, 331.
Prahl, J. W., Mandy, W. J., David, G. S., Steward, M. W. and Todd, C. W. 1970. *Protides of the Biological Fluids.* Proc. 17th Colloq. Bruges. Pergamon, Oxford, p. 125.
Prescott, D. M., Stevens, A. R. and Lauth, M. R. 1971. Exp. Cell Res. *64*, 145.
Press, E. M. and Hogg, N. M. 1969. Nature *223*, 807.
Press, E. M., Piggot, P. J. and Porter, R. R. 1966. Biochem. J. *99*, 356.
Prichard, P. M., Gilbert, J. M., Shafritz, D. A. and Anderson, W. F. 1970. Nature *226*. 511.
Primakoff, P. and Berg, P. 1970. Cold Spring Harbor Symp. Quant. Biol. *35*, 391.
Printz, D. B. and Gross, S. R. 1967. Genetics *55*, 451.
Ptashne, M. 1967(a). Proc. Natl. Acad. Sci. *57*, 306.
Ptashne, M. 1967(b). Nature *214*, 232.
Ptashne, M. and Hopkins, N. 1968. Proc. Natl. Acad. Sci. *60*, 1282.
Puca, G. A. and Bresciani, F. 1968. Nature *218*, 967.
Pucci-Minafra, I., Minafra, S. and Collier, J. R. 1969. Exp. Cell Res. *57*, 167.
Pulitzer, J. F. 1970. J. Mol. Biol. *49*, 473.
Pulitzer, J. F. and Geiduschek, E. P. 1970. J. Mol. Biol. *49*, 489.
Putnam, F. W. 1967. *Gamma Globulins, Structure and Control of Biosynthesis.* Nobel Symposium *3*. Editor J. Killander, Almqvist and Wiksell, Stockholm, p. 45.
Putnam, F. W. 1969. Science *163*, 633.
Quagliarotti, G. and Ritossa, F. M. 1968. J. Mol. Biol. *36*, 57.
Rabinovitz, M., Freedman, M. L., Fisher, J. M. and Maxwell, C. R. 1969. Cold Spring Harbor Symp. Quant. Biol. *34*, 567.
Rabinowitz, M., Getz, G. S., Casey, J. and Swift, H. 1969. J. Mol. Biol. *41*, 381.
Rabovsky, D. and Konrad, M. 1970. Virology *40*, 10.
Rachkus, Y. A., Kupriyanova, N. S., Timofeeva, M. Y. and Kafiani, K. A. 1969(a). Molek. Biol. *3* (*Trans.*), 486.
Rachkus, Y. A., Timofeeva, M. Y., Kupriyanova, N. S. and Kafiani, K. A. 1969(b). Molek. Biol. *3* (*Trans.*), 338.
Radding, C. M. and Kaiser, A. D. 1963. J. Mol. Biol. *7*, 225.
Rae, P. M. M. 1970. Proc. Natl. Acad. Sci. *67*, 1018.
Rahamimoff, H. and Arnstein, H. R. V. 1969. Biochem. J. *115*, 113.
Rakhimbekova, L. S. and Gaitskhoki, V. S. 1969. Molek. Biol. *3* (*Trans.*), 315.
Ramming, K. P. and Pilch, Y. H. 1970. Science *168*, 492.
Raška, K., Frohwirth, D. H. and Schlesinger, R. W. 1970. J. Virol. *5*, 464.
Rasmussen, P. S., Murray, K. and Luck, J. M. 1962. Biochemistry *1*, 79.
Ravel, J. M. 1967. Proc. Natl. Acad. Sci. *57*, 1811.
Ravel, J. M., Shorey, R. L., Froehner, S. and Shive, W. 1968. Arch. Biochem. Biophys. *125*, 514.
Ravel, J. M., Shorey, R. L. and Shive, W. 1967. Biochem. Biophys. Res. Commun. *29*, 68.
Ravel, J. M., Shorey, R. L. and Shive, W. 1968. Biochem. Biophys. Res. Commun. *32*, 9.
Ravin, A. W. 1965. *The Evolution of Genetics.* Academic Press, New York.
Rawson, J. R. and Stutz, E. 1969. Biochim. Biophys. Acta *190*, 368.
Recheigl, M. and Heston, W. E. 1967. Biochem. Biophys. Res. Commun. *27*, 119.
Rechler, M. M. and Martin, R. G. 1970. *226*, 908.
Redman, C. M. 1969. J. Biol. Chem. *244*, 4308.
Reel, J. R., Lee, K. L. and Kenney, F. T. 1970. J. Biol. Chem *245*, 5800.

Reger, B. J., Fairfield, S. A., Epler, J. L. and Barnett, W. E. 1970. Proc. Natl. Acad. Sci. *67,* 1207.
Rennert, O. M. 1970. Life Sciences *9,* Part *II,* 277.
Retèl, J., Van Den Bos, R. C. and Planta, R. J. 1969. Biochim. Biophys. Acta *195,* 370.
Revel, H. R. and Luria, S. E. 1963. Cold Spring Harbor Symp. Quant. Biol. *28,* 403.
Revel, M., Aviv, H., Groner, Y. and Pollack, Y. 1970. FEBS Letters *9,* 213.
Revel, M., Greenshpan, H. and Herzberg, M. 1970. Europ. J. Biochem. *16,* 117.
Revel, M. and Gros, F. 1967. Biochem. Biophys. Res. Commun. *27,* 12.
Revel, M., Herzberg, M., Becarevic, A. and Gros, F. 1968. J. Mol. Biol. *33,* 231.
Revel, M., Herzberg, M. and Greenshpan, H. 1969. Cold Spring Harbor Symp. Quant. Biol. *34,* 261.
Revel, M., Hiatt, H. H. and Revel, J. P. 1964. Science *146,* 1311.
Rho, H. M. and DeBusk, A. G. 1971. Biochem. Biophys. Res. Commun. *42,* 319.
Rho, J. H. and Bonner, J. 1961. Proc. Natl. Acad. Sci. *47,* 1611.
Rich, A. 1967. *Organizational Biosynthesis.* Editors H. J. Vogel, J. O. Lampen and V. Bryson, Academic Press, New York, p. 143.
Rich, A., Eikenberry, E. F. and Malkin, L. I. 1966. Cold Spring Harbor Symp. Quant. Biol. *31,* 303.
Rich, A., Warner, J. R. and Goodman, H. M. 1963. Cold Spring Harbor Symp. Quant. Biol. *28,* 269.
Richards, B. M. and Pardon, J. F. 1970. Exp. Cell Res. *62,* 184.
Richardson, B. J., Czuppon, A. B. and Sharman, G. B. 1971. Nature New Biology *230,* 154.
Richardson, J. P. 1966(a). J. Mol. Biol. *21,* 83.
Richardson, J. P. 1966(b). J. Mol. Biol. *21,* 115.
Richardson, J. 1969. Prog. Nucleic Acid Res. *9,* 75.
Richardson, J. P. 1970. Nature *225,* 1109.
Richmond, M. H. 1967. J. Mol. Biol. *26,* 357.
Richmond, M. H. 1968. *Essays in Biochemistry.* Editors P. N. Campbell and G. D. Greville, Academic Press, London, Vol. *4,* p. 105.
Richmond, M. H. 1969. Biochem. J. *113,* 225.
Richter, D. and Lipmann, F. 1970. Biochemistry *9,* 5065.
Rickenberg, H. V., Cohen, G. N., Buttin, G. and Monod. J. 1956. Ann. Inst. Pasteur *91,* 829.
Rickenberg, H. V., Hsie, A. W. and Janeček, J. 1968. Biochem. Biophys. Res. Commun. *31,* 603.
Rickenberg, H. V. and Lester, G. 1955. J. Gen. Microbiol. *13,* 279.
Rifkin, M. R., Wood, D. D. and Luck, D. J. L. 1967. Proc. Natl. Acad. Sci. *58,* 1025.
Rifkind, R. A., Danon, D. and Marks, P. A. 1964. J. Cell Biol. *22,* 599.
Riggs, A. D. and Bourgeois, S. 1968. J. Mol. Biol. *34,* 361.
Riggs, A. D., Bourgeois, S. and Cohn, M. 1970. J. Mol. Biol. *53,* 401.
Riggs, A. D., Bourgeois, S., Newby, R. F. and Cohn, M. 1968. J. Mol. Biol. *34,* 365.
Riggs, A. D., Newby, R. F. and Bourgeois, S. 1970. J. Mol. Biol. *51,* 303.
Riggs, A. D., Suzuki, H. and Bourgeois, S. 1970. J. Mol. Biol. *48,* 67.
Riggsby, W. S. 1969. Biochemistry *8,* 222.
Rinaldi, A. M. and Monroy, A. 1969. Dev. Biol. *19,* 73.
Ringborg, U., Daneholt, B., Edström, J. E., Egyházi, E. and Lambert, E. 1970(a). J. Mol. Biol. *51,* 327.
Ringborg, U., Daneholt, B., Edström, J. E., Egyházi, E. and Rydlander, L. 1970(b). J. Mol. Biol. *51,* 679.

Ringertz, N. R. and Bolund, L. 1969. Exp. Cell Res. *55,* 205.
Ritossa, F. M. 1968. Proc. Natl. Acad. Sci. *59,* 1124.
Ritossa, F. M. and Atwood, K. C. 1966. Proc. Natl. Acad. Sci. *56,* 496.
Ritossa, F. M., Atwood, K. C., Lindsley, D. L. and Spiegelman, S. 1966. Natl. Cancer Inst. Monograph *23,* 449.
Ritossa, F. M., Atwood, K. C. and Spiegelman, S. 1966. Genetics *54,* 819.
Ritossa, F. M. and Scala, G. 1969. Genetics *61,* Supp. *1,* 305.
Ritossa, F. M. and Spiegelman, S. 1965. Proc. Natl. Acad. Sci. *53,* 737.
Riva, S., Cascino, A. and Geiduschek, E. P. 1970(a). J. Mol. Biol. *54,* 85.
Riva, S., Cascino, A. and Geiduschek, E. P. 1970(b). J. Mol. Biol. *54,* 103.
Robbins, E. and Borun, T. W. 1967. Proc. Natl. Acad. Sci. *57,* 409.
Robert, R. and Kroeger, H. 1965. Experientia *21,* 326.
Roberts, J. W. 1969(a). Nature *223,* 480.
Roberts, J. W. 1969(b). Nature *224,* 1168.
Robertson, H., Webster, R. E. and Zinder, N. D. 1968. Nature *218,* 533.
Robertson, H. D. and Zinder, N. D. 1968. Nature *220,* 69.
Robertson, H. D. and Zinder, N. D. 1969. J. Biol. Chem. *244,* 5790.
Robison, G. A., Butcher, R. W. and Sutherland, E. W. 1968. Ann. Rev. Biochem. *37,* 149.
Roeder, R. G. and Rutter, W. J. 1969. Nature *224,* 234.
Roeder, R. G. and Rutter, W. J. 1970(a). Proc. Natl. Acad. Sci. *65,* 675.
Roeder, R. G. and Rutter, W. J. 1970(b). Biochemistry *9,* 2543.
Rogers, M. E., Loening, U. E. and Fraser, R. S. S. 1970. J. Mol. Biol. *49,* 681.
Romanoff, A. L. 1960. *The Avian Embryo.* Macmillan, New York.
Romanoff, A. L. 1967. *Biochemistry of the Avian Embryo.* John Wiley and Sons, Inc., New York.
Roodyn, D. B., Reis, P. J. and Work, T. S. 1961. Biochem. J. *80,* 9.
Roodyn, D. B., Suttie, J. W. and Work, T. S. 1962. Biochem. J. *83,* 29.
Rose, J. K., Mosteller, R. D. and Yanofsky, C. 1970. J. Mol. Biol. *51,* 541.
Rosenberg, A. H. and Gefter, M. L. 1969. J. Mol. Biol. *46,* 581.
Roth, J. R. and Ames, B. N. 1966. J. Mol. Biol. *22,* 325.
Roth, J. R., Antón, D. N. and Hartman, P. E. 1966. J. Mol. Biol. *22,* 305.
Roth, J. R., Silbert, D. F., Fink, G. R., Voll, M. J., Antón, D., Hartman, P. E. and Ames, B. N. 1966. Cold Spring Harbor Symp. Quant. Biol. *31,* 383.
Roth, J. S. 1958. J. Biol. Chem. *231,* 1085.
Roth, R., Ashworth, J. M. and Sussman, M. 1968. Proc. Natl. Acad. Sci. *59,* 1235.
Rotman, B. 1958. J. Bact. *76,* 1.
Rotman, B. 1959. Biochim. Biophys. Acta *32,* 599.
Rotman, B. 1961. Proc. Natl. Acad. Sci. *47,* 1981.
Rotman, B. and Spiegelman, S. 1954. J. Bact. *68,* 419.
Rotman, B., Zderic, J. A. and Edelstein, M. 1963. Proc. Natl. Acad. Sci. *50,* 1.
Roufa, D. J., Doctor, B. P. and Leder, P. 1970. Biochem. Biophys. Res. Commun. *39,* 231.
Roufa, D. J., Skogerson, L. E. and Leder, P. 1970. Nature *227,* 567.
Roulland-Dussoix, D. and Boyer, H. W. 1969. Biochim. Biophys. Acta *195,* 219.
Rowbury, R. J. and Woods, D. D. 1966. J. Gen. Microbiol. *42,* 155.
Rowley, D., Cooper, P. D., Roberts, P. W. and Smith, E. L. 1950. Biochem. J. *46,* 157.
Rowley, P. T. and Morris, J. A. 1967. J. Biol. Chem. *242,* 1533.
Roy, K. L. and Söll, D. 1968. Biochim. Biophys. Acta *161,* 572.
Rubin, A. D. 1968. Nature *220,* 196.
Rudkin, G. T. and Woods, P. S. 1959. Proc. Natl. Acad. Sci. *45,* 997.

Rudland, P. S., Whybrow, W. A., Marcker, K. A. and Clarck, B. F. C. 1969. Nature *222,* 750.
Rudzik, M. B. and Imsande, J. 1970. J. Biol. Chem. *245,* 3556.
Rush, E. A. and Starr, J. L. 1970. Biochim. Biophys. Acta *199,* 41.
Russell, D. H. 1971. Proc. Natl. Acad. Sci. *68,* 523.
Ryan, F. J. 1952. J. Gen. Microbiol. *7,* 69.
Ryskov, A. P. and Georgiev, G. P. 1970. FEBS Letters *8,* 186.
Ryskov, A. P., Mantieva, V. L., Avakian, E. R. and Georgiev, G. P. 1971. FEBS Letters *12,* 141.
Sabatini, D. D. and Blobel, G. 1970. J. Cell Biol. *45,* 146.
Sabatini, D. D., Tashiro, Y. and Palade, G. E. 1966. J. Mol. Biol. *19,* 503.
Sabol, S., Sillero, M. A. G., Iwasaki, K. and Ochoa, S. 1970. Nature *228,* 1269.
Sadowski, P. D. and Howden, J. A. 1968. J. Cell Biol. *37,* 163.
Sadowski, P. and Hurwitz, J. 1969(a). J. Biol. Chem. *244,* 6182.
Sadowski, P. and Hurwitz, J. 1969(b). J. Biol. Chem. *244,* 6192.
Saedler, H., Gullon, A., Fiethen, L. and Starlinger, P. 1968. Molec. Gen. Genetics *102,* 79.
Saedler, H. and Starlinger, P. 1967(a). Molec. Gen. Genetics *100,* 178.
Saedler, H. and Starlinger, P. 1967(b). Molec. Gen. Genetics *100,* 190.
Sagik, B. P., Green, M. H., Hayashi, M. and Spiegelman, S. 1962. Biophys. J. *2,* 409.
Sala, F. and Küntzel, H. 1970. Europ. J. Biochem. *15,* 280.
Salas, J. and Green, H. 1971. Nature New Biology *229,* 165.
Salas, M., Hille, M. B., Last, J. A., Wahba, A. J. and Ochoa, S. 1967. Proc. Natl. Acad. Sci. *57,* 387.
Salb, J. M. and Marcus, P. I. 1965. Proc. Natl. Acad. Sci. *54,* 1353.
Salser, W. 1969. Molec. Gen. Genetics *105,* 125.
Salser, W. 1970. Cold Spring Harbor Symp. Quant. Biol. *35,* 19.
Salser, W., Bolle, A. and Epstein, R. 1970. J. Mol. Biol. *49,* 271.
Salser, W., Gesteland, R. F. and Bolle, A. 1967. Nature *215,* 588.
Salser, W., Gesteland, R. F. and Ricard, B. 1969. Cold Spring Harbor Symp. Quant. Biol. *34,* 771.
Samarina, O. P., Asrijan, I. S. and Georgiev, G. P. 1965. Dokl. Akad. Nauk. SSSR. *163,* 1510.
Samarina, O. P., Krichevskaya, A. A. and Georgiev, G. P. 1966. Nature *210,* 1319.
Samarina, O. P., Krichevskaya, A. A., Molnar, J., Bruskov, V. I. and Georgiev, G. P. 1967(a). Molek. Biol. *1* (*Trans.*), 110.
Samarina, O. P., Lerman, M. I., Tumanyan, V. D., Anan'eva, L. N. and Georgiev, G. P. 1965. Biokhimiya *30* (*Trans.*), 755.
Samarina, O. P., Lukanidin, E. M. and Georgiev, G. P. 1967. Biochim. Biophys. Acta *142,* 561.
Samarina, O. P., Lukanidin, E. M. and Georgiev, G. P. 1968. Molek. Biol. *2* (*Trans.*), 61.
Samarina, O. P., Lukanidin, E. M., Molnar, J. and Georgiev, G. P. 1968. J. Mol. Biol. *33,* 251.
Samarina, O. P., Molnar, J., Lukanidin, E. M., Bruskov, V. I., Krichevskaya, A. A. and Georgiev, G. P. 1967(b). J. Mol. Biol. *27,* 187.
Samuel, C. E., D'Ari, L. and Rabinowitz, J. C. 1970. J. Biol. Chem. *245,* 5115.
Sarabhai, A. S., Stretton, A. O. W., Brenner, S. and Bolle, A. 1964. Nature *201,* 13.
Sarkar, N. K. 1969. FEBS Letters *4,* 37.
Sarkar, S. and Luria, S. 1963. Biochim. Biophys. Acta *68,* 505.
Sarkar, S. and Thach, R. E. 1968. Proc. Natl. Acad. Sci. *60,* 1479.
Sassa, S. and Granick, S. 1970. Proc. Natl. Acad. Sci. *67,* 517.

Satake, K., Rasmussen, P. S. and Luck, J. M. 1960. J. Biol. Chem. *235,* 2801.
Sato, K. and Matsushiro, A. 1965. J. Mol. Biol. *14,* 608.
Scaife, J. and Beckwith, J. R. 1966. Cold Spring Harbor Symp. Quant. Biol. *31,* 403.
Scarano, E. and Augusti-Tocco, G. 1967. *Comprehensive Biochemistry.* Editors M. Florkin and E. H. Stotz, Elsevier, Amsterdam, Vol. *28,* p. 55.
Schalet, A. 1969. Genetics *63,* 133.
Scharff, M. D. 1967. *Gamma Globulins. Structure and Control of Biosynthesis.* Nobel Symposium *3.* Editor J. Killander, Almqvist and Wiksell, Stockholm, p. 385.
Scharff, M. D. and Robbins, E. 1966. Science *151,* 992.
Schaup, H. W., Best, J. B. and Goodman, A. B. 1969. Nature *221,* 864.
Schedl, P. D., Singer, R. E. and Conway, T. W. 1970. Biochem. Biophys. Res. Commun. *38,* 631.
Scherberg, N. H. and Weiss, S. B. 1970. Proc. Natl. Acad. Sci. *67,* 1164.
Scherrer, K., Latham, H. and Darnell, J. E. 1963. Proc. Natl. Acad. Sci. *49,* 240.
Scherrer, K. and Marcaud, L. 1968. J. Cell. Phys. *72,* Supp. *1,* 181.
Scherrer, K., Marcaud, L., Zajdela, F., Breckenridge, B. and Gros, F. 1966(a). Bull. Soc. Chim. Biol. *48,* 1037.
Scherrer, K., Marcaud, L., Zajdela, F., London, I. M. and Gros, F. 1966(b). Proc. Natl. Acad. Sci. *56,* 1571.
Schiefer, H. G. 1969. Zeit. Physiol. Chem. *350,* 921.
Schimke, R. T. 1966. Bull. Soc. Chim. Biol. *48,* 1009.
Schimke, R. T., Sweeney, E. W. and Berlin, C. M. 1965. J. Biol. Chem. *240,* 322.
Schlesinger, M. J. 1968. J. Bact. *96,* 727.
Schlesinger, S. and Magasanik, B. 1964. J. Mol. Biol. *9,* 670.
Schlesinger, S. and Magasanik, B. 1965. J. Biol. Chem. *240,* 4325.
Schlesinger, S. and Nester, E. W. 1969. J. Bact. *100,* 167.
Schmid, W., Gallwitz, D. and Sekeris, C. E. 1967. Biochim. Biophys. Acta *134,* 80.
Schmoyer, I. R. and Lovett, J. S. 1969. J. Bact. *100,* 854.
Schneider, J. A. and Weiss, M. C. 1971. Proc. Natl. Acad. Sci. *68,* 127.
Scholnick, P. L., Hammaker, L. E. and Marver, H. S. 1969. Proc. Natl. Acad. Sci. *63,* 65.
Schreier, M. X. and Noll, H. 1970. Nature *227,* 128.
Schreier, M. H. and Noll, H. 1971. Proc. Natl. Acad. Sci. *68,* 805.
Schubert, D. 1970. J. Mol. Biol. *51,* 287.
Schubert, D. and Cohn, M. 1968. J. Mol. Biol. *38,* 273.
Schubert, D. and Cohn, M. 1970. J. Mol. Biol. *53,* 305.
Schultz, J. 1952. Exp. Cell Res., Supp. *2,* 17.
Schultz, J. 1965. Brookhaven Symp. Biol. *18,* 116.
Schütz, G., Gallwitz, D. and Sekeris, C. E. 1968. Europ. J. Biochem. *4,* 149.
Schwartz, D. and Beckwith, J. 1970. *The Lactose Operon.* Editors J. R. Beckwith and D. Zipser, Cold Spring Harbor Laboratory, New York, p. 417.
Schwartz, E. 1970. Science *167,* 1513.
Schwartz, J. H., Meyer, R., Eisenstadt. J. M. and Brawerman, G. 1967. J. Mol. Biol. *25,* 571.
Schwartz, T., Craig, E. and Kennell, D. 1970. J. Mol. Biol. *54,* 299.
Schweet, R., Lamfrom, H. and Allen, E. 1958. Proc. Natl. Acad. Sci. *44,* 1029.
Scolnick, E. M. and Caskey, C. T. 1969. Proc. Natl. Acad. Sci. *64,* 1235.
Scolnick, E., Milman, G., Rosman, M. and Caskey, T. 1970. Nature *225,* 152.
Scolnick, E., Tompkins, R., Caskey, T. and Nirenberg, M. 1968. Proc. Natl. Acad. Sci. *61,* 768.
Scott, R. B. and Bell, E. 1965. Science *147,* 405.
Scott, W. A. and Mitchell, H. K. 1969. Biochemistry *8,* 4282.

Seed, R. W. and Goldberg, I. H. 1965. J. Biol. Chem. *240,* 764.
Seifart, K. H. 1970. Cold Spring Harbor Symp. Quant. Biol. *35,* 719.
Seifert, W. 1970. Lepetit Colloq. Biol. Med. *1,* 158.
Seifert, W., Qasba, P., Walter, G., Palm, P., Schachner, M. and Zillig, W. 1969. Europ. J. Biochem. *9,* 319.
Sekiguchi, M. and Cohen, S. S. 1963. J. Biol. Chem. *238,* 349.
Sekiya, T., Takeishi, K. and Ukita, T. 1969. Biochim. Biophys. Acta *182,* 411.
Seligy, V. and Miyagi, M. 1969. Exp. Cell Res. *58,* 27.
Seligy, V. L. and Neelin, J. M. 1970. Biochim. Biophys. Acta *213,* 380.
Seligy, V. L. and Neelin, J. M. 1971. Cited in Seligy and Neelin (1970).
Sels, A., Fukuhara, H., Péré, G. and Slonimski, P. 1965. Biochim. Biophys. Acta *95,* 486.
Selvig, S. E., Gross, P. R. and Hunter, A. L. 1970. Dev. Biol. *22,* 343.
Senior, A. E. and MacLennan, D. H. 1970. J. Biol. Chem. *245,* 5086.
Shaeffer, J. R. 1967. Biochem. Biophys. Res. Commun. *28,* 647.
Shaeffer, J. R., Trostle, P. K. and Evans, R. F. 1969. J. Biol. Chem. *244,* 4284.
Shaffer, B., Rytka, J. and Fink, G. R. 1969. Proc. Natl. Acad. Sci. *63,* 1198.
Shafritz, D. A. and Anderson, W. F. 1970. Nature *227,* 918.
Shafritz, D. A., Laycock, D. G. and Anderson, W. F. 1971. Proc. Natl. Acad. Sci. *68,* 496.
Shapiro, A. L., Scharff, M. D., Maizel, J. V. and Uhr, J. W. 1966(a). Proc. Natl. Acad. Sci. *56,* 216.
Shapiro, A. L., Scharff, M. D., Maizel, J. V. and Uhr, J. W. 1966(b). Nature *211,* 243.
Shapiro, J. A. 1969. J. Mol. Biol. *40,* 93.
Shapiro, J. A. and Adhya, S. L. 1969. Genetics *62,* 249.
Shapiro, J., MacHattie, L., Eron, L., Ihler, G., Ippen, K. and Beckwith, J. 1969. Nature *224,* 768.
Sharma, O. K. and Borek, E. 1970(a). J. Bact. *103,* 705.
Sharma, O. K. and Borek, E. 1970(b). Biochemistry *9,* 2507.
Sharma, S. K. and Talwar, G. P. 1970. J. Biol. Chem. *245,* 1513.
Sharman, G. B. 1971. Nature *230,* 231.
Shatkin, A. J. 1962. Biochim. Biophys. Acta *61,* 310.
Shearer, R. W. and McCarthy, B. J. 1967. Biochemistry *6,* 283.
Shearer, R. W. and McCarthy, B. J. 1970. J. Cell. Phys. *75,* 97.
Shearn, A. and Horowitz, N. H. 1969. Biochemistry *8,* 304.
Sheinin, R. and Crocker, B. F. 1961. Can. J. Biochem. Physiol. *39,* 63.
Shelton, K. R. and Allfrey, V. G. 1970. Nature *228,* 132.
Shen, T. C. 1969. Plant Physiol. *44,* 1650.
Shepherd, G. R., Noland, B. J. and Hardin, J. M. 1971. Biochim. Biophys. Acta *228,* 544.
Sheppard, D. E. and Englesberg, E. 1967. J. Mol. Biol. *25,* 443.
Sherr, C. J. and Uhr, J. W. 1969. Proc. Natl. Acad. Sci. *64,* 381.
Shields, R. and Korner, A. 1970. Biochim. Biophys. Acta *204,* 521.
Shih, A., Eisenstadt, J. and Lengyel, P. 1966. Proc. Natl. Acad. Sci. *56,* 1599.
Shih, T. Y. and Bonner, J. 1969. Biochim. Biophys. Acta *182,* 30.
Shin, D. H. and Moldave, K. 1966. J. Mol. Biol. *21,* 231.
Shiokawa, K. and Yamane, K. 1967. Dev. Biol. *16,* 368.
Shiokawa, K. and Yamana, K. 1969. Exp. Cell Res. *55,* 155.
Shirley, M. A. and Wainwright, S. D. 1972. In preparation.
Shorey, R. L., Ravel, J. M., Garner, C. W. and Shive, W. 1969. J. Biol. Chem. *244,* 4555.
Shortman, K. 1962. Biochim. Biophys. Acta *61,* 50.

Shugart, L., Chastain, B. H., Novelli, G. D. and Stulberg, M. P. 1968. Biochem. Biophys. Res. Commun. *31,* 404.
Shugart, L., Novelli, G. D. and Stulberg, M. P. 1968. Biochim. Biophys. Acta *157,* 83.
Shuster, R. C. and Weissbach, A. 1969. Nature *223,* 852.
Shyamala, G. and Gorski, J. 1969. J. Biol. Chem. *244,* 1097.
Sibatani, A., de Kloet, S. R., Allfrey, V. G. and Mirsky, A. E. 1962. Proc. Natl. Acad. Sci. *48,* 471.
Siddiqui, M. A. Q. and Hosokawa, K. 1968. Biochem. Biophys. Res. Commun. *32,* 1.
Siddiqui, M. A. Q. and Hosokawa, K. 1969. Biochem. Biophys. Res. Commun. *36,* 711.
Sidebottom, E. and Harris, H. 1969. J. Cell. Sci. *5,* 351.
Siekevitz, P. 1952. J. Biol. Chem. *195,* 549.
Siersma, P. W. and Lederberg, S. 1970. J. Bact. *101,* 398.
Silagi, S., Darlington, G. and Bruce, S. A. 1969. Proc. Natl. Acad. Sci. *62,* 1085.
Silbert, D. F., Fink, G. R. and Ames, B. N. 1966. J. Mol. Biol. *22,* 335.
Siler, J. G. and Fried, M. 1968. Biochem. J. *109,* 185.
Siler, J. and Moldave, K. 1969. Biochim. Biophys. Acta *195,* 123.
Silverstone, A. E., Arditti, R. R. and Magasanik, B. 1970. Proc. Natl. Acad. Sci. *66,* 773.
Silverstone, A. E., Magasanik, B., Reznikoff, W. S., Miller, J. H. and Beckwith, J. R. 1969. Nature *221,* 1012.
Simoni, R. D., Levinthal, M., Kundig, F. D., Kundig, W., Anderson, B., Hartman, P. E. and Roseman, S. 1967. Proc. Natl. Acad. Sci. *58,* 1963.
Simpson, M. V. 1953. J. Biol. Chem. *201,* 143.
Singer, M. F. and Tolbert, G. 1965. Biochemistry *4,* 1319.
Siniscalco, M., Klinger, H. P., Eagle, H., Koprowski, H., Fujimoto, W. Y. and Seegmiller, J. E. 1969. Proc. Natl. Acad. Sci. *62,* 793.
Sirbasku, D. A. and Buchanan, J. M. 1970(a). J. Biol. Chem. *245,* 2679.
Sirbasku, D. A. and Buchanan, J. M. 1970(b). J. Biol. Chem. *245,* 2693.
Sirlin, J. L. and Jacob, J. 1960. Exp. Cell Res. *20,* 283.
Sirlin, J. L., Kato, K. and Jones, K. W. 1961. Biochim. Biophys. Acta *48,* 421.
Sirlin, J. L. and Loening, U. E. 1968. Biochem. J. *109,* 375.
Sivolap, Y. M. and Bonner, J. 1971. Proc. Natl. Acad. Sci. *68,* 387.
Skiff, D. D. 1970. Editor, Cold Spring Harbor Symp. Quant. Biol. Vol. 35.
Skogerson, L. and Moldave, K. 1968. Arch. Biochem. Biophys. *125,* 497.
Skogerson, L., Roufa, D. and Leder, P. 1971. Proc. Natl. Acad. Sci. *68,* 276.
Skold, Ö. and Buchanan, J. M. 1964. Proc. Natl. Acad. Sci. *51,* 553.
Skoultchi, A., Ono, Y., Moon, H. M. and Lengyel, P. 1968. Proc. Natl. Acad. Sci. *60,* 675.
Skoultchi, A., Ono, Y., Waterson, J. and Lengyel, P. 1970. Biochemistry *9,* 508.
Skutelsky, E. and Danon, D. 1967. J. Cell. Biol. *33,* 625.
Slater, D. W. and Spiegelman, S. 1966(a). Proc. Natl. Acad. Sci. *56,* 164.
Slater, D. W. and Spiegelman, S. 1966(b). Biophys. J. *6,* 385.
Slater, D. W. and Spiegelman, S. 1968. Biochim. Biophys. Acta *166,* 82.
Slater, D. W. and Spiegelman, S. 1970. Biochim. Biophys. Acta *213,* 194.
Slonimski, P. 1953. *La formation des enzymes respiratoires chez la levure.* Masson et Cie, Paris.
Slonimski, P., Archer, R., Péré, G., Sels, A. and Somlo, M. 1963. *Méchanismes de régulation des activités cellulaires chez les organismes.* Symp. Int., Marseille. Centre National de la Recherche Scientifique, Paris, p. 435.
Sluyser, M. and Snellen-Jurgens, N. H. 1970. Biochim. Biophys. Acta *199,* 490.

Smith, A. E. and Marcker, K. A. 1970. Nature *226,* 607.
Smith, D. W. E. 1968. J. Biol. Chem. *243,* 3361.
Smith, D. W. E. and Russell, N. L. 1970. Biochim. Biophys. Acta *209,* 171.
Smith, F. L. and Haselkorn, R. 1969. Cold Spring Harbor Symp. Quant. Biol. *34,* 91.
Smith, I., Dubnau, D., Morell, P. and Marmur, J. 1968. J. Mol. Biol. *33,* 123.
Smith, J. A., Martin, L., King, R. J. B. and Vértes, M. 1970. Biochem. J. *119,* 773.
Smith, J. D., Abelson, J. N., Clark, B. F. C., Goodman, H. M. and Brenner, S. 1966. Cold Spring Harbor Symp. Quant. Biol. *31,* 479.
Smith, J. D. and Dunn, D. B. 1959. Biochem. J. *72,* 294.
Smith, K. D., Church, R. B. and McCarthy, B. J. 1969. Biochemistry *8,* 4271.
Smith, L. D. 1965. Proc. Natl. Acad. Sci. *54,* 101.
Smith, L. D., Ecker, R. E. and Subtelny, S. 1966. Proc. Natl. Acad. Sci. *56,* 1724.
Smithies, O. 1967. Science *157,* 267.
Smithies, O. 1970. Science *169,* 882.
Smulson, M. 1970. Biochim. Biophys. Acta *199,* 537.
Smulson, M. E. and Thomas, J. 1969. J. Biol. Chem. *244,* 5309.
Snyder, L. and Geiduschek, E. P. 1968. Proc. Natl. Acad. Sci. *59,* 459.
Soeiro, R., Birnboim, H. C. and Darnell, J. E. 1966. J. Mol. Biol. *19,* 362.
Soeiro, R. and Darnell, J. E. 1970. J. Cell Biol. *44,* 467.
Soeiro, R., Vaughan, M. H. and Darnell, J. E. 1968. J. Cell Biol. *36,* 91.
Soeiro, R., Vaughan, M. H., Warner, J. R. and Darnell, J. E. 1968. J. Cell. Biol. *39,* 112.
Soffer, R. L. 1970. J. Biol. Chem. *245,* 731.
Sokawa, Y., Nakao-Sato, E. and Kaziro, Y. 1970. Biochim. Biophys. Acta *199,* 256.
Sokoloff, L. 1968. *Regulatory Mechanisms for Protein Synthesis in Mammalian Cells.* Editors A. S. Pietro, M. R. Lamborg and F. T. Kenney, Academic Press, New York, p. 345.
Söll, D., Jones, D. S., Ohtsuka, E., Faulkner, R. D., Lohrmann, R., Hayatsu, H., Khorana, H. G., Cherayil, J. D., Hampel, A. and Bock, R. M. 1966. J. Mol. Biol. *19,* 556.
Soll, L. and Berg, P. 1969(a). Proc. Natl. Acad. Sci. *63,* 392.
Soll, L. and Berg, P. 1969(b). Nature *223,* 1340.
Solomon, A. and McLaughlin, C. L. 1969. J. Biol. Chem. *244,* 3393.
Somerville, R. L. and Yanofsky, C. 1965. J. Mol. Biol. *11,* 747.
Southern, E. M. 1970. Nature *227,* 794.
Spahr, P. F. 1962. J. Mol. Biol. *4,* 395.
Spahr, P. F., Farber, M. and Gesteland, R. F. 1969. Nature *222,* 455.
Spelsberg, T. C. and Hnilica, L. S. 1969. Biochim. Biophys. Acta *195,* 63.
Spelsberg, T. C. and Hnilica, L. S. 1970. Biochem. J. *120,* 435.
Spelsberg, T. C. and Hnilica, L. S. 1971(a). Biochim. Biophys. Acta *228,* 202.
Spelsberg, T. C. and Hnilica, L. S. 1971(b). Biochim. Biophys. Acta *228,* 212.
Spelsberg, T. C., Hnilica, L. S. and Ansevin, A. T. 1971. Biochim. Biophys. Acta *228,* 550.
Spencer, D. and Whitfeld, P. R. 1967. Biochem. Biophys. Res. Commun. *28,* 538.
Spiegelman, S. and Hayaishi, M. 1963. Cold Spring Harbor Symp. Quant. Biol. *28,* 161.
Spiegelman, S., Lindegren, C. C. and Hedgecock, L. 1944. Proc. Natl. Acad. Sci. *30,* 13.
Spiegelman, S., Lindegren, C. C. and Lindegren, G. 1945. Proc. Natl. Acad. Sci. *31,* 95.
Spiegelman, S., Sussman, M. and Taylor, B. 1950. Fed. Proc. *9,* 120.

Spiegelman, W. G. 1971. Virology *43,* 16.
Spirin, A. S. 1966. *Current Topics in Developmental Biology.* Editors A. A. Moscona and A. Monroy, Academic Press, New York, Vol. *1,* p. 1.
Spirin, A. S. 1969(a). Europ. J. Biochem. *10,* 20.
Spirin, A. S. 1969(b). Cold Spring Harbor Symp. Quant. Biol. *34,* 197.
Spirin, A. S., Belitsina, N. Y. and Aitkhozhin, M. A. 1964. Zhur. Obsh. Biol. *25,* 321.
Spirin, A. S. and Gavrilova, L. P. 1969. *The Ribosome.* Molecular Biology, Biochemistry and Biophysics, Vol. 4, Springer-Verlag, Berlin.
Spirin, A. S. and Nemer, M. 1965. Science *150,* 214.
Spohr, G., Granboulan, N., Morel, C. and Scherrer, K. 1970. Europ. J. Biochem. *17,* 296.
Spratt, N. T. 1964. *Introduction to Cell Differentiation.* Reinhold, New York.
Srinivasan, P. R. and Borek, E. 1963. Proc. Natl. Acad. Sci. *49,* 529.
Srinivasan, P. R. and Borek, E. 1964(a). Science *145,* 548.
Srinivasan, P. R. and Borek, E. 1964(b). Biochemistry *3,* 616.
Staehelin, T., Wettstein, F. O., Oura, H. and Noll, H. 1964. Nature *201,* 264.
Stanier, R. Y. 1947. J. Bact. *54,* 339.
Stanier, R. Y. 1950. Bact. Rev. *14,* 179.
Stanier, R. Y. 1954. *Aspects of Synthesis and Order in Growth.* Editor D. Rudnick, Growth Symposium *13,* 43.
Stanier, R. Y., Hegeman, G. D. and Ornston, L. N. 1965. Colloq. Intern. Centre Nat. Rech. Sci. (Paris), p. 227.
Starbuck, W. C., Mauritzen, C. M., Taylor, C. W., Saroja, I. S. and Busch, H. 1968. J. Biol. Chem. *243,* 2038.
Stavy, L. and Gross, P. R. 1967. Proc. Natl. Acad. Sci. *57,* 735.
Stavy, L. and Gross, P. R. 1969(a). Biochim. Biophys. Acta *182,* 193.
Stavy, L. and Gross, P. R. 1969(b). Biochim. Biophys. Acta *182,* 203.
Stebbins, G. L. 1966. Science *152,* 1463.
Stedman, E. and Stedman, E. 1950. Nature *166,* 780.
Steggles, A. W., Spelsberg, T. C., Glasser, S. R. and O'Malley, B. W. 1971. Proc. Natl. Acad. Sci. *68,* 1479.
Stein, G. and Baserga, R. 1970. Biochem. Biophys. Res. Commun. *41,* 715.
Stein, H. and Hausen, P. 1969. Science *166,* 393.
Stein, H. and Hausen, P. 1970(a). Europ. J. Biochem. *14,* 270.
Stein, H. and Hausen, P. 1970(b). Cold Spring Harbor Symp. Quant. Biol. *35,* 709.
Steinberg, D., Vaughan, M. and Anfinsen, C. B. 1956. Science *124,* 389.
Steinberg, R. A. and Ptashne, M. 1971. Nature New Biology *230,* 76.
Steitz, J. A. 1969. Nature *224,* 957.
Steitz, J. A., Dube, S. K. and Rudland, P. S. 1970. Nature *226,* 824.
Stellwagen, R. H. and Cole, R. D. 1968. J. Biol. Chem. *243,* 4456.
Stellwagen, R. H. and Cole, R. D. 1969(a). Ann. Rev. Biochem. *38,* 951.
Stellwagen, R. H. and Cole, R. D. 1969(b). J. Biol. Chem. *244,* 4878.
Stent, G. S. 1963. *Molecular Biology of Bacterial Viruses.* W. H. Freeman and Co., San Francisco.
Stent, G. S. 1964. Science *144,* 816.
Stent, G. S. 1966. Proc. Roy. Soc. B, *164,* 181.
Stent, G. S. 1967. *Organizational Biosynthesis.* Editors H. J. Vogel, J. O. Lampen and V. Bryson, Academic Press, New York, p. 99.
Stent, G. S. and Brenner, S. 1961. Proc. Natl. Acad. Sci. *47,* 2005.
Stephenson, J. R., Axelrad, A. A., McLeod, D. L. and Shreeve, M. W. 1971. Proc. Natl. Acad. Sci. *68,* 1542.

Stephenson, M. 1937. Ergebn. Enzymforsch. *6,* 139.
Stephenson, M. and Stickland, L. H. 1933. Biochem. J. *27,* 1528.
Stephenson, M. and Yudkin, J. 1936. Biochem. J. *30,* 506.
Stern, C. 1964. In discussion of paper by B. Ephrussi, *Somatic Cell Genetics,* Editor R. S. Krooth, University of Michigan Press, Ann Arbor, p. 253.
Stern, R., Gonano, F., Fleissner, E. and Littauer, U. Z. 1970. Biochemistry *9,* 10.
Stevely, W. S. and Stocken, L. A. 1968. Biochem. J. *110,* 187.
Stévenin, J., Mandel, P. and Jacob, M. 1969. Proc. Natl. Acad. Sci. *62,* 490.
Stévenin, J., Mandel, P. and Jacob, M. 1970. Bull. Soc. Chim. Biol. *52,* 703.
Stevens, A. 1970. Biochem. Biophys. Res. Commun. *41,* 367.
Stevenson, I. L. and Mandelstam, J. 1965. Biochem. J. *96,* 354.
Steward, F. C. 1970. Proc. Roy. Soc. B, *175,* 1.
Steward, F. C., Mapes, M. O., Kent, A. E. and Holsten, R. D. 1964. Science *143,* 20.
Steward, J. W., Sherman, F., Shipman, N., Thomas, F. L. X. and Cravens, M. 1969. Fed. Proc. *28,* 597.
Stewart, M. J. and Corrance, M. H. 1969. Cancer Res. *29,* 1642.
Stewart, T. S., Roberts, R. J. and Strominger, J. L. 1971. Nature *230,* 36.
Stockdale, F. E., Juergens, W. G. and Topper, Y. J. 1966. Dev. Biol. *13,* 266.
Stockdale, F. E. and Topper, Y. J. 1966. Proc. Natl. Acad. Sci. *56,* 1283.
Stubbs, J. D. and Hall, B. D. 1968. J. Mol. Biol. *37,* 289.
Stulberg, M. P., Isham, K. R. and Stevens, A. 1969. Biochim. Biophys. Acta *186,* 297.
Sturtevant, A. H. 1925. Genetics *10,* 117.
Sturtevant, A. H. 1951. *Genetics in the 20th Century.* Editor L. C. Dunn, Macmillan Co., New York, p. 101.
Stutz, E. and Noll, H. 1967. Proc. Natl. Acad. Sci. *57,* 774.
Subak-Sharpe, H. and Hay, J. 1965. J. Mol. Biol. *12,* 924.
Subramanian, A. R. and Davis, B. D. 1970. Nature *228,* 1273.
Subramanian, A. R., Davis, B. D. and Beller, R. J. 1969. Cold Spring Harbor Symp. Quant. Biol. *34,* 223.
Subramanian, A. R., Ron, E. Z. and Davis, B. D. 1968. Proc. Natl. Acad. Sci. *61,* 761.
Subtelny, S. 1965(a). J. Exp. Zool. *159,* 47.
Subtelny, S. 1965(b). J. Exp. Zool. *159,* 59.
Suda, M., Hayaishi, O. and Oda, Y. 1949. Symp. on Enzyme Chemistry (Tokyo) *1,* 73.
Sueoka, N. and Cheng, T. Y. 1962. J. Mol. Biol. *4,* 161.
Sueoka, N. and Kano-Sueoka, T. 1970. Prog. Nucleic Acid Res. *10,* 23.
Sugiura, M., Okamoto, T. and Takanami, M. 1969. J. Mol. Biol. *43,* 299.
Sugiyama, T. and Nakada, D. 1968. J. Mol. Biol. *31,* 431.
Summers, W. C. 1970. J. Mol. Biol. *51,* 671.
Summers, W. C. 1971. Nature New Biology *230,* 208.
Summers, W. C. and Siegel, R. B. 1969. Nature *223,* 1111.
Summers, W. C. and Siegel, R. B. 1970. Nature *228,* 1160.
Sundararajan, T. A. and Thach, R. E. 1966. J. Mol. Biol. *19,* 74.
Sung, M. T. and Dixon, G. H. 1970. Proc. Natl. Acad. Sci. *67,* 1616.
Sunshine, G. H., Williams, D. J. and Rabin, B. R. 1971. Nature New Biology *230,* 133.
Sussman, A. J. and Gilvarg, C. 1969. J. Biol. Chem. *244,* 6304.
Sussman, M. 1966. *Current Topics in Developmental Biology.* Editors A. A. Moscona and A. Monroy, Academic Press, New York, Vol. *1,* p. 61.
Sussman, M. 1970. Nature *225,* 1245.
Sussman, M. and Sussman, R. 1969. Symp. Soc. Gen. Microbiol. *19,* 403.

Svensson, I., Björk, G. R. and Lundahl, P. 1969. Europ. J. Biochem. *9,* 216.
Swan, D., Sander, G., Bermek, E., Krämer, W., Kreuzer, T., Arglebe, C., Zöllner, R., Eckert, K. and Matthaei, H. 1969. Cold Spring Harbor Symp. Quant. Biol. *34,* 179.
Swaneck, G. E., Chu, L. L. H. and Edelman, I. S. 1970. J. Biol. Chem. *245,* 5382.
Szego, C. M. and Davis, J. S. 1967. Proc. Natl. Acad. Sci. *58,* 1711.
Szybalski, W. 1969. Can. Cancer Conf. *8,* 183.
Szybalski, W. 1970. Lepetit Colloq. Biol. Med. *1,* 209.
Szybalski, W., Bøvre, K., Fiandt, M., Guha, A., Hradecna, Z., Kumar, S., Lozeron, H. A., Maher, V. M., Nijkamp, H. J. J., Summers, W. C. and Taylor, K. 1969. J. Cell. Phys. *74,* Supp. *1,* 33.
Szybalski, W., Bøvre, K., Fiandt, M., Hayes, S., Hradecna, Z., Kumar, S., Lozeron, H. A., Nijkamp, H. J. J. and Stevens, W. F. 1970. Cold Spring Harbor Symp. Quant. Biol. *35,* 341.
Takagi, M., Tanaka, T. and Ogata, K. 1970. Biochim. Biophys. Acta *217,* 148.
Takaku, F., Nakao, K., Ono, T. and Terayama, H. 1969. Biochim. Biophys. Acta *195,* 396.
Takanami, M. 1963. Biochim. Biophys. Acta *61,* 432.
Takanami, M. and Okamoto, T. 1963. J. Mol. Biol. *7,* 323.
Takanami, M., Yan, Y. and Jukes, T. H. 1965. J. Mol. Biol. *12,* 761.
Takanami, M. and Zubay, G. 1964. Proc. Natl. Acad. Sci. *51,* 834.
Takeda, Y. 1969(a). Biochim. Biophys. Acta *182,* 258.
Takeda, Y. 1969(b). J. Biochem. *66,* 345.
Takeda, Y. and Yura, T. 1968. Virology *36,* 55.
Takeishi, K., Sekiya, T. and Ukita, T. 1970. Biochim. Biophys. Acta *199,* 559.
Talwar, G. P., Segal. S. J., Evans, A. and Davidson, O. W. 1964. Proc. Natl. Acad. Sci. *52,* 1059.
Tan, C. H. and Miyagi, M. 1970. J. Mol. Biol. *50,* 641.
Tanner, M. J. A. 1967. Biochemistry *6,* 2686.
Tao, M. and Huberman, A. 1970. Arch. Biochem. Biophys. *121,* 236.
Tao, M., Salas, M. L. and Lipmann, F. 1970. Proc. Natl. Acad. Sci. *67,* 408.
Tao, M. and Schweiger, M. 1970. J. Bact. *102,* 138.
Tartof, K. D. 1971. Science *171,* 294.
Tata, J. R. 1967. Biochem. J. *104,* 1.
Tatum, E. L. and Lederberg, J. 1947. J. Bact. *53,* 673.
Tavill, A. S., Grayzel, A. I., London, I. M., Williams, M. K. and Vanderhoff, G. A. 1968. J. Biol. Chem. *243,* 4987.
Taylor, A. L. and Trotter, C. D. 1967. Bact. Rev. *31,* 332.
Taylor, K., Hradecna, Z. and Szybalski, W. 1967. Proc. Natl. Acad. Sci. *57,* 1618.
Taylor, M. W., Buck, C. A., Granger, G. A. and Holland, J. J. 1968. J. Mol. Biol. *33,* 809.
Teng, C. S. and Hamilton, T. H. 1968. Proc. Natl. Acad. Sci. *60,* 1410.
Teng, C. S. and Hamilton, T. H. 1969. Proc. Natl. Acad. Sci. *63,* 465.
Teng, C. S. and Hamilton, T. H. 1970(a). Biochem. Biophys. Res. Commun. *40,* 1231.
Teng, C. S. and Hamilton, T. H. 1970(b). Biochem. J. *118,* 341.
Teng, C. S., Teng, C. T. and Allfrey, V. G. 1971(a). Cited in Teng, Teng and Allfrey (1970).
Teng, C. S., Teng, C. T. and Allfrey, V. G. 1971(b). Cited in Teng, Teng and Allfrey (1970).
Teng, C. T., Teng, C. S. and Allfrey, V. G. 1970. Biochem. Biophys. Res. Commun. *41,* 690.
Terman, S. A. 1970. Proc. Natl. Acad. Sci. *65,* 985.

Thach, S. S. and Thach, R. E. 1971. Nature New Biology *229,* 219.
Thiebe, R. and Zachau, H. G. 1968. Europ. J. Biochem. *5,* 546.
Thomas, C. 1970. Biochim. Biophys. Acta *224,* 99.
Thompson, E. B., Granner, D. K. and Tomkins, G. M. 1970. J. Mol. Biol. *54,* 159.
Thompson, L. R. and McCarthy, B. J. 1968. Biochem. Biophys. Res. Commun. *30,* 166.
Thompson, L. R. and McCarthy, B. J. 1970. Personal communication.
Thompson, P., English, D. S. and Bleecker, W. 1969. Genetics *63,* 183.
Thuriaux, P., Ramos, F., Wiame, J. M., Grenson, M. and Bechet, J. 1968. Arch. Internat. Physiol. Biochim. *76,* 955.
Tidwell, T., Allfrey, V. G. and Mirsky, A. E. 1968. J. Biol. Chem. *243,* 707.
Tiedemann, H. 1966. *Current Topics in Developmental Biology.* Editors A. A. Moscona and A. Monroy, Academic Press, New York, Vol. *1,* p. 85.
Till, J. E., McCulloch, E. A., Phillips, R. A. and Siminovitch, L. 1967. Cold Spring Harbor Symp. Quant. Biol. *32,* 461.
Tissières, A., Watson, J. D., Schlessinger, D. and Hollingworth, B. R. 1959. J. Mol. Biol. *1,* 221.
Tocchini-Valentini, G. P. and Crippa, M. 1970. Nature *228,* 993.
Tocchini-Valentini, G. P., Stodolsky, M., Aurisicchio, A., Sarnat, M., Graziosi, F., Weiss, S. B. and Geiduschek, E. P. 1963. Proc. Natl. Acad. Sci. *50,* 935.
Todd, C. W. 1963. Biochem. Biophys. Res. Commun. *11,* 170.
Toft, D. and Gorski, J. 1966. Proc. Natl. Acad. Sci. *55,* 1574.
Toft, D., Shyamala, G. and Gorski, J. 1967. Proc. Natl. Acad. Sci. *57,* 1740.
Tomkins, G. M. 1968. *Regulatory Mechanisms for Protein Synthesis in Mammalian Cells.* Editors A. S. Pietro, M. R. Lamborg and F. T. Kenney, Academic Press, New York, p. 269.
Tomkins, G. M., Garren, L. D., Howell, R. R. and Peterkofsky, B. 1965. J. Cell. Comp. Phys. *66,* Supp. *1,* 137.
Tomkins, G. M., Gelehrter, T. D., Granner, D., Martin, D., Samuels, H. H. and Thompson, E. B. 1969. Science *166,* 1474.
Tonoue, T., Eaton, J. and Frieden, E. 1969. Biochem. Biophys. Res. Commun. *37,* 81.
Torriani, A. M. 1956. Biochim. Biophys. Acta *19,* 224.
Torriani, A. 1968. J. Bact. *96,* 1200.
Torriani, A. and Rothman, F. 1961. J. Bact. *81,* 835.
Tóth, M. and Mányai, S. 1968. Acta Biochim. et Biophys. Acad. Sci. Hung. *3,* 337.
Traut, R. R., Delius, H., Ahmad-Zadeh, C., Bickle, T. A., Pearson, P. and Tissières, A. 1969. Cold Spring Harbor Symp. Quant. Biol. *34,* 25.
Travers, A. A. 1969. Nature *223,* 1107.
Travers, A. A. 1970(a). Nature *225,* 1009.
Travers, A. 1970(b). Cold Spring Harbor Symp. Quant. Biol. *35,* 241.
Travers, A. 1971. Nature New Biology *229,* 69.
Travers, A. A. and Burgess, R. R. 1969. Nature *222,* 537.
Travers, A., Kamen, R. and Cashel, M. 1970. Cold Spring Harbor Symp. Quant. Biol. *35,* 415.
Travers, A. A., Kamen, R. I. and Schleif, R. F. 1970. Nature *228,* 748.
Trevithick, J. R. and Wainwright, S. D. 1970. Can. J. Biochem. *48,* 833.
Trinkaus, J. P. 1969. *Cells into Organs.* Prentice-Hall Inc., Englewood Cliffs, New Jersey.
Truffa-Bachi, P. and Cohen, G. N. 1968. Ann. Rev. Biochem. *37,* 79.
Tsanev, R. and Sendov, B. 1971. J. Theoret. Biol. *30,* 337.
Tschudy, D. P., Marver, H. S. and Collins, A. 1965. Biochem. Biophys. Res. Commun. *21,* 480.

Ts'o P. O. P. and Sato, C. S. 1959. Exp. Cell. Res. *17,* 237.
Tsugita, A., Inouye, M., Imagawa, T., Nakanishi, T., Okada, Y., Emrich, J. and Streisinger, G. 1969. J. Mol. Biol. *41,* 349.
Tsukada, K., Majumdar, C. and Liebermann, I. 1966. Biochem. Biophys. Res. Commun. *25,* 181.
Tsutsui, E., Srinivasan, P. R. and Borek, E. 1966. Proc. Natl. Acad. Sci. *56,* 1003.
Turkington, R. W. 1968(a). *Current Topics in Developmental Biology.* Editors A. A. Moscona and A. Monroy. Academic Press, London, Vol. *3,* p. 199.
Turkington, R. W. 1968(b). Endocrinol. *82,* 540.
Turkington, R. W. 1968(c). Endocrinol. *82,* 575.
Turkington, R. W. 1968(d). Biochim. Biophys. Acta *158,* 274.
Turkington, R. W. 1969(a). Exp. Cell Res. *58,* 296.
Turkington, R. W. 1969(b). J. Biol. Chem. *244,* 5140.
Turkington, R. W. 1970(a). Biochim. Biophys. Acta *213,* 484.
Turkington, R. W. 1970(b). J. Biol. Chem. *245,* 6690.
Turkington, R. W. 1970(c). Biochem. Biophys. Res. Commun. *41,* 1362.
Turkington, R. W. 1971. *Sex Steroids: Molecular Mechanisms.* Editor K. W. McKerns, Appleton-Century-Crofts, New York. In press.
Turkington, R. W., Brew, K., Vanaman, T. C. and Hill, R. L. 1968. J. Biol. Chem. *243,* 3382.
Turkington, R. W. and Hill, R. L. 1969. Science *163,* 1458.
Turkington, R. W., Lockwood, D. H. and Topper, Y. J. 1967. Biochim. Biophys. Acta *148,* 475.
Turkington, R. W. and Riddle, M. 1969. J. Biol. Chem. *244,* 6040.
Turkington, R. W. and Riddle, M. 1970(a). Cancer Res. *30,* 650.
Turkington, R. W. and Riddle, M. 1970(b). J. Biol. Chem. *245,* 5145.
Turkington, R. W. and Topper, Y. J. 1967. Endocrinol. *80,* 329.
Turkington, R. W. and Ward, O. T. 1969. Biochim. Biophys. Acta *174,* 291.
Twu, J. S. and Bretthauer, R. K. 1971. Biochemistry *10,* 1576.
Tyler, A. 1963. Amer. Zool. *3,* 109.
Tyler, A. 1967. *Control Mechanisms in Developmental Processes.* Editor M. Locke, Academic Press, New York, p. 170.
Tyler, B. and Magasanik, B. 1969. J. Bact. *97,* 550.
Tyler, B. and Magasanik, B. 1970. J. Bact. *102,* 411.
Tyler, B., Wishnow, R., Loomis, W. F. and Magasanik, B. 1969. J. Bact. *100,* 809.
Udaka, S. 1970. Nature *228,* 336.
Udenfriend, S. 1966. Science *152,* 1335.
Ullmann, A. and Monod, J. 1968. FEBS Letters *2,* 57.
Van Dijk-Salkinoja, M. S. and Planta, R. J. 1970. Arch. Biochem. Biophys. *141,* 477.
Van Dijk-Salkinoja, M. S. and Planta, R. J. 1971. J. Bact. *105,* 20.
Van Furth, R., Schuit, H. R. E. and Hijmans, W. 1967. *Gamma Globulins. Structure and Control of Biosynthesis.* Nobel Symposium *3.* Editor J. Killander. Almqvist and Wiksell, Stockholm, p. 443.
Varmus, H. E., Perlman, R. L. and Pastan, I. 1970(a). J. Biol. Chem. *245,* 2259.
Varmus, H. E., Perlman, R. L. and Pastan, I. 1970(b). J. Biol. Chem. *245,* 6366.
Varmus, H. E., Perlman, R. L. and Pastan, I. 1971. Nature New Biology *230,* 41.
Vaughan, M. H., Soeiro, R., Warner, J. R. and Darnell, J. E. 1967. Proc. Natl. Acad. Sci. *58,* 1527.
Vaughan, M. H., Warner, J. R. and Darnell, J. E. 1967. J. Mol. Biol. *25,* 235.
Vendrely, R. and Picaud, M. 1968. Exp. Cell Res. *49,* 13.
Venetianer, P. 1968. Biochem. Biophys. Res. Commun. *33,* 959.
Venetianer, P., Berberich, M. A. and Goldberger, R. F. 1968. Biochim. Biophys. Acta *166,* 124.

Vesco, C. and Colombo, B. 1970. J. Mol. Biol. *47,* 335.
Vesco, C. and Penman, S. 1969. Proc. Natl. Acad. Sci. *62,* 218.
Vickers, J. D. and Logan, D. M. 1970. Biochem. Biophys. Res. Commun. *41,* 741.
Vidali, G., Gershey, E. L. and Allfrey, V. G. 1968. J. Biol. Chem. *243,* 6361.
Vidali, G. and Neelin, J. M. 1968. Europ. J. Biochem. *5,* 330.
Vilček, J. 1969. *The Interferons.* Virology Monographs. Springer-Verlag, New York.
Villee, C. A., Hagerman, D. D. and Joel, P. B. 1960. Recent Progress in Hormone Research *16,* 49.
Vincent, W. S. 1957. Science *126,* 306.
Vincent. W. S., Halvorson, H. O., Chen, H. R. and Shin, D. 1969. Exp. Cell Res. *57,* 240.
Vittorelli, M. L., Caffarelli-Mormino, I. and Monroy, A. 1969. Biochim. Biophys. Acta *186,* 408.
Vogel, H. J. 1953. Proc. Natl. Acad. Sci. *39,* 578.
Vogel, H. J. 1957(a). *The Chemical Basis of Heredity.* Editors W. D. McElroy and B. Glass, Johns Hopkins Press, Baltimore, p. 276.
Vogel, H. J. 1957(b). Proc. Natl. Acad. Sci. *43,* 491.
Vogel, H. J. 1960. Proc. Natl. Acad. Sci. *46,* 488.
Vogel, H. J. 1961. Cold Spring Harbor Symp. Quant. Biol. *26,* 163.
Vogel, H. J. 1967. Can. Cancer Conf. *7,* 133.
Vogel, H. J., Bacon, D. F. and Baich, A. 1963. *Informational Macromolecules.* Editors H. J. Vogel, V. Bryson and J. O. Lampen, Academic Press, New York, p. 293.
Vogel, H. J., Baumberg, S., Bacon, D. F., Jones, E. E., Unger, L. and Vogel, R. H. 1967. *Organizational Biosynthesis.* Editors H. J. Vogel, J. O. Lampen and V. Bryson, Academic Press, New York, p. 223.
Vogel, H. J. and Bonner, D. M. 1956. J. Biol. Chem. *218,* 97.
Vogel, H. J. and Davis, B. D. 1952. Fed. Proc. *11,* 485.
Vogel, Z., Zamir, A. and Elson, D. 1968. Proc. Natl. Acad. Sci. *61,* 701.
Vogel, Z., Zamir, A. and Elson, D. 1969. Biochemistry *8,* 5161.
Vogt, V. 1969. Nature *223,* 854.
Vold, B. S. and Sypherd, P. S. 1968. Proc. Natl. Acad. Sci. *59,* 453.
Volkers, S. A. S. and Taylor, M. W. 1971. Biochemistry *10,* 488.
Volkin, E. and Astrachan, L. 1956(a). Virology *2,* 149.
Volkin, E. and Astrachan, L. 1956(b). Virology *2,* 433.
Volkin, E., Astrachan, L. and Countryman, J. L. 1958. Virology *6,* 545.
von Ehrenstein, G. 1966. Cold Spring Harbor Symp. Quant. Biol. *31,* 705.
Voytovich, A. E., Owens, I. S. and Topper, Y. J. 1969. Proc. Natl. Acad. Sci. *63,* *213.*
Vyas, S. and Maas, W. K. 1963. Arch. Biochem. Biophys. *100,* 542.
Waddington, C. H. 1969(a). Nature *224,* 269.
Waddington, C. H. 1969(b). Science *166,* 639.
Wagner, C., Payne, A. N. and Briggs, W. 1969. Exp. Cell Res. *55,* 330.
Wagner, E. K., Penman, S. and Ingram, V. M. 1967. J. Mol. Biol. *29,* 371.
Wagner, E. K. and Roizman, B. 1969. Proc. Natl. Acad. Sci. *64,* 626.
Wahba, A. J., Chae, Y. B., Iwasaki, K., Mazumder, R., Miller, M. J., Sabol, S. and Sillero, M. A. G. 1969. Cold Spring Harbor Symp. Quant. Biol. *34,* 285.
Wainfan, E. 1968. Virol. *35,* 282.
Wainfan, E., Srinivasan, P. R. and Borek, E. 1965. Biochemistry *4,* 2845.
Wainwright, S. D. 1953. Arch. Biochem. Biophys. *47,* 445.
Wainwright, S. D. 1970. Can. J. Biochem. *48,* 393.
Wainwright, S. D. 1971(a). Cancer Res. *31,* 694.
Wainwright, S. D. 1971(b). *Symposium Erythropoieticum.* Charles University, Prague. In press.

Wainwright, S. D. and Bonner, D. M. 1959. Can. J. Biochem. Physiol. *37*, 741.
Wainwright, S. D. and Thompson, J. C. 1972. In preparation.
Wainwright, S. D., Thompson, J. C., Prchal, J. F. and Tsay, H. M. 1972. In preparation.
Wainwright, S. D. and Wainwright, L. K. 1966. Can J. Biochem. *44*, 1543.
Wainwright, S. D. and Wainwright, L. K. 1967(a). Can J. Biochem. *45*, 255.
Wainwright, S. D. and Wainwright, L. K. 1967(b). Can J. Biochem. *45*, 344.
Wainwright, S. D. and Wainwright, L. K. 1967(c). Can. J. Biochem. *45*, 1483.
Wainwright, S. D. and Wainwright, L. K. 1969. Can. J. Biochem. *47*, 1089.
Wainwright, S. D. and Wainwright, L. K. 1970. Can. J. Biochem. *48*, 400.
Wainwright, S. D. and Wainwright, L. K. 1972(a). In preparation.
Wainwright, S. D. and Wainwright, L. K. 1972(b). In preparation.
Wainwright, S. D. and Wainwright, L. K. 1972(c). In preparation.
Wainwright, S. D. and Wainwright, L. K. 1972(d). In preparation.
Wainwright, S. D. and Wainwright, L. K. 1972(e). In preparation.
Wainwright, S. D., Wainwright, L. K. and Tsay, H. M. 1972. In preparation.
Walker, P. B. M. 1968. Nature *219*, 228.
Wallace, H., Morray, J. and Langridge, W. H. R. 1971. Nature New Biol. *230*, 201.
Wallenfels, K. and Malhotra, O. P. 1961. Advances in Carbohydrate Chem. *16*, 239.
Walsh, D. A., Perkins, J. P. and Krebs, E. G. 1968. J. Biol. Chem. *243*, 3763.
Walter, G., Seifert, W. and Zillig, W. 1968. Biochem. Biophys. Res. Commun. *30*, 240.
Wang, A. C., Wilson, S. K., Hopper, J. E., Fudenberg, H. H. and Nisonoff, A. 1970. Proc. Natl. Acad. Sci. *66*, 337.
Wang, T. Y. 1966. J. Biol. Chem. *241*, 2913.
Wang, T. Y. 1968. Arch. Biochem. Biophys. *127*, 235.
Ward, G. A. and Plagemann, P. G. W. 1969. J. Cell. Phys. *73*, 213.
Ward, R., Konings, R. H. and Hofschneider, P. H. 1970. Europ. J. Biochem. *17*, 106.
Waring, M. and Britten, R. J. 1966. Science *154*, 791.
Warner, H. R., Snustad. P., Jorgensen, S. E. and Koerner, J. F. 1970. J. Virol. *5*, 700.
Warner, J. R. 1966. J. Mol. Biol. *19*, 383.
Warner, J. R., Girard, M., Latham, H. and Darnell, J. E. 1966. J. Mol. Biol. *19*, 373.
Warner, J. R., Knopf, P. M. and Rich, A. 1963. Proc. Natl. Acad. Sci. *49*, 122.
Warner, J. R. and Rich, A. 1964(a). J. Mol. Biol. *10*, 202.
Warner, J. R. and Rich, A. 1964(b). Proc. Natl. Acad. Sci. *51*, 1134.
Warner, J. R., Rich, A. and Hall, C. E. 1962. Science *138*, 1399.
Warner, J. R. and Soeiro, R. 1967. Proc. Natl. Acad. Sci. *58*, 1984.
Waterson, J., Beaud, G. and Lengyel, P. 1970. Nature *227*, 34.
Waxman, H. S., Freedman, M. L. and Rabinovitz, M. 1967. Biochim. Biophys. Acta *145*, 353.
Waxman, H. S. and Rabinovitz, M. 1966. Biochim. Biophys. Acta *129*, 369.
Weatherall, D. J. 1968. Brit. J. Haematol. *15*, 1.
Weatherall, D. J., Clegg, J. B. and Naughton, M. A. 1965. Nature *208*, 1061.
Weber, G., Singhal, R. L., Stamm, N. B., Lea, M. A. and Fisher, E. A. 1966. Adv. in Enzyme Regulation *4*, 59.
Webster, R. E., Engelhardt, D. L., Zinder, N. D. and Konigsberg, W. 1967. J. Mol. Biol. *29*, 27.
Webster, R. E. and Zinder, N. D. 1969. J. Mol. Biol. *42*, 425.
Wegman, J. and Crawford, I. P. 1968. J. Bact. *95*, 2325.

Weigert, M. G., Gallucci, E., Lanka, E. and Garen, A. 1966. Cold Spring Harbor Symp. Quant. Biol. *31,* 145.
Weiler, A. 1965. Proc. Natl. Acad. Sci. *54,* 1765.
Weinberg, R. A., Loening, U., Willems, M. and Penman, S. 1967. Proc. Natl. Acad. Sci. *58,* 1088.
Weinberg, R. and Penman, S. 1969. Biochim. Biophys. Acta *190,* 10.
Weinberg, R. A. and Penman, S. 1970. J. Mol. Biol. *47,* 169.
Weisberg, R. A. and Gallant, J. A. 1966. Cold Spring Harbor Symp. Quant. Biol. *31,* 374.
Weisberg, R. A. and Gallant, J. A. 1967. J. Mol. Biol. *25,* 537.
Weisblum, B., Cherayil, J. D., Bock, R. M. and Söll, D. 1967. J. Mol. Biol. *28,* 275.
Weisblum, B., Gonano, F., von Ehrenstein, G. and Benzer, S. 1965. Proc. Natl. Acad. Sci. *53,* 328.
Weiss, J. F. and Kelmers, A. D. 1967. Biochemistry *6,* 2507.
Weiss, S. B. 1960. Proc. Natl. Acad. Sci. *46,* 1020.
Weiss, S. B. 1963. *Informational Macromolecules.* Editors H. J. Vogel, V. Bryson and J. O. Lampen, Academic Press, New York, p. 61.
Weiss, S. B., Hsu, W. T., Foft, J. W. and Scherberg, N. H. 1968. Proc. Natl. Acad. Sci. *61,* 114.
Weissbach, H., Miller, D. L. and Hachmann, J. 1970. Arch. Biochem. Biophys. *137,* 262.
Weissbach, H. and Redfield, B. 1967. Biochem. Biophys. Res. Commun. *27,* 7.
Weissbach, H., Redfield, B. and Hachmann, J. 1970. Arch. Biochem. Biophys. *141,* 384.
Wellings, S. R., Cooper, R. A. and Rivera, E. M. 1966. J. Nat. Cancer Inst. *36,* 657.
Wellings, S. R. and Deome, K. B. 1961. J. Biophys. Biochem. Cytol. *9,* 479.
Wellings, S. R., Deome, K. B. and Pitelka, D. R. 1960. J. Nat. Cancer Inst. *25,* 393.
Welshons, W. J. 1965. Science *150,* 1122.
Wettstein, F. O., Staehelin, T. and Noll, H. 1963. Nature *197,* 430.
Wettstein, F. O. and Stent, G. S. 1968. J. Mol. Biol. *38,* 25.
Wevers, W. F., Baguley, B. C. and Ralph, R. K. 1966. Biochim. Biophys. Acta *123,* 503.
Wheelis, M. L. and Stanier, R. Y. 1970. Genetics *66,* 245.
White, B. N. and Bayley, S. T. 1971. Personal communication.
White, M. J. D. 1951. *Genetics in the 20th Century,* Editor L. C. Dunn, Macmillan Co., New York, p. 333.
Whitehouse, H. L. K. 1967(a). J. Cell. Sci. *2,* 9.
Whitehouse, H. L. K. 1967(b). Nature *215,* 371.
Whiteley, H. R. and Hemphill, E. 1970. Biochem. Biophys. Res. Commun. *41,* 647.
Whitfield, H. J., Gutnick, D. L., Margolies, M. N., Martin, R. G., Rechler, M. M. and Voll, M. J. 1970. J. Mol. Biol. *49,* 245.
Whitfield, H. J., Smith, D. W. E. and Martin, R. G. 1964. J. Biol. Chem. *239,* 3288.
Whitt, D. D. and Carlton, B. C. 1968(a). Biochem. Biophys. Res. Commun. *33,* 636.
Whitt, D. D. and Carlton, B. C. 1968(b). J. Bact. *96,* 1273.
Wicks, W. D. 1968. *Regulatory Mechanisms for Protein Synthesis in Mammalian Cells.* Editors A. S. Pietro, M. R. Lamborg and F. T. Kenney, Academic Press, New York, p. 143.
Wicks, W. D. 1969. J. Biol. Chem. *244,* 3941.
Widnell, C. C. and Tata, J. R. 1964. Biochim. Biophys. Acta *87,* 531.
Widnell, C. C. and Tata, J. R. 1966. Biochim. Biophys. Acta *123,* 478.
Wigle, D. T. and Dixon, G. H. 1970. Nature *227,* 676.
Wijesundra, S. and Woods, D. D. 1953. Biochem. J. *55,* viii.

Wilhelm, F. X., Champagne, M. H. and Daune, M. P. 1970. Europ. J. Biochem. *15,* 321.
Wilhelm, J. M. and Haselkorn, R. 1970. Proc. Natl. Acad. Sci. *65,* 388.
Wilhelm, J. M. and Haselkorn, R. 1971. Virology *43,* 198.
Wilhelm, R. C. 1966. Cold Spring Harbor Symp. Quant. Biol. *31,* 496.
Wilkinson, J. M. 1969. Biochem. J. *112,* 173.
Willems, M., Penman, M. and Penman, S. 1969. J. Cell Biol. *41,* 177.
Willems, M., Wagner, E., Laing, R. and Penman, S. 1968. J. Mol. Biol. *32,* 211.
Williams, L. and Freundlich, M. 1969. Biochim. Biophys. Acta *186,* 305.
Williamson, A. R. and Askonas, B. A. 1968. Arch. Biochem. Biophys. *125,* 401.
Williamson, R. 1970. J. Mol. Biol. *51,* 157.
Willson, C. and Gros, F. 1964. Biochim. Biophys. Acta *80,* 478.
Willson, C., Perrin, D., Cohn, M., Jacob, F. and Monod, J. 1964. J. Mol. Biol. *8,* 582.
Wilson, D. B. 1969. *Medicine in the University and Community of the Future.* Proceedings of the Scientific Sessions Dalhousie University Faculty of Medicine Centennial. Whynot and Associates, Ltd., Halifax, p. 209.
Wilson, D. B. and Dintzis, H. 1969. Cold Spring Harbor Symp. Quant. Biol. *34,* 313.
Wilson, D. B. and Dintzis, H. M. 1970. Proc. Natl. Acad. Sci. *66,* 1282.
Wilson, D. B. and Hogness, D. S. 1969. J. Biol. Chem. *244,* 2143.
Wilson, D. L. and Geiduschek, E. P. 1969. Proc. Natl. Acad. Sci. *62,* 514.
Wilt, F. H. 1962. Proc. Natl. Acad. Sci. *48,* 1582.
Wilt, F. H. 1965(a). J. Mol. Biol. *12,* 331.
Wilt, F. H. 1965(b). Science *147,* 1588.
Wilt, F. H. 1966. Amer. Zool. *6,* 67.
Wilt, F. H. 1967. Advances in Morphogenesis. Editors M. Abercrombie and J. Brachet, Academic Press, New York, Vol. *6,* p. 89.
Wilt, F. H. 1968. Biochem. Biophys. Res. Commun. *33,* 113.
Wilt, F. H. and Hultin, T. 1962. Biochem. Biophys. Res. Commun. *9,* 313.
Wimber, E. and Steffensen, D. M. 1970. Science *170,* 639.
Wimhurst, J. M. and Manchester, K. L. 1970. FEBS Letters *8,* 91.
Winslow, R. M. and Ingram, V. M. 1966. J. Biol. Chem. *241,* 1144.
Winterhalter, K. H. 1966. Nature *211,* 932.
Woese, C. 1970. Nature *226,* 817.
Woese, C., Naono, S., Soffer, R. and Gros, F. 1963. Biochem. Biophys. Res. Commun. *11,* 435.
Wollman, E., Jacob, F. and Hayes, W. 1956. Cold Spring Harbor Symp. Quant. Biol. *21,* 141.
Wolpert, L. and Gingell, D. 1970. J. Theoret. Biol. *29,* 147.
Wong, J. T. F. and Nazard, R. N. 1970. J. Biol. Chem. *245,* 4591.
Wong, T. W., Weiss, S. B., Eliceiri, G. L. and Bryant, J. 1970. Biochemistry *9,* 2376.
Wood, D. D. and Luck, D. J. L. 1969. J. Mol. Biol. *41,* 211.
Wood, W. B. and Berg, P. 1962. Proc. Natl. Acad. Sci. *48,* 94.
Woodland, H. R. and Graham, C. F. 1969. Nature *221,* 327.
Woodland, H. R. and Gurdon, J. B. 1968. J. Emb. Exp. Morph. *19,* 363.
Woodland, H. R. and Gurdon, J. B. 1969. Dev. Biol. *20,* 89.
Woods, P. S. 1959. Brookhaven Symp. Biol. *12,* 153.
Woods, P. S. and Zubay, G. 1965. Proc. Natl. Acad. Sci. *54,* 1705.
Wool, I. G., Martin, T. E. and Low, R. B. 1968. *Regulatory Mechanisms for Protein Synthesis in Mammalian Cells.* Editors A. S. Pietro, M. R. Lamborg and F. T. Kenney, Academic Press, New York, p. 323.

Wortmann, J. 1882. Ztschr. f. physiol. Chem. *6,* 287.
Wright, D. A. and Subtelny, S. 1971. Dev. Biol. *24,* 119.
Wu, R. and Kaiser, A. D. 1967. Proc. Natl. Acad. Sci. *57,* 170.
Yamamura, H., Takeda, M., Kumon, A. and Nishizuka, Y. 1970. Biochem. Biophys. Res. Commun. *40,* 675.
Yamasaki, M. and Arima, K. 1969. Biochem. Biophys. Res. Commun. *37,* 430.
Yang, P. C., Hamada, K. and Schweet, R. 1968. Arch. Biochem. Biophys. *125,* 506.
Yang, S. S. and Comb, D. G. 1968. J. Mol. Biol. *31,* 139.
Yang, S. and Criddle, R. S. 1969. Biochem. Biophys. Res. Commun. *35,* 429.
Yang, S. S. and Sanadi, D. R. 1969. J. Biol. Chem. *244,* 5081.
Yang, W. K., Hellman, A., Martin, D. H., Hellman, K. B. and Novelli, G. D. 1969. Proc. Natl. Acad. Sci. *64,* 1411.
Yang, W. K. and Novelli, G. D. 1968. Biochem. Biophys. Res. Commun. *31,* 534.
Yaniv, M., Favre, A. and Barrell, B. G. 1969. Nature *223,* 1331.
Yankofsky, S. A. and Spiegelman, S. 1962(a). Proc. Natl. Acad. Sci. *48,* 1069.
Yankofsky, S. A. and Spiegelman, S. 1962(b). Proc. Natl. Acad. Sci. *48,* 1466.
Yankofsky, S. A. and Spiegelman, S. 1963. Proc. Natl. Acad. Sci. *49,* 538.
Yanofsky, C. 1960. Bact. Rev. *24,* 221.
Yanofsky, C. 1967. Harvey Lectures *61,* 145.
Yanofsky, C., Carlton, B. C., Guest, J. R., Helinski, D. R. and Henning, U. 1964. Proc. Natl. Acad. Sci. *51,* 266.
Yanofsky, C., Cox, E. C. and Horn, V. 1965. Proc. Natl. Acad. Sci. *55,* 274.
Yanofsky, C., Ito, J. and Horn, V. 1966. Cold Spring Harbor Symp. Quant. Biol. *31,* 151.
Yanofsky, C. and Lennox, E. S. 1959. Virology *8,* 425.
Yarmolinsky, M. B., Jordon, E. and Weismeyer, H. 1961. Cold Spring Harbor Symp. Quant. Biol. *26,* 217.
Yasmineh, W. G. and Yunis, J. J. 1970. Exp. Cell Res. *59,* 69.
Yasmineh, W. G. and Yunis, J. J. 1971. Exp. Cell Res. *64,* 41.
Yčas, M. 1969. *The Biological Code, Frontiers of Biology.* North-Holland Publishing Co., Amsterdam.
Yegian, C. D. and Stent, G. S. 1969(a). J. Mol. Biol. *39,* 45.
Yegian, C. D. and Stent, G. S. 1969(b). J. Mol. Biol. *39,* 59.
Yelton, D. B. and Thorne, C. B. 1970. J. Bact. *102,* 573.
Yeung, D. and Oliver, I. T. 1968. Biochemistry *7,* 3231.
Yoneyama, Y., Sawada, H., Takeshita, M. and Sugita, Y. 1969. Lipids *4,* 321.
Yoshida, A., Watanabe, S. and Morris, J. 1970. Proc. Natl. Acad. Sci. *67,* 1600.
Yoshida, M. and Shimura, K. 1970. J. Biochem. *67,* 507.
Yoshikawa, H. and Maruo, B. 1961. Biochim. Biophys. Acta *45,* 270.
Yoshikawa-Fukada, M. 1966. Biochim. Biophys. Acta *123,* 91.
Yoshikawa-Fukada, M. 1967. Biochim. Biophys. Acta *145,* 651.
Young, E. T. 1970. J. Mol. Biol. *51,* 591.
Young, R. J. 1968. Biochemistry *7,* 2263.
Yudkin, J. 1938. Biol. Rev. *13,* 93.
Yudkin, M. D. 1969. Biochem. J. *114,* 307.
Yudkin, M. D. 1970. Biochem. J. *118,* 741.
Yura, T., Marushige, K. and Imai, M. 1963. Biochim. Biophys. Acta *76,* 442.
Zabin, I. 1963. J. Biol. Chem. *238,* 3300.
Zabin, I., Kepes, A. and Monod, J. 1962. J. Biol. Chem. *237,* 253.
Zachau, H. G. 1968. Europ. J. Biochem. *5,* 559.
Zalokar, M. 1959. Nature *183,* 1330.

Zalta, J. P., Lachurie, F. and Osono, S. 1960. Comptes Rend. Acad. Sci. (Paris) *251,* 814.
Zamecnik, P. C. and Keller, E. B. 1954. J. Biol. Chem. *209,* 337.
Zeikus, J. G., Taylor, M. W. and Buck, C. A. 1969. Exp. Cell Res. *57,* 74.
Zillig, W., Fuchs, E., Palm, P., Rabussay, D. and Zechel, K. 1970(a). Lepetit Colloq. Biol. Med. *1,* 151.
Zillig, W., Zechel, K., Rabussay, D., Schachner, M., Sethi, V. S., Palm, P., Heil, A. and Seifert, W. 1970(b). Cold Spring Harbor Symp. Quant. Biol. *35,* 47.
Zimmerman, E. F. 1968. Biochemistry 7, 3156.
Zimmerman, E. F., Hackney, J., Nelson, P. and Arias, I. M. 1969. Biochemistry *8,* 2636.
Zinder, N. D., Engelhardt, D. L. and Webster, R. E. 1966. Cold Spring Harbor Symp. Quant. Biol. *31,* 251.
Zipser, D. 1969. Nature *221,* 21.
Zubay, G. and Doty, P. 1959. J. Mol. Biol. *1,* 1.
Zubay, G., Schwartz, D. and Beckwith, J. 1970. Proc. Natl. Acad. Sci. *66,* 104.
Zucker, W. V. and Schulman, H. M. 1967. Biochim. Biophys. Acta *138,* 400.
Zuckerkandl, E. 1964. J. Mol. Biol. *8,* 128.
Zylber, S. and Penman, S. 1969. J. Mol. Biol. *46,* 201.
Zylber, S., Vesco, C. and Penman, S. 1969. J. Mol. Biol. *44,* 195.

BIBLIOGRAPHY FOR ADDENDUM

Abe, H. and Yamana, K. 1971. Biochim. Biophys. Acta *240,* 392.
Adler, F. L., Fishman, M. and Dray, S. 1966. J. Immunol. *97,* 554.
Artman, M. and Roth, J. S. 1971. J. Mol. Biol. *60,* 291.
Baglioni, C., Bleiberg, I. and Zauderer, M. 1971. Nature New Biology *232,* 8.
Baril, E. F., Brown, O. E., Jenkins, M. D. and Laslo, J. 1971. Biochemistry *10,* 1981.
Bell, C. and Dray, S. 1969. J. Immunol. *103,* 1196.
Bell, E. 1969. Nature *224,* 326.
Bell, E. 1971. Science *174,* 603.
Berissi, H., Groner, Y. and Revel, M. 1971. Nature New Biology *234,* 44.
Blasi, F., Barton, R. W., Kovach, J. S. and Goldberger, R. F. 1971. J. Bact. *106,* 508.
Bothwell, A. L. M. and Apirion, D. 1971. Biochem. Biophys. Res. Commun. *44,* 844.
Chesterton, C. J. and Butterworth, P. H. W. 1971(a). FEBS Letters *13,* 275.
Chesterton, C. J. and Butterworth, P. H. W. 1971(b). FEBS Letters *15,* 181.
Coffman, R. L., Norris, T. E. and Koch, A. L. 1971. J. Mol. Biol. *60,* 1.
Conde, F., Del Campo, F. F. and Ramirez, J. M. 1971. FEBS Letters *16,* 156.
Crick, F. 1971. Nature *234,* 25.
Crippa, M. and Tocchini-Valentini, P. 1971. Proc. Natl. Acad. Sci. *68,* 2769.
Darnell, J. E., Philipson, L., Wall, R. and Adesnik, M. 1971. Science *174,* 507.
Davidson, E. H. and Britten, R. J. 1971. J. Theoret. Biol. *32,* 123.
de Crombrugghe, B., Chen, B., Anderson, W., Nissley, P., Gottesman, M., Pastan, I. and Perlman, R. 1971. Nature New Biology *231,* 139.
Dofuku, R., Tettenborn, U. and Ohno, S. 1971. Nature New Biology *232,* 5.
Duke, L. J. and Harshman, S. 1971. Immunochemistry *8,* 431.
Edmonds, M., Vaughan, M. H. and Nakazato, H. 1971. Proc. Natl. Acad. Sci. *68,* 1336.
Eron, L. and Block, R. 1971. Proc. Natl. Acad. Sci. *68,* 1828.
Fabricant, R. and Kennell, D. 1971. J. Virol. *6,* 772.
Faiferman, I., Cornudella, L. and Pogo, A. O. 1971. Nature New Biology *233,* 234.

Ficq, A. and Brachet, J. 1971. Proc. Natl. Acad. Sci. *68,* 2774.
Ford, P. J. 1971. Nature *233,* 561.
Gehring, U., Tomkins, G. M. and Ohno, S. 1971. Nature New Biology *232,* 106.
Greenblatt, J. and Schleif, R. 1971. Nature New Biology *233,* 166.
Groner, Y. and Revel, M. 1971. Eur. J. Biochem. *22,* 144.
Gurdon, J. B., Lane, C. D., Woodland, H. R. and Marbaix, G. 1971. Nature New Biology *233,* 177.
Herzog, A., Ghysen, A. and Bollen, A. 1971. Molec. Gen. Genetics *110,* 211.
Hill, R. J., Poccia, D. L. and Doty, P. 1971. J. Mol. Biol. *61,* 445.
Holmes, R. K. and Singer, M. F. 1971. Biochem. Biophys. Res. Commun. *44,* 837.
Hong, J. S., Smith, G. R. and Ames, B. N. 1971. Proc. Natl. Acad. Sci. *68,* 2258.
Horgen, P. A. and Griffin, D. H. 1971. Nature New Biology *234,* 17.
Hradec, J., Dušek, E., Bermek, E. and Matthaei, H. 1971. Biochem. J. *123,* 959.
Hunter, T. 1971. Cited by a correspondent, Nature New Biology *233,* 255.
Ishihama, A., Murakami, S., Fukuda, R., Matsukage, A. and Kameyama, T. 1971. Molec. Gen. Genetics *111,* 66.
Jacquet, M. and Kepes, A. 1971. J. Mol. Biol. *60,* 453.
Kanehisha, T., Fujitani, H., Sano, M. and Tanaka, T. 1971. Biochim. Biophys. Acta *240,* 46.
Kedinger, C., Nuret, P. and Chambon, P. 1971. FEBS Letters *15,* 169.
Koch, J. and Von Pfeil, H. 1971. FEBS Letters *17,* 312.
Kosaganov, Y. N., Zarudnaja, M. I., Lazurkin, Y. S., Frank-Kamenetskii, M. D., Beabealashvilli, R. S. and Savochkina, L. P. 1971. Nature New Biology *231,* 212.
Küntzel, H. and Schäfer, K. P. 1971. Nature New Biology *231,* 265.
Kuwano, M., Schlessinger, D. and Morse, D. E. 1971. Nature New Biology *231,* 214.
Lane, C. D., Marbaix, G. and Gurdon, J. B. 1971. J. Mol. Biol. *61,* 73.
Lebowitz, P., Weissman, S. M. and Radding, C. M. 1971. J. Biol. Chem. *246,* 5120.
Lee, S. Y., Mendecki, J. and Brawerman, G. 1971. Proc. Natl. Acad. Sci. *68,* 1331.
Lerner, R. A., Meinke, W. and Goldstein, D. A. 1971. Proc. Natl. Acad. Sci. *68,* 1212.
Lim, L. and Canellakis, E. S. 1970. Nature *227,* 710.
Lockard, R. E. and Lingrel, J. B. 1971. Nature New Biology *233,* 204.
Mandel, J. L. and Chambon, P. 1971. FEBS Letters *15,* 175.
Mans, R. J. and Walter, T. J. 1971. Biochim. Biophys. Acta *247,* 113.
Maroun, L. E., Driscoll, B. F. and Nardone, R. M. 1971. Nature New Biology *231,* 270.
Mathews, M. B., Osborn, M. and Lingrel, J. B. 1971. Nature New Biology *233,* 206.
Melcher, U. 1971. Biochim. Biophys. Acta *246,* 216.
Moar, V. A., Gurdon, J. B., Lane, C. D. and Marbaix, G. 1971. J. Mol. Biol. *61,* 93.
Naora, H., Kodaira, K. and Pritchard, M. J. 1971. Biochim. Biophys. Acta *246,* 280.
Nissley, S. P., Anderson, W. B., Gottesman, M. E., Perlman, R. L. and Pastan, I. 1971. J. Biol. Chem. *246,* 4671.
Ohno, S. 1971. Nature *234,* 134.
Parks, J. S., Gottesman, M., Perlman, R. L. and Pastan, I. 1971(a). J. Biol. Chem. *246,* 2419.
Parks, J. S., Gottesman, M., Shimada, K., Weisberg, R. A., Perlman, R. L. and Pastan, I. 1971(b). Proc. Natl. Acad. Sci. *68,* 1891.
Philipson, L., Wall, R., Glickman, G. and Darnell, J. E. 1971. Proc. Natl. Acad. Sci. *68,* 2806.
Phillips, S. L. 1971. J. Mol. Biol. *59,* 461.
Raacke, I. D. 1971. Proc. Natl. Acad. Sci. *68,* 2357.
Reichardt, L. and Kaiser, A. D. 1971. Proc. Natl. Acad. Sci. *68,* 2185.
Roodman, S. T. and Greenberg, G. R. 1971(a). J. Biol. Chem. *246,* 2609.

Roodman, S. T. and Greenberg, G. R. 1971(b). J. Biol. Chem. *246,* 4853.
Seifert, W., Rabussay, D. and Zillig, W. 1971. FEBS Letters *16,* 175.
Sudilovsky, O., Pestana, A., Hinderaker, P. H. and Pitot, H. C. 1971. Science *174,* 142.
Tanaka, N., Lin, Y. C. and Okuyama, A. 1971. Biochem. Biophys. Res. Commun. *44,* 477.
Thach, S. S. and Thach, R. E. 1971. Proc. Natl. Acad. Sci. *68,* 1791.
Turkington, R. W., Majumder, G. C. and Riddle, M. 1971. J. Biol. Chem. *246,* 1814.
Turkington, R. W. and Spielvogel, R. L. 1971. J. Biol. Chem. *246,* 3835.
Twu, J. S. and Bretthauer, R. K. 1971. Biochemistry *10,* 1576.
Weissbach, H., Redfield, B. and Brot, N. 1971. Arch. Biochem. Biophys. *144,* 224.
Wetekam, W., Staack, K. and Ehring, R. 1971. Molec. Gen. Genetics *112,* 14.
Wilcox, G., Clemetson, K. J., Santi, D. V. and Englesberg, E. 1971. Proc. Natl. Acad. Sci. *68,* 2145.
Yudkin, M. D. 1969. Biochem. J. *114,* 307.
Zagorska, L., Dondon, J., Lelong, J. C., Gros, F. and Grunberg-Manago, M. 1971. Biochimie *53,* 63.
Zubay, G., Gielow, L. and Englesberg, E. 1971. Nature New Biology *233,* 164.

INDEX